Tree Physiology

Volume 9

Series Editor

Ülo Niinemets, Estonian University of Life Sciences
Tartu, Estonia

Our perceptions of forests has dramatically changed in the past decades, and forests are now considered central in mitigating climate change, providing vital ecosystem services, and maintaining human health.

Fundamental to understanding our global forest reserves, and managing them according to our current needs, is the need to understand and predict the physiological responses of trees to their abiotic and biotic environment.

Springer's Tree Physiology series takes a broad approach to address this need drawing together expertise from around the world to address issues and present findings, spanning molecular biology, biochemistry, biophysics, ecophysiology and atmospheric sciences, ranging from the cellular to the landscape scale. Providing state-of-the-art analyses on key topics, the volumes constitute an invaluable resource for researchers and advanced students involved in both pure and applied fields – including forestry, ecology, conservation, biodiversity and pest and disease management.

Daniel M. Johnson • Ülo Niinemets
Jean-Christophe Domec
Editors

Physiology and Ecology of Pines Worldwide

Springer

Editors
Daniel M. Johnson
Warnell School of Forestry
and Natural Resources
University of Georgia
Athens, USA

Ülo Niinemets
Institute of Agricultural
and Environmental Sciences
Estonian University of Life Sciences
Tartu, Estonia

Jean-Christophe Domec
Department of Forestry
Bordeaux Sciences Agro
UMR INRAE-ISPA 1391
Gradignan, France

ISSN 1568-2544
Tree Physiology
ISBN 978-3-032-23000-3 ISBN 978-3-032-23001-0 (eBook)
https://doi.org/10.1007/978-3-032-23001-0

This Springer imprint is published by the registered company Springer Nature Switzerland AG
The registered company address is: Gewerbestrasse 11, 6330 Cham, Switzerland

Photo courtesy of Joe Landsberg

This book is dedicated to Sune Linder (3 May 1943–1 February 2026) whose studies of pine trees reshaped for many of us our understanding of pine ecophysiology, and whose legacy of teaching inspires each of us.

Preface

The genus *Pinus* is naturally occurring throughout the Northern Hemisphere where it often dominates forest landscapes from subarctic to tropical regions, being particularly competitive in moderately open habitats with infertile soils. Due to their high adaptability and low nutrient and water requirements, pines have been planted in forest plantations throughout the world. In natural ecosystems, pines are often dominant species, and natural and planted pine forests are a cornerstone of modern forestry, providing a myriad of economic and ecological benefits. These include providing wildlife habitat (e.g., Smith et al. 2017), improving local hydrology (Hovis et al. 2021; Younger et al. 2023), CO_2 sequestration (Aspinwall et al. 2012; Jörgensen et al. 2021), and many forest products used by humans. Economically, pine forestry represents a major sector of the global forest industry; for example, in the southeastern USA alone, intensively managed pine plantations contribute tens of billions of dollars annually to the regional economy (e.g., Pelkki and Sherman 2020).

Pines have played a central role in human societies for millions of years, initially as sources of heat and subsequently managed by fire to influence grazing areas, hunting patterns, and the regeneration of other species. Humans have intentionally cultivated pine trees for over 2000 years where they have been used for medicine, religious ceremonies, food (e.g., edible pine nuts in SW USA as recently as ~10,000 YA), timber, resin, naval construction (Le Maitre 2000), and ornamental purposes since the pre-Roman times (Scarascia-Mugnozza and Paganini 2024). These early practices mark the transition from opportunistic use of natural pine stands to organized silviculture, laying the foundation for the widespread pine plantation forestry that characterizes many regions of the world today.

Currently, pine forests are under threat: over 40% of the ~110 *Pinus* species (Gernandt et al. 2005) are near-threatened or worse with almost 20% formally listed as endangered (International Conifer Conservation Program, Royal Botanical Gardens Edinburgh). Evergreen conifers, including pines, are disproportionately impacted by drought compared to angiosperm tree species in general (Anderegg et al. 2019). Given that many pine-dominated forests already colonize relatively dry habitats, future increase of drought severity and drought/heat stress combinations are expected to be particularly devastating in pine forests. Widespread drought-induced mortality has already been observed in *Pinus*-dominated ecosystems (Floyd et al. 2009; McDowell et al. 2010; Hereş et al. 2012) and climate projections indeed indicate that these regions are likely to experience increasingly severe and prolonged droughts in the future (e.g., Seager et al. 2007; Dai 2011). Moreover, recent studies have shown that natural *Pinus* regeneration in the western US has declined dramatically, and in some cases has completely failed over the past decade (Kemp et al. 2019; Davis et al. 2019, 2023; Bailey et al. 2021).

This book was undertaken based on these two converging realities: the ecological and economic importance of pines worldwide and the current level of threats to pines across the globe. The following chapters address specific aspects of pine ecology and physiology. They explore the evolutionary history and global distribution of pines, the variation in their needle and shoot structure and concomitant modifications in light harvesting and carbon capture, their physiological responses to environmental stresses, their roles in fire-adapted ecosystems, and the management practices that influence their growth, productivity, and resilience. Collectively, these chapters aim to provide a comprehensive synthesis of the ecology, ecophysiology and physiology of pines, the state of pine forests, the challenges they face under climate change, and the adaptive strategies needed to ensure their future sustainability.

Specifically, Brodribb and Jordan (2026) explore biogeographical and evolutionary patterns in the Pinaceae as compared to the Podocarpaceae. While members of the Pinaceae tend to be found in the Northern Hemisphere, Podocarps are more restricted to Southern Hemisphere locations. The authors also evaluate the impacts of temperature and moisture availability and each family's adaptations (or lack thereof) to cold temperatures, shady habitats, fire and moisture limitations on the current biogeographical patterns observed.

Domec and Johnson (2026) bridge the evolutionary-physiological gap by focusing on pine anatomy and how it dictates physiological function. The genus *Pinus* is divided into two subgenera that have diverged ~150 MYA. These two subgenera have distinct anatomical characteristics, including major differences in wood structure. The authors explore how differences in anatomy between the two subgenera may underpin their differences in physiological and ecological performance.

Many pines are fire-tolerant and/or fire-dependent, and Hood et al. (2026) explore the numerous adaptations that pines have to fire. Anatomical, morphological, and physiological traits are discussed that have resulted in pines being dominant in certain fire-prone habitats. Additionally, Hood et al. (2026) summarize the relatively new field of pyroecophysiology and how fires impact pine carbon and water relations as well as growth and mortality.

Insects and fire interact with pines in most pine-dominated ecosystems. Pines have not only evolved to tolerate fires but have also co-evolved with herbivorous insects. Both wildfires and prescribed fires can change pine-insect dynamics. Additionally, human-dependent fire suppression and a changing climate will likely alter these interactions. Curbing fires potentially alters the range of fire-dependent pines that characteristically dominate open habitats and will be out-competed as the vegetation becomes denser once the fires are suppressed. Gandhi and Hood (2026) provide a framework for understanding the multiple interactions between pines, fire, and insects.

The next four chapters focus on the physiology of pines from carbon, water, and light absorbance perspectives. Tor-Ngern et al. (2026) analyze how pine leaves and shoots are coordinated to efficiently absorb photons for photosynthesis while also providing hydraulic supply to evaporative surfaces inside the leaf. They also develop a model linking shoot structure, stand density, and light interception and present several case studies to examine how effective or ineffective establishing pine species beyond their native ranges would be.

Pine needle photosynthesis is difficult to characterize due to complex three-dimensional leaf shape and associated light capture properties. Furthermore, there is inherent variation in pine needle longevities and associated suite of structural and chemical needle traits. Niinemets and Zhang (2026) explore needle longevities across pines worldwide, develop a pine needle economics spectrum framework for pines, and look at photosynthesis per investment in nitrogen and biomass. They further analyze the extent to which needle photosynthesis is controlled by biochemistry, mesophyll, and stomatal diffusion and analyze changes in needle photosynthesis as driven by needle age, whole-plant age and size, and environmental acclimation. Additionally, they put forth a framework for the expression of pine photosynthesis per unit leaf area, discuss the benefits and drawbacks of different ways to express it, and call for more rigor in the research community in reporting pine photosynthesis data.

Plant hydraulics is a research field that has grown dramatically over the past three decades; however, pine hydraulics has been understudied. Augustine and McCulloh (2026) explore global patterns in pine hydraulics and water relations and how they vary as compared to other coniferous trees. Additionally, they link the hydraulics and underlying anatomy in pines to their ecology, particularly as it relates to drought tolerance.

Sun et al. (2026) synthesize our understanding of the hydrology of pine forests and evaluate hydrological budgets for pine forests across the globe from wet to dry environments. They also study how stand management influences stand hydrology and its components. Additionally, they use an existing framework to explore water versus energy limitations on pine forest evapotranspiration.

Multiple manipulative experiments have been performed in pine forests, most of which were designed to understand the potential impacts of climate change on functioning of these forests. Elevated CO_2, warming, and manipulation of soil water availability and air humidity affect carbon uptake, water loss, and growth. Maier et al. (2026) synthesized the results of these experiments from leaf to canopy scales and developed a model to simulate changes in biomass, leaf area, and hydrology with changes in interacting environmental variables. Increases or decreases in growth and evaporation rate with elevated CO_2 are site/species-specific and can be altered with changes in temperature as well as soil and atmospheric moisture availability.

Pine mortality has now been observed worldwide both in pines in their native ranges and pines planted in plantations outside their native range. Hammond et al. (2026) explore five case studies of high pine mortality and the mechanisms underlying pine mortality events. They conclude that "hotter droughts" are driving many of the observed large-scale pine mortality events globally and propose a set of research questions to guide future investigations in this area.

Pines are key ecological and economic components of forests across the Northern Hemisphere and in plantations worldwide, supplying multiple essential ecosystem services, including provision of timber, resin, bioenergy, wildlife habitat, carbon sequestration, and hydrological regulation. Despite their adaptability, pine forests face growing threats, including drought-induced mortality, declining natural regeneration, pest pressures, and climate-driven stressors, with over 40% of species near-threatened or worse. This book synthesizes current knowledge on pine anatomy, ecology, physiology, and ecophysiology, covering evolutionary history, fire and insect adaptations, carbon and water relations, photosynthesis, stand hydrology, and responses to climate change. By integrating these perspectives, it highlights mechanisms underlying pine performance and mortality and identifies adaptive management strategies to sustain productivity, resilience, and ecosystem services under current and future environmental challenges. The book also highlights key research priority areas with major knowledge gaps, including physiology and ecology of tropical pines in their native range and in plantations, invasions of planted pines and their threats to local biodiversity, invasions of pine pests and host changes of pine pests, and sensitivity and adaptability of pine forests to global change-driven enhancement of the severity of environmental stresses.

Athens, GA, USA — Daniel M. Johnson
Tartu, Estonia — Ülo Niinemets
Gradignan, France — Jean-Christophe Domec

References

Anderegg WR, Anderegg LD, Kerr KL, Trugman AT (2019) Widespread drought-induced tree mortality at dry range edges indicates that climate stress exceeds species' compensating mechanisms. Glob Chang Biol 25(11):3793–3802

Aspinwall MJ, McKeand SE, King JS (2012) Carbon sequestration from 40 years of planting genetically improved loblolly pine across the southeast United States. For Sci 58(5):446–456

Augustine SP, McCulloh KA (2026) Doing just Pine, thank you very much: the hydraulics of Pinus. In: Johnson DM, Domec J-C, Niinemets Ü (eds) Physiology and ecology of pines worldwide. Springer

Bailey SN, Elliott GP, Schliep EM (2021) Seasonal temperature–moisture interactions limit seedling establishment at upper treeline in the Southern Rockies. Ecosphere 12(6):e03568

Brodribb TJ, Jordan GJ (2026) North-south rivalry—Pinaceae versus Podocarpaceae. In: Johnson DM, Domec J-C, Niinemets Ü (eds) Physiology and ecology of pines worldwide. Springer

Dai A (2011) Drought under global warming: a review. Wiley Interdiscip Rev Clim Chang 2(1):45–65
Davis KT, Dobrowski SZ, Higuera PE, Holden ZA, Veblen TT, Rother MT, Parks SA, Sala A, Maneta MP (2019) Wildfires and climate change push low-elevation forests across a critical climate threshold for tree regeneration. Proc Natl Acad Sci 116(13):6193–6198
Davis KT, Robles MD, Kemp KB, Higuera PE, Chapman T, Metlen KL, Peeler JL, Rodman KC, Woolley T, Addington RN, Buma BJ (2023) Reduced fire severity offers near-term buffer to climate-driven declines in conifer resilience across the western United States. Proc Natl Acad Sci 120(11):e2208120120
Domec JC, Johnson DM (2026) The anatomy of pine wood and leaves. In: Johnson DM, Domec J-C, Niinemets Ü (eds) Physiology and ecology of pines worldwide. Springer
Floyd ML, Clifford M, Cobb NS, Hanna D, Delph R, Ford P, Turner D (2009) Relationship of stand characteristics to drought-induced mortality in three Southwestern piñon–juniper woodlands. Ecol Appl 19(5):1223–1230
Gandhi KJK, Hood SM (2026) Eco-evolutionary interactions between pines, fire, and herbivorous insects. In: Johnson DM, Domec J-C, Niinemets Ü (eds) Physiology and ecology of pines worldwide. Springer
Gernandt DS, López GG, García SO, Liston A (2005) Phylogeny and classification of *Pinus*. Taxon 54(1):29–42
Hammond WM, Kant ME, Mantova M, Sapes G, Fernández ME, Gyenge J, Gatica G, Rodríguez RL, Preisler Y, Saenz-Romero C, Wion AP, Allen CD (2026) Pine death in the Anthropocene. In: Johnson DM, Domec J-C, Niinemets Ü (eds) Physiology and ecology of pines worldwide. Springer
Hereş AM, Martínez-Vilalta J, López BC (2012) Growth patterns in relation to drought-induced mortality at two Scots pine (*Pinus sylvestris* L.) sites in NE Iberian Peninsula. Trees 26(2):621–630
Hood SM, Bär A, Battipaglia G, Mayr S, Shearman T, Varner JM, Gandhi KJK (2026) Fire ecology and ecophysiology of Pinus species. In: Johnson DM, Domec J-C, Niinemets Ü (eds) Physiology and ecology of pines worldwide. Springer
Hovis M, Hollinger JC, Cubbage F, Shear T, Doll B, Kurki-Fox JJ, Line D, Fox A, Baldwin M, Klondike T, Lovejoy M (2021) Natural infrastructure practices as potential flood storage and reduction for farms and rural communities in the North Carolina coastal plain. Sustainability 13(16):9309
Jörgensen K, Granath G, Lindahl BD, Strengbom J (2021) Forest management to increase carbon sequestration in boreal Pinus sylvestris forests. Plant Soil 466(1):165–178
Kemp KB, Higuera PE, Morgan P, Abatzoglou JT (2019) Climate will increasingly determine post-fire tree regeneration success in low-elevation forests, Northern Rockies, USA. Ecosphere 10(1):e02568
Le Maitre DC (2000) Pines in cultivation: a global view. In: Ecology and biogeography of Pinus. Cambridge University Press, p 407
Maier CA, Medvigy D, Oren R, Palmroth S, Schäfer KVR (2026) Hypothetical responses of pine forests to climate change based on manipulation experiments. In: Johnson DM, Domec J-C, Niinemets Ü (eds) Physiology and ecology of pines worldwide. Springer
McDowell NG, Allen CD, Marshall L (2010) Growth, carbon-isotope discrimination, and drought-associated mortality across a Pinus ponderosa elevational transect. Glob Chang Biol 16(1):399–415
Niinemets Ü, Zhang Z (2026) Pine photosynthesis: needle longevity, structural and physiological controls, needle and tree age dependent changes and plasticity. In: Johnson DM, Domec J-C, Niinemets Ü (eds) Physiology and ecology of pines worldwide. Springer
Pelkki M, Sherman G (2020) Forestry's economic contribution in the United States, 2016. For Prod J 70(1):28–38
Scarascia-Mugnozza G, Paganini R (2024) The pine forests of Rome: an ecosystem and a scenic beauty under threat. European Forest Institute, pp 1–6

Seager R, Ting M, Held I, Kushnir Y, Lu J, Vecchi G, Huang HP, Harnik N, Leetmaa A, Lau NC, Li C (2007) Model projections of an imminent transition to a more arid climate in southwestern North America. Science 316(5828):1181–1184

Smith LL, Cox JA, Conner LM, McCleery RA, Schlimm EM (2017) Management and restoration for wildlife. In: Ecological restoration and management of longleaf pine forests. CRC Press, pp 233–252

Sun G, Aguilos M, Liu N, Boggs J, Amatya D, Zhang L, Hao L (2026) Evapotranspiration of pine forests: a global synthesis. In: Johnson DM, Domec J-C, Niinemets Ü (eds) Physiology and ecology of pines worldwide. Springer

Tor-Ngern P, Palmroth S, Lim H, Oren R (2026) Shoot architecture—a crown attribute regulating biosphere-atmosphere exchanges via coordinated light absorption and water delivery. In: Johnson DM, Domec J-C, Niinemets Ü (eds) Physiology and ecology of pines worldwide. Springer

Younger SE, Cannon JB, Brantley ST (2023) Impacts of longleaf pine (*Pinus palustris* Mill.) on long-term hydrology at the watershed scale. Sci Total Environ 902:165999

Acknowledgments

DMJ: I would like to thank my mentors, Dr. William K. Smith and Dr. Frederick C. Meinzer, for their support and encouragement to go study coniferous trees. I would also like to acknowledge Dr. J. Dan Pittillo, who inspired me to study coniferous tree species and to pursue a career path in research. I thank the Warnell School at the University of Georgia for their support and the National Science Foundation and US Department of Agriculture for research funding. Finally, I would like to thank my family who have always encouraged me in my pursuit of spending time in the woods.

ÜN: I am very grateful to my mentors Dr. Olevi Kull and Dr. John D. Tenhunen. Olevi pioneered in developing tools to measure conifer shoot silhouette to total surface area and photosynthesis. We have climbed many trees, built towers in forests, and taken thousands of hemispheric photographs in tree canopies. John is a global leader in modeling plant photosynthesis from biochemistry to leaves, stands, and landscapes. John introduced me to modeling tree photosynthesis, and it was in Bayreuth where I learned what it means to work hard.

JCD: First and foremost, I wish to thank Dan for sowing the first seeds of this book in our minds. Although none of us expected it to grow into such a wild forest of pages, it turned out to be a pretty great place to wander. Gratitude also goes to the coffee that has been fueling my research for the last 25 years and the multiple forests that tolerated my clumsy presence. I owe special thanks to Barbara Lachenbruch, my PhD advisor, who patiently taught me wood anatomy and proved that staring at tiny cells under a microscope was in fact both fun and a full-time lifestyle choice. Most of all, I thank my wife for reminding me that life exists outside data, and my daughter for ensuring I never take myself too seriously.

Acknowledgements

[illegible]

[illegible]

[illegible]

Contents

About the Editors

Daniel M. Johnson is Professor of tree physiology and forest ecology in the Warnell School of Forestry and Natural Resources at the University of Georgia. He received his undergraduate and master's degrees from North Carolina State University with a focus on the biochemistry of plants. He received his PhD from Wake Forest University where he focused on the physiology and ecology of high-altitude conifers. After finishing his PhD, he moved to Corvallis, Oregon, to do a postdoc focused on conifer tree water relations. He eventually made his way to the University of Idaho where he continued his work on conifer tree physiology and ecology. After four cold years in Idaho, he moved to the University of Georgia. He currently teaches undergraduate and graduate classes, mentors graduate students, and explores his scientific passions: tree water relations, tree responses to abiotic stress, and impacts of climate change on forests.

Jean-Christophe Domec is a Distinguished Professor and forest scientist whose work explores how trees cope with drought, climate stress, and management practices. He earned his engineering degree in forestry from Bordeaux Sciences Agro and a master's in wood science from the University of Bordeaux/ENGREF, before venturing across the Atlantic to complete a PhD in forest ecology and plywood science at Oregon State University (go Beavers). His career reflects a successful case of assisted migration between France (Bordeaux Sciences Agro) and the United States (Duke University's Nicholas School of the Environment), mixing the best of both research ecosystems and proving that adaptation isn't just for trees. Actively involved in the *Integrated Carbon Observation System (ICOS)*, he contributes to Europe's long-term monitoring of greenhouse gases and ecosystem fluxes. He leads research within the ISPA EcoFun and AGROFOR teams at INRAE Bordeaux, supervises a forest's worth of students, and teaches courses on wood anatomy, tree function, forest ecology, and the art of turning science into publishable prose. His integrative and collaborative approach unites wood traits, field ecology, and eco-friendly management to guide sustainable forestry practices and climate-change adaptation strategies; all while reminding us that even trees have their limits when the world heats up.

Ülo Niinemets is Head of the Department of Crop Science and Plant Biology at the Estonian University of Life Sciences and member of the Estonian Academy of Sciences. His work broadly addresses the question of what determines vegetation distribution on the globe, emphasizing plant adaptation and acclimation to their environment in current and future conditions and functional relationships and trade-offs between leaf structure and function. He has extensively worked on conifer responses to light and nutrient availability, looking at how acclimation alters needle structure, chemical composition, and photosynthetic activity, and what traits drive conifer species differentiation across environmental gradients. He has further advanced woody species shade, drought and waterlogging tolerance rankings for the entire woody flora of the Northern Hemisphere, and linked the tolerance scores to leaf structure/function relationships and contributed to the development of the concept of stress tolerance biomes. This information has made it possible to demonstrate that global change-dependent changes in species composition constitute a major way of community acclimation, and that assessments of plant productivity in future environments must include species ecological potentials and associated suites of structural, chemical, and physiological traits.

North-South Rivalry: Pinaceae Versus Podocarpaceae

Gregory Jordan and Tim Brodribb

Abstract Pinaceae and Podocarpaceae are the two most diverse families of conifers, but have strikingly different ecological and geographic ranges. Pinaceae are almost exclusively Northern Hemisphere. Podocarpaceae are widespread across the Southern Hemisphere but extend into the north, and even overlap with Pinaceae in several areas. This chapter investigates the ecology and history of these families, and how they led to the families' current distributions.

We use bioclimatic analyses combined with existing knowledge of phylogeny, fossil record, and physiology to argue that the patterns primarily derive from slow rates of evolution of functionally significant characteristics (evolutionary niche conservatism). The evidence indicates that very limited evolution of tolerance of frost and freezing climates have constrained the success of Podocarpaceae in the Northern Hemisphere, while evolution of traits allowing Pinaceae to compete with angiosperms in tropical rainforest has been very limited. Even Podocarpaceae, with their convergence towards the broad leaves and animal seed dispersal common in angiosperms, have little capacity to compete in lowland tropical rainforest. Important exceptions to these patterns demonstrate that, even though this conservatism has restricted the geography of the family for over tens of millions of years, it is not absolute. This evolutionary conservatism is stronger in precipitation-related dimensions of niches than in thermal dimensions, especially for Podocarpaceae. Although the tropically derived Podocarpaceae of the Northern Hemisphere have failed to colonise freezing climates, temperate Southern Hemisphere Podocarpaceae have evolved into most extreme freezing climates that are compatible with woody plant growth within that region. Dispersal limitations may play a secondary role: tropical rainforest may have provided too large a barrier for Pinaceae to disperse naturally into temperate Southern Hemisphere, and extensive deserts may have constrained the expansion of Podocarpaceae within the Northern Hemisphere.

G. Jordan (✉) · T. Brodribb
Biological Sciences, University of Tasmania, Hobart, Australia
e-mail: greg.jordan@utas.edu.au

D. M. Johnson et al. (eds.), *Physiology and Ecology of Pines Worldwide*, Tree Physiology 9, https://doi.org/10.1007/978-3-032-23001-0_1

Keywords Biogeography · Evolution · Fossil record · Niche conservatism · Phylogeny · Pinaceae · Podocarpaceae

1 Introduction

The division between conifer clades that dominate northern and southern regions of the globe is one of the most striking patterns in plant biogeography, and one for which the drivers remain uncertain. Conifer clades whose native ranges dominate north of the equator derive from the families Pinaceae, Sciadopityaceae, Taxaceae (excluding *Austrotaxus*), Cephalotaxaceae and Cupressaceae (excluding *Athrotaxis* and the Callitroid clade). These trees and shrubs occur almost exclusively in the Northern Hemisphere, with only *Pinus merkusii* and *Juniperus procera* reaching the Southern Hemisphere tropics (Farjon and Filer 2013). Conifer clades that dominate south of the equator derive exclusively from Podocarpaceae, Araucariaceae, *Austrotaxus*, *Athrotaxis*, and the Callitroid clade of Cupressaceae. Unlike their northern counterparts, these southern conifers extend across the equator, overlapping with the northern conifers extensively in south-east and east Asia; in Central America, Mexico, and the Caribbean; and to a small degree in north-eastern Africa.

This chapter will focus on the two largest conifer families Pinaceae and Podocarpaceae, which apart from some parts of North Africa span almost the whole geographic and environmental ranges of the northern and southern conifers, respectively (Fig. 1). They are the most successful families of conifers in their respective regions and include the two most species rich genera—*Pinus* with ~113 species and

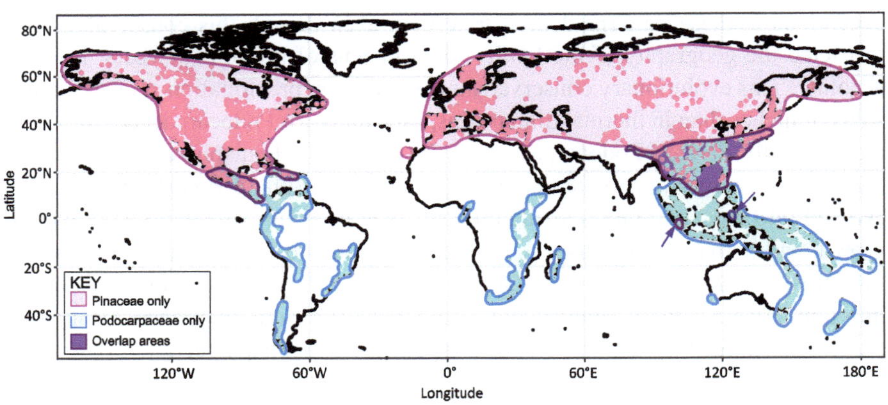

Fig. 1 Geographic distribution of Pinaceae and Podocarpaceae. The small circles are point record data (pink = Pinaceae; blue = Podocarpaceae) from Brown et al. (2020), with downsampling of heavily sampled areas (Chinese Taipei, Mexico, New Caledonia, New Zealand, Slovenia, U.S.A., Vanuatu). The curved shapes indicate the approximate ranges of the families. Regions where Pinaceae and Podocarpaceae co-occur are indicated by purple lines and shading (two of these regions are highlight by arrows). Note, even though no Pinaceae overlaps with Podocarpaceae in North Africa, *Juniperus procera* (a member of northern Cupressaceae) does

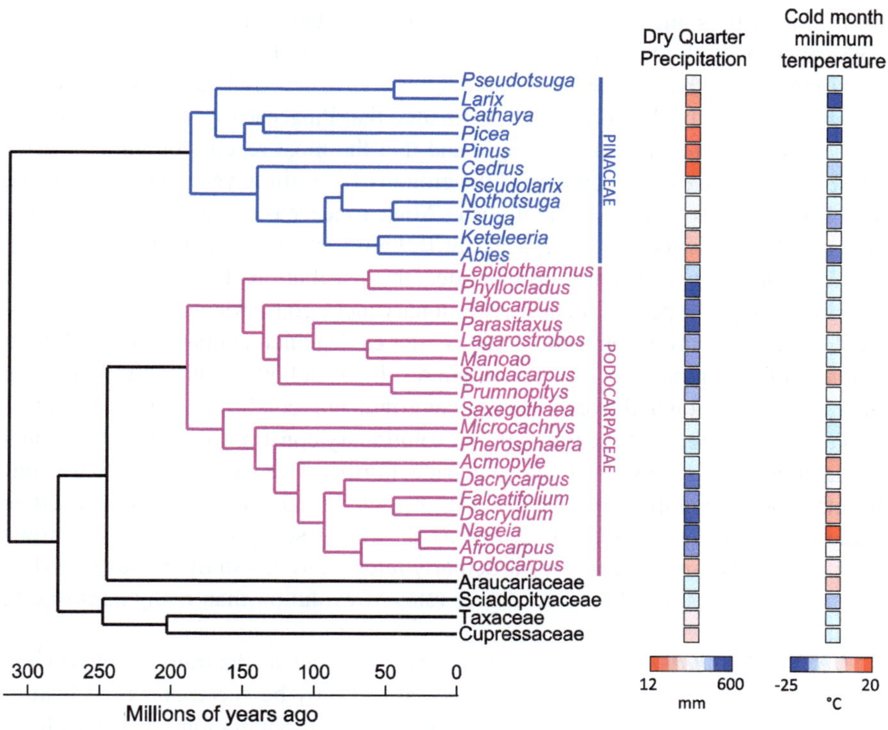

Fig. 2 Summary of the dated phylogeny of (Leslie et al. 2018), with Pinaceae and Podocarpaceae reduced to genera, and other groups to family level. Estimated median values (medians of the medians of species) of the mean dry quarter precipitation and cold month mean minimum temperature occupied by the groups are shown on the right, based on the data in Fig. 1. Note that most genera of Pinaceae occupy drier and more frosty climates than Podocarpaceae

Podocarpus with ~97 species (Farjon 2010). Our study is constructed around strong fossil records, phylogenetic and biogeographic knowledge and a suite of physiological studies for both families. Combining information about the distribution and climatic range with fossil and physiological data for both families, this chapter considers possible explanations for the peculiar global differentiation of conifers. Although both families are ancient (Fig. 2), we focus on processes that have occurred over the Cenozoic, the last 66 million years (Cohen et al. 2013).

1.1 *What Kinds of Factors Underpin This North/South Polarization?*

Expansion of Pinaceae from its likely area of dominance in the Northern Hemisphere into the Southern Hemisphere and greater expansion of the Podocarpaceae through the Northern Hemisphere in the Cenozoic (the last 66 million years) could have been limited by some combination of dispersal limitation and phylogenetic niche conservatism. Both of these factors should be considered in the light of the extensive time frames that has been available for evolving new traits and dispersing into suitable territory. Leslie et al. (2018) estimated that Pinaceae diverged from the rest of the conifers 314 million years ago, and that the most recent common ancestor (the *crown*) of the extant members of the group is 187 million years old (Fig. 2). The split between Podocarpaceae and its sister Araucariaceae is estimated at 245 million years, and the crown of Podocarpaceae at 189 million (Leslie et al. 2018).

Slow rates of evolution of traits that affect the distribution of species—and resulting slow change in species' niches: evolutionary niche conservatism (Losos 2008)—can lead to extirpation from some regions after environmental change and failure to colonise other regions. Such conservatism results from both intrinsic limitations on the rate of evolution and from a lack of environmental conditions that create selection to drive change in these traits. Both evolutionary constraints on physiology and interactions with other species can contribute. Furthermore, because niches are multidimensional, evolutionary changes within a clade may be more constrained in some niche dimensions than in others, as shown for southern conifers by Brown et al. (2021). These concepts are particularly relevant to this study because conifers are generally considered to have slower rates of evolution than many angiosperm families.

Dispersal limitation is related to dispersal traits, such as the traits enabling wind or animal dispersal. However, such dispersal can only be successful (i.e., lead to establishment in new areas or gene flow into existing populations) if the niche is available in the new area. Thus, traits related to habitat preferences (and therefore niches) can be more important than traits related to the movement of propagules in determining rates of successful long-distance dispersal (Jordan 2001).

1.2 Past Distributions

The fossil record shows that the families have been present across much of their overall current geographic distributions for much of the Cenozoic—the last 66 million years. Thus, both Pinaceae and Podocarpaceae have had considerable time to extend their ranges, and these ranges must have been restricted by either niche conservatism or dispersal traits. The simplest explanation of the families' broad distributions through the Cenozoic, is that (1) Pinaceae occupied more-or-less its full current range throughout that time and (2) Podocarpaceae occurred predominantly in the Southern Hemisphere but expanded into the northern hemisphere through the mid to late Cenozoic expansion of multiple clades during that time.

The Pinaceae have a strong Northern Hemisphere fossil record that extends back into the Mesozoic. The major patterns during the Cenozoic are that each genus of Pinaceae was widespread across the northern hemisphere and either contracted markedly or remained widespread (Akkemik 2021; Ding et al. 2021; Doweld 2018; Erwin and Schorn 2005; Klymiuk and Stockey 2012; Lakhanpal 1958; LePage

2001, 2003, 2009; LePage and Basinger 1995a, b; Wu et al. 2020; Xiang et al. 2007). Thus, *Pinus*, *Picea*, *Abies*, and *Larix* retained widespread distributions across the Northern Hemisphere, whereas *Pseudolarix*, *Nothotsuga*, *Keteleeria*, and *Cathaya* contracted to East Asia and *Pseudotsuga*, *Cedrus*, and *Tsuga* contracted to highly fragmented distributions. During periods of warmer and wetter climates in the Cenozoic, several genera of Pinaceae occurred much further north than they now do. For example, the now subtropical *Keteleeria* and temperate *Pseudolarix* occurred on Axel Heiberg Island at 79° N in the Middle Eocene along with *Tsuga*, *Larix* and *Picea*.

The fossil record of Podocarpaceae is also ancient and strong, and is mostly from the Southern Hemisphere. Molecular phylogenetic evidence indicates that the Northern Hemisphere Podocarpaceae fall into 16 clades from seven genera (Table 1), providing an indication of how many times Podocarpaceae may have migrated across the Equator. Nine of the 16 clades overlap geographically with Pinaceae. Several of the overlapping genera (*Dacycarpus*, *Nageia*, *Dacrydium* and *Podocarpus*) have good fossil records in Asia/south-east Asia, broadly within these clades' current range (Jin et al. 2010; Morley 2011; Wu et al. 2021a, b). The records of *Dacrycarpus* and *Podocarpus* are consistent with arrival after the impact of Australia with south-east Asia at the end of the Oligocene (Wu et al. 2021b). However, the records for *Nageia* and *Dacrydium* appear to be older (Eocene and mid Oligocene, respectively) (Jin et al. 2010; Morley 2011). All of these ages, apart from that of *Nageia*, are consistent with the molecular dating of the northern clades (Table 1). Morley (2000) argued that *Podocarpus* sensu lato (i.e., including *Nageia*, *Afrocarpus* and *Retrophyllum*) arrived in Asia much earlier via the Gondwanan fragment, India. However, this inference is inconsistent with molecular dates and was based on the record of the pollen form genus, *Podocarpidites*. *Podocarpidites* is also characteristic of *Prumnopitys* sensu lato (Macphail 2007), which is phylogenetically distant from *Podocarpus*. Thus, the early records of *Podocarpidites* from Asia may represent the pollen of one or more extinct lineages of Podocarpaceae. The first appearance of *Phyllocladus* in the northern hemisphere is relatively recent (early Pleistocene) (Morley 2011). Thus, the Podocarpaceae arrived in the northern hemisphere many times, and some genera have been present for tens of millions of years. As in Pinaceae, many of the genera of Podocarpaceae were much more widespread within the Southern Hemisphere than they are now (Hill and Brodribb 1999; Wilf et al. 2013). Fossils of mesic genera such as *Microcachrys*, *Lagarostrobos*, *Dacrycarpus*, *Halocarpus*, and *Dacrydium* can be found in now arid Central Australia (Macphail 2007), while fossils of extant tropical genera such as *Acmopyle* and *Retrophyllum* can be found at high palaeolatitudes (Hill and Brodribb 1999; Wilf et al. 2017).

The Cenozoic Northern Hemisphere fossil record of Podocarpaceae outside Asia is weak, but molecular dating is consistent with Podocarpaceae arriving in the Northern Hemisphere of Africa and the Americas in the Neogene (the last 23 million years) (Table 1). Only two Podocarpaceae clades are present in tropical Africa, and these clades are both estimated by Leslie et al. (2018) as being less than four million years old. Four clades are in Northern Hemisphere in the Americas, with

Table 1 Clades of Podocarpaceae present in the Northern Hemisphere

Region	Clade	Location	Notes	Clade age: stem (crown)
Asia	*Dacrycarpus imbricatus* clade (5 species)	Vanuatu to southern China	Overlaps with Pinaceae. Sister to *D. dacrydioides* from New Zealand.	20.9 (10.7) Ma
	Dacrycarpus cumingii	Sumatera to Philippines	Overlaps with Pinaceae. Sister to the rest of the genus	55.2 (25.0) Ma
	Dacrydium nidulum	Moluccas to Fiji	No overlap with Pinaceae. Embedded in a New Caledonian/Fijian clade	6.7 Ma
	Dacrydium elatum clade (10 out of 15 species)	New Guinea to China	Five species overlap with Pinaceae. Southern Hemisphere records are from Indonesia, New Guinea and Borneo	29.8 (20.3) Ma
	Falcatifolium falciforme clade (3 species)	Moluccas to Philippines	Two species overlap with Pinaceae. Uncertainty about the monophyly of this group—only one of these species is in the phylogeny of (Leslie et al. 2018)	25.37 Ma
	Nageia (5 species)	New Guinea to India and China	All five species overlap with Pinaceae. One species (*N. wallichiana*) reaches New Guinea	26.1 (10.9) Ma
	Phyllocladus hypophyllus	New Guinea to Borneo	No overlap with Pinaceae.	4.6 Ma
	Podocarpus neriifolius clade (27 out of ~40 species)	Fiji to China	Nineteen species overlap with Pinaceae. Southern Hemisphere records are tropical, from Malesia and Pacific islands. Possibly more than one immigration into the Northern Hemisphere.	24.6 (21.3) Ma
	Podocarpus polystachyus clade (3 out of 4 species)	Australia to China	Three species overlap with Pinaceae. *P. elatus* (Australia) is embedded in this clade	14.1 (10.5) Ma
	Sundacarpus amara	Australia to Philippines	Overlaps with Pinaceae.	46.1 Ma

Americas	*Podocarpus matudae* clade (21 of ~25 species)	Americas as far north as Mexico	Six species overlap with Pinaceae. Comprises *Podocarpus* sections *Pratensis*, *Lanceolatus*, *Pumilis*, and *Nemoralis*. Southern Hemisphere records are from tropical South America. Possibly more than one immigration into the Northern Hemisphere.	26.4 (16.7) Ma
	Prumnopitys montana/ standlyi	Northern South America	No overlap with Pinaceae. Sister to *P. andina* from Chile and Argentina.	5.0 Ma
	Prumnopitys harmsiana	Northern South America	No overlap with Pinaceae. Sister to *P. exigua* from Bolivia	1.1 Ma
	Retrophyllum rospigliosii	Northern South America	No overlap with Pinaceae. Sister to *R. piresia* from tropical Brazil	7.6 Ma
Africa	*Podocarpus milanjianus*	Nigeria to Zimbabwe	No overlap with Pinaceae. Embedded in an African and Madagascan clade	3.3 Ma
	Afrocarpus mannii clade (2 species)	Ethiopia to Tanzania; Sao Tomé	No overlap with Pinaceae. Sister to a central and southern African clade.	3.9 (1.1) Ma

Ages are those estimated by Leslie et al. (2018). The stem age (the age of the divergence of the clade from its sister group) can be taken as an estimate of the maximum age of the clade. The crown age represents the age of the first divergence within the clade three of them being less than ten million years old. The fourth (a large tropical clade of *Podocarpus*) is estimated to have split from its sister group approximately 26.4 million years ago (Table 1), and is the only one to overlap geographically with the range of Pinaceae. Immigration into the Northern Hemisphere in this clade could be younger than that age, given that the crown age of the clade is 16.7 million years and several of the species involved occur in both hemispheres.

1.3 This Chapter

This chapter therefore investigates the habitat characteristics of Pinaceae and Podocarpaceae, and how these relate to niche conservatism, dispersal and habitat suitability. It will focus on climatic ranges and the physiological traits underpinning them, but will also consider the roles of competition, forest structure, and fire. Specifically it will assess some key existing ideas, with the proviso that these explanations are not mutually exclusive.

1. It is widely held that a lack of frost tolerance was a major factor in the failure of Podocarpaceae to occupy more of the northern hemisphere (Brodribb 2011). This view is based on the widespread occurrence of Pinaceae in areas with climates exposed to severe freezing, and the absence of Podocarpaceae from such climates (Figs. 1, 2, and 3). Climatic niche conservatism could explain these patterns: with Pinaceae either having existing frost tolerance or a greater capacity to evolve it than Podocarpaceae.
2. Another view is that Podocarpaceae are physiologically limited to relatively mesic environments (Brodribb and Hill 2004), whereas tolerance of dry climates is widespread in Pinaceae. Extensive areas with arid climates in the "Horse latitudes," regions with latitudes close to 30° N and to 30° S, would therefore provide substantial barriers to Podocarpaceae.
3. A third view is that Podocarpaceae are largely closed forest species, whereas many Pinaceae—especially *Pinus*—lack the shade tolerance required to be competitive with angiosperm trees in these environments. As a result the extensive rainforests of the tropics could then pose a barrier to Pinaceae (Brodribb and Feild 2008). This view is also related to fire tolerance, with a general lack of fire tolerance in Podocarpaceae (Ladd and Enright 2011) and fire tolerance in many Pinaceae. High fire frequencies in the monsoon tropics and Mediterranean climate regions (Bond 2005) would then pose barriers to Podocarpaceae but not Pinaceae.

Fig.3 (continued) of the driest quarter. Note the non-linear (fourth root) scale for (**c**) and (**d**). This analysis is based on the point record data shown in Fig. 1. The Worldclim2 30″ grid (Fick and Hijmans 2017) was used to estimate average climates for each point record. Available climate space was inferred from climatic estimates made for each grid cell in a global 3 min grid for latitudes from 54.6° S to 72° N. Trendlines were fitted using a generalised additive model implemented in mgcv (Wood 2018) in R (R Core Team 2014)

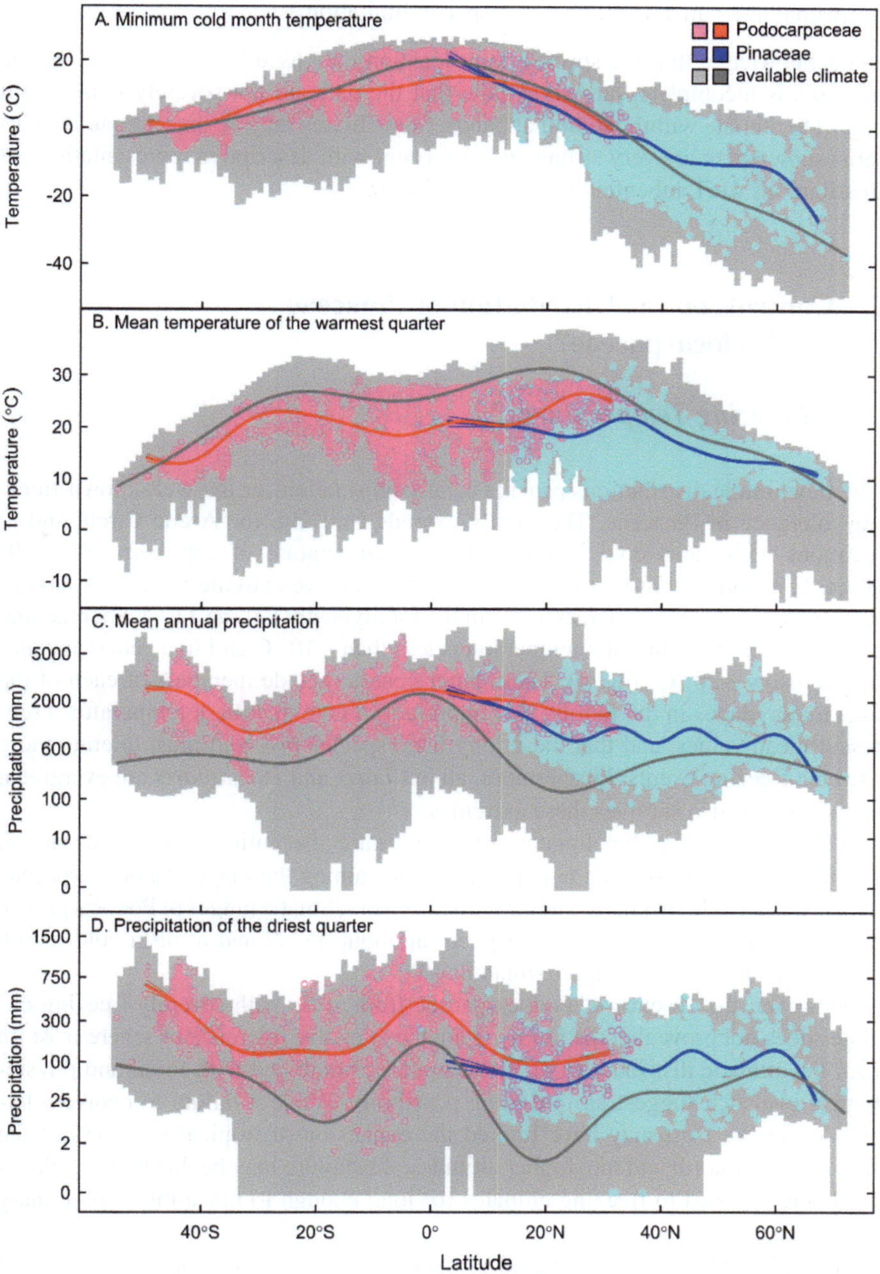

Fig. 3 Latitudinal trends in climatic ranges of Podocarpaceae (pink = point records; red = trend line) and Pinaceae (pale blue = point records; dark blue = trend line), with available climates for the latitudes (pale grey = range in 1° latitudinal bands; dark grey = trend line for median values for 0.1° latitudinal bands). Fine lines show the trend plus or minus standard error (note that these are mostly too close to the trend to be visible). (**a**) Mean minimum temperature of the coldest month; (**b**) mean temperature of the warmest quarter; (**c**) mean annual precipitation; (**d**) mean precipitation

4. A final alternative is the role of dispersal limitation.

We acknowledge that the species point location data used here in the bioclimatic analyses is incomplete, but we assume that the data are not severely biased with regard to climate within latitudinal bands. Some biases may be present, but the patterns shown here are very strong and consistent with descriptive representations of distributions (de Laubenfels 1984; Farjon 2010).

2 Controls on the Distribution of Pinaceae and Podocarpaceae?

2.1 *Freezing Temperatures?*

The bioclimatic distributions provide clear confirmation for the well-known freezing tolerance in Pinaceae. The family extends well into the Arctic Circle and to locations with mean cold month minimum temperature of approximately −40° (Figs. 1, 2, and 3). Furthermore, the capacity to evolve extreme freezing tolerance appears to be a general characteristic in the family, with 7 of the 11 genera reaching mean cold month minimum temperatures less than −10 °C and five genera extending to below −20 °C (Fig. 4). These latter genera include members of each of the four major clades in the family. It is important to note that these temperatures represent mean values, and that these long lived species involved must tolerate more extreme freezing events. Furthermore, all but *Larix* and *Pseudolarix* are evergreen, so the foliage also survives these extremes.

Biogeographic and bioclimatic distributions are also entirely consistent with an inability to tolerate freezing temperatures constraining the range of Podocarpaceae in the Northern Hemisphere. The latitudinal and bioclimatic ranges of Podocarpaceae are truncated in the northern hemisphere at about 35° N and a mean cold month minimum temperature of approximately −3°.

However, the absence of Podocarpaceae from areas with severely freezing climates does not prove that the family is unable to evolve tolerance of severe frost. In fact, the climatic distributions of Podocarpaceae combined with fossil and physiological evidence suggest an intriguing possibility. While a failure to evolve tolerance of severe frost may have limited the expansion of tropical members of the family further north and into higher altitudes, this failure may be due to these clades not being exposed to freezing climates for long enough to make this evolutionary transition.

Several southern temperate clades of Podocarpaceae occur in areas with cold winters where they have evolved significant frost tolerance, suggesting that frost tolerance in this family may not be limited by adaptive capacity, but by the availability of appropriate climates. Thus, Podocarpaceae extend to much colder climates in the Southern Hemisphere (mean cold month minimum temperature of approximately −7.6°) than in the Northern Hemisphere. The Podocarpaceae

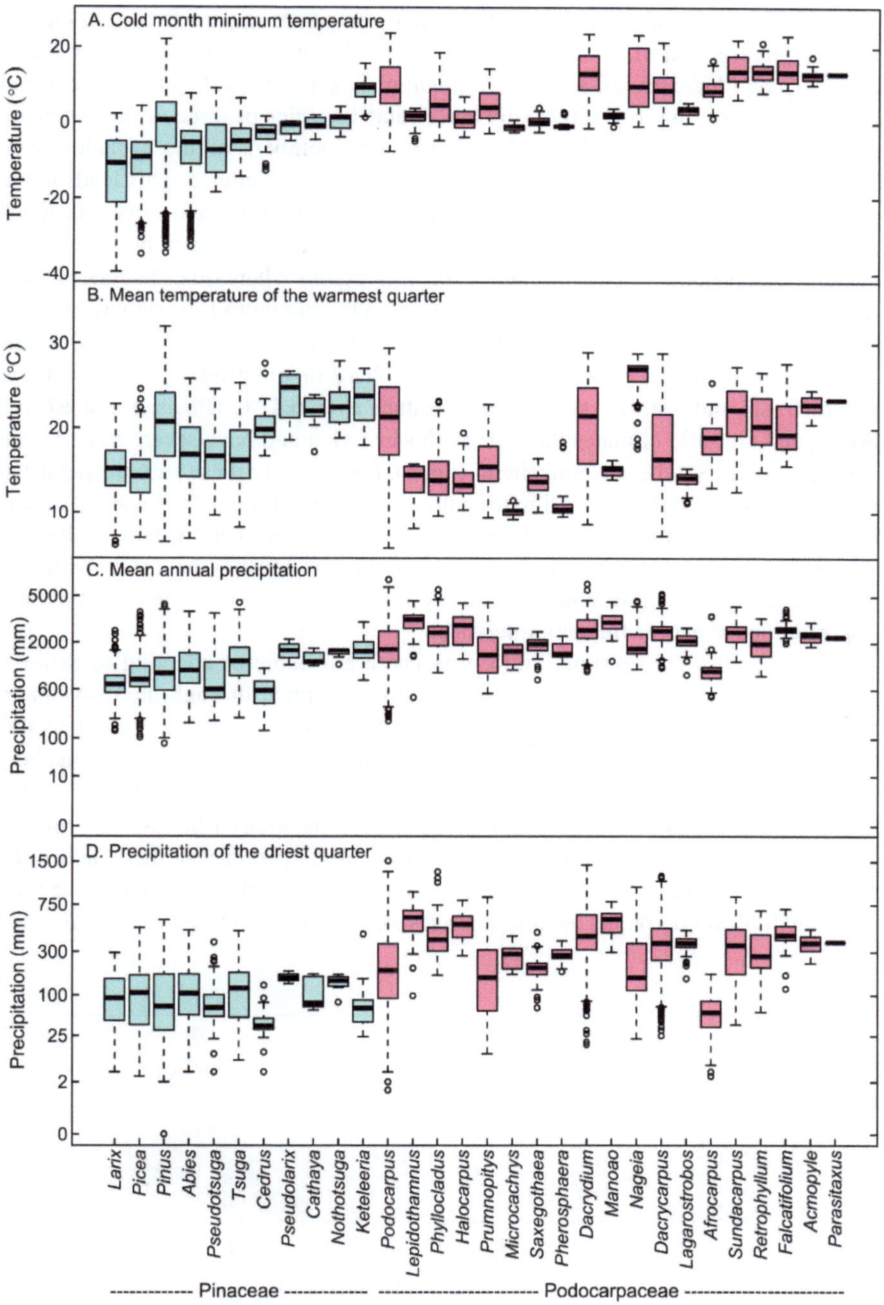

Fig. 4 Climatic ranges of genera of Podocarpaceae (pink) and Pinaceae (pale blue), arranged by minimum values of mean minimum temperature of the coldest month. (**a**) Mean minimum temperature of the coldest month; (**b**) mean temperature of the warmest quarter; (**c**) mean annual precipitation; (**d**) mean precipitation of the driest quarter. The boxes represent the inter-quartile range, the horizontal bar is the median, the whiskers represent the maximum and minimum excluding outliers, and circles are outliers. This analysis is based on the point record data shown in Fig. 1

occupying the coldest climates are all montane temperate species from four clades: *Podocarpus* section *Australis* (a clade of small leaved Australasian species), *Halocarpus*, *Phyllocladus*, and *Lepidothamnus*. Sakai and Wardle (1978) measured tolerance of freezing to ~−20 °C in New Zealand species of three of these genera, and Feild and Brodribb (2001) found that two members of these clades from Tasmania also showed moderately high levels of freeze tolerance. This leads to the possibility that Podocarpaceae has the capacity to evolve extreme freezing tolerance, but has failed to do so partly because of niche conservatism, but also partly because of a paucity of these climates except in regions where other factors exclude these species. Indeed, severe freezing climates in the Southern Hemisphere are very rare, except in conditions with summers that are too cool to support woody plants. Thus, the altitudinal treeline occurs in areas with a 3 month growing season without severe frosts and with an average temperature over 6.4 °C (Paulsen and Körner 2014), and non-arid climates that satisfy this criterion and have severe freezing are very rare in the Southern Hemisphere (Fig. 5). Furthermore these patterns persisted if the criteria were relaxed to climates with mean temperatures of the warmest quarter as low as 5 °C to allow for woody growth above treelines. Although climates were colder during the many glacial periods of the Pleistocene (the last 2.58 million years), considerably drier climates at those times may have restricted evolutionary opportunities to evolve freezing tolerance.

It may be that, given time, Podocarpaceae are capable of adapting to extreme cold, however the clades of Podocarpaceae that migrated through the tropics into the Northern Hemisphere may have been warm-adapted taxa that have not had sufficient time to evolve significant frost tolerance. These Northern Hemisphere Podocarpaceae clades appear to be Cenozoic arrivals (Table 1), and most of the Cenozoic has been considerably warmer than the Quaternary (the last 2.58 million years) (Zachos et al. 2001). As a result the tropical Podocarpaceae that reached the Northern Hemisphere have only been exposed to climates that are compatible with the growth of woody plants but also have severe freezing events since they migrated

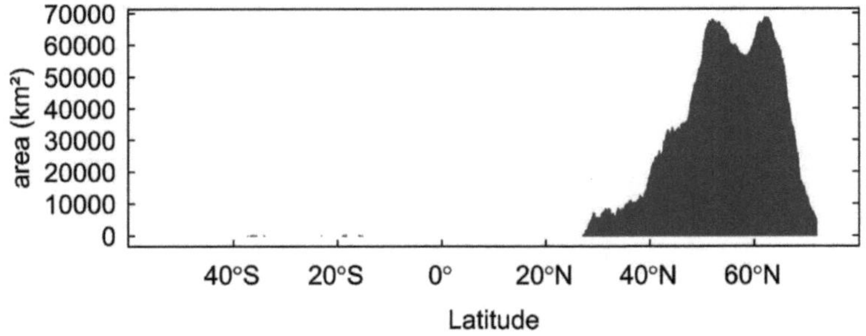

Fig. 5 Latitudinal variation in areas with climate that is both suitable for tree growth and exposed to substantial freezing temperatures. The criteria used were that the mean temperature of the warmest quarter is more than 6.4 °C, mean minimum temperatures of the coldest month is less than −10 °C, and mean annual precipitation is greater than 250 mm. This analysis was based on the point record data shown in Fig. 1

to northern parts of their range. Such climates are common in the Northern Hemisphere at latitudes above 25° N, but are largely absent from the tropics (Fig. 5). Furthermore, given that freezing tolerance comes with some costs (Hacke and Sperry 2001) then either selection for competitive ability in tropical forests (see below) or drift may have eliminated the genetic capacity of the tropical podocarps to evolve severe frost tolerance.

2.2 *Hydroclimate*

The bioclimatic ranges of the species indicate that hydroclimate has significant impacts on both families, with both Podocarpaceae and Pinaceae occupying wetter climates than the median available climate at all latitudes (Fig. 3). Thus, Podocarpaceae has failed to occupy the driest climates across its latitudinal range, and although Pinaceae reach some of the driest environments at some latitudes, the family is more likely to be found in relatively mesic climates across its range. These patterns could be direct consequences of physiology, but may also reflect the impacts of other factors such as fire and competition as discussed below.

However, even though the climatic ranges of the families overlap substantially across all latitudes, Podocarpaceae mostly occupy wetter climates than Pinaceae. Thus, for both mean annual temperature and precipitation of the driest quarter the median values of most genera of Podocarpaceae are higher than those of any genus of Pinaceae (Fig. 4), and in latitudes where both families occur, the median values for Podocarpaceae are substantially higher than those for Pinaceae (Fig. 4) indicating that the differences between family are not direct effects of latitudinal patterns in available climates. Furthermore, many Pinaceae occur in dry habitats (Figs. 3 and 4), with nearly 80 species occurring in areas with less than 450 mm mean annual precipitation. These species are phylogenetically diverse, including species from 7 of the 12 genera (Fig. 4), including members of each of the subfamilies recognised by Farjon (2010). Thus, Pinaceae in general appears to have the capacity to evolve tolerance of dry climates.

All but a few species of Podocarpaceae are restricted to relatively mesic environments—thus median mean annual precipitation is greater than 1500 mm for all genera (Fig. 4). However, there is evidence that Podocarpaceae can evolve tolerance of dry climates, although this appears to be even rarer than the evolution of tolerance of freezing climates. One species (*Podocarpus elongatus* from southern Africa) mostly occurs in areas with less than 450 mm mean annual precipitation, and extends down to ~200 mm. In addition, some other podocarps occupy seasonally dry climates, especially species of *Afrocarpus*, *Podocarpus*, and *Prumnopitys*, growing in monsoonal tropical climates with mean precipitation of the driest quarter less than 25 mm, and in some cases less than 5 mm (Fig. 3). Most of these species experience high summer rainfall, but must have mechanisms to tolerate the severe dry season. These mechanisms are not well known but may involve having xylem with high resistance to cavitation and/or being deeply rooted. Importantly, the genera in seasonally dry climates are phylogenetically diverse: they include members of both of the major clades in the family—*Prumnopitys* is a member of the "prumnopityoid" clade whereas *Afrocarpus* and *Podocarpus* are members of the "tropical" clade (Leslie et al. 2018). Thus, the capacity to potentially evolve into such climates appears to be present across the family.

However, the number of Podocarpaceae species that have become successful in dry climates is small, and there has been abundant opportunity for such evolution to happen. Thus, the niche conservatism in precipitation-related dimensions in Podocarpaceae is greater than the conservatism in thermal dimensions—frost tolerance has evolved to a significant degree in the family in the temperate Southern Hemisphere areas in spite of limited evolutionary opportunities (Figs. 4 and 5), but the evolution into arid climates in Podocarpaceae has been very rare relative to the major opportunities for it to happen. Large areas of arid climates occur in the mid-latitudes of the Southern Hemisphere (Fig. 2), and these areas can be found in all the major regions containing Podocarpaceae in Australia, South America and Africa, and even to a degree in New Zealand.

Many of the currently dry regions of the Southern Hemisphere have experienced arid or semi-arid climates for millions of years. For example, climates in Australia became drier through the Neogene, such that an extensive semi-arid zone was present by ~7 million years ago (Byrne et al. 2008, 2011; Macphail 2007). Furthermore, many (and perhaps all) of these areas were occupied by mesic members of the family in the past. Thus, there is fossil clear evidence of multiple mesic genera growing in regions of Australia (Macphail 2007), New Zealand (Jordan et al. 2011) and Patagonia (Wilf et al. 2013) that now have mean annual precipitation far lower than in the extant range of those genera. This indicates that these genera failed to evolve to drying climates in the Neogene or Quaternary. Some genera, such as *Willungia* and *Sigmaphyllym*, went extinct (Hill and Brodribb 1999), but most contracted to or migrated to mesic areas. The very limited colonization of drier climates by Podocarpaceae contrasts markedly with the success of another southern conifer clade, *Callitris*, in Australia (Brodribb et al. 2013).

This idea of greater niche conservatism in precipitation-related variables than temperature variables in Podocarpaceae is consistent with an analysis of fossil assemblages which indicated much stronger temporal conservatism in precipitation-related variables than temperature variables in the southern conifers (Brown et al. 2021). Clear physiological evidence that water stress is a major factor in the restriction of Podocarpaceae to mesic environments supports the concept of strong niche conservatism in hydroclimate. One possible limitation among Podocarpaceae is that they tend to produce leaves that are particularly leaky to water vapour during stress. The leaves of species of Podocarpaceae that live in dry climates are less leaky than species that live in wet climates: specifically, the mean precipitation of the driest quarter is positively correlated with the minimum conductance of the epidermis to water vapour (Brodribb et al. 2015). In other words, there is extreme, although not absolute, niche conservatism in precipitation-related dimensions in Podocarpaceae.

2.3 *Indirect Impacts: Forest Structure, Fire, and Competition*

Although the low temperatures and dry climates appear to constrain both major conifer families directly through lethal damage or severe reduction of growth, the most important impacts of summer temperatures and mesic climates may be indirect. The main effects appear to derive from competition with other species (especially angiosperms) and from fire. Thus, Pinaceae in the tropics occupy considerably more open vegetation than Podocarpaceae, as indicated by Pinaceae occurring areas with lower estimated leaf area index than Podocarpaceae (Fig. 6), consistent with very limited occupation of closed forest by Pinaceae.

The bioclimatic evidence supports the idea that Pinaceae may be constrained by a limited ability to compete with angiosperms in tropical rainforest (Brodribb and Feild 2008). Furthermore it also supports the idea that Podocarpaceae performs well in upland tropical rainforest but is much less competitive in the lowlands, as suggested by de Laubenfels (1984). Both Pinaceae and Podocarpaceae occur across a wide climatic range in the tropics (Fig. 2), demonstrating that the direct effects of warm and mesic climates do not limit the distributions of these families. However, the pattern is very different when one considers the families' performance relative

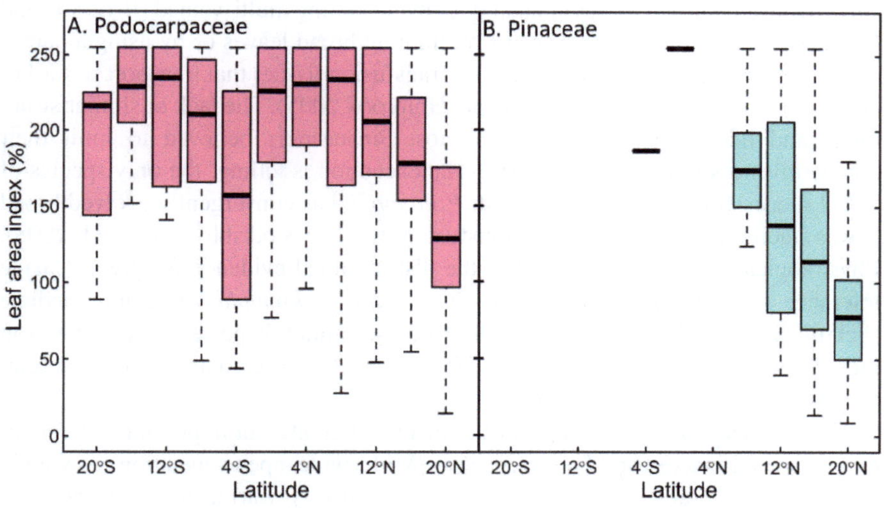

Fig. 6 Boxplots of records of Podocarpaceae (**a**), Pinaceae (**b**) showing leaf area index (LAI) according to 4° latitudinal band in tropical areas. Note the higher LAI for Podocarpaceae than Pinaceae both overall and in the region where the families overlap. This data has considerable uncertainty—LAI is estimated from satellite data for 0.1° grid cells, and vegetation turnover is at much finer scales than this. However, the broad trends are likely to be informative, but some extreme values may represent errors. For example some extreme values may be cases where species occupy small areas of open vegetation within cells dominated by closed forest. The boxes represent the inter-quartile range, the horizontal bar is the median, the whiskers represent the maximum and minimum excluding outliers. Data is from NASA (https://neo.gsfc.nasa.gov/view.php?datasetId=MOD15A2_M_LAI)

to the amount of available environment. Figure 7 shows the relationship of Podocarpaceae and Pinaceae to summer temperature in tropical areas with mesic climates (i.e. areas of mesic tropical rainforest). Although over 90% of the land with such climates has mean summer temperatures above 23.5°, Podocarpaceae and Pinaceae appear to be rare in such climates. Podocarpaceae show strong preference for areas with mean summer temperatures between 9 and 15 °C and, although Pinaceae are most common in that same range, are generally much rarer than Podocarpaceae at these low latitudes (Fig. 7). Thus, in the tropics Podocarpaceae are observed in rainforest much more often than Pinaceae (Fig. 6), and much more often observed in the cooler upland forest than in the lowlands.

The success of Podocarpaceae and failure of *Pinus* in rainforest are likely to be linked to leaf characteristics. Wider leaves are more competitive in warm and mesic climates (Wright et al. 2017), but the leaves of all Pinaceae and most Podocarpaceae have only one vein—the midrib- posing a significant limitation to the maximum leaf width. The reason for this is that water outside the leaf xylem encounters a very large resistance to flow as it moves through cell walls or between cells, rather than through the highly efficient tubes of xylem tissue (Brodribb et al. 2007). Thus, for the stomata and photosynthetic mesophyll cells to remain hydrated they must remain very close to specialized water conducting cells. This constrains the maximum width of conifer needles. However, Podocarpaceae have a diverse array of ways to develop broad leaves or analogues by becoming multi-veined (Brodribb and Hill 1997), by employing leaf arrays resembling broad leaves or by using an array of radially arranged tracheids (accessory transfusion tissue) that transport water laterally out from the midrib (Brodribb and Holbrook 2005). The lack of similar structures, and hence broader leaves, in most Pinaceae is believed to limit their competitiveness in tropical rainforest. The exception is telling: the only species of *Pinus* successful in tropical rainforest, *P. krempfii*, has convergently evolved broad leaves facilitated by accessory transfusion tissue (Brodribb and Feild 2008). Physiological measurements support the distributional evidence for greater shade tolerance in Podocarpaceae than in *Pinus*, with *P. krempfii* being intermediate (Schmiege et al. 2021). The evolution of fleshy, animal-dispersed female cones in most Podocarpaceae (Contreras et al. 2017) may also have contributed to the relative success of the family in rainforest.

Fire is a second indirect impact of climate that creates an important differentiation between Pinaceae and Podocarpaceae. World-wide, open vegetation growing in conditions favourable for plant growth (i.e. relatively warm, non-arid and non-oligotrophic conditions), are typically associated with fire (Bond 2005). This contributes to the success of Pinaceae and lack of success of Podocarpaceae in these regions. Fire adaptations including serotiny and thick bark are widespread and apparently ancient in *Pinus* (He et al. 2012), and this is reflected in the widespread incidence of the genus in fire prone areas. Fire adaptation also occurs in some species of other genera of Pinaceae, including *Picea*, *Larix*, *Abies*, and *Pseudotsuga* (He et al. 2012). However, other genera in the family appear to have late successional niches, and some species are characteristic of mature forests. In contrast, most species of Podocarpaceae lack any adaptations to fire and are restricted to

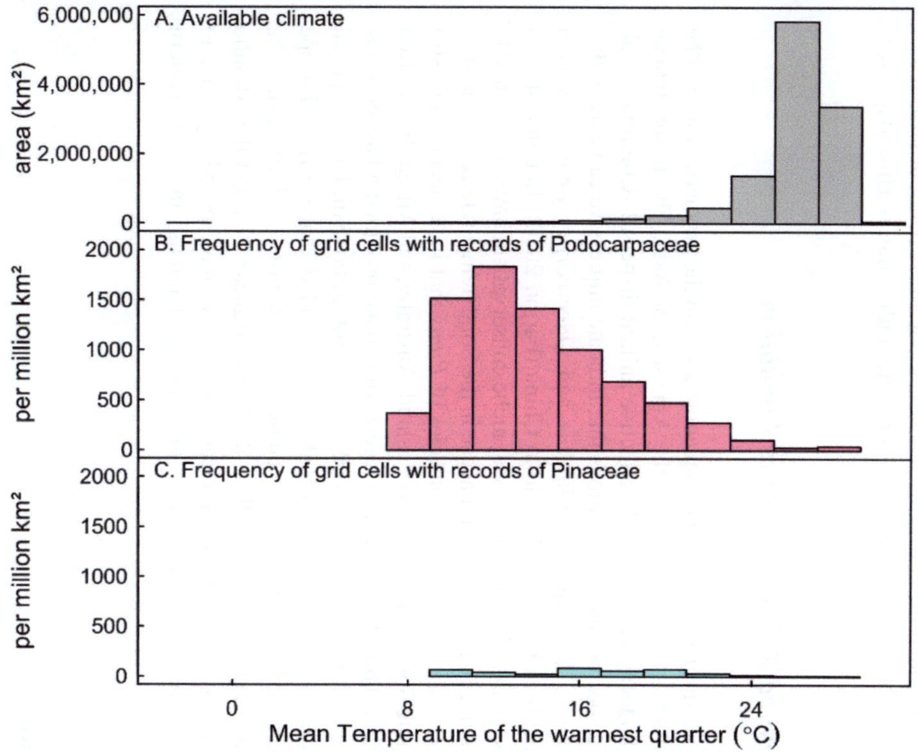

Fig. 7 A comparison of Pinaceae and Podocarpaceae in areas suitable for the growth of mesic tropical rainforest. (**a**) Tropical land area (between 23.4° S and 23.4° N) with high levels of precipitation both overall (mean annual precipitation greater than 1500 mm) and in the dry season (mean precipitation of the driest quarter greater than 100 mm) versus mean temperature of the warmest quarter. The climate limits were set to correspond to the distribution of rainforest in Brazil and Central Africa. (**b**, **c**) Frequency of 3′ (~2 × 2 km) grid cells containing Podocarpaceae (**b**) and Pinaceae (**c**) relative to the areas from panel (**a**). Note, the data on which this is based (Fig. 1), is likely to under-represent the incidence of Podocarpaceae and Pinaceae in each band, but there is no reason to suspect major biases among bands or between families

rainforests or other mesic areas with few fires (Farjon 2010; Jordan et al. 2016). The main exceptions are that *Podocarpus drouynianus* and *P. spinulosus* from Australia and *P. elongatus* from southern Africa can resprout from the base of the plant after fire and occur in fire-prone sclerophyll plant communities (Farjon 2010; Ladd and Enright 2011). This capacity may have evolved in response to Neogene fires—according to Leslie et al. (2018) the former two species form a clade with a crown age of 16.2 million years, and the latter has a stem age of 6 million years. Animal-dispersal of seed allows some species (e.g. *Phyllocladus aspleniifolius*) to colonise areas after relatively small fires (Kirkpatrick 1977), but this is better described as a generalised dispersal trait that allows use of burnt sites rather than as a fire adaptation.

2.4 The Limited Role of Dispersal Limitation

Dispersal traits are unlikely to be important constraints on the northern limits on the distribution of Podocarpaceae in either the Americas or Asia. The fleshy female cones of most Podocarpaceae provide effective animal dispersal (Contreras et al. 2017), and some species have distributions that indicate impressive capacity to disperse. For example, *Podocarpus neriifolius* and *Dacrycarpus imbricatus* both extend from islands in the South Pacific to China (Farjon 2010). In contrast, it is possible that dispersal traits, combined with niche conservatism have contributed to the absence of native populations of Pinaceae in the South temperate zone. The suitability of habitat is shown by multiple species of *Pinus* that have become invasive weeds in temperate South America, New Zealand, Australia, and Africa (Richardson and Nsikani 2021). Thus, the equatorial forests may have provided a barrier to the southward movement of Pinaceae (Brodribb and Feild 2008). Similarly, dispersal limitations may have contributed to the absence of Podocarpaceae from Europe because of the biogeographic barrier formed by the arid regions of North Africa, the Middle East and central Asia, as well as the Mediterranean Sea and the Atlantic Ocean. It is also possible that the Mexican deserts may have posed a barrier to Podocarpaceae dispersing northwards into mesic habitats in south-eastern United States.

3 Conclusions

The major implication of the combination of fossil, phylogenetic, physiological, and bioclimatic evidence is that phylogenetic niche conservatism underpins the failures of Podocarpaceae to expand more widely in the Northern Hemisphere and of Pinaceae to invade the Southern Hemisphere. A failure of tropical Podocarpaceae to evolve tolerance to arid or extreme freezing climates and a failure of Pinaceae to compete in closed forests drove this lack of expansion. However, these do not represent absolute limits on these families. Thus Pinaceae has evolved broad leaves at least once that allowed effective colonization of tropical rainforest, and one Podocarpaceae has evolved substantial tolerance of arid climates. In addition, the lack of Podocarpaceae in severely freezing environments in the southern temperate areas may be due more to an almost complete lack of places suitable for woody plant growth with freezing winters than to underlying constraints on physiology. Evolution of traits that allow the colonization of substantially different habitats (Crisp et al. 2009) is extremely slow, but is possible and is contingent on the availability of that habitat.

Several areas of further study would enhance our understanding of the constraints on and evolutionary potential of Pinaceae and Podocarpaceae. These include studies to refine our understanding of the interaction of freezing tolerance and drought tolerance; the impacts of greater temperature on water stress; the control of drought tolerance in Pinaceae, which, unlike in other conifers, is not well predicted by vulnerability to cavitation (Brodribb et al. 2015); and the mechanisms underpinning freezing tolerance, especially in Podocarpaceae.

A history of slow evolution and climate-sensitive niche boundaries in Pinaceae and Podocarpaceae explains their overrepresentation among the endangered species catalogues such as the IUCN Redlist. Other conifer groups show similar patterns, highlighting the need for very close observation of conifer health and distribution under the current conditions of very fast climate change.

References

Akkemik Ü (2021) A new fossil *Cedrus* species from the early Miocene of northwestern Turkey and its possible affinities. Palaeoworld 30(4):746–756

Brodribb TJ (2011) A functional analysis of podocarp ecology. Smithsonian contributions to botany. 95:165–173.

Bond WJ (2005) Large parts of the world are brown or black: a different view on the "Green World" hypothesis. J Veg Sci 16(3):261–266. https://doi.org/10.1111/j.1654-1103.2005.tb02364.x

Brodribb TJ, Feild TS (2008) Evolutionary significance of a flat-leaved Pinus in Vietnamese rainforest. New Phytol 178(1):201–209. https://doi.org/10.1111/j.1469-8137.2007.02338.x

Brodribb TJ, Hill RS (1997) Light response characteristics of a morphologically diverse group of southern hemisphere conifers as measured by chlorophyll fluorescence. Oecologia 110(1):10–17

Brodribb TJ, Hill RS (2004) The rise and fall of Podocarpaceae in Australia – a physiological explanation. In: Hemsley AR, Poole IR (eds) The evolution of plant physiology from whole plants to ecosystems. Elsevier Academic Press, Amsterdam, pp 381–400

Brodribb TJ, Holbrook NM (2005) Water stress deforms tracheids peripheral to the leaf vein of a tropical conifer. Plant Physiol 137(3):1139–1146. https://doi.org/10.1104/pp.104.058156

Brodribb TJ, Feild TS, Jordan GJ (2007) Leaf maximum photosynthetic rate and venation are linked by hydraulics. Plant Physiol 144(4):1890–1898. https://doi.org/10.1104/pp.107.101352

Brodribb TJ, Bowman DMJS, Grierson PF, Murphy BP, Nichols S, Prior LD (2013) Conservative water management in the widespread conifer genus Callitris. AoB Plants 5:plt052. https://doi.org/10.1093/aobpla/plt052

Brodribb TJ, McAdam SAM, Jordan GJ, Martins SCV (2015) Conifer species adapt to low-rainfall climates by following one of two divergent pathways. Proc Natl Acad Sci USA 111(40):14489–14493. https://doi.org/10.1073/pnas.1407930111

Brown MJM, Holland BR, Jordan GJ (2020) Hyperoverlap: detecting biological overlap in n-dimensional space. Methods Ecol Evol 11(4):513–523. https://doi.org/10.1111/2041-210X.13363

Brown MJM, Brodribb TJ, Jordan GJ (2021) No-analogue associations in the fossil record of southern conifers reveal conservatism in precipitation, but not temperature axes. Glob Ecol Biogeogr 30(12):2455–2466. https://doi.org/10.1111/geb.13398

Byrne M, Yeates DK, Joseph L et al (2008) Birth of a biome: insights into the assembly and maintenance of the Australian arid zone biota. Mol Ecol 17(20):4398–4417

Byrne M, Steane DA, Joseph L et al (2011) Decline of a biome: evolution, contraction, fragmentation, extinction and invasion of the Australian mesic zone biota. J Biogeogr 38(9):1635–1656. https://doi.org/10.1111/j.1365-2699.2011.02535.x

Cohen KM, Finney SC, Gibbard PL, Fan J-X (2013) The ICS international chronostratigraphic chart. Episodes 36(3):199–204

Contreras DL, Duijnstee IAP, Ranks S, Marshall CR, Looy CV (2017) Evolution of dispersal strategies in conifers: functional divergence and convergence in the morphology of diaspores. Perspect Plant Ecol Evol Syst 24:93–117. https://doi.org/10.1016/j.ppees.2016.11.002

Crisp MD, Arroyo MTK, Cook LG et al (2009) Phylogenetic biome conservatism on a global scale. Nature 458(7239):754–U90. https://doi.org/10.1038/nature07764

de Laubenfels DJ (1984) Coniferales. Flora Malesiana-Series 1. Spermatophyta 10(1):337–347

Ding S-T, Chen S-Y, Ruan S-C et al (2021) First fossil record of *Nothotsuga* (Pinaceae) in China: implications for palaeobiogeography and palaeoecology. Hist Biol 33(12):3617–3624

Doweld AB (2018) On *Cathaya*, living and fossil (Pinaceae). Taxon 67(1):196–202

Erwin DM, Schorn HE (2005) Revision of the conifers from the Eocene Thunder Mountain flora, Idaho, USA. Rev Palaeobot Palynol 137(3–4):125–145

Farjon A, Filer D (2013) An atlas of the world's conifers. An analysis of their distribution, biogeography, diversity and conservation status. Brill, Leiden.

Farjon A (2010) A handbook of the world's conifers. Brill, Leiden

Feild TS, Brodribb T (2001) Stem water transport and freeze-thaw xylem embolism in conifers and angiosperms in a Tasmanian treeline heath. Oecologia 127(3):314–320

Fick SE, Hijmans RJ (2017) WorldClim 2: new 1-km spatial resolution climate surfaces for global land areas. Int J Climatol 37(12):4302–4315. https://doi.org/10.1002/joc.5086

Hacke UG, Sperry JS (2001) Functional and ecological xylem anatomy. Perspect Plant Ecol Evol Syst 4(2):97–115

He T, Pausas JG, Belcher CM, Schwilk DW, Lamont BB (2012) Fire-adapted traits of Pinus arose in the fiery Cretaceous. New Phytol 194(3):751–759

Hill RS, Brodribb TJ (1999) Southern conifers in time and space. Aust J Bot 47(5):639–696. https://doi.org/10.1071/bt98093

Jin J, Qiu J, Zhu Y, Kodrul TM (2010) First fossil record of the genus *Nageia* (Podocarpaceae) in south China and its phytogeographic implications. Plant Syst Evol 285(3):159–163. https://doi.org/10.1007/s00606-010-0267-4

Jordan GJ (2001) An investigation of long-distance dispersal based on species native to both Tasmania and New Zealand. Aust J Bot 49(3):333–340. https://doi.org/10.1071/BT00024

Jordan GJ, Carpenter RJ, Bannister JM, Lee DE, Mildenhall DC, Hill RS (2011) High conifer diversity in Oligo-Miocene New Zealand. Aust Syst Bot 24(2):121–136. https://doi.org/10.1071/SB11004

Jordan GJ, Harrison PA, Worth JRP, Williamson GJ, Kirkpatrick JB (2016) Palaeoendemic plants provide evidence for persistence of open, well-watered vegetation since the Cretaceous. Glob Ecol Biogeogr 25(2):127–140. https://doi.org/10.1111/geb.12389

Kirkpatrick JB (1977) Native vegetation of the west coast region of Tasmania. In: Banks MR, Kirkpatrick JB (eds) Landscape and man, Papers and proceedings of the Royal Society of Tasmania. Royal Society of Tasmania, Hobart, pp 55–80

Klymiuk AA, Stockey RA (2012) A lower Cretaceous (Valanginian) seed cone provides the earliest fossil record for *Picea* (Pinaceae). Am J Bot 99(6):1069–1082

Ladd PG, Enright NJ (2011) Ecology of fire-tolerant podocarps in temperate Australian forests. Smithson Contrib Bot 95:141–156

Lakhanpal RN (1958) The Rujada flora of west central Oregon, University of California publications in geological sciences, vol 35. University of California Press, Berkeley, pp 1–65

LePage BA (2001) New species of *Picea* A. Dietrich (Pinaceae) from the middle Eocene of Axel Heiberg Island, Arctic Canada. Bot J Linn Soc 135(2):137–167

Lepage BA (2003) A new species of *Tsuga* (Pinaceae) from the middle Eocene of Axel Heiberg Island, Canada, and an assessment of the evolution and biogeographical history of the genus. Bot J Linn Soc 141(3):257–296

LePage BA (2009) Earliest occurrence of *Taiwania* (Cupressaceae) from the early Cretaceous of Alaska: evolution, biogeography, and paleoecology. Proc Acad Natl Sci Phila 158:129–158

LePage BA, Basinger J (1995a) The evolutionary history of the genus *Larix* (Pinaceae). USDA For Serv Int Res Stn GTR-INT 319:19–29

LePage BA, Basinger JF (1995b) Evolutionary history of the genus *Pseudolarix* Gordon (Pinaceae). Int J Plant Sci 156(6):910–950

Leslie AB, Beaulieu J, Holman G et al (2018) An overview of extant conifer evolution from the perspective of the fossil record. Am J Bot 105(9):1531–1544. https://doi.org/10.1002/ajb2.1143

Losos JB (2008) Phylogenetic niche conservatism, phylogenetic signal and the relationship between phylogenetic relatedness and ecological similarity among species. Ecol Lett 11(10):995–1003

Macphail MK (2007) Australian palaeoclimates: Cretaceous to tertiary. A review of palaeobotanical and related evidence to the year 2000. CRC LEME, Bentley

Morley RJ (2000) Origin and evolution of tropical rain forests. Wiley

Morley RJ (2011) Dispersal and paleoecology of tropical podocarps. Smithson Contrib Bot 95:21–40

Paulsen J, Körner C (2014) A climate-based model to predict potential treeline position around the globe. Alp Bot 124(1):1–12

R Core Team (2014) R: a language and environment for statistical computing. R Foundation for Statistical Computing, Vienna

Richardson DM, Nsikani MM (2021) Mediterranean pines as invasive species in the Southern Hemisphere. In: Ne'eman G, Nsikani MM (eds) Pines and their mixed forest ecosystems in the Mediterranean Basin, Managing forest ecosystems. Springer Nature Switzerland, pp 83–99

Sakai A, Wardle P (1978) Freezing resistance of New Zealand trees and shrubs. N Z J Ecol 1:51–61

Schmiege SC, Buckley BM, Stevenson D, Cuong TQ, Nam LC, Griffin KL (2021) Contrasting physiological traits of shade tolerance in *Pinus* and Podocarpaceae native to a tropical Vietnamese forest: insight from an aberrant flat-leaved pine. Tree Physiol 41(2):223–239

Wilf P, Cúneo NR, Escapa IH, Pol D, Woodburne MO (2013) Splendid and seldom isolated: the paleobiogeography of Patagonia. Annu Rev Earth Planet Sci 41:561–603

Wilf P, Donovan MP, Cúneo NR, Gandolfo MA (2017) The fossil flip-leaves (*Retrophyllum*, Podocarpaceae) of southern South America. Am J Bot 104(9):1344–1369. https://doi.org/10.3732/ajb.1700158

Wood S (2018) Package "mgcv". R package version 1.8

Wright IJ, Dong N, Maire V et al (2017) Global climatic drivers of leaf size. Science 357(6354):917–921. https://doi.org/10.1126/science.aal4760

Wu M-X, Huang J, Su T, Leng Q, Zhou Z-K (2020) *Tsuga* seed cones from the late Paleogene of southwestern China and their biogeographical and paleoenvironmental implications. Palaeoworld 29(3):617–628

Wu J-Y, Chen H, Ruan S-C et al (2021a) Fossil leaves of *Podocarpus* subgenus *Foliolatus* (Podocarpaceae) from the Pliocene of southwestern China and biogeographic history of *Podocarpus*. Rev Palaeobot Palynol 287:104380

Wu XK, Zavialova NE, Kodrul TM et al (2021b) Northern Hemisphere megafossil of *Dacrycarpus* (Podocarpaceae) from the Miocene of South China and its evolutionary and paleoecological implications. J Syst Evol 59(2):352–374. https://doi.org/10.1111/jse.12534
Xiang X, Cao M, Zhou Z (2007) Fossil history and modern distribution of the genus *Abies* (Pinaceae). Front For China 2(4):355–365
Zachos J, Pagani H, Sloan L, Thomas E, Billups K (2001) Trends, rhythms, and aberrations in global climate 65 Ma to present. Science 292(5517):686–693

The Anatomy of Pine Wood and Leaves

Jean-Christophe Domec and Daniel M. Johnson

Abstract The genus *Pinus* represents ~110 species with a global distribution. The genus is divided into two subsections: subgenus *Pinus*, or the hard pines, and subgenus *Strobus*, or the soft pines. These subgenera differ in their ecology, physiology, and anatomy. In this chapter, we compare the anatomy of wood and leaves in subgenera *Pinus* and *Strobus*. In general, species in *Pinus* have two to three needles per fascicle with each having two vascular bundles per leaf, whereas species in *Strobus* tend to have four to five needles per fascicle (but sometimes one or two) with each needle having one vascular bundle. Additionally, leaves of *Strobus* tend to have greater amounts of mesophyll tissue per volume than do those from subgenus *Pinus*. The wood of soft pines tends to be less dense than wood in hard pines and soft pines appear to be more vulnerable to embolism than hard pines. Other wood differences between hard and soft pines include maximum hydraulic conductivities, earlywood-latewood transition, ray parenchyma pitting patterns, and the amount of parenchyma (as a percentage of total wood volume) which may be related to the amounts of resin produced and associated vulnerability to pathogens. These anatomical differences are placed in the contexts of both physiology and ecology of this economically and ecologically important genus.

J.-C. Domec (✉)
Department of Forestry, Bordeaux Sciences Agro, UMR INRAE-ISPA 1391, Gradignan, France

Nicholas School of the Environment, Duke University, Durham, NC, USA
e-mail: jc.domec@agro-bordeaux.fr

D. M. Johnson
Warnell School of Forestry and Natural Resources, University of Georgia, Athens, GA, USA

D. M. Johnson et al. (eds.), *Physiology and Ecology of Pines Worldwide*, Tree Physiology 9, https://doi.org/10.1007/978-3-032-23001-0_2

1 Introduction

The genus *Pinus* represents between 105 and 124 species of evergreen conifers of the pine family (Pinaceae; Gernandt et al. 2008). The earliest fossil record of *Pinus* is from 130 Ma (late Jurassic and early Cretaceous; Millar 1998) corresponding to what phylogenetic analyses have recently estimated (155.3 Ma; Jin et al. 2021). Pines diverged ~300 Ma in Laurasia, one of the two ancient supercontinents produced by the first split of Pangaea (see chapter "North-South Rivalry: Pinaceae Versus Podocarpaceae"). By mid-Cretaceous, the pines had separated into two lineages, which are further divided into sections and subsections (Keeley 2012). Both subgenera are thought to have diverged during the late Jurassic (Stull et al. 2021) into the subgenus *Pinus* (formerly *Diploxylon*), or hard pines, and subgenus *Strobus* (formerly *Haploxylon*), or soft pines, characterized by different morphological but also anatomical features. Soft pines have needles in fascicles (anatomically short shoots) of four to five (or occasionally one or two), and stalked cones with scales lacking prickles. The sheaths of the leaf clusters are deciduous, and the leaves contain a single fibrovascular bundle. Hard pines have needles in fascicles of two or three, cone scales with prickles, and a large amount of resin. The sheaths of the leaf clusters are persistent, and the leaves have two fibrovascular bundles. However, contemporary pines have diverged and diversified during the Oligocene and Miocene 25–40 Ma into species adapted to extreme climatic sites (arid or cold) and to areas prone to regular fires (Millar 1998; Willyard et al. 2007). These changes had an immediate impact on anatomical traits of those pines adapted to environmental extremes, and drove the high pine speciation rates observed during this epoch (López et al. 2002). With the exception of Sumatran pine (*Pinus merkusii*) that crosses the equator, all current pines are native to the Northern Hemisphere; yet as important forestry species, they are now distributed throughout the world (Richardson et al. 1994). In this chapter, our goal is to provide a synthesis of pine wood and needle anatomical features and potentially ways to differentiate *Pinus* species from each other. Because wood and leaf traits of needle-leaved gymnosperms, including pines, strongly differ from angiosperm anatomical traits, we start with basic information about similarities and differences in wood and leaf traits between the major pine species. Then we focus on the properties that are most important for tree structure and function according to hydraulic and mechanical principles. Finally, we present the morphology and anatomy of pine leaves and discuss the functional significance of wood and needle anatomy for water transport, carbon gain and species ecological potentials.

2 Macroscopic Differences Between Wood of Soft and Hard Pines

Eastern white pine (*Pinus strobus*), sugar pine (*Pinus lambertiana*), western white pine (*Pinus monticola*), and limber pine (*Pinus flexilis*) are the principal and economically important soft pine species. Differences in macroscopic wood structure between soft pine species are difficult to differentiate visually or even with a hand lens. The use of a microscope and analyses of radial and tangential sections are usually required. Pine stem cross-section is made of distinct dead cells of dark tone defined as heartwood (i.e., the wood closest to the center of the tree) and of hydroactive cells defined as sapwood (i.e., the wood closest to the bark of a tree transporting xylem sap). The proportion of sapwood within a tree decreases with tree age and is indistinguishable between hard and soft pines. However, unlike in hard pines, soft pine latewood zone is always narrow and the growth ring boundaries inconspicuous, often undulated, especially in the proximity of latewood resin canals (Fig. 1). They can also be distinguished from the hard pines by having uniformly thin-walled tracheids and smaller resin canals. The hard pine group is somewhat opposite of the soft pines, not only in physical characteristics such as hardness and wood density,

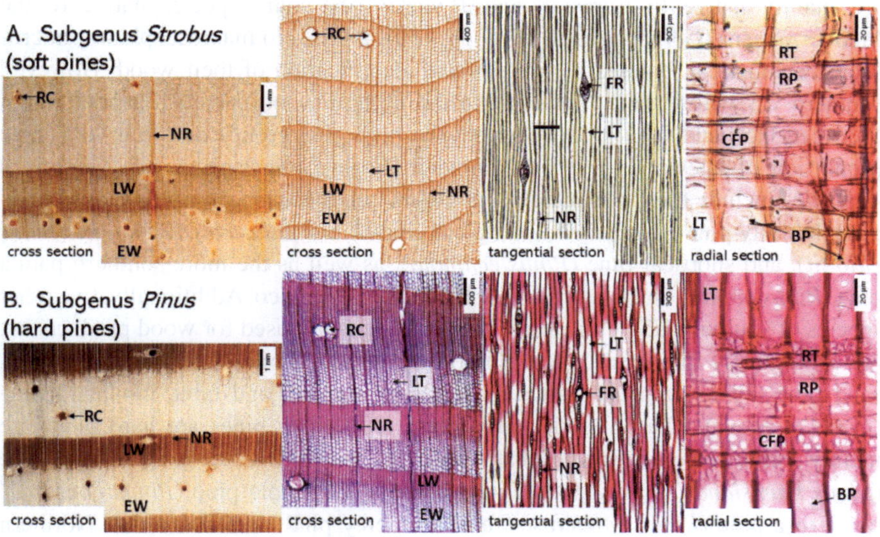

Fig. 1 Abrupt transition between earlywood (EW) and latewood (LW) of soft pines (**a**, top images from *Pinus strobus*) and gradual transition of hard pines (**b**, bottom images from *Pinus virginiana*). The cross sections (transverse views) are dominated by rows of longitudinal tracheids (LT); also visible are the narrow rays (NR) and the resin canals (RC). In tangential sections, some rays, called fusiform rays (FR), have a horizontal resin canal through their center. In radial sections, ray parenchyma (RP) appears as brick-like bands of cells surrounded by ray tracheids (RT). Bordered pits (BP) connecting two adjacent LT, and crossfield pits connecting RP to LT are also visible. (Images taken from InsideWood. 2004-onwards. http://insidewood.lib.ncsu.edu/search [November 2024]; Wheeler (2011))

but also in regards to earlywood to latewood transition. Hard pines in general tend to have a distinct abrupt transition from earlywood to latewood (and occasionally false growth rings), thus leading to an irregular wood grain appearance (though there can be some exceptions such as longleaf pine, *Pinus palustris*).

Wood is a porous material, made of empty cells characterized by a matrix of solid wood fibers (cell walls); the fibers without air space have a density of about 1.5 g cm^{-3} (oven dry basis; Siau 1984; Wilkinson et al. 2015). Wood density is therefore a measure of the total amount of solid-wood substance in a given volume of wood and in pines, it is driven by the proportion of early- and latewood (Panshin and de Zeeuw 1980). Mean wood density and density variation can be used as a good proxy for wood quality and suitability for some end uses (Zobel and Van Buijtenen 1989). Unlike hard pines, soft pine wood is characterized by a low wood density (ranging between 0.30 and 0.43 g cm^{-3} at 12% moisture content), fine texture and an even grain owing to a gradual earlywood to latewood transition (Panshin and de Zeeuw 1980). Overall, average wood density (at 12% moisture content) for hard pine species ranges between 0.38 and 0.65 g cm^{-3} (Zanne et al. 2009). Although the hard pines of great commercial importance can further be divided in four subgroups corresponding to the phylogeny subsections *Australes* (the southern US yellow as well as species native to central America and the Caribbean), *Contortae and Ponderosae* (the northern and western US yellow pines), *Pinus* (the red pines that are mainly native to Eurasia), and *Pinaster* (the main species native to the Mediterranean) (Gernandt et al. 2008; Jin et al. 2021). No macroscopic nor microscopic anatomical features permit positive identification of their wood. However, southern US yellow pine species have the highest wood densities at maturation (between 0.58 and 0.65 g cm^{-3}) within the hard pine groups due to a very large proportion of latewood. The six commercially important species of subsections *Australes* include the five southern US yellow pines Caribbean pine (*Pinus caribaea*), loblolly pine (*Pinus taeda*), longleaf pine (*Pinus palustris*), slash pine (*Pinus elliottii*), and shortleaf pine (*Pinus echinata*), as well as the more southern patula pine (*Pinus patula*) that is native to the highlands of Mexico. Additionally, five other species of southern US yellow pines are less frequently used for wood products but are ecologically important: pitch pine (*Pinus rigida*), pond pine (*Pinus serotina*), sand pine (*Pinus clausa*), spruce pine (*Pinus glabra*), and table mountain pine (*Pinus pungens*). The northern and western US yellow pines are hard pines are mainly comprised of six species that are Virginia pine (*Pinus virginiana*), Jack pine (*Pinus banksiana*), Jeffrey pine (*Pinus jeffreyi*), lodgepole pine (*Pinus contorta*), ponderosa pine (*Pinus ponderosa*), and Monterey pine (*Pinus radiata*). Northern and western US yellow pines have a smaller proportion of latewood transitions and therefore tend to have lower wood density ranging from 0.41 to 0.45 g cm^{-3}, and have a more even grain than southern US yellow pines (Schimleck et al. 2024). Note that although *P. ponderosa, P. jeffreyi and P contorta* are structurally indistinguishable, and so no commercial distinction is made between the lumber of the two species, the former tends to have slightly larger resin canals than the latter (Wiedenhoeft et al. 2003). Native to coastal California, today *P. radiata* is grown almost exclusively in plantations, mostly in the southern hemisphere where native pines are essentially absent (Chile, Australia, and New Zealand). This species, followed by *P. taeda* and *P. Patula*, has been the most widely planted pine in the world for the last 70 years (Lavery and Mead 1998). The red pines subgroup is comprised of one species in the United States (red pine, *Pinus resinosa*) and two in Europe (Austrian pine, *Pinus nigra* and Scots pine, *Pinus sylvestris*) that are sister species in a distant clade (Geada López et al. 2002), and thus cannot be distinguished on the basis of their wood anatomy. Finally, the *Pinaster* subgroup is composed of several species from Mediterranean basin with the most important species being maritime pine (*Pinus pinaster*), sadly the unique pine species of one of the largest man-made forests in the world (870,000 ha) that is located on the Atlantic coast of SW France. In addition, Aleppo pine (*Pinus halepensis*) that is a wide-spread species, and to some extent umbrella pine (*Pinus pinea*) that is planted for its edible seeds are also commercially important species of the *Pinaster* subgroup.

3 Anatomical Features Common to All Pine Species

3.1 Radial Wood Variation Patterns

Wood density within tree stems varies according to species, following typical radial and longitudinal patterns that are strongly correlated with tree age, size and growth (Lachenbruch et al. 2011). The typical radial pattern of wood variation in conifers in general and in pines in particular constitutes a suite of properties that change asymptotically from the first growth ring outward (Fig. 2). Wood scientists refer to the "transition age" as the dividing zone between the juvenile wood and mature wood (Zobel and Sprague 1998). Wood density is often used as a proxy to show this radial pattern of wood properties and to estimate the transition age from juvenile to mature wood (Fig. 2). The wood formed during the first 5–10 years of tree life is generally dominated by the so-called juvenile wood. In general, juvenile wood tends to have high variability among species in physical and anatomical properties (Zobel and Van Buijtenen 1989). Wood density is often negatively correlated with annual ring width, but no consistent relationship between these variables has been reported in several pine species (Megraw 1985; Peltola et al. 2009). This may be because concurrent changes in wood density and ring width as cambium ages reflect the physiological responses to growing environmental conditions rather than evidence of a causal relationship between the two variables (Schimleck et al. 2024). In pines, three types of wood density radial patterns from pith to bark can be visible (Zobel and van Buijtenen 1989; Schimleck et al. 2025). The first one characteristic of southern hard pines and some European hard pines such as *P. pinaster* denotes a rapid wood density increase with increasing cambial age due to a latewood becoming more prominent in later rings, and then a leveling off or a decline in the rate of increase of wood density (Fig. 2a). The soft pines also follow that pattern (Thor 1995). The second one found in northern and western hard pines in the North

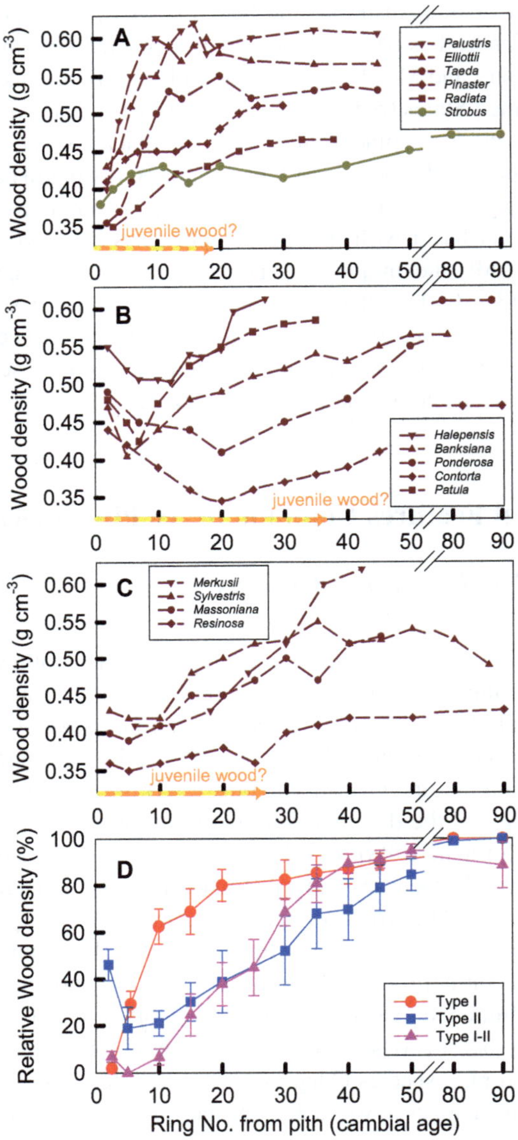

Fig. 2 Average wood density from pith to bark (cambial age) and estimated range of juvenile wood at breast height in second-growth commercially important pine species. Wood juvenility was determined empirically through examining the rate of change in wood density. The influence of cambial age on the radial patterns in wood density was described by different patterns. The first two, referred to as Type I and Type II wood density variations (Panshin and de Zeeuw 1980; Schimleck et al. 2025) are characteristic of pine species and described in panels (**a**) and (**b**), respectively. (**a**) An increase in wood density for some number of years with a levelling off that is characteristic of most southern hard pines plus *Pinus pinaster* from southern Europe; (**b**) A high density near the pith with a fast decrease for several growth rings and then a slow increase to a constant value that is characteristic of most western North America hard pines plus *Pinus halepensis* from the Mediterranean basin and *Pinus patula* from Mexico. Our literature survey identified

American but also in *P. Patula* and in *P. halepensis* is more complex and reveals a high wood density near the pith followed by a dip in density for several growth rings and a final increase with increasing cambial age before reaching more stable values in the outer growth rings (Fig. 2b). The wood closest to the pith of those hard pines is of high density but this apparent advantage is offset by large fibril angle that contributes to low strength and stiffness and greater longitudinal shrinkage (Josza and Middleton 1994). Finally, a third and intermediate pattern exists among pine species characterized by a short and flat trend in wood density near the pith followed by a steady increase before reaching maximum values (Fig. 2c, d).

The increased proportion of juvenile wood relative to mature wood has quality implications for lumber production. Juvenile wood can shrink excessively along the grain because of large fibril angles, or twist because of spiral grain. Therefore, in pine species where the difference in wood density between juvenile wood and mature wood is relatively large (Fig. 2), a higher proportion of juvenile wood can result in warping problems and reduced lumber strength. The transition age is typically taken as 5–20 years, depending on the pine species and the definition used. Juvenile and mature wood in pines differ in their properties, often by a large amount—with in some pine species a three- to fourfold increase in tracheid length (Fig. 3). The extensive reviews on wood variation from pith to bark consider mostly pine species used for timber and in construction and therefore are taxonomically biased toward the hard pines (e.g., Zobel and van Buijtenen 1989; Kennedy 1995; Zobel and Sprague 1998; Larson et al. 2001; Woodcock and Shier 2002). Although much of the literature considered here has used the terms "juvenile vs. mature wood," the terms "corewood vs. outerwood" have been instead proposed to avoid terminology that mixes up wood structure and cambium maturity (Burdon et al. 2004; Lachenbruch et al. 2011).

Fig. 2 (continued) another intermediate trend (Type I–II) shown in panel (**c**). This intermediate pattern is characterized by a short and flat trend in wood density near the pith followed by a steady increase before reaching maximum values. The mean standardized values (±SE) of these three wood density patterns shown in (**a**)–(**c**) are represented in (**d**). For each species, standardized density values were calculated as the difference between the density at a given cambial age and the minimum density of the data series divided by the differences between the maximum and the minimum density of the data series. In (**a**), the species are *Pinus palustris* (Eberhardt et al. 2018); *Pinus elliottii* (Clark and Saucier 1989; Strickland and Goddard 1966); *Pinus taeda* (Dahlen et al. 2021; Tasissa and Burkhart 1998); *P. pinaster* (Bouffier et al. 2008; Wilkinson et al. 2015); *Pinus radiata* (Kimberley et al. 2015; McConchie 1999) and the only white pine for which we could fine data: *Pinus strobus* (Lawhon and Woods 1976; Thor 1995; Paul 1963). In (**b**), the species are *P. halepensis* (Foti et al. 2018); *Pinus banksiana* (Park et al. 2009; Schneider et al. 2008); *Pinus ponderosa* (Bouffier et al. 2003); *Pinus contorta* (Mansfield et al. 2007) and *P. patula* (Kamala et al. 2013). In (**c**), the species are *Pinus merkusii* (Darmawan et al. 2018); *Pinus sylvestris* (Auty et al. 2014; Sauter et al. 1999); *Pinus massoniana* (Deng et al. 2014; Zhang et al. 2012); and *Pinus resinosa* (Deresse 1998; Larocque and Marshall 1995; Zhu et al. 2007). If multiple sources are provided for a specific species in panels (**a**–**c**), the average values are displayed

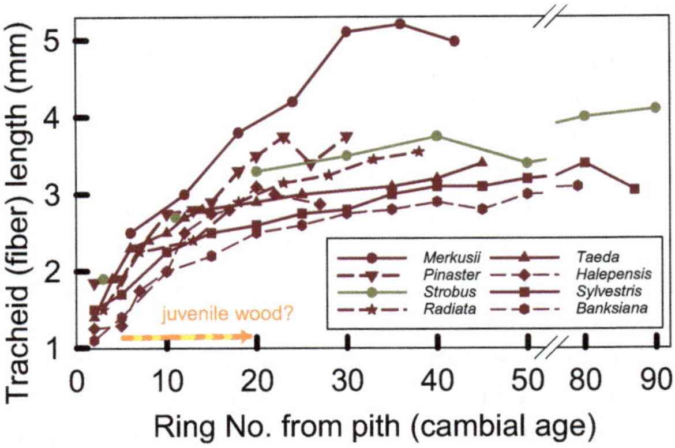

Fig. 3 Average tracheid (fiber) length from pith to bark (cambial age) at breast height in commercially important pine species. Species are *Pinus merkusii* (Darmawan et al. 2018); *Pinus strobus* (Thor 1995); *Pinus radiata* (McConchie 1999); *Pinus pinaster* (Bisset et al. 1951; Louzada 2000); *Pinus taeda* (Dahlen et al. 2021); *Pinus halepensis* (Foti et al. 2018); *Pinus sylvestris* (Fabisiak et al. 2020); *Pinus banksiana* (Kribs 1928). If multiple sources are provided for a specific species, the average values are displayed

3.2 *Epithelial Cells, Resin Canals, and Bordered Pits*

All pines have heterocellular rays made of 5–15 cells stacked on the top of each other. Longitudinal parenchyma is sometimes present in strands adjacent to the epithelial cells of vertical resin canals. Its presence is, for the most part, overlooked in the literature and it never represents more than 1% of the wood volume (Panshin and de Zeeuw 1980). Note that another cell type that is sometimes present in pine species is axial parenchyma. Axial parenchyma cells are similar in size and shape to ray parenchyma cells, but they are vertically oriented and stacked one on top of the other to form a parenchyma strand (Wheeler and Baas 1998).

Among conifers, resin canals are exclusively found within the Pinaceae family. The genus *Pinus* is separated anatomically from all other conifers by bearing resin canal complexes (resin ducts) made of thin-walled specialized epithelial cells (Phillips 1948; García Esteban et al. 2021). Resin canals in species from the genera *Cathaya*, *Keteleeria*, *Larix*, *Picea*, and *Pseudotsuga* are made of thick-walled epithelium cells. Moreover, of all the conifers, the pines have the largest and most numerous resin canals (Wiedenhoeft and Miller 2002). Resin canals are technically not cells, but voids surrounded by interconnected epithelial cells forming a closed continuous system that secrete oleoresin. Horizontal (or radial) resin canals are embedded in fusiform rays (Fig. 1a, b). During heartwood formation, those epithelial cells can divide and protrude into the resin canals, thus clogging them by the formation of tylosoids (Spicer 2005). The diurnal and seasonal dynamics of resin production is controlled by changes in environmental conditions, in particular by soil and tree water status (Rissanen et al. 2019).

Both in soft (Fig. 1a) and hard pines (Fig. 1b) long and narrow longitudinal parenchyma (rectangular in shape) surrounds longitudinal epithelial cells forming resin canals (Esau 1965; Wheeler 2011). Parenchyma measured as a fraction of wood volume has been shown to be lower in the subgenus *Strobus* (soft pines). For example, in *P. lambertiana*, *P. albicaulis* and *P. monticola*, ray parenchyma fraction represents less than 4% of the whole wood volume, and up to 6% in most hard pines, i.e., species of the subgenus *Pinus* (Godfrey et al. 2020). It has been hypothesized that low ray parenchyma fraction may correlate with lower resin production and delivery for successful isolation of xylem wound and pathogen spread (Trapp and Crouteau 2001; Morris et al. 2016). This hypothesis could in fact explain why the subgenus *Strobus* is generally more susceptible to pests and diseases than the subgenus *Pinus* (Gibson et al. 2008). In *Pinus*, the formation of elongated traumatic resin ducts represents an additional way to increase oleoresin production in response to a range of stimuli, including mechanical wounding, fire, insects, or pathogen attack (Hudgins et al. 2004). However, there is limited knowledge about the differences that could exist in the formation of those resin canals and the production of traumatic resinosis between hard and soft pines.

Bordered pits are also a key anatomical features common to all pine species. Bordered pits in pines are mostly located on radial cell walls and are generally uniseriate (Dute 2015). Bordered pits on tangential cell walls are mainly found at the tracheid ends of the earlywood section of a growth ring. Even though studies on border-pits patterning differentiation between tangential and radial tracheid walls are still nonexistent, when present, tangential bordered pits are of the same size and shape as radial pits (Mauseth 1988). Those conifer pits are composed of a central circular or oblong thickening on the inner surface of the cell wall, called the torus, surrounded by a thin, elastic membrane called the margo. The torus/margo system is a specialized structure acting as a valve that helps prevent the spread of air embolisms from one tracheid to another and thus maintains water transport through the xylem (Domec et al. 2006; Delzon et al. 2010).

4 Consistent Differences Among Specific Anatomical Features Between Soft and Hard Pines

All conifer species bearing resin canals also have several rows of specialized horizontal tracheids surrounding the ray parenchyma cells as seen in a radial section (Fig. 1). Ray tracheids also occur in a few species that do not have resin canals, such as *Tsuga* spp. and *Chamaecyparis* spp. (IAWA Committee 2004; García Esteban et al. 2021). In general, the wood of subgenus *Strobus* can be distinguished by the presence of smooth-walled (nondentate) ray tracheids with rare cell-wall thickenings (Fig. 4a), in contrast to the dentate or reticulate ray tracheids found in the subgenus *Pinus* (Fig. 4b). These ray tracheids have bordered pits like axial

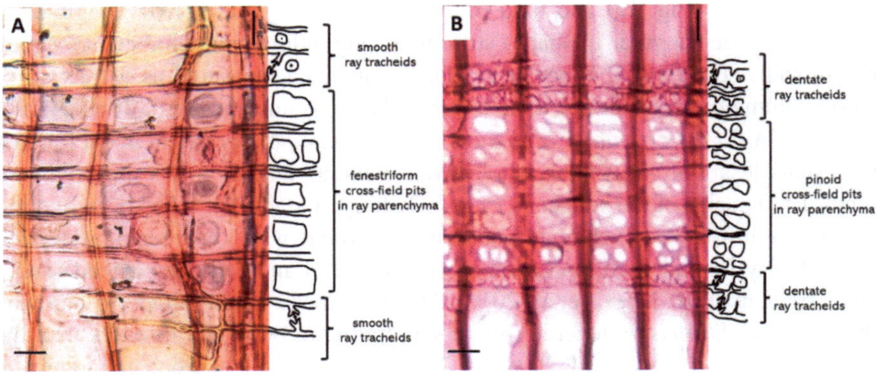

Fig. 4 Differences in wood ray tracheids and cross-field pitting areas between soft (**a**, *Pinus strobus*) and hard pines (**b**, *Pinus virginiana*). All cross-field pits are half-bordered pits. Soft pines are characterized by smooth non-dentate ray tracheids and by fenestriform pits, that are large, borderless, rectangular "window-like" pits (usually two to four fenestriform pits per cross field). Hard pines are characterized by several rows of dentate ray tracheids on the upper and lower margins of ray parenchyma (here two are visible in this (*P. virginiana*) radial section). Except in red pine, pinoid pittings occur along with dentate ray tracheids. The five layers of ray parenchyma show one to two pinoid-like cross-field pittings that are elliptical or shaped as irregular polygons and that vary in size. When the border of the pinoid pittings are visible they are narrow and usually wider on one side than the other. Scale bars = 20 µm. (Images taken from InsideWood. 2004-onwards. Published on the Internet. http://insidewood.lib.ncsu.edu/search [November 2022]; Wheeler (2011))

tracheids, but are much shorter and narrower. In general, ray parenchyma walls remain non-lignified, but ray tracheid walls are lignified (Saka and Goring 1985). In addition, spiral thickenings are also present in hard pines latewood and usually occur sporadically along the tracheid length (Howard and Manwiller 1968).

Sometimes axial tracheid and ray parenchyma cell-walls intersect and connects to form an area called a cross-field pit. Those specialized pits connect the dead and purely hydraulic part of the xylem to its living and storage system. The shape and size of these cross-field pits are different between the wood of subgenus *Strobus* and *Pinus* and can be used for wood identification (Wheeler and Baas 1998). Soft pines have rays with distinctly one (rarely up to 3) large fenestriform cross-field pits, whereas hard pines near pinoid pittings (Akkemik and Yaman 2012); the latter generally occur with dentate ray tracheids (Fig. 4a). In hard pines, red pine has dentate ray tracheids with fenestriform pitting (Fig. 4b). Within the soft pines, if fewer than three pits per cross-field are always present in early wood tracheids, then the species is likely to be *P. monticola* (Kellogg et al. 1982). Further, if three or more pits per cross-field are observed, then the species is likely *P. lambertiana*, and if crystals are common in the longitudinal resin canals, then the species probably is *P. flexilis* (Hoadley 1990; IAWA Committee 2004).

5 Do Differences in Anatomical Properties Between Pines Species Explain Wood Hydraulic Differences?

Pinus species are drought tolerant, and the potential plasticity of their xylem anatomical structures might play a role in enabling them to thrive in stressful habitats (Vilalta et al. 2004; Creese et al. 2011; Hacke et al. 2015; Augustine and McCulloh 2024). Xylem area-specific conductivity (k_S, in kg m^{-1} s^{-1} MPa^{-1}) for plant ecophysiologists (equivalent to wood permeability for wood scientists) is a measure of the ease with which liquids (or gas) can flow through a xylem sample under a given pressure gradient. One of the most important questions in tree physiology and wood technology concerns how k_S is related to wood structure and in particular to tracheid diameter and end-wall bordered-pit characteristics (Tyree and Ewers 1991; Sperry et al. 2006; Pittermann et al. 2006) (see chapter "Doing Just Pine, Thank You Very Much: The Hydraulics of Pinus"). These relationships impact our understanding of areas as diverse as how trees cope with drought, how tall trees can become, or how silvicultural regimes that affect tree form will affect tree growth and wood quality. Additionally, k_S also affects kiln drying schedules, and appropriate methods to impregnate timber with wood preservatives and fire retardants (Milota et al. 1994; Lehringer et al. 2009).

The type of wood cells formed during a given year depends on the changing hydraulic and mechanical requirements of the tree, which reflect the site and environmental conditions (Domec and Gartner 2003; Bouche et al. 2014). Although it is not yet fully clear what the specific advantages of juvenile wood are, a wood scientist, who typically has more training in structural engineering than plant physiologists, would be likely to assume that juvenile wood plays an important role in the strength of a tree, given that the modulus of elasticity (ε) is low in juvenile wood and high in mature wood. Mechanical performance (as judged by flexural stiffness, which is ε times the second moment of area, I) increases greatly as the stem diameter gets larger, but this change is dominated by the increase in I, that scales as a function of stem diameter to the fourth power. For example, as a tree increases from 1 to 10 cm in radius, its ε may increase by a factor of 3–4, but its I will increase by a factor of 10,000 (10^4) and the contribution of ε to $\varepsilon \times I$ drops by a factor of 2500. Thus, from the mechanical viewpoint, the low ε and torsional stiffness conferred by the high microfibril angles in young stems may be adaptive (facilitating bending of thin axes in the wind), but the higher ε in older stems may have no adaptive value. Low ε together with smaller diameter, results in much lower flexural stiffness of young tree stems than older trees stems, which along with low torsional stiffness could be seen as giving them more mechanical flexibility to contend with wind stresses (Niklas 1998). However, several lines of evidence indicate that the purpose of juvenile wood is improved hydraulic conductance to support fast growth rather than tree mechanics (Domec and Gartner 2002; Rosner et al. 2008; Rosner 2013). In mature pine wood, there is a universal decrease in torsional rigidity as driven by lower microfibril angle and by larger stem thickness with number of years from the pith (Gartner 2006), and there is no good functional explanation for this pattern. A more convincing explanation for many of the juvenile and mature wood differences in pines is that the radial changes seen in Fig. 3 are driven by a need for higher water transport (i.e., higher k_s) (Schimleck et al. 2024). The higher water transport is accomplished by wider and longer conduits (Fig. 3), which just happen to have a lower microfibril angle, which is probably a developmental consequence of the increase in tracheid size with cambial age (Wardrop and Dadswell 1951; Megraw 1985). There is a universal pattern of increase in length and diameter of tracheids with radial position throughout the juvenile wood zone (Zobel and van Buijtenen 1989). The longer the tracheid, the higher k_s, because longer tracheids bear fewer pits per unit length (Pothier et al. 1989; Hacke et al. 2015) and because in longer wider tracheids, water can move faster (the flow rate is proportional to the four power of conduit radius; Tyree and Ewers 1991). These facts are consistent with the hypothesis that trees have evolved to increase k_s as they increase in age and diameter (Domec and Gartner 2003; Mauriat et al. 2014). Therefore, it appears that xylem hydraulic conductivity is the main driver of the decrease in microfibril angle as well as cell dimensions with radial position. Limited data show that the increase in tracheid size from pith to bark (Fig. 3) actually leads to increased k_s as trees grow (Domec et al. 2006; Wilkinson et al. 2015; Held et al. 2021). Even at the whole-tree level, compensating changes to increase overall hydraulic efficiency between short and tall pine trees occur more through an increased in k_s, rather than through an increase in sapwood to leaf area ratio (Huber value) (Whitehead et al. 1984; Delzon et al. 2004; Domec et al. 2012).

As xylem tension (water potential) in the water column of a tree decreases, tracheids become embolized (air-filled), and k_s declines (Fig. 5a). This process is not a "cavitation," in which the water column literally breaks; it is a gas bubble that enters the conduit, rapidly expands, then the conduit becomes increasingly air-filled as gasses diffuse into the conduit at near vacuum (Johnson et al. 2022). The embolisms are thought to be formed by air becoming "seeded" into the conduit through the bordered-pit margo pores. In conifers, the exact tension at which the bubble is pulled in is a function of pore size and geometry. It occurs when this tension is sufficiently large to pull the air through the air/water interface directly through the pores, or between the torus and the pit border once the membrane is deflected (Hacke et al. 2004; Domec et al. 2006; Delzon et al. 2010). However, with increasing conduit lumen diameter, the risk of hydraulic failure can potentially increase (Rosner et al. 2016; Losso et al. 2018). Because of overall lower mature wood density and generally longer cell size in the pine subgenus *Strobus* (Figs. 3a and 4), it is then logical to expect the wood of soft pines to be more conductive and have a higher k_s. We conducted a literature survey of more than 55 studies looking at pine hydraulics, and for comparisons we were able to select seven soft pines and seven hard pines growing in the same ecological niche. There is some evidence that the gain in embolism resistance associated with increasing wood density in hard pine is associated with a cost in terms of reduced k_s (Fig. 5a). Our literature survey shows that k_s of hard pine species are marginally lower than k_s of soft pine species ($p = 0.08$ paired samples t-test; Fig. 5a).

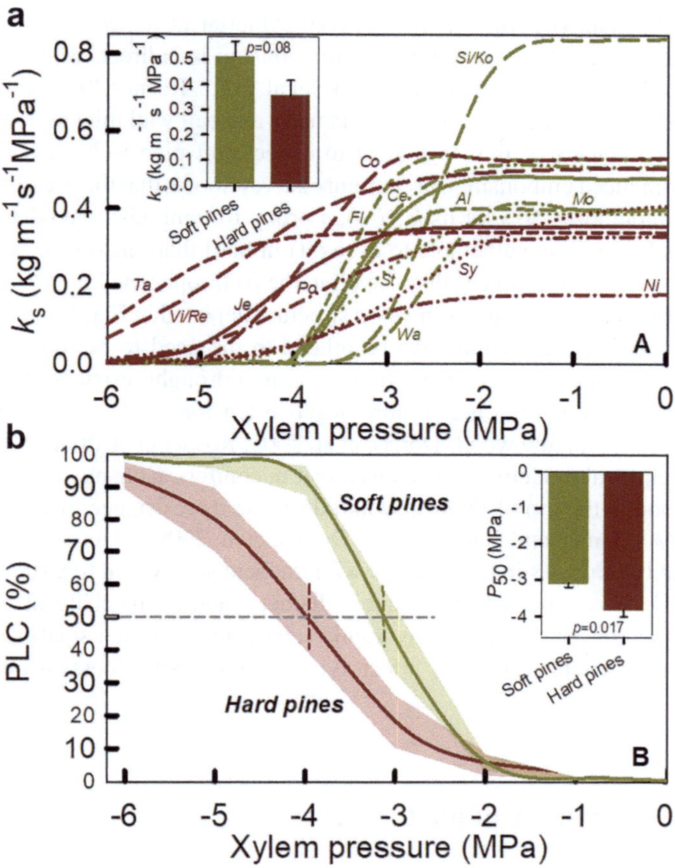

Fig. 5 (**a**) Xylem area-specific hydraulic conductivity (k_s) and (**b**) vulnerability curves showing the percent loss of k_s (PLC) as a function of xylem pressure for soft and hard pines. In (**a**) soft pine species (subgenus *Strobus*) are *Mo* = *Pinus monticola*, *St* = *Pinus strobus*, *Fl* = *Pinus flexilis*, *Ce* = *Pinus cembra*, *Wa* = *Pinus wallichiana*, *Al* = *Pinus albicaulis* and *Si*/*Ko* = *Pinus sibirica*/*Pinus koraiensis*. In the same order these seven soft pine species were compared to the respective seven hard pine species (subgenus *Pinus*) from the same ecological niche: *Je* = *Pinus jeffreyi*, *Vi*/*Re* = *Pinus virginiana*/*Pinus Resinosa*, *Co* = *Pinus contorta*, *Ta* = *Pinus taeda*, *Sy* = *Pinus sylvestris*, *Ni* = *Pinus nigra*, and *Po* = *Pinus ponderosa*. Insets in (**a**) and (**b**) represent k_s and the pressure at which 50% loss of k_s occurred (P_{50}) for the seven species per pine group, respectively. *p*-values represent the statistical outputs from an unpaired t-test between the mean difference in k_s or P_{50} between the species groups. In (**b**) Means and standard errors (shaded areas) are shown for the seven species per pine group. The data were taken from the global vulnerability of plant xylem database (https://xylemfunctionaltraitsxylemfunctionaltraits.org), and Wubble (2010) and Augustine and McCulloh (2024) for *P. strobus* and from Zhang et al. (2024) for *P. virgininia*

The "vulnerability curve" is a plot of percentage loss of k_s as a function of tension in the water column (Fig. 5b), and it is important to understand species drought tolerance (Sperry 1995) and plant hydraulic architecture (Tyree and Ewers 1991). Different parts of plants, ages of plants, and plants from different environments have different characteristic vulnerability curves (Mauriat et al. 2014; Johnson et al. 2022; Zhang et al. 2024). Although the relationship is indirect, wood density is a good proxy for embolism resistance (Hacke et al. 2001; Rosner 2013), reflecting the dependence of wood density on lumen diameter and cell wall thickness. Because of higher wood density, it is then logical to expect soft pine to be less resistant to cavitation-induced embolism. Our literature survey shows that the pressure at which a wood sample loses 50% of its k_s (P_{50}), a proxy for embolism and drought resistance, was lower (less vulnerable to drought) in hard than in soft pines ($p < 0.01$; Fig. 5b), but also see chapter "Pine Death in the Anthropocene".

Furthermore, radial changes in wood structure depicted in Fig. 2 and the delay in mature wood formation could also be related to the need to be more resistant to embolism. Juvenile wood has been shown to confer drought resistance to stems, i.e., juvenile wood can sustain much more negative tensions before the water column breaks than does mature wood (Domec et al. 2005; Rosner et al. 2008; Mauriat et al. 2014). Reciprocal modeling also showed that in conifers, a change from juvenile to mature wood at the top of the tree would decrease its resistance to water stress by almost 50% (Domec and Gartner 2002; Domec et al. 2009a). However, few studies have looked at changes in pit structure with tracheid size within tree rings and from pith to bark. It is expected that tracheid diameter and pit mechanical functioning should act in concert to increase hydraulic safety and thus even small variations in pit dimensions could potentially affect xylem hydraulics (Sviderskaya et al. 2021; Held et al. 2021).

6 Leaf Anatomy Unique to Pines

As previous sections have covered the structure that supplies water to the photosynthetic tissues of the plant (i.e., xylem), this section will cover the organs that are the sink for that water: the leaves. Probably the most obvious leaf anatomical property of pines is the acicular needle shape with circular, triangular or hemicircular cross-section (with the exception of the rare flat-leaved *Pinus krempfii*; Brodribb and Feild 2008). Additionally, pines have unique stomatal distributions (Ghimire et al. 2015, see also below), mesophyll tissue that is not differentiated into spongy and palisade mesophyll (i.e., plicate mesophyll, Esau 1965; Fig. 6) and a vascular bundle containing a single vein that is surrounded by an endodermis (bundle sheath) with Casparian-strip-like properties (impregnation of lignin and or/ suberin in cell walls; for review see Esau (1965), Lersten (1997), and references therein). The vascular bundle also contains transfusion tissue (Mai et al. 2024) including transfusion tracheids and transfusion parenchyma.

The closest relatives to *Pinus*, *Cathaya*, and *Picea* (Gernandt et al. 2008; Ran et al. 2018), share many leaf anatomical features with *Pinus*, including a vascular bundle surrounded by an endodermis as well as plicate mesophyll. Plicate mesophyll is only found in this clade of conifers and not in other clades (Pant and Basu 1977). Additionally, pines tend to have a smaller mesophyll volume and smaller

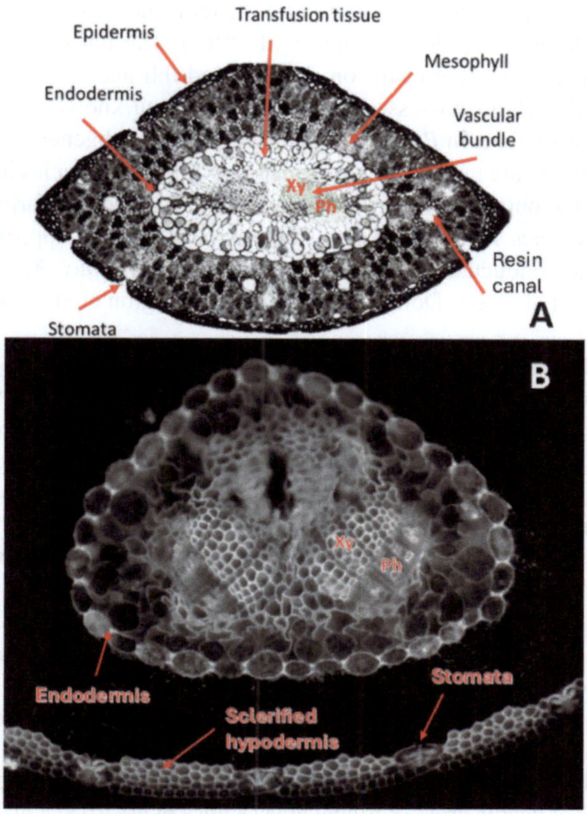

Fig. 6 Cross sections of *Pinus taeda* needles indicating anatomical traits discussed in the text. Panel (**a**) was imaged using light microscopy and panel (**b**) was imaged using autofluorescence imaging to highlight lignified and/or suberized tissues. Xy = xylem, Ph = phloem. Sclerified hypodermis, sometimes referred to as sclerenchyma cells, are specialized dead cells providing mechanical support and strength. Epidermis in B is located just below the sclerenchyma cells and is represented by the last single-celled layer. (Image in Panel (**a**) reproduced from Domec et al. 2016; image in Panel (**b**) DMJ unpublished). Terminology for abaxial (upper surface) and adaxial (lower surface) follows Leyton and Armitage (1968) and Lin et al. (2001), and refers to the phloem or xylem side of the vascular bundle, respectively

mesophyll intercellular airspace volumes (Wang et al. 2019; Trueba et al. 2022) than other conifers.

Although all gymnosperms contain transfusion tissue, transfusion tissue in pines is distinct in its morphology as compared to other gymnosperms (Mai et al. 2024). In *Pinus*, transfusion tissue is nearly evenly distributed throughout the vascular bundle with slightly higher abundance on the abaxial side of the bundle (Fig. 6). However, in other gymnosperm genera these tissues may be reduced, or located more on either the vascular abaxial (phloem side) or adaxial (xylem side) side of the bundle (Leyton and Armitage 1968; Lin et al. 2001). Additionally, these tissues may extend beyond the vascular bundle into the mesophyll (termed "accessory transfusion tissue") in other gymnosperm genera such as *Taxus*, *Cycas*, and *Podocarpus* (Hu and Yao 1981; Zhang et al. 2014). The flat-leaved pine species, *P. krempfii*, has accessory transfusion tissue (Brodribb and Feild 2008; Brodribb 2011) but no other *Pinus* possesses this tissue type to our knowledge.

Stomatal distributions in *Pinus* tend to differ between subgenera. Species in subgenus *Pinus* species are usually amphistomatous while most species in *Strobus* have no stomata on the outer (curved, abaxial) surface of the needle (Farjon 2021). This may be why species in the subgenus *Pinus* appear to have approximately 20% greater stomatal conductance than species in subgenus *Strobus* (Maier and Teskey 1992; Hubbard et al. 2001; Domec et al. 2009b; Addington et al. 2006; Asbjornsen et al. 2021).

Tissues in leaves of *Pinus* can be broken down into two categories: mechanical tissues and photosynthetic tissues (Kuusk et al. 2018; Wang et al. 2019). In general, most of the information we have about tissue partitioning come from the *Pinus* subgenus. Within the *Pinus* subgenus, there is large variation in mesophyll volume fractions with ~40–70% being the observed range. This value changes with leaf age, becoming smaller in older leaves as more volume is dedicated to mechanical tissues (Kuusk et al. 2018). The volume fraction of mesophyll also changes with leaf length, with mesophyll volume decreasing and mechanical tissues, such as sclerenchyma cells (Fig. 6b), increasing with leaf length (Wang et al. 2019). There is much less data available for tissue fractions in leaves of the subgenus *Strobus* but Trueba et al. (2022) showed that mesophyll tissue fractions were 0.67 (SE = 0.01) and 0.56 (SE = 0.01) in *Strobus* and *Pinus*, respectively and they were significantly different ($p < 0.0001$).

Additionally, resin ducts, which are neither photosynthetic or mechanical, can make up 1–8% of needle cross sectional area (Kuusk et al. 2018; Trueba et al. 2022) but do not differ in tissue fraction between subgenera. However, resin ducts in in subgenus *Pinus* are typically arranged symmetrically whereas in *Strobus* they are most often asymmetrical (Farjon 2021). In addition, the volume of resin ducts per volume of leaf appears to decline with increasing needle maturation (Kuusk et al. 2018).

7 Leaf Anatomy Impacts on Physiological Function

The unique leaf anatomy of pines (needle-shaped with a single longitudinal vein) constrains photosynthesis and transpiration due to a lack of a vein network and resultant lower leaf hydraulic conductance (Zwieniecki et al. 2006; Brodribb et al. 2007). However, conifers, including *Pinus*, can adjust mesophyll proportions and stomatal distribution in response to needle nitrogen content with concomitant alterations in leaf hydraulics and gas exchange (Domec et al. 2016; Wang et al. 2019; Trueba et al. 2022). More research is needed to understand the constraints of having a single vein and the compensating mechanisms as they relate to both liquid and gas phase mass transfer in pine leaves.

The functional significance of transfusion tracheids and other transfusion tissues within the vascular bundles of pine leaves is poorly understood but recent work has shown them to be involved in both water and sugar transport between the mesophyll and vascular tissues (Gao et al. 2023; Mai et al. 2024). In other conifers, transfusion tracheids have been observed to collapse and or/embolize during periods of water stress. For example, Zhang et al. (2014) found reversible collapse of transfusion tracheids with water stress in *Taxus*. Additionally, reversible collapse of accessory transfusion tissues (outside of the vascular bundle) with moisture stress in *Podocarpus* has also been observed (Brodribb and Holbrook 2005). It is not known how widespread transfusion tissue collapse is in *Pinus*, but there is evidence that it may occur in multiple pine species (Cochard et al. 2004). Transfusion tracheid embolism has also observed in pine species under water stress (Johnson et al. 2009, 2012). If transfusion tracheid embolism and/or collapse is indeed common in *Pinus*, it might suggest that transfusion tracheids are acting as a sort of hydraulic "circuit breaker" to limit embolism in the xylem by releasing water inito the transpiration stream and thus decreasing water potential gradients (Zhang et al. 2014).

The function of the vascular bundle sheath in *Pinus* is also poorly understood. One important yet unanswered question is how does the hydraulic resistance of the bundle sheath compare to the rest of the hydraulic transport pathway in *Pinus* leaves (see Gao et al. 2023)? One may speculate that because water is forced to travel symplastically through the bundle sheath, the resistance can be much greater than that of either the vascular or mesophyll hydraulic pathways. For example, Buckley et al. (2015) estimated the outside-xylem conductance to leaf water transport and found that outside-xylem conductance was approximately 50% lower in leaves with suberized bundle sheaths as compared to leaves with unsuberized bundle sheaths. But, Laur and Hacke (2014) did observe up- and downregulation of aquaporins in *Picea* needles in response to changing environmental conditions. One aquaporin in particular (PIP1), appeared to be localized to the bundle sheath and needle phloem. It is possible that the bundle sheath in *Pinus*, assuming that similar aquaporin expression patterns are present in *Pinus*, could serve as a point of regulation of water flow between the leaf xylem and the mesophyll tissue.

Additionally, needle anatomy in several pine species has been shown to vary with needle length, reflecting adjustments in the capacity for mechanical support and leaf hydraulic efficiency (Zwieniecki et al. 2006; Wang et al. 2019). For example, needle length was positively correlated with overall leaf hydraulic conductance and gas exchange (Wang et al. 2019). However, in another study, the area of xylem and the number of tracheids decreased moving from the base of the needle to the tip which resulted in predicted leaf xylem permeability decreasing from base to tip (Zwieniecki et al. 2006). Questions that arise from this research are (1) are there gradients of water potential along the length of *Pinus* leaves? and (2) does gas exchange decrease along *Pinus* leaves going from base to tip? (but see Jain et al. 2024).

Another research question that has received very little attention is what is the impact of plicate mesophyll on photosynthesis, transpiration and light processing?

Since plicate mesophyll appears to have less air space than spongy mesophyll (Trueba et al. 2022), should we expect reduced light scattering and less diffuse light inside the mesophyll of species with plicate mesophyll? It is also unclear if plicate mesophyll has a greater hydraulic resistance (or resistance to CO_2 movement) than differentiated mesophyll. It may be possible that plicate mesophyll is a xerophytic adaptation since many pines live in dry areas, but this has not been explored, to our knowledge.

Much research on *Pinus* wood anatomy has been reported previously, likely due to the economic and ecological importance of the genus. However, there are obvious differences in both wood and leaves of the two *Pinus* subgenera. In the subgenus *Pinus*, wood density is greater, and the wood is more resistant to embolism compared to the subgenus *Strobus* (Fig. 5b). There is also a much more abrupt transition from earlywood to latewood in *Strobus* versus *Pinus*. Leaves have much less studied as compared to wood, but leaves of the two subgenera differ in numbers of needle per fascicle, number of vascular bundles per vein, stomatal distributions, and volume of tissues (e.g., mesophyll).

Although much previous work exists on Pine anatomy, especially wood anatomy, much work on needle anatomy, development and structure and their relations to physiological functions is still needed. For example, in Scots pines, plasticity in needle longevity has been shown to compensate for reduced needle production so that a constant foliar biomass per shoot and thus carbon uptake was maintained (Pensa and Jalkanen 2005). Similarly, across several pine species, a negative correlation between mesophyll volume and structural tissues was observed, revealing a trade-off between needle biomass investments in mechanical support and photosynthetic functions (Kuusk et al. 2018). Clearly, this study showed that juvenile needles maximize carbon gain in seedlings, while adult needles enhance environmental stress tolerance in established plants. Furthermore, our understanding of water movement through the living tissues that water must travel through on its path from the soil to the atmosphere is as of yet, underdeveloped. These tissues, the endodermis in the root and the outside-xylem pathway for water movement in the leaf (Enstone et al. 2001; Wang et al. 2019; Gao et al. 2023), may be critical for understanding water use, growth, and survival under a changing climate. However, there is a large knowledge gap in our understanding of these tissues as compared to xylem. Future work should aim to elucidate structures and mechanisms in these living tissues and integrate this information with existing knowledge about apoplastic pathways for a more integrated, whole-plant view of physiological function in *Pinus*.

Acknowledgements JCD was supported in by the Department of Energy, USA (No. DE-SC0023309), from the French ANR (grant PHYDRAUCC 21-CE02-0033-02) and the Nlle Aquitaine Region (Projects VITIPIN AAPR2020-2020-8596610, PSGAR Multirisque Forêt GRIFFON and IMPACTS). DMJ was supported by the USDA National Institute for Food and Agriculture program under award numbers 2023-67013-39411 and 2024-67013-42313.

References

Addington RN, Donovan LA, Mitchell RJ, Vose JM, Pecot SD, Jack SB, Hacke UG, Sperry JS, Oren R (2006) Adjustments in hydraulic architecture of Pinus palustris maintain similar stomatal conductance in xeric and mesic habitats. Plant Cell Environ 29(4):535–545

Akkemik Ü, Yaman B (2012) Wood anatomy of eastern Mediterranean species. Verlag Kessel, Remagen. 31p

Asbjornsen H, McIntire CD, Vadeboncoeur MA, Jennings KA, Coble AP, Berry ZC (2021) Sensitivity and threshold dynamics of Pinus strobus and Quercus spp. in response to experimental and naturally occurring severe droughts. Tree Physiol 41(10):1819–1835

Augustine SP, McCulloh KA (2024) Physiological trait coordination and variability across and within three *Pinus* species. New Phytol 244:451–463

Auty D, Achim A, Macdonald E, Cameron AD, Gardiner BA (2014) Models for predicting wood density variation in Scots pine. Forestry 87:449–458

Bisset IJ, Dadswell HE, Wardrop B (1951) Factors influencing tracheid length in conifers stems. Aust For 15:17–30

Bouche PS, Larter M, Domec JC, Burlett R, Gasson P, Jansen S et al (2014) A broad survey of hydraulic and mechanical safety in the xylem of conifers. J Exp Bot 65:4419–4431

Bouffier LA, Gartner BL, Domec JC (2003) Wood density and hydraulic properties of ponderosa pine from the Willamette Valley vs. the Cascade Mountains. Wood Fiber Sci 35:217–233

Bouffier L, Rozenberg P, Raffin A, Kremer A (2008) Wood density variability in successive breeding populations of maritime pine. Can J For Res 38:2148–2158

Brodribb TJ (2011) A functional analysis of podocarp ecology. In: Turner BL, Cernusak LA (eds) Ecology of the podocarpaceae in tropical forests. Smithsonian Press, Washington, DC

Brodribb TJ, Feild TS (2008) Evolutionary significance of a flat-leaved Pinus in Vietnamese rainforest. New Phytol 178:201–209

Brodribb TJ, Holbrook NM (2005) Water stress deforms tracheids peripheral to the leaf vein of a tropical conifer. Plant Physiol 137:1139–1146

Brodribb TJ, Feild TS, Jordan GJ (2007) Leaf maximum photosynthetic rate and venation are linked by hydraulics. Plant Physiol 144:1890–1898

Buckley TN, John GP, Scoffoni C, Sack L (2015) How does leaf anatomy influence water transport outside the xylem? Plant Physiol 168:1616–1635

Burdon RD, Kibblewhite RP, Walker JCF, Megraw RA, Evans R, Cown DJ (2004) Juvenile versus mature wood: a new concept, orthogonal to corewood versus outerwood, with special reference to *Pinus radiata* and *P. taeda*. For Sci 50:399–415

Clark A, Saucier JR (1989) Influence of initial planting density, geographic location, and species on juvenile wood formation in southern pine. For Prod J 39:42–48

Cochard H, Froux F, Mayr S, Coutand C (2004) Xylem wall collapse in water-stressed pine needles. Plant Physiol 134:401–408

Creese C, Benscoter AM, Maherali H (2011) Xylem function and climate adaptation in *Pinus*. Am J Bot 98:1437–1445

Dahlen J, Nabavi M, Auty D, Schimleck L, Eberhardt TL (2021) Models for predicting the within-tree and regional variation of tracheid length and width for plantation loblolly pine. Forestry 94:127–140

Darmawan W, Nandika D, Hasna Afaf BD, Rahayu I, Lumongga D (2018) Radial variation in selected wood properties of Indonesian Merkusii pine. J Korean Wood Sci Technol 46:323–327

Delzon S, Sartore M, Burlett R, Dewar R, Loustau D (2004) Hydraulic responses to height growth in maritime pine trees. Plant Cell Environ 27:1077–1087

Delzon S, Douthe C, Sala A, Cochard H (2010) Mechanism of water-stress induced cavitation in conifers: bordered pit structure and function support the hypothesis of seal capillary-seeding. Plant Cell Environ 33:2101–2111

Deng Z, Zhang L, Xiang P, Yan W (2014) Variations of wood basic density with tree age and social classes in the axial direction within *Pinus massoniana* stems in Southern China. Ann For Sci 71:505–516

Deresse T (1998) The influence of age and growth rate on selected properties of Maine-grown red pine. PhD thesis, University of Maine, Orono, ME, 177p

Domec J-C, Gartner BL (2002) Age- and position-related changes in hydraulic vs. mechanical dysfunction of xylem: inferring the design criteria for Douglas-fir wood structure. Tree Physiol 22:91–104

Domec J-C, Gartner BL (2003) Relationship between growth rates and xylem hydraulic characteristics in young, mature and old-growth ponderosa pine trees. Plant Cell Environ 26:471–483

Domec J-C, Pruyn M, Gartner BL (2005) Axial and radial xylem conductivities, water storage and native embolism in trunks of young and old ponderosa pine trees. Plant Cell Environ 28:1103–1113. http://onlinelibrary.wiley.com/doi/10.1111/j.1365-3040.2005.01347.x/pdf

Domec J-C, Lachenbruch B, Meinzer FC (2006) Bordered pit structure and function determine spatial patterns of air-seeding thresholds in xylem of Douglas-fir (*Pseudotsuga menziesii*; Pinaceae) trees. Am J Bot 93:1588–1600

Domec J-C, Warren JM, Meinzer FC, Lachenbruch B (2009a) Safety factors for xylem failure byimplosion and air-seeding within roots, trunks and branches of young and old conifer trees. IAWA J 30:100–120

Domec J-C, Palmroth S, Ward E, Maier CA, Thérézien M, Oren R (2009b) Acclimation of leaf hydraulic conductance and stomatal conductance of *Pinus taeda* (loblolly pine) to long-term growth in elevated CO2 (free-air CO2 enrichment) and N-fertilization. Plant Cell Environ 32(11):1500–1512

Domec J-C, Pruyn M, Lachenbruch B, Spicer R (2012) Effects of age-related increases in sapwood area, leaf area, and xylem conductivity on height-related hydraulic costs in two contrasting coniferous species. Ann For Sci 69:17–27. https://doi.org/10.1007/s13595-011-0154-3

Domec J-C, Palmroth S, Oren R (2016) Effects of *Pinus taeda* leaf anatomy on vascular and extravascular leaf hydraulic conductance as influenced by N-fertilization and elevated CO_2. J Plant Hydraul 3:e007. https://doi.org/10.20870/jph.2016.e007

Dute RR (2015) Development, structure, and function of torus–margo pits in conifers, ginkgo and dicots. In: Functional and ecological xylem anatomy. Springer, Berlin, Heidelberg, pp 77–102

Eberhardt TL, So C-L, Leduc DJ (2018) Wood variability in mature longleaf pine: differences related to cardinal direction for a softwood in a humid subtropical climate. Wood Fiber Sci 50:323–333

Enstone DE, Peterson CA, Hallgren SW (2001) Anatomy of seedling tap roots of loblolly pine (Pinus taeda L.). Trees 15:98

Esau K (1965) Plant anatomy, 2nd edn. Wiley, New York

Fabisiak E, Fabisiak B, Krauss A (2020) Variation in tracheid lengths in dominant trees of selected coniferous species. Bioresources 15:7330–7341

Farjon A (2021) Pines: drawings and descriptions of the genus Pinus, 2nd revised edn. Brill. 236 pp

Foti D, Passialis C, Voulgaridis E, Skaltsoyiannes A, Tsaktsira M (2018) Wood density and tracheid length of Aleppo pine (Pinus halepensis Mill.) grafts in relation to cambium age and growth rate. J For Sci 64:101–107

Gao C, Marker SJ, Gundlach C, Poulsen HF, Bohr T, Schulz A (2023) Tracing the opposing assimilate and nutrient flows in live conifer needles. J Exp Bot 74(21):6677–6691

García Esteban L, de Palacios P, García-Iruela A, García-Fernández F, García-Esteban L, González de Vega D (2021) Comparative wood anatomy in Pinaceae with reference to its systematic position. Forests 12:1706. https://doi.org/10.3390/f12121706

Gartner BL (2006) Prediction of wood structural patterns in trees using ecological models of plant water relations. In: Stokke DD, Groom LH (eds) Characterization of the cellulosic cell wall. Blackwell Publishing, Ames

Geada López G, Kamiya K, Harada K (2002) Phylogenetic relationships of diploxylon pines (subgenus *Pinus*) based on plastid sequence data. Int J Plant Sci 163:737–747

Gernandt DS, Magallón S, Geada López G, Zerón FO, Willyard A, Liston A (2008) Use of simultaneous analyses to guide fossil-based calibrations of Pinaceae phylogeny. Int J Plant Sci 169:1086–1099

Ghimire B, Lee C, Yang J, Heo K (2015) Comparative leaf anatomy of native and cultivated Pinus (Pinaceae) in Korea: implication for the subgeneric classification. Plant Syst Evol 301:531–540

Gibson K, Skov K, Kegley S, Jorgensen C, Smith S, Witcosky J (2008) Mountain pine beetle impacts in high-elevation five-needle pines: current trends and challenges. US Department of Agriculture Forest Service, Forest Health Protection Note R1-08-020. US Department of Agriculture Forest Service, Northern Region, Missoula

Godfrey JM, Riggio J, Orozco J, Guzmán-Delgado P, Chin ARO, Zwieniecki MA (2020) Ray fractions and carbohydrate dynamics of tree species along a 2750 m elevation gradient indicate climate response, not spatial storage limitation. New Phytol 225:2314–2330

Hacke UG, Sperry JS, Pockman WT, Davis SD, McCulloh KA (2001) Trends in wood density and structure are linked to prevention of xylem implosion by negative pressure. Oecologia 2001(126):457–461

Hacke UG, Sperry JS, Pittermann J (2004) Analysis of circular bordered pit function II. Gymnosperm tracheids with torus-margo pit membranes. Am J Bot 91:386–400

Hacke UG, Lachenbruch B, Pittermann J, Mayr S, Domec J-C, Schulte PJ (2015) The hydraulic architecture of conifers. In: Functional and ecological xylem anatomy. Springer, Berlin, Heidelberg, New York, pp 39–75

Held M, Ganthaler A, Lintunen A, Oberhuber W, Mayr S (2021) Tracheid and pit dimensions hardly vary in the xylem of *Pinus sylvestris* under contrasting growing conditions. Front Plant Sci 12:786593. https://doi.org/10.3389/fpls.2021.786593

Hoadley RB (1990) Identifying wood: accurate results with simple tools. Taunton Press, Newtown, p 223

Howard ET, Manwiller FG (1968) Anatomical characteristics of southern pine stemwood. Wood Sci 2:77–86

Hu YS, Yao BJ (1981) Transfusion tissue in gymnosperm leaves. Bot J Linn Soc 83:263–272

Hubbard RM, Ryan MG, Stiller V, Sperry JS (2001) Stomatal conductance and photosynthesis vary linearly with plant hydraulic conductance in ponderosa pine. Plant Cell Environ 24(1):113–121

Hudgins JW, Christiansen E, Franceschi VR (2004) Induction of anatomically based defense responses in stems of diverse conifers by methyl jasmonate: a phylogenetic perspective. Tree Physiol 24:251–264

IAWA Committee (2004) IAWA list of microscopic features of softwood identification. IAWA J 25:1–70

Jain P, Huber AE, Rockwell FE, Sen S, Holbrook NM, Stroock AD (2024) Localized measurements of water potential reveal large loss of conductance in living tissues of maize leaves. Plant Physiol 194:2288–2300. https://doi.org/10.1093/plphys/kiad679

Jin WT, Gernandt DS, Wehenkel C, Xia X, Wei X, Wang X (2021) Phylogenomic and ecological analyses reveal the spatiotemporal evolution of global pines. Proc Natl Acad Sci 118:e2022302118. https://doi.org/10.1073/pnas.2022302118

Johnson DM, Meinzer FC, Woodruff D, McCulloh KA (2009) Leaf xylem embolism, detected acoustically and by cryo-SEM, corresponds to decreases in leaf hydraulic conductance in four evergreen species. Plant Cell Environ 32:828–836

Johnson DM, McCulloh KA, Meinzer FC, Woodruff DR (2012) Evidence for leaf xylem embolism as a primary factor in dehydration-induced declines in leaf hydraulic conductance. Plant Cell Environ 35:760–769

Johnson DM, Katul GG, Domec J-C (2022) Catastrophic hydraulic failure and tipping points in plants. Plant Cell Environ 45:2231–2266

Josza LA, Middleton GR (1994) Wood quality attributes and their practical implications. Forintek Canada Corp., Vancouver

Kamala FD, Sakagami H, Oda K, Matsumura J (2013) Wood density and growth ring structure of *Pinus patula* planted in Malawi, Africa. IAWA J 34:61–70

Keeley JE (2012) Ecology and evolution of pine life histories. Ann For Sci 69:445–453

Kellogg RM, Rowe S, Koeppen RC, Miller RB (1982) Identification of the wood of the soft pines of western North America. IAWA Bull 3:95–101

Kennedy RW (1995) Coniferous wood quality in the future: concerns and strategies. Wood Sci Technol 29:321–338

Kimberley MO, Cown DJ, McKinley RB et al (2015) Modelling variation in wood density within and among trees in stands of New Zealand-grown radiata pine. N Z J For Sci 45:22. https://doi.org/10.1186/s40490-015-0053-8

Kribs DA (1928) Length of tracheids in Jack pine in relation to their position in the vertical and horizontal position axes of the trees. Technical bulletin, vol 54. University of Minnesota, University Farm, St Paul, 9pp

Kuusk V, Niinemets Ü, Valladares F (2018) A major trade-off between structural and photosynthetic investments operative across plant and needle ages in three Mediterranean pines. Tree Physiol 38:543–557. https://doi.org/10.1093/treephys/tpx139

Lachenbruch B, Moore JR, Evans R (2011) Radial variation in wood structure and function in woody plants, and hypotheses for its occurrence. In: Meinzer FC, Lachenbruch B, Dawson TE (eds) Size- and age-related changes in tree structure and function. Springer-Verlag, Berlin, pp 121–164

Larocque GR, Marshall PL (1995) Wood relative density development in red pine (*Pinus resinosa* Ait.) stands as affected by different initial spacings. For Sci 41:709–728

Larson PR, Kretschmann DE, Clark A III, Isebrands JG (2001) Formation and properties of juvenile wood in southern pines: a synopsis. General technical report FPL-GTR-129. U.S. Department of Agriculture, Forest Service, Forest Products Laboratory, Madison, 42p

Laur J, Hacke UG (2014) Exploring *Picea glauca* aquaporins in the context of needle water uptake and xylem refilling. New Phytol 203:388–400

Lavery PB, Mead DJ (1998) Pinus radiata: a narrow endemic from North America takes on the world. In: Richardson DM (ed) Ecology and biogeography of Pinus. Cambridge University Press, pp 432–449

Lawhon WT, Woods FW (1976) Radial growth and wood density of white pine in relation to fossil-fired power plant operations. In: Dochinger LS, Seliga TA (eds) Proceedings of the first international symposium on acid precipitation and the forest ecosystem; General technical report NE-23. U.S. Department of Agriculture, Forest Service, Northeastern Forest Experiment Station, Upper Darby, pp 1025–1026

Lehringer C, Richter K, Schwarze FW, Militz H (2009) A review on promising approaches for liquid permeability improvement in softwoods. Wood Fiber Sci 41:373–385

Lersten NR (1997) Occurrence of endodermis with a casparian strip in stem and leaf. Bot Rev 63:265–272

Leyton L, Armitage IP (1968) Cuticle structure and water relations of the needles of Pinus radiata (D. Don). New Phytol 67:31–33

Lin J, Jach ME, Ceulemans R (2001) Stomatal density and needle anatomy of Scots pine (Pinus sylvestris) are affected by elevated CO2. New Phytol 150:665–674

López GG, Kamiya K, Harada K (2002) Phylogenetic relationships of *Diploxylon* pines (subgenus *Pinus*) based on plastid sequence data. Int J Plant Sci 163:737–747

Losso A, Anfodillo T, Ganthaler A, Kofler W, Markl Y, Nardini A et al (2018) Robustness of xylem properties in conifers: analyses of tracheid and pit dimensions along elevational transects. Tree Physiol 38:212–222

Louzada J (2000) Variação fenotípica e genética em características estruturais na madeira de Pinus pinaster Ait. O comprimento das fibras e a densidade até aos 80 anos de idade das árvores. Parâmetros genéticos na evolução juvenil-adulto das componentes da densidade da madeira, Série didáctica, Ciências aplicadas, n° 143. UTAD, Vila Real. 293p

Mai M, Gao C, Bork PAR, Holbrook NM, Sculz A, Bohr T (2024) Relieving the transfusion tissue traffic jam: a network model of radial transport in conifer needles. New Phytol 244:2183–2196. https://doi.org/10.1111/nph.20189

Maier CA, Teskey RO (1992) Internal and external control of net photosynthesis and stomatal conductance of mature eastern white pine (Pinus strobus). Can J For Res 22(9):1387–1394

Mansfield SD, Parish R, Goudie JW, Kang K-Y, Ott P (2007) The effects of crown ratio on the transition from juvenile to mature wood production in lodgepole pine in western Canada. Can J For Res 37:1450–1459

Mauriat M, Le Provost G, Rozenberg P, Delzon S, Bréda N, Clair B, Coutand C, Domec JC, Fourcaud T, Grima-Pettenati J et al (2014) Wood formation in trees. In: Ramawat KG, Mérillon JP, Ahuja MR (eds) Tree biotechnology. CRC Press, Boca Raton. 550p

Mauseth JD (1988) Plant anatomy. Benjamin/Cummings Publishing Company, California

McConchie D (1999) An update on within-ring internal checking including a procedure to identify logs prone to form checks during drying. In: Emerging technologies for evaluating wood quality for processing, Proceedings of the 3rd wood quality symposium, Rotorua, New Zealand, 30 November. Forest Industries Engineering Association/Forest Research. 19p

Megraw RA (1985) Wood quality factors in loblolly pine: the influence of tree age, position in tree, and cultural practice on wood specific gravity, fiber length, and fibril angle. Tappi Press, Atlanta. 89p

Millar CI (1998) Early evolution of pines. In: Richardson DM (ed) Ecology and biogeography of *Pinus*. Cambridge University Press, Cambridge, pp 69–91

Milota M, Tschernitz JL, Verrill SP, Mianowski T (1994) Gas permeability of plantation loblolly pine. Wood Fiber Sci 27:34–40

Morris H, Broderson C, Schwarze FWM, Jansen S (2016) The parenchyma of secondary xylem and its critical role in tree defense against fungal decay in relation to the CODIT model. Front Plant Sci 7:1665

Niklas KJ (1998) The influence of gravity and wind on land plant evolution. Rev Palaeobot Palynol 102:1–14

Panshin AJ, de Zeeuw C (1980) Textbook of wood technology, 4th edn. McGraw-Hill, New York

Pant DD, Basu N (1977) A comparative study of the leaves of *Cathaya argyrophylla* Chun & Kuang and three species of *Keteleeria* Carriere. Bot J Linn Soc 75:271–282

Park YI, Koubaa A, Brais S, Mazerolle MJ (2009) Effects of cambial age and stem height on wood density and growth of jack pine grown in boreal stands. Wood Fiber Sci 41:346–358

Paul BH (1963) The application of silviculture in controlling the specific gravity of wood, Technical bulletin no. 1288. U.S. Department of Agriculture, Madison

Peltola H, Gort J, Pulkkinen P, Zubizarreta Gerendiain A, Karppinen J, Ikonen V-P (2009) Differences in growth and wood density traits in Scots pine (*Pinus sylvestris* L.) genetic entries grown at different spacing and sites. Silva Fenn 43:339–354

Pensa M, Jalkanen R (2005) Variation in needle longevity is related to needle-fascicle production rate in Pinus sylvestris. Tree Physiol 25:1265–1271

Phillips EWJ (1948) Identification of softwoods by their microscopic structure. For Prod Res Bull (Lond) 22:1–56

Pittermann J, Sperry JS, Hacke UG, Wheeler JK, Sikkema EH (2006) Inter-tracheid pitting and the hydraulic efficiency of conifer wood: the role of tracheid allometry and cavitation protection. Am J Bot 93:1265–1273

Pothier D, Margolis HA, Poliquin J, Waring RH (1989) Relation between the permeability and the anatomy of jack pine sapwood with stand development. Can J For Res 19:1564–1570

Ran JH, Shen TT, Wu H, Gong X, Wang XQ (2018) Phylogeny and evolutionary history of Pinaceae updated by transcriptomic analysis. Mol Phylogenet Evol 129:106–116

Richardson DM, Williams PA, Hobbs RJ (1994) Pine invasions in the Southern Hemisphere: determinants of spread and invadability. J Biogeogr 21:511–527. https://doi.org/10.2307/2845655

Rissanen K, Hölttä T, Barreira LFM, Hyttinen N, Kurtén T, Bäck J (2019) Temporal and spatial variation in scots pine resin pressure and composition. Front For Glob Change 2:23. https://doi.org/10.3389/ffgc.2019.00023

Rosner A (2013) Hydraulic and biomechanical optimization in Norway spruce trunk wood – a review. IAWA J 34:365–390

Rosner S, Klein A, Muller U, Karlsson B (2008) Tradeoffs between hydraulic and mechanical-stress responses of mature Norway spruce trunk wood. Tree Physiol 28:1179–1188

Rosner S, Svìtlík J, Andreassen K, Børja I, Dalsgaard L, Evans R et al (2016) Novel hydraulic vulnerability proxies for a boreal conifer species reveal that opportunists may have lower survival prospects under extreme climatic events. Front Plant Sci 7:831. https://doi.org/10.3389/fpls.2016.00831

Saka S, Goring DAI (1985) Localisation of lignins in wood cell walls. In: Higuchi T (ed) Biosynthesis and biodegradation of wood components. Academic Press Inc., New York, pp 51–62

Sauter UH, Mutz R, Munro BD (1999) Determining juvenile-mature wood transition in Scots pine using latewood density. Wood Fiber Sci 31:416–425

Schimleck LR, McCulloh KA, Dahlen J, Auty D (2024) New perspectives on radial profiles of specific gravity in North American conifers. Can J For Res 54:940–955. https://doi.org/10.1139/cjfr-2023-0235

Schimleck LR, Dahlen J, Mora C, Antony F, Eberhardt TL, Onakporma I (2025) Maps depicting the within-tree wood property variation of some North American conifers: a review. Can J For Res 55:1–14. https://doi.org/10.1139/cjfr-2024-0142

Schneider R, Zhang SY, Swift DE, Bégin J, Lussier J-M (2008) Predicting selected wood properties of jack pine following commercial thinning. Can J For Res 38:2030–2043

Siau JF (1984) Wood structure and chemical composition. In Transport processes in wood (pp. 35–72). Berlin, Heidelberg: Springer Berlin Heidelberg

Sperry JS (1995) Limitations of stem water transport and their consequences. In: Gartner BL (ed) Plant stems: physiology and functional morphology. Academic Press, San Diego, pp 95–104

Sperry JS, Hacke UG, Pittermann J (2006) Size and function in conifer tracheids and angiosperm vessels. Am J Bot 93:1490–1500

Spicer R (2005) Senescence in secondary xylem: heartwood formation as an active developmental program. In: Michele Holbrook N, Zwieniecki MA (eds) Physiological ecology, vascular transport in plants. Academic Press, pp 457–475. https://doi.org/10.1016/B978-012088457-5/50024-1

Strickland RK, Goddard RE (1966) Correlation studies of slash pine tracheid length. For Sci 12:54–62. https://doi-org.proxy.lib.duke.edu/10.1093/forestscience/12.1.54

Stull GW, Qu XJ, Parins-Fukuchi C, Yang YY, Yang JB, Yang ZY, Hu Y, Ma H, Soltis PS, Soltis DE, Li DZ (2021) Gene duplications and phylogenomic conflict underlie major pulses of phenotypic evolution in gymnosperms. Nat Plants 7(8):1015–1025

Sviderskaya IV, Vaganov EA, Fonti MV, Fonti P (2021) Isometric scaling to model water transport in conifer tree rings across time and environments. J Exp Bot 72:2672–2685. https://doi.org/10.1093/jxb/eraa595

Tasissa G, Burkhart HE (1998) Modeling thinning effects on ring specific gravity of loblolly pine (Pinus taeda L.). For Sci 44:212–223

Thor E (1995) Variation in some wood properties of eastern white pine. For Sci 11:451–455

Trapp S, Crouteau R (2001) Defensive resin biosynthesis in conifers. Annu Rev Plant Physiol Plant Mol Biol 52:689–724

Trueba S, Théroux-Rancourt G, Earles JM, Buckley TN, Love D, Johnson DM, Brodersen C (2022) The three-dimensional construction of leaves is coordinated with water use efficiency in conifers. New Phytol 233(2):851–861

Tyree MT, Ewers FW (1991) The hydraulic architecture of trees and other woody plants. New Phytol 119:345–360

Vilalta JM, Sala A, Pinol J (2004) The hydraulic architecture of Pinaceae – a review. Plant Ecol 171:3–13

Wang N, Palmroth S, Maier CA, Domec J-C, Oren R (2019) Anatomical changes with needle length affect leaf structural and physiological traits across five Pinus species. Plant Cell Environ 42:1690–1704. https://doi.org/10.1111/pce.13516

Wardrop AB, Dadswell HE (1951) Helical thickenings and micellar orientation in the secondary wall of conifer tracheids. Nature 168:610–613
Wheeler EA (2011) InsideWood – a web resource for hardwood anatomy. IAWA J 32(2):199–211
Wheeler EA, Baas P (1998) Wood identification: a review. IAWA J 19(3):241–264
Whitehead D, Edwards WRN, Jarvis PG (1984) Relationships between conducting sapwood area, foliage area and permeability in mature *Picea stichensis* and *Pinus contorta* trees. Can J For Res 14:940–947
Wiedenhoeft AC, Miller RB, Theim JB (2003) Analysis of three microscopic characters for separating the wood of *Pinus contorta* and *P-ponderosa*. IAWA J 24(3):257–267
Wilkinson S, Ogée J, Domec J-C, Rayment M, Wingate L (2015) Biophysical modelling of intra-ring variations in tracheid features and wood density of *Pinus pinaster* trees exposed to seasonal droughts. Tree Physiol 35:305–318. https://doi.org/10.1093/treephys/tpv010
Willyard A, Syring J, Gernandt DS, Liston A, Cronn R (2007) Fossil calibration of molecular divergence infers a moderate mutation rate and recent radiations for *Pinus*. Mol Biol Evol 24:90–101
Woodcock DW, Shier AD (2002) Wood specific gravity and its radial variations: the many ways to make a tree. Trees 16:437–443
Wubble J (2010) Tree species distribution in relation to stem hydraulic traits and soil moisture in a mixed hardwood forest in central Pennsylvania. Master of Science, Horticulture, The Pennsylvania State University, p 39
Zanne AE, Lopez-Gonzalez G, Coomes DA, Ilic J, Jansen S, Lewis S et al (2009) Data from: towards a worldwide wood economics spectrum. Dryad, Dataset. https://doi.org/10.5061/dryad.234
Zhang S, Fei B, Yu Y, Wang H (2012) Fundamental properties of Masson pine (*Pinus massoniana* Lamb.) wood from plantation. In: Proceedings of the 55th international convention of Society of Wood Science and Technology, 27–31 Aug 2012, Beijing, China
Zhang YJ, Rockwell FE, Wheeler JK, Holbrook NM (2014) Reversible deformation of transfusion tracheids in *Taxus baccata* is associated with a reversible decrease in leaf hydraulic conductance. Plant Physiol 165(4):1557–1565
Zhang H, Domec JC, Maier CA, Chainey A, Wang N, Palmroth S, Oren R (2024) Similar response of canopy conductance to increasing vapor pressure deficit and decreasing soil conductivity with drought among five co-occurring pine species of a large range in crown architecture. Agric For Meteorol 366:110479
Zhu JY, Scott CT, Myers GC (2007) Effects of plantation density on wood density and anatomical properties of red pine (*Pinus resinosa* Ait). Wood Fiber Sci 39:502–512
Zobel BJ, Sprague JR (1998) Characteristics of juvenile wood. In: Juvenile wood in forest trees. Springer, pp 21–55
Zobel BJ, Van Buijtenen JP (1989) Wood variation – its causes and control. In: Timell TE (ed) Springer series in wood science. Springer, Berlin, p 363
Zwieniecki MA, Stone HA, Leigh A, Boyce CK, Holbrook NM (2006) Hydraulic design of pine needles: one-dimensional optimization for single-vein leaves. Plant Cell Environ 29(5):803–809

Fire Ecology and Ecophysiology of *Pinus* Species

Sharon M. Hood, Andreas Bär, Giovanna Battipaglia, Stefan Mayr, Timothy Shearman, J. Morgan Varner, and Kamal J. K. Gandhi

Abstract Fire has played a driving role in the speciation and evolution of the genus *Pinus* (pines) and continues to exert dominant influences on *Pinus*-dominated forests around the world. This chapter is divided into two major sections: fire ecology and fire ecophysiology. In the fire ecology section, we review fire-adapted traits of *Pinus* that allow persistence and recovery after fire, including how traits cluster to form fire syndromes. Fire regimes are described as an emergent property of fire syndromes and climate. In the fire ecophysiology section, we review the effects of fire on *Pinus* carbon relations, water relations, defense, growth, and mortality.

Keywords Fire regimes · Nonstructural carbohydrates · Plant defense · Post-fire tree mortality · Syndromes · Traits · Tree hydraulics

---

S. M. Hood (✉)
USDA Forest Service, Rocky Mountain Research Station, Fire Sciences Laboratory, Missoula, MT, USA
e-mail: sharon.hood@usda.gov

A. Bär · S. Mayr
Department of Botany, University of Innsbruck, Innsbruck, Austria

G. Battipaglia
Department of Environmental Biological and Pharmaceutical Sciences and Technologies, University of Campania Luigi Vanvitelli, Caserta, Italy

T. Shearman
College of Forestry, Wildlife and Environment, Auburn University, Auburn, AL, USA

J. M. Varner
Tall Timbers, Tallahassee, FL, USA

K. J. K. Gandhi
D.B. Warnell School of Forestry and Natural Resources, University of Georgia, Athens, GA, USA

D. M. Johnson et al. (eds.), *Physiology and Ecology of Pines Worldwide*, Tree Physiology 9, https://doi.org/10.1007/978-3-032-23001-0_3

1 Fire Ecology

Fire ecology is the study of fire as an ecosystem process, including its relationships to living organisms and the physical environment (Kobziar et al. 2024). In this section, we describe traits in pines that allow persistence—via survival or via regeneration—in fire-prone environments and evolutionary pressures that shaped these traits. We then describe fire syndromes, life history strategies based on several correlated traits and their associated fire regimes. Numerous papers and reviews exist on this topic, and we refer readers to these for more details, especially for non-*Pinus* species (Keeley and Zedler 1998; Keeley et al. 2011; He et al. 2012; Keeley 2012; Archibald et al. 2018; Stevens et al. 2020; Keeley and Pausas 2022).

1.1 Fire-Adapted Traits

Fire and the genus *Pinus* have a long-intertwined history. Fires, because they can be lethal across life history stages from seed to adult, have served as a strong selective force on the speciation and phenotypic traits of *Pinus* (Keeley and Zedler 1998; Schwilk and Ackerly 2001; Jin et al. 2021) and continue to play an essential role in many pine-dominated forests. He et al. (2012) hypothesized that fire in the Cretaceous period may have been a primary selecting force that initiated the divergence of the two pine lineages—subgenus *Pinus* (the diploxylon or hard pines) and subgenus *Strobus* (the haploxylon or soft pines), with fire playing a driving role in pine diversification beginning in the Oligocene (Jin et al. 2021). Fire-adapted species are more prevalent in the *Pinus* subgenus than the *Strobus* lineage (Badik et al. 2018; Jin et al. 2021). These studies align with Bond and Scott (2010) that changing fire regimes in the Cretaceous promoted the spread of angiosperms. The diversity of fires, from frequent surface fires to rare crown fires, also resulted in a diversity of fire-adapted traits across the *Pinus* genus.

Pinus species have a variety of traits that correspond to their past or extant fire regimes (Fig. 1). Because fire generates heat of lethal temperatures to plant tissues, pines and other woody species have developed traits to avoid or endure heat across their life histories. Protective traits, such as rapid development of thick bark, "grass stage" seedling physiognomy, self-pruning of lower branches, and the "basal crook" of seedlings are all beneficial for protecting vital meristematic tissues from heat damage in frequent fire regimes (Keeley and Zedler 1998; Pausas 2015a; Jin et al. 2021). Other traits provide an advantage in recolonizing the post-fire environment, such as cone serotiny and basal or epicormic resprouting (Pausas and Keeley 2014). Still other traits, such as highly flammable litter, help ensure the continuity of fuels that allow surface fires to spread throughout the landscape and result in selective mortality of competitors (Keeley and Pausas 2022; Varner et al. 2022). The fire ecology and fire resistance has been reviewed for *Pinus* in Mexico (Rodríguez-Trejo and Fulé 2003) and Europe (Fernandes et al. 2008). However, there are still many

Fig. 1 Examples of fire-adapted traits in *Pinus*. (**a**) Bud survival from fire in *P. plaustris*; (**b**) epicormic resprouting after fire in *P. canariensis*; (**c**) basal resprouting after fire in *P. echinata*; (**d**) *P. ponderosa* self-pruning with frequent fire; (**e**) *P. contorta* regeneration after fire from serotinous cones; (**f**) black arrow pointing to thick bark in *P. brutia*

knowledge gaps about fire resistance traits for numerous pine species, and most of what is known is disproportionately from the United States and Europe. Identifying morphological traits that arose due to the selective pressure of fire regimes is difficult and not without criticism. Many traits that are considered "fire-adapted" can also be argued to be exaptations that arose from other selective pressures (Snyder 1984). In this section, we review fire-adapted traits of *Pinus* and how combinations of these traits form fire-adaptive syndromes.

1.1.1 Bark Thickness

A primary defense against heat from wildland fires is bark, providing protection of underlying phloem, vascular cambium and buds. There have been many studies on the protective nature of bark on trees in relation to fire (Hare 1965; Vines 1968; Hengst and Dawson 1994; Stephens and Libby 2006). Bark thickness, moisture content, bark density, thermal conductivity, and surface structure influence heat insulation capacity, but bark thickness has by far the most influence on thermal protection (Dickinson and Johnson 2004; Bär and Mayr 2020). Bark protection scales exponentially with bark thickness (Hare 1965). Thus, many post-fire tree mortality models include some measure of bark thickness (or tree size) as a variable when predicting mortality (Woolley et al. 2012). Although bark serves multiple functions in addition to fire protection (Rosell 2016), the development of thick bark is presumed to have evolved due to the selective force of historical fire regimes

(Pausas 2015b). Pine species occurring in frequently burned savannas and woodlands tend to develop thicker bark than those found in closed canopy forests that experience less frequent fire (Pellegrini et al. 2017).

Since bark generally increases in thickness with tree diameter, large trees of many species can have thick, fire protective bark regardless of whether they are considered "pyrophytic" (i.e., fire tolerant). For this reason, many studies have emphasized the need to examine the rate at which bark thickness accumulates in younger trees with relation to the fire regime (Jackson et al. 1999; Shearman et al. 2018). For instance, Jackson et al. (1999) found that bark development can change over an individual tree's lifetime with some species investing in thick bark at an early age, and reducing the investment later in life. Additionally, bark development differs along tree height, with rapid accumulation at the base where surface heating would be most intense (Shearman et al. 2018; Rodman et al. 2021). Fire's selective pressure is greatest on smaller trees that need to develop thick bark during the fire-free period to survive the next fire and ultimately to recruit into the canopy. As a juvenile, when bark is thin across most species, bark thickness may indeed be paramount. For species that develop thick bark with age, the relative importance of bark becomes less critical for surviving surface fires. For juvenile seedlings and saplings, accruing sufficient bark to resist cambial injury in the "fire trap" stage is critical (Hammond et al. 2015; Rodman et al. 2021). In *P. pinea*, Battipaglia et al. (2016) observed that bark plates twisted and flaked off during combustion, which reduced surface stem temperatures.

Bark in pines, as in other genera, varies among species and closely correlates with fire regime (Keeley and Zedler 1998; Schwilk and Ackerly 2001; Stevens et al. 2020). He et al. (2012) suggest that thick bark originated in the *Pinus* genus ca. 126 Ma. The two subgenera, competing with angiosperms, spread to inhabit different environmental conditions: the *Strobus* subgenus generally inhabit stressful alpine and desert conditions, while the *Pinus* subgenus adapted to frequent fire environments (Keeley and Zedler 1998; Keeley 2012). Building on this hypothesis, Varner et al. (2022) linked bark thickness development in pines of North America to their flammability and their hypothesized fire return interval. They found that species in subgenus *Pinus* that develop thick bark early also have litter that is highly flammable, consistent with the hypothesis that these pines have a strong evolutionary history with fire.

1.1.2 Self-Pruning

Another fire-adapted trait is the ability of a species to shed lower branches, so termed self-pruning, natural pruning, or branch shedding. Self-pruning is affected by multiple species-specific physiology and environmental conditions, including shade tolerance, neighborhood competition/light intensity, insects, soil, and weather (Millington and Chaney 1973). Pines seasonally abscise short shoots, or the needle fascicles, at the point of attachment to the long shoot (i.e., branch), but self-pruning of branches in pines does not involve abscission (i.e., programmed cell death resulting in shedding of tissue). Rather, branch death is a slow process that occurs when the leaves on branches have low photosynthetic rates that do not meet maintenance demands and become susceptible to insects and disease (Millington and Chaney 1973; Sprugel et al. 1991). As pines are highly resinous, they are able to effectively seal dead branches from the rest of the tree, preventing disease spread. The degree of natural pruning of dead branches varies greatly by species and seems to be genetically and environmentally determined (Millington and Chaney 1973; Pallardy 2010). Self-pruning has been reported as a binary trait (He et al. 2012) and relativized scales (Keeley and Zedler 1998; Moris et al. 2022). These subjective measurements lead to uncertainty in the literature. For example, Keeley and Zedler (1998) rank *P. balfouriana* as a 2 (i.e., poor pruning ability), while He et al. (2012) suggests that it does shed branches (on a binary scale). There is clearly much more to learn about how pines shed branches and how these differ among species.

Self-pruning is relevant to fire ecology because mortality of mature *Pinus* tends to be driven primarily by injuries to the crown, thus increasing the distance from radiative and convective heat by pruning is advantageous (Varner et al. 2021). Beyond protection from heating from surface fires, this trait reduces ladder fuels that could otherwise allow fire to spread into the crown. Schwilk (2003) experimentally manipulated the crown of a non-pine (*Adenostoma fasciculatum*) to test whether crown architecture (pruning vs. non-pruning) affected fire behavior and found that non-pruned individuals had higher temperatures and heat release than the pruned treatments. In pines, Schwilk and Ackerly (2001) found positive correlation between pine species that have the self-pruning trait, thick bark, and tall height, suggesting that species with these traits are typically found in frequent surface fire regimes. In contrast, the retention of dead branches (i.e., no self-pruning) are hypothesized to increase crown flammability and to be correlated with cone serotiny (Keeley and Zedler 1998; Schwilk and Ackerly 2001).

1.1.3 Height and Height Growth

Being tall is advantageous in evading radiative and convective heating, and the rate at which species ascend to fire-safe heights matters. Pines vary widely in maximum heights, varying from shrub stature (<5 m tall) species to emergent canopy trees like *P. lambertiana* that can exceed 75 m. The rate of height growth enables seedlings and saplings to abbreviate the period of fire-sensitivity, as virtually all seedlings and saplings are susceptible to surface fire (but see Sect. 1.1.6). Several pines undergo rapid height growth during the "fire trap" (sensu Hoffmann et al. 2012) period where the crown is vulnerable to even low-intensity surface fires. Stevens et al. (2020) found that tall conifers (*P. jeffreyi*, *P. ponderosa*, and *P. lambertiana* included) had other fire-adapted traits that collectively enabled them to resist fire injury better than short-statured conifers.

1.1.4 Crown Traits

Crown injury, especially damage to foliage and buds, is the primary cause of mortality across pines (Hood et al. 2018; Cansler et al. 2020a). Self-pruning enables pines to evade surface heat, but the geometry and allometry of crowns matters. Savanna and woodland pines reflect these contrasting selective pressures via a typical physiognomy of flat-topped crowns that are pruned from below, but also spread widely. Pine crowns are differentially porous, with lacunarity varying widely across species. This lacunarity results in ventilation that allows heat to escape, and thus bypass sensitive tissues in the crown (Rowell et al. 2022). Other crown traits that can influence tree survival from fire include bud size, needle length and mass, and bud growth patterns (Fernandes et al. 2008). Large buds have higher heat tolerance and long, thick needles can shield buds from heat and mitigate meristem death (Varner et al. 2021). Pines can also have fixed or preformed buds that undergo one growth flush per year or exhibit continuous or multiple growth flushes per year (Businský 2016). Species with multiple growth flushes can recover more quickly from needle scorching if meristems are not killed compared to species that only have one growth flush per year (Wade and Johansen 1986). While some research reporting crown traits exists (Fernandes et al. 2008; Paula et al. 2009; Bär et al. 2021), measurements are lacking for many species, limiting understanding of how trait variation results in differential heating.

1.1.5 Cone Serotiny

A common trait in pines is the retention of a seed bank within crowns via cones that remain closed until they are exposed to heating. This trait, known as serotiny, is suggestive of a high intensity, high severity fire regime that enables the species to quickly reseed in the post-fire environment while mineral soil is exposed and competition is reduced (Lamont et al. 1991; Keeley et al. 2011; Pausas and Keeley 2014). Serotinous cones remain closed due to a resin seal that holds the scales together (Johnson and Gutsell 1993). When the cone is subjected to sufficient heat for enough time, the resin seal weakens and breaks, resulting in the scales reflexing away from the central axis of the cone and releasing the seeds. Serotinous cones differ anatomically from non-serotinous cones. Aside from the resin seal, serotinous cones are better insulators, with higher sclereid cell diameter and sclereid wall thickness (Moya et al. 2008). He et al. (2012) estimates that serotiny emerged in the *Pinus* subgenus approximately 89 Ma.

Cone serotiny is often reported across species as a binary trait but this is too simplistic, as there is notable polymorphism, with some individuals producing both serotinous and non-serotinous cones and differences in serotiny levels among populations (Muir and Lotan 1985b; Tinker et al. 1994; Climent et al. 2004; Radeloff et al. 2004; Greene et al. 2024). The degree of serotiny can vary based on age, landscape position, and fire frequency (McMaster and Zedler 1981; Tapias et al. 2001). For example, studies on *P. halepensis* showed that younger trees had cone opening temperatures about 5 °C higher than older individuals and that opening temperatures between species varied from 45 to 60 °C (Gauthier et al. 1996; Tapias et al. 2001). Some pine species that display serotiny have dwarf growth forms that tend to have a higher degree of serotiny and occur in more frequently burned environments. Dwarf pitch pines (*P. rigida*) in the New Jersey Pine Plains, USA (which were subject to more frequent fires) have a higher percentage of serotinous cones than their taller counterparts in areas of less frequent fires (Ledig and Fryer 1972). Similarly, the shrub variety of *P. yunnanensis* in the Yunnan Province of China has a higher percentage of serotinous cones and experiences more frequent fires than the tree variety (Pausas et al. 2021).

Serotiny is a complex polymorphic trait and considerable research is needed to understand evolutionary pressures and environmental factors influencing its prevalence. While fire and herbivory are selective pressures favoring serotiny (Smith 1970; He et al. 2012), wind-dispersed pine species have ample opportunity for gene flow between serotinous and non-serotinous individuals that reduce selection pressure (Muir and Lotan 1985b). Serotiny also develops with age, potentially further reducing selection pressure. Numerous studies in pines have found the degree of serotiny in a population is most related to the fire regime, with areas of large, high-severity fire having higher proportions of serotiny and areas with complex, finer grain fires of mixed-severity having higher proportions of non-serotinous trees (Muir and Lotan 1985a; Gauthier et al. 1996; Radeloff et al. 2004). Serotiny is likely a bet-hedging strategy by which a species can reproduce under unpredictable disturbance regimes (Muir and Lotan 1985b). In a large study comparing cone and seed traits in serotinous versus non-serotinous conifer species, Greene et al. (2024) found little differences between the two cone types and demonstrated that seeds from non-serotinous cones can survive fire and contribute to post-fire regeneration. Collectively, this research highlights that while cone serotiny can certainly increase regeneration after fire, serotinous species are not entirely dependent on this trait to be resilient to fire.

Given the pressure of fire on the vulnerable reproductive period, there are likely other fire-cued or fire-adapted reproductive-related traits in pines. Strobili, pollen cones, and pollen all have vulnerable periods within pine crowns. Seeds of many pines lack thick protection once they disperse from cones and thus are vulnerable before germination and as juvenile seedlings. There has been scant work on investigating how these critical stages of pine life history may be cued to fire timing to avoid or evade heat via crown architecture or phenology more broadly.

1.1.6 Resprouting and Grass Stage

The ability to resprout after fire is rare in the *Pinus* genus, occurring in approximately 15 of all extant pines. For pines with this trait, resprouting can occur from basal or underground buds or higher up along the bole. Some species such as *P. echinata*, *P. rigida*, and presumably others develop a unique "basal crook" at the root collar where the stem grows horizontally at or below the mineral soil before turning vertical (Little and Somes 1956). This morphological trait allows buds that develop in the crook to be protected from the heat of fire and sprout after fire. Evidence of the selective ability of fire on the basal crook trait have been demonstrated in studies of *P. echinata*, the less fire adapted *P. taeda* that lacks the basal crook and rarely resprouts after fire, and a hybrid between the two species that has a less pronounced crook and intermediate resprouting capability (Stewart et al. 2015; Bradley et al. 2016). *P. echinata* was able to persist in frequently burned sites, whereas *P. taeda* and the majority of hybrid individuals were eliminated (Stewart et al. 2015). Crown and epicormic resprouting are also notable, enabling scorched and consumed crowns to refoliate via stem sprouts on the bole or in the crown. Examples of epicormic resprouting include *P. serotina*, *P. echinata*, and *P. canariensis*. Resprouting, as with other traits, is often viewed and analyzed as a binary trait, ignoring the wide variation in resprouting and its vigor within species and across the *Pinus* genus (Lilly et al. 2012; Pausas and Keeley 2014).

An even rarer trait is the "grass stage" juvenile form, occurring in only nine (<8%) pine species (e.g., *P. devoniana*, *P. merkusii*, and *P. palustris*) (Mirov 1967; Earle 2023). The grass stage is a juvenile form where no vertical growth occurs, instead the seedling invests growth in roots and radial growth. Beyond a threshold that may transpire in a year or over decades, the grass stage seedling rapidly ascends from its grass stage. This rapid juvenile height growth enables the seedling to exceed the height of injury from surface fires in a short period, perhaps taking advantage of stored carbohydrates to ascend beyond a fire-sensitive height. Recent investigations have better evaluated the grass stage (Pile et al. 2017; Knapp et al. 2018), but much is left to understand from this curious trait in *Pinus*. While the grass stage is rare in pines, this strategy of investing heavily in roots during juvenile development is relatively common in angiosperms as a strategy to allow rapid height growth after disturbance. For example, many species in tropical and subtropical savannas also have this life history strategy of early heavy belowground investment followed by a developmental shift to rapid height growth that can allow escape from grass fires (Schutz et al. 2009; Werner and Prior 2013).

1.1.7 Litter Flammability

Surface fire regimes in woodlands and forests are primarily driven by senesced foliage as surface litter. The flammability of litter has been documented for many tree species, pines a notable fraction of these (Fonda 2001; Varner et al. 2022). Mutch (1970) hypothesized that litter flammability is a selected trait in fire-dependent ecosystems, with species that occur in these systems having more flammable litter than species that have not experienced a long history with fire. Thus, pyrogenic species not only have adapted traits to protect them from fire, but also have adapted increased flammability to exclude species without defensive traits (Mutch 1970; Williamson and Black 1981; Bond and Midgley 1995; Gagnon et al. 2010). The Mutch (1970) hypothesis introduced somewhat of a paradox—fire is damaging to all plants to some degree, so the idea that a trait that increases injury to an individual is beneficial in fire-prone environments seems counterintuitive to natural selection at the individual level (Snyder 1984). Snyder (1984) criticized this hypothesis stating that flammability could have evolved due to completely unrelated selective forces (i.e., exaptation; e.g., an increase in phenolic compounds that are highly flammable could be the result of selective pressure from herbivory).

Numerous hypotheses have since been proposed to explain how the flammability trait could evolve in a species. Bond and Midgley (1995) showed through simulation that it is possible for increased flammability to be a fitness benefit if the flammable species also has some of the other traits listed above and the nearby non-flammable species does not have any fire defensive or post-fire regenerative traits. Gagnon et al. (2010) hypothesized that increased flammability could actually benefit the individual plants because highly flammable litter would have less soil heating due to low flame residence times, and therefore, belowground roots and meristems would be protected. These hypotheses all assume that species already possess some traits that confer advantage to survive fire or to colonize the post-fire environment. However, Kerr et al. (1999) modeled a scenario that suggested the opposite: anatomical and reproductive traits that improve fitness in fire-prone environments may have evolved as a result of the evolution of flammability. Schwilk and Kerr (2002) proposed that increased flammability as a trait could spread through a population without any direct fitness benefit by creating localized gaps that are likely to be colonized by propagules possessing the high flammability trait, assuming mating and seed dispersal are local processes. Alternatively, Midgley (2013) argued that flammability is an emergent property conferring no advantage to individual plants, and therefore, is not a selected trait. Instead, anti-flammability is the trait that could be selected for and evolved with time. Regardless of whether adaptation or exaptation, flammability and the associated morphological and physiological traits are clearly linked in ecosystems that are maintained by recurring fire which occur in all pine-dominated ecosystems.

Pines vary tremendously in their flammability. In a synthesis of North American pine litter flammability, Varner et al. (2022) found species with rapid, high intensity flaming combustion (e.g., *P. palustris*, *P. ponderosa*, and *P. serotina*) and others with poor flammability (*P. contorta* varieties, *P. clausa* varieties, and many high-elevation pines). Flammability in those species was related to their phylogeny, with subgenus *Pinus* exceeding subgenus *Strobus*, and to needle length, with longer-needled pines having greater flammability than short-needled pines. Past work has also illustrated the chemical nature of flammability, as terpene composition and content is correlated with ignition and intensity (Ormeño et al. 2009). Work remains to be conducted on flammability of pines beyond the U.S. and Mediterranean regions, where there exists substantial diversity of pine species and fire regimes.

1.2 Fire Syndromes and Regimes

1.2.1 Syndromes

Several fire syndromes have been described for pine species with common combinations of traits (Agee 1998; Keeley and Zedler 1998; Schwilk and Ackerly 2001; Jin et al. 2021; Keeley and Pausas 2022). Fire tolerators or resisters typically survive low-intensity fires due to thick bark and self-pruning. Fire embracers and endurers are easily killed by fire but have traits that allow regeneration and recovery, such as cone serotiny and ability to resprout. Fire evaders have few traits to survive or regenerate after fire and can persist through unburned areas or fire refugia. Fire avoiders also have few traits to survive fire, but occupy environments that rarely burn (cold, wet, or desert). These categorical approaches suffer from confounding issues but are convenient to describe the diverse suites of traits that enable survival and fuel local fire regimes.

To illustrate how fire-adapted traits are associated with the entire *Pinus* genus, we constructed a classification of all 118 extant pines including subspecies *P. contorta* ssp. *contorta*, *bolanderi*, and *murrayana*, as well as varieties *P. elliottii* vars. *elliottii* and *densa* (Fig. 2). We used nonmetric multidimensional ordination (NMDS) on trait metrics with a Gower distance metric. The ordination includes

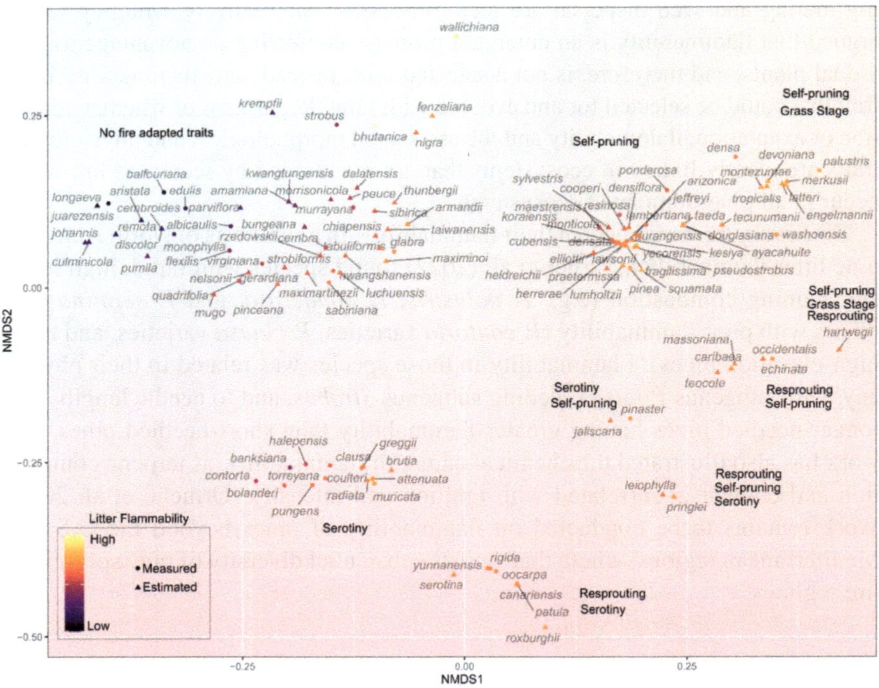

Fig. 2 Nonmetric Multidimensional Ordination (NMDS) showing relationship of *Pinus* species and fire-adapted traits

measured litter flammability for species as well as estimated flammability based on the relationship to leaf length found in Varner et al. (2022). Other traits include tree height and binary measures of serotiny, resprouting, grass stage, and pruning. Binary measurements from He et al. (2012) were used with some corrections and the additional species were estimated from various resources (Fire Effects Information System (FEIS); Perry 1991; Keeley and Zedler 1998; Saladin 2013). The binary measurements have a high influence on the ordination of species, which cluster into regions that relate to their hypothesized syndrome. The NMDS1 axis is correlated with response to fire, with fire avoiders clustered together with species that lack any of the fire-adapted traits, while fire resisters cluster in regions with species that have self-pruning and grass stage physiognomy. The NMDS2 axis separates species with serotiny versus self-pruning. The ordination also highlights species that have seemingly conflicting traits to any one syndrome. For example, serotiny and self-pruning are suggested to be negatively correlated (Schwilk and Ackerly 2001), yet the ordination suggests that some pines possess both of these traits. One possibility is that these species are using a "bet hedging" strategy for multiple fire regimes. Regardless, it is an indicator of how more studies on fire-adapted traits are needed.

1.2.2 Fire Regimes

Fire disturbance regimes are useful emergent properties that help describe general domains of fire dynamics over space and time (Agee 1993, 1998). Understanding historical fire regimes is essential to understanding how ecosystems evolved with wildland fire and for anticipating how they will respond to management, alterations to disturbance regimes, and climate change (Keeley and Pausas 2022). While individual plant traits influence relative resistance and responses to fire, in fire-affected systems, it is the distribution of fires *over time* (i.e., the fire regime) interacting with climate and topography, that determines the structure and composition of ecosystems (Moritz et al. 2005; Keeley et al. 2011). Fire regimes can include many characteristics, such as fire frequency, intensity (i.e., energy release), severity (i.e., fire effects such as tree mortality or soil heating), seasonality, type (i.e., ground, surface, or crown fire), extent (i.e., average size), geometry, or synergies with other disturbances (Agee 1993; Turner 2010). In practice, fire regimes are often reduced to simplified classifications of fire frequency and severity (Guyette et al. 2012; Blankenship et al. 2021) or fire type (Keeley and Pausas 2022). These groupings describe the dominant, or most likely fire attributes, but fires of any severity, frequency, and type can occur over smaller areas.

As fire exerts a large influence in many pine-dominated ecosystems, understanding the dominant fire regime is essential to understanding the system's ecology and departures from historical regimes (Falk et al. 2011; Greenberg and Collins 2021). Pines with broad distributions may have different species associations and, therefore, can be associated with different fire regimes. A fire regime is an emergent ecosystem property where a given climate and distribution of fire behavior favors convergence of a fire-adaptation syndrome and associated species traits and forest characteristics (Table 1). Pine ecosystems with frequent, low-severity fire regimes are comprised of *tolerator* or *resistor* species with traits that allow survival of low-intensity surface fires, such as developing thick bark at an early age, self-pruning branches, and large or protected buds. Forests typically are open and have a low density of multiple age cohorts, but with little overlapping vertical strata. Infrequent, high-severity fire regimes are dominated by *embracer* species that are easily killed by fire and forest characteristics that typically support high-intensity crown fire behavior such as dense forests with continuous forest canopies. Although *embracers* and *endurers* are easily killed or top-killed by fire, they have traits to allow rapid recolonization of the site after fire, such as serotinous cones, light-wind born seed, or the ability to resprout. Mixed-severity fire regimes, as the name implies, have high spatial variability of fire intensities, allowing for *evaders* to survive by escaping the fire (i.e., *fire refugia*); *tolerators* and *embracers* can also occur. In areas where fire is rare, such as low productivity arid or cold environments, *avoiders* are not adapted to fires of any kind.

Fire regimes are governed by both top-down climatic factors and bottom-up factors of topography, human population density, and fine-scale forest mosaics (Falk et al. 2011; Guyette et al. 2012). Climate and its effect on fire vary greatly by geographic location, with synoptic circulation patterns, such as the El Niño-Southern

Table 1 Relationships between fire regimes and fire-adaptation syndromes, with associated fire-adapted traits and forest characteristics

Fire regime	Fire-adaptation syndrome	Associated fire-adapted species Traits	Associated forest characteristics	Example *Pinus* species
Frequent, low-severity	Tolerators[a], Resisters[b, c], Survivors[d]	Thick bark Self-pruning branches Open crown architecture Protected buds Grass stage	Low density Patchy, discontinuous vertical strata	*P. ponderosa* *P. palustris* *P. echinata* *P. jeffreyi* *P. pinaster* *P. nigra* *P. sylvestris*
Infrequent, high-severity	Embracers[a, d], Endurers[b, c]	Serotiny Thin bark at maturity	High density Connecting overstory crowns	*P. contorta* *P. clausa* *P. banksiana*
Mixed-severity	Evaders[b, c], Fire-refugia[d]		Complex topography to allow fire escape. Horizontal and vertical heterogeneity	*P. albicaulis* *P. uncinata*
Rare, little-to-no fire	Avoiders[a, c]	None	Low productivity with little understory	*P. edulis* *P. longeva* *P. mugo*

Footnotes are primary citations describing the fire-adaptation syndromes in more detail
[a]Keeley (2012)
[b]Agee (1998)
[c]Jin et al. (2021)
[d]Schwilk and Ackerly (2001)

Oscillation (ENSO) or the Pacific Decadal Oscillation (PDO), exerting large influences on regional fire activity (Swetnam and Betancourt 1998). In contrast to regional fire regime patterns, fire severity and frequency vary at finer scales as a function of forest vegetation, topography, and cultural fire use practices, and land use (Falk et al. 2011). As fire is sensitive to climate, anthropogenic climate change is now altering contemporary fire regimes (Jones et al. 2022). Changes in climate and fire regimes have ecological consequences for postfire resilience and recovery which may affect distributions of *Pinus* species and community associations.

2 Fire Ecophysiology

Plant ecophysiology is the study of a plant's physiological mechanisms interacting with its physical, chemical, and biological environment to determine survival, reproduction, and distribution (Lambers et al. 2008). In this section, we describe the effects of fire on the metabolic functions of carbon, hydraulics, and defenses that lead to growth and survival or tree death. This is an active area of research, with more unknowns than Sect. 1; therefore, we go into more detail in this section when research is available then the previous section. In the carbon and water relations sections we describe and summarize the evidence supporting the various hypothesized mechanisms of fire-caused tree death (Fig. 3).

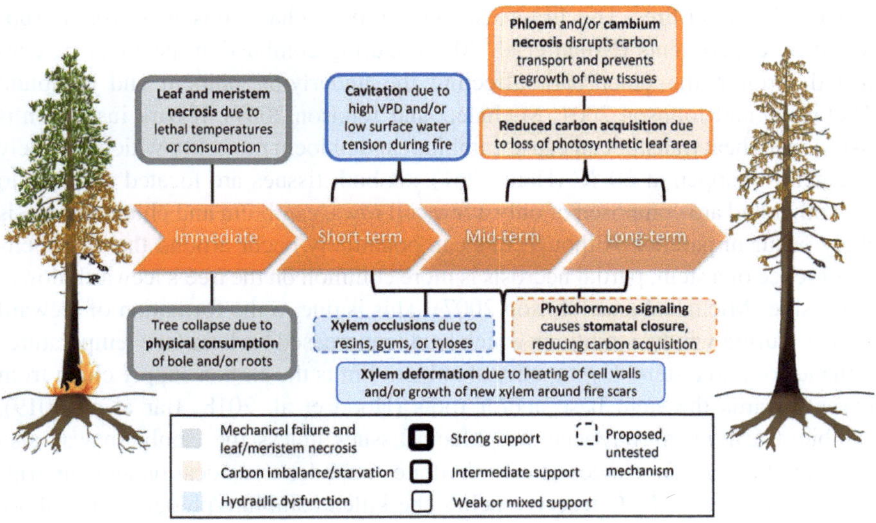

Fig. 3 Hypothesized mechanisms of fire-caused tree death illustrating the differential timing of the mechanisms and their level of support in the literature. VPD = vapor preesure deficit (Reprinted from Nolan et al. (2024), by permission of Oxford University Press)

2.1 Fire Effects on Carbon Relations

Fires can affect three important processes associated with carbon relations: acquisition, translocation, and storage. Pools of non-structural carbohydrates (NSCs; mainly sugars and starch) are essential for all plant metabolic processes (Hartmann and Trumbore 2016). They are involved with maintaining hydraulic function, phloem transport, plant defenses, cold tolerance, and buffering against stress (Dietze et al. 2014; Villar-Salvador et al. 2015; Martínez-Vilalta et al. 2016; Erbilgin et al. 2021).

First, post-fire carbon acquisition can be impaired due to heat-induced foliage loss, and there is intermediate supporting evidence that this can lead to tree mortality (Fig. 3) (Nolan et al. 2024). The amount of heat transferred to the crown and subsequent foliage death depends on fireline intensity, residence time, tree architecture, and crown component traits (Michaletz and Johnson 2007; Varner et al. 2021; see also Sect. 1.1.4). Necrotic foliage reduces the leaf area of affected trees limiting their overall photosynthetic capacity, but the unaffected foliage may increase photosynthetic efficiency to compensate (Wallin et al. 2003; Sayer et al. 2020). Especially when levels of leaf necrosis are high, the remaining foliage may not able to sustain whole tree carbohydrate demands, which consequently can limit foliage reestablishment, growth and carbon-based defense. Additionally, heat may have damaged photosystems of leaves that survived fire (Sparks et al. 2023). This can further curtail carbohydrate assimilation until the photosynthetic apparatus is repaired.

Second, the translocation of NSCs from source to sink organs can be limited or even blocked after fires. This fire-caused mortality mechanism is also strongly supported by experiments (Nolan et al. 2024). During combustion, heat can be conducted through the outer bark affecting the underlying phloem and cambium (Dickinson and Johnson 2001; Michaletz and Johnson 2007). If bark insulation is insufficient, heat transfer can cause cambium and phloem mortality which is widely assumed to happen at 60 °C (Hare 1961). As both tissues are located adjacent to each other and are composed of only a few cell rows, cambium and phloem necrosis likely occur in parallel. Although this can potentially occur around the whole circumference of a stem, partial necrosis is more common on the tree's leeward (downwind) side (Michaletz and Johnson 2007). This is due to the formation of leeward vortices during windy conditions which lead to increased bark surface temperatures in the lee of a tree stem. Dysfunctional phloem limits the carbon supply chain from leaves towards the stem base and/or roots (Hood et al. 2018; Bär et al. 2019). Cambium activity may regenerate phloem tissues unless the cambium was also killed by the heat. Blockages of carbohydrates impedes translocation and can trigger the depletion of NSC reserves below the killed cambium/phloem. Once those reserves are exhausted, tissues are considered to suffer from carbon starvation which ultimately may lead to dieback. Hoffmann et al. (2024) recently proposed another mechanism that could impact NSC acquisition through phytohormone signaling that causes stomatal closure. Further research is needed to determine if this could lead to reduced NSCs and eventual tree death in fire-injured trees.

Third, trees may exploit NSCs which were stored before the fire to ensure carbon assimilation (Smith et al. 2017) and water uptake (Varner et al. 2009) after a fire. The importance of NSCs as an actively regulated storage pool required for plant growth and respiration maintenance is increasingly recognized, with strong evidence that plant death occurs if NSCs decline below a minimum threshold (Martínez-Vilalta et al. 2016; Barker Plotkin et al. 2021). Exact thresholds are unclear; however, and seem to vary across species, functional groups, and tissue types (Adams et al. 2017). NSCs are more tightly linked to hydraulic function and drought resistance in gymnosperms than angiosperms, with reductions in NSCs associated with lower resistance to xylem embolism (Adams et al. 2017). NSC stores may also be used post-fire to repair the photosynthetic apparatus, regrow buds/leaves or fine roots.

The effects of fire-induced impairments of carbon acquisition, translocation, and storage are often assessed indirectly by examining growth limitations (see below and Sect. 2.4) as a final consequence, and inferring conclusions on NSC dynamics. However, only a few studies have directly quantified the effects of fire on NSC pools in obligate seeders such as *Pinus* species; most research has focused on resprouting angiosperms (Piper and Paula 2020). In a study on *P. palustris*, NSC concentrations after long-duration smoldering duff fires declined in coarse roots but not in fine roots, probably because carbon reserves of coarse roots were re-allocated to rebuild damaged fine roots (Varner et al. 2009). Fall burning of *P. palustris* saplings caused reduced shoot starch concentrations for several months, as well as reduced radial growth, and the authors attributed these declines to crown scorch and high NSC demand to replenish foliage (Sayer et al. 2020). In laboratory-burned 2-year-old *P. ponderosa*, NSCs of stems and roots of burned trees were lower than unburned controls as soon as 1 week after exposure to lethal fire (Partelli-Feltrin et al. 2022). The authors hypothesized that fire-induced phloem death likely impeded the translocation of NSCs. Reed and Hood (2024) measured NSCs pre-fire and six timesteps after a spring prescribed fire (4 days through 16 months) in young *P. ponderosa* that experienced varying levels of crown scorch. No trees were immediately killed by the fire and all trees showed some degree of bud survival and new needle growth. NSC concentrations declined quickly in burned trees relative to unburned controls over the same post-fire period. The decline was strongest for trees that eventually died, with surviving trees recovering to unburned levels within 14 months post-fire. Two months post-fire, phloem NSCs in the main stem were strongly negatively related to crown scorch levels. Taken together, these studies support the importance of NSCs for tree survival and recovery post-fire and suggest that post-fire NSC depletion may occur due to reduced photosynthetic leaf area, phloem death, and use of remaining NSC pools for repair and regrowth processes.

The cumulative effects of fire on NSCs can also influence subsequent long-term tree growth. *P. palustris* trees with longer duration smoldering suffered growth declines as much as a decade following burning, with the effects most severe in drought years (Slack et al. 2016). Niccoli et al. (2023a) found a delay in cambial phenology of burned *P. pinaster* individuals with a lower number of mature xylem cells during the growing season compared to unburned stands. Generally, in conifers, the formation of new xylem cells is fueled by the photosynthates of the current year rather than by the stored carbon (Hansen and Beck 1994). Hence, the post-fire decline in overall photosynthetic capacity for burned trees likely compromised the ability to support required carbohydrate acquisition (Valor et al. 2018), causing severe limitations in cambium productivity (Gričar et al. 2020) and regrowth of killed needles (Sayer et al. 2020; Reed and Hood 2024). In fact, in cases of severe defoliation, only a small part of the carbohydrate reserves is allocated to growth and foliage because the majority is used primarily to ensure cell respiration (Jacquet et al. 2014).

2.2 *Fire Effects on Water Relations*

Water relations are a major critical physiological component in plant life and rely on sufficient water uptake, water transport, and storage, as well as transpirational control, the latter obtained by the combination of hydrophobic shields (cuticula and tissues containing suberin) and stomatal adjustment. Water transport is based on a meta-stable state of water (Tyree and Zimmermann 2002), which is under tension (i.e., negative water potential) due to transport resistances (under transpirational conditions), limited soil humidity, and gravitational forces (which are relevant in trees), and thus at risk for xylem embolism and/or impairments of extra-vascular pathways.

During a forest fire, heat may affect all components (water uptake, transport and storage, and transpiration) and organs involved in tree water relations, though with varying heat transfer rates (Michaletz and Johnson 2007; Hood et al. 2018). The water uptake system (i.e., root system) is affected by heat conduction from fuels burning above and within the soil. The aboveground water transport system (i.e., stem and axes system) is exposed to heat by primarily via radiation and convection processes, although conduction can occur at the base of the stem through smoldering combustion. When fires move through forest stands, stem bases are often exposed to high temperatures, which may affect main sections of the xylem and thus respective water transport and storage capacities. Exposure of leaves to heat (by radiation and convection; Michaletz and Johnson 2007; Varner et al. 2021) can empty leaf water reservoirs and dramatically increase transpiration. Longer-term, fire-caused loss of foliage may also reduce transpiration.

Plant water relations are based on optimized physical and chemical properties of the water phase and of cellular structures in all organs and tissues (Bär et al. 2019). Heat can affect both the stability of water bodies (non-structurally, by reduction of the surface tension and/or of xylem water potential), as well as the structural integrity (e.g., by altering physical xylem structures in conduits). Symplastic structures may be repaired, while apoplastic components, like the xylem conduits, have to be replaced and thus the damage of meristematic tissues during fires is also hydraulically relevant. In the following, mechanisms underlying fire and post-fire damage of plant hydraulic systems and potential counter strategies are discussed.

Structural alterations of the xylem can occur when the heat arriving at stem surfaces is conducted towards the stem or root center (Michaletz and Johnson 2007). The amount of heat able to penetrate the xylem depends on the energy dose released during the fire and on the bark's thermal insulation capability—which is determined mainly by bark thickness, as well as density and moisture (Pausas 2015a; Bär and Mayr 2020). Without sufficient bark insulation, phloem, vascular cambium and xylem can reach critical temperatures (Michaletz et al. 2012). Xylem conduit walls are composed of viscoelastic polymers (lignin, cellulose, and hemicellulose) which start to soften above threshold temperatures ("glass transition point") transitioning from a solid to a viscous state. Xylem softening has been demonstrated to occur between 60 and 90 °C with variations among species, as exact transition temperatures are affected by molecular structures of the cell wall matrix and present moisture content (Hillis and Rozsa 1985; Olsson and Salmén 2004; Börcsök and Pásztory 2021). During this softening phase, alterations of the conduit cell walls (e.g., deformations and/or ruptures) can occur, which become permanent after the xylem cools. Such physical changes of conduit structures can potentially impair both the hydraulic efficiency (e.g., clogging of pits by ruptured wall particles) and the hydraulic vulnerability (e.g., modification of the pit architecture) of affected xylem.

Support for heat-induced structural alterations and reduced hydraulic efficiency is weak or mixed (Fig. 3), largely due to limitations in experiments that do not mimic the conditions that occur during fire (Nolan et al. 2024). Various heat experiments, as well as after a wildfire, have shown that reduced hydraulic efficiency can occur and that it may be due to xylem cell alterations (Michaletz et al. 2012; West et al. 2016; Bär et al. 2018). Alterations of secondary cell walls after exposure to 90 °C were also detected for *P. sylvestris* (Bär et al. 2018), however without any permanent effects on hydraulic efficiency. When small *P. ponderosa* were exposed to controlled surface fires (Partelli-Feltrin et al. 2020, 2022), the xylem did not reach temperatures provoking physical alterations of conduits and accordingly, no reduction in hydraulic conductivity was observed. Similarly, no hydraulic limitations were found in the main stem of mature *P. pinea* after a prescribed fire (Battipaglia et al. 2016). These results indicate that conifer tracheids possess a high mechanical resistance against heat-induced destabilization of the overall wall geometry. This can be attributed to their high ratio between wall thickness and lumen size, which was attributed to higher hydraulic safety (Hacke and Sperry 2001), but also may prevent major wall ruptures even under comparatively high heat transfer rates to the xylem. Thus, limitations of the hydraulic efficiency by heat-induced wall deformations and/or ruptures seem to be unlikely in pines, while other mechanisms may play a role. Hydraulic conductivity can also be reduced by traumatic resinosis as response to fire-induced wounding (Smith et al. 2016; Partelli-Feltrin et al. 2020). Post-fire resin soaking can extend from the heat-killed area radially towards the pith plugging pit connections and impairing the water flow through the xylem. Non-conductive xylem close to fire wounds in *P. ponderosa* has previously been associated with resin depositions using staining techniques (Partelli-Feltrin et al. 2020). However, the exact effect of resin blockages on hydraulic conductivity is unevaluated (Hoffmann et al. 2024; Nolan et al. 2024), as the amount of resin released after injuries can vary widely within the genus *Pinus* (Hood and Sala 2015).

Heat-initiated alterations of conduit structures can also affect the xylem's hydraulic vulnerability post-fire. As vulnerability to drought-induced embolism is closely linked to pit architecture (Tyree and Zimmermann 2002), any physical irregularities of pit structures can facilitate embolism formation and therefore permanently shift vulnerability thresholds (Plavcová et al. 2013). It is assumed that thermal softening of lignin and hemicelluloses can result in pit modifications and enable movements of cellulose microfibrils, which both favor torus sealing failures (Michaletz et al. 2012; Bär et al. 2019). Heat-induced changes in hydraulic vulnerability were demonstrated for *P. sylvestris* (Bär et al. 2018) and *P. ponderosa* (Partelli-Feltrin et al. 2020). Shifts of 1.3 MPa (from −3.6 to −2.3 MPa) and 0.6 MPa (from −3.4 to −2.8 MPa) in the water potential inducing 50% loss of conductivity were detected in branches of *P. sylvestris* after heat treatments and after a wildfire, respectively (Fig. 4). Experimental heat exposure using water baths may not always reflect heat regimes during fires and changes in vulnerabilities thus vary. While Bär et al. (2018) assigned changes of thresholds to structural alterations which occurred during xylem heating, Partelli-Feltrin et al. (2020) demonstrated on *P. ponderosa* saplings that the development of traumatic xylem elements post-fire can also increase vulnerability. They

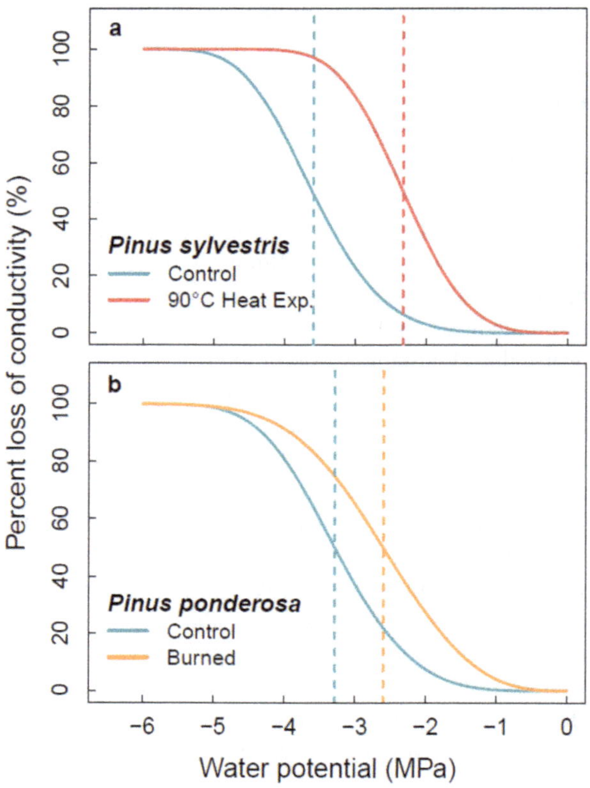

Fig. 4 Hydraulic vulnerability curves as a function of loss of hydraulic conductivity and water potential for (**a**) *P. sylvestris* control branches and exposed to a heated water bath and (**b**) burned and control *P. ponderosa* stems 21 months post-fire. (Adapted from Bär et al. (2018) and Partelli-Feltrin et al. (2020), respectively)

found newly formed tracheids in the vicinity of the fire scar to be disorganized and deformed, which facilitated drought-induced embolism formation.

Besides xylem structures, fire can also affect the water phase by enhancing the likelihood of embolism formation, and there is intermediate support for these hypotheses (Nolan et al. 2024). First, heat transferred to the xylem also raises the temperature of the sap which is associated with a reduction of its surface tension (Lodge et al. 2018). Sap surface tension impacts the vulnerability to embolism formation (Michaletz et al. 2012; West et al. 2016; Losso et al. 2017). Rapid threshold shifts can potentially lead to embolism formation, especially when tensions in the xylem are already enhanced due to pre-fire drought stress, or when atmospheric conditions during the fire add additional stress on the water columns (as described in the following).

It is assumed that the heat plume of hot air associated with fire can provoke extremely low water potentials in the xylem network (Kavanagh et al. 2010; Midgley et al. 2011; West et al. 2016; Hoffmann et al. 2021). Abrupt shifts of air temperature and humidity above flames lead to a high leaf-to-air vapor pressure deficit (VPD). This "heat plume hypothesis" assumes that stomatal response is slower than the heat plume passage, and therefore high VPDs can trigger extensive water loss and decrease xylem water potentials to critical levels and thus embolism formation. In vessel-bearing species, high water loss rates and extensive embolism formation have been documented after exposing branches to experimental fires and to simulated heat-plumes, respectively (West et al. 2016; Hoffmann et al. 2021). In contrast, no heat plume-related conductivity losses were detected in stems of *P. ponderosa* saplings burned under controlled conditions (Partelli-Feltrin et al. 2020, 2022). This suggests that the valve-like pit structures of conifers provide additional hydraulic protection during fires. The sudden rise of VPD creates a steep water potential gradient in the xylem which probably triggers pit closure (aspiration of the torus sealing the pit aperture; Tyree and Zimmermann 2002). Aspirated pits isolate tracheids and prevent air entry unless the water potential reaches critical levels. Then, this safety mechanism fails, whereas the subsequent propagation of embolism in conifer xylem depends on the time of exposure to pressure gradients (Mayr et al. 2006). Especially during fast moving fires with short residence times, it is likely that embolism therefore may not spread far into the xylem of proximal branch parts or the main stem. However, when residence times are longer, VPD-driven embolism can also affect hydraulic efficiency in stem xylem of conifers. This was demonstrated in experiments with *Sequoia sempervirens* where conductivity losses of 23% and higher were detected when branches were exposed to heat plumes of 70 °C for 6 min and longer (Salladay and Pittermann 2023). Further studies are required to better understand the underlying physical mechanisms and their interplay with the hydraulic architecture of trees. Conifers, with their needles, might substantially differ from broadleaved-angiosperms due to lower specific leaf area, higher cuticular protection, and xylem hydraulic segmentation.

Studies indicate that pines benefit from xylem structures resistant to fire and thus to hydraulic impairments (although the number of studies is limited). The high mechanical resistance and the sealing mechanism of tracheids may help prevent extensive conductivity losses, especially during fast moving surface fires. Many pine species also profit from bark insulation and develop relatively thick bark with advancing age, which provides additional thermal protection of the xylem. However, if the amount of heat released during the fire is high enough to induce wounding of stem internal tissues, hydraulic functions can be long-term impaired, e.g., by physical alterations or the formation of traumatic xylem and resinosis. In addition, conductive sapwood may be directly consumed in recurrent fires. Affected xylem needs to be replaced in order to restore full hydraulic capacity. Replacement times can vary from few years up to decades depending on species and its number of active sapwood rings. Long-term limitations of hydraulic efficiency and vulnerability can have important impacts on post-fire tree physiology (Bär et al. 2019). On one hand, reduced conductivity curtails the water supply to more distal tree parts, which can cause low water potentials in the xylem and leaves, as well as affect leaf turgor. In consequence, stomatal regulation may result in reductions in assimilation and productivity. On the other, fire-caused increases of hydraulic vulnerability imply that affected trees have to cope with reduced hydraulic "safety margins." Trees will be less tolerant to embolism formation during future drought events, which particularly affects species operating at water potentials close to hydraulic failure (Choat et al. 2012).

2.3 *Fire Effects on Defenses*

Wildland fire can affect resin-based defenses in pines, including resin flow, chemistry, and ducts (i.e., canals), but the effects attenuate with time and are dependent on burn severity and level of fire-caused injuries (Hood 2020; Fettig et al. 2022). See chapter 4 by Gandhi and Hood (2026) for a more detailed discussion of pine defenses. Immediately after fire and in trees with high levels of injury, fire-induced susceptibility may increase colonization by bark and woodboring beetles (Coleoptera: Buprestidae, Cerambycidae, Curculionidae), leading to mortality in trees that may have otherwise survived fire's initial injury (e.g., McHugh et al. 2003; Lerch et al. 2016; Westlind and Kelsey 2019; Valor et al. 2021). In trees with high levels of crown scorch and basal cambium kill from fires, the injuries can limit phloem transport, photosynthetic potential, and reduce resources necessary for compartmentalizing basal wounds and mounting an energetically demanding defense response (Hood et al. 2018; Bär et al. 2019). However, trees with high burn injury levels are also poor hosts for bark beetle brood development due to low substrate quality (Lombardero and Ayres 2011; Powell et al. 2012), suggesting a unimodal response of beetles to fire-caused injury levels, with attacks most common in moderately-injured trees.

The pulse of bark beetle-caused mortality after fire in numerous pines species (e.g., *P. radiata*, *P. pinaster*, *P. ponderosa*, *P. contorta*) is typically short-lived, persisting only 1 or 2 years and not leading to outbreaks outside of the burned area (Schwilk et al. 2006; Lombardero and Ayres 2011; Davis et al. 2012; Powell et al. 2012; Tabacaru et al. 2016). More research is needed for other pine species and to determine the duration of susceptibility and how the response is modulated by other factors (e.g., insect species, tree injury level, drought, host characteristics such as size, and bark beetle predation). For example, in contrast to the ephemeral short-term increase in susceptibility to bark beetles from burning that is often reported in pines, *P. lambertiana* mortality from *Dendroctonus ponderosae* was much higher after low-intensity prescribed burns compared to unburned areas (Maloney et al. 2008), with the increased susceptibility lasting over 15 years and exacerbated by extreme drought (Steel et al. 2021). *Ips sexdentatus* preferentially attacked *P. sylvestris* over *P. nigra* after prescribed burning, but for both species, trees with faster growth, higher resin duct area, and larger resin ducts were more likely to survive attacks (Valor et al. 2021).

Likely mechanisms for the oft-observed switch in surviving pines from increased susceptibility to increased resistance are increases in resin flow and resin ducts. These increases in defenses can contribute to declines in attack success and begin within a month to a year after fire. The exact timing of this switch depends on the specific bark beetle and host tree species involved, growing season length, and post-fire climate. There is widespread evidence of increased resin flow after fire in several pine species, including *P. resinosa* (Santoro et al. 2001; Lombardero et al. 2006), *P. ponderosa* (Wallin et al. 2004; Six and Skov 2009; Davis et al. 2011; Perrakis et al. 2011), *P. pungens*, *P. rigida*, *P. strobus*, *P. virginiana* (Knebel and Wentworth 2007), and *P. nigra* (Cannac et al. 2009). Wildfire resulted in higher resin yields in *P. merkusii* after tapping for resin production compared to unburned trees; the response was highest in trees with stem charring compared to trees where the fire did not affect the stem, as well as for trees with crown scorch (Della Prasetya et al. 2017). In contrast, prescribed fire did not affect resin production of tapped *P. pinaster* relative to unburned tapped trees (Rodríguez-García et al. 2018). The increases in resin flow after fire declined over time, such that 4 years post-fire resin flows were no different than unburned pines (Cannac et al. 2009; Perrakis et al. 2011). Increased resin flow after fire may also help with compartmentalizing fire-cause basal wounds and limiting wood decay (Verrall 1938).

Low-severity fire can induce resin ducts, which, as the sites of resin synthesis, storage, and delivery, are strongly positively correlated with tree resistance to bark beetles (reviewed in Vázquez-González et al. 2020). Because resin ducts remain functional for years (Hood and Sala 2015), the 1- or 2-year increase in resin ducts after fire is consistent with the long-term increase in resin flow found in previous studies. Initial increases in resin flow could occur by *de novo* resin synthesis in existing ducts and then continue to increase if fire induces increases in resin ducts, as axial resin duct density is positively related to resin flow (Blanche et al. 1992; Hood and Sala 2015). Fire increased resin duct defenses in surviving *P. ponderosa* 1–2 years after wildfire, and those defenses declined when frequent fire ceased (Hood et al. 2015). Studies of duct responses after prescribed fire report conflicting results. In *P. ponderosa*, fall prescribed burns showed no increases in duct responses, even when accounting for fire intensity (Sparks et al. 2017), while spring prescribed burns were associated with small increases in total duct area, but the magnitude was very small compared to the effects of mechanical thinning (Hood et al. 2016) and not associated with crown scorch levels (Vilà-Vilardell et al. 2024). Prescribed burning reduced resin duct area and size in *P. palustris*, where smoldering time of duff was negatively related to resin duct size in the dry years post-fire (Slack et al. 2016). *P. lambertiana* resin ducts increased after fall prescribed burning (Bernal et al. 2023). In this study, defense response interacted with local competition in the burned treatments, with higher likelihood of mortality in trees with high competition and lower resin duct density and relative duct area. The competition × defense response was not observed in unburned treatments. Prescribed burning combined with tapping for resin production in *P. pinaster* only increased resin duct frequency and size near the tapped site, but the response did not lead to an increase in resin flow (Rodríguez-García et al. 2018).

Fire can also affect resin chemistry. *P. ponderosa* that had experienced a prescribed burn had different phloem resin chemical composition 10 years after burning, with lower levels of several specific monoterpenes known to benefit *D. ponderosae* attacks (Raffa 2014), including (−)- α- pinene, myrcene, and 3-carene (Hood et al. 2016). Wildfire reduced induced phloem resin concentrations in *P. contorta* and altered relative composition of resin monoterpenes (Powell and Raffa 2011). In contrast to Hood et al. (2016), burned trees had increased proportions of (−)- α- pinene compared to unburned trees 1 year post-fire, but both studies found increased limonene, a monoterpene highly toxic to bark beetles, in burned trees. Much more research is needed about the extent to which fire severity alters resin chemistry and how responses affect resistance to bark beetles.

While increased resin duct production after fire is likely a direct response to wounding by fire, changes in microclimate due to frequent, low-severity fire may also enhance resin duct production via changes in stand structure (Hood et al. 2016; Bernal et al. 2023). Low-severity fire can reduce tree density, which can reduce water stress in surviving trees, but also increase surface temperatures due to increased solar radiation. Resin duct area is positively correlated with both warmer and wetter conditions and with ring width (Rigling et al. 2003; Hood et al. 2015). Hence, increases in tree radial growth in response to higher water availability, combined with warmer temperatures may also further stimulate resin duct production beyond the effects of fire alone. Favorable environmental conditions from fire or thinning may increase duct production and resin flow in the long-term (Zausen et al. 2005), while low-severity fire can provide a pulsed increase in resin duct-related defenses. Top-down control of bark beetles by competitors and predators also increase after fire and can further decrease bark beetle attacks (Tabacaru and Erbilgin 2015). Therefore, frequent, low-severity fire can increase pine forests' resistance to bark beetles via these mechanisms—direct, physiological stimulation of defenses, indirect changes in stand structure, and bark beetle predation (Fig. 5). As resin duct traits are heritable (Westbrook et al. 2015; Fabián-Plesníková et al. 2021), low-severity fire could favor evolution of increased pine defenses via bark beetles using a window of susceptibility after fire to attack weakened trees until the fire-induced defenses increase to bolster resistance (Lombardero et al. 2006; Hood et al. 2015). This hypothesized relationship with fire and defense likely only applies to resister or tolerator pine species that can survive low-intensity fire.

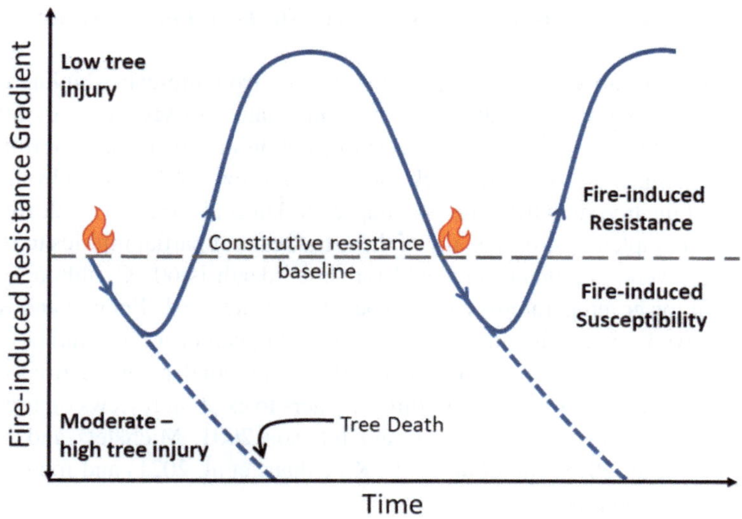

Fig. 5 Hypothesized conceptual model of fire-induced tree defenses and forest-level resistance from repeated fire in pines adapted to frequent, low-intensity fire regimes. Solid line—For trees with low levels of injury, fire causes a brief (days—1 growing season) reduction in resistance to bark beetle attacks before induced resin ducts form, followed by a period of increased resistance that may last for several years before returning to constitutive levels as induced resin ducts lose connectively due to annual tree growth. Repeated low-severity fires foster low-density forests that reinforce resistance to bark beetles. Dashed line—Fire increases susceptibility to attack and death for trees with moderate to high levels of injury. (Adapted from Hood et al. (2015))

2.4 Fire Effects on Tree Mortality and Growth

By killing trees, fire controls forest structure, tree density and species dominance, with cascading positive and negative impacts to biodiversity, carbon stores, hydrologic processes, and economic and social services (Bond and Keeley 2005). Tree sensitivity to fire varies by species, size, and phenology, as well as indirect interactions with insects, diseases, and climate (Hood et al. 2018). In fire-adapted ecosystems where trees have traits to survive or regenerate after fire (see Sect. 1), climate-driven alterations to fire regimes are an emerging threat, with increasing fire size, frequency, and severity potentially having long-term consequences via fire-induced tree mortality (Seidl et al. 2016). Climate-mediated increases in fire severity and frequency are projected to cause large decreases in carbon stocks through loss in forested area (Liang et al. 2017)—changes that can shift forests to nonforested states (Nolan et al. 2021). Understanding these direct, indirect, and long-term effects of fire on tree mortality and tree growth is essential to predicting fire effects from local to global scales.

2.4.1 Direct and Short-Term Indirect Fire Effects on Pine Mortality

Post-fire tree mortality can be caused either directly from fire-caused injury or from indirect effects, with tree death occurring immediately to several years after fire (Fig. 3). Direct tree death from fire is due to cambium necrosis via heat transfer to the crown, stem, and root tissue (Dickinson and Johnson 2001) or the physical consumption of the tree stem that causes collapse (Nolan et al. 2021). Tree death results from either complete necrosis to any of these tissues or if partial injuries to multiple tissues are severe. A general threshold for tissue death is 60 °C, although longer exposure at lower temperatures can also be lethal (Hare 1961; Pingree and Kobziar 2019). Delayed tree death in the years following fire can occur as a direct result of high levels of fire-caused tissue injuries or through secondary interactions of injuries with climate or bark beetles. We direct readers to existing reviews on fire effects on trees for more details (Dickinson and Johnson 2001; Michaletz and Johnson 2007; Hood et al. 2018; Bär et al. 2019; Kleynhans et al. 2021) and focus the following sections on pines.

Post-fire tree mortality has been extensively studied in many pine species due to their wide distribution range and high commercial and ecological value, with a large body of research related to predicting mortality from fire. The Fire and Tree Mortality (FTM) database of fire-caused tree injury and mortality data in the USA, contains records for 98,100 pines from 24 species, representing 58% of the total database (Cansler et al. 2020b). Post-fire tree mortality models are overwhelming based on empirical data using logistic regression (i.e., binary outcome of the tree surviving or dying from fire). Common important predictive variables include species, some measure of crown injury, and some measure of resistance such as tree diameter or bark thickness (Woolley et al. 2012; Cansler et al. 2020a). Mortality models have been developed for numerous pine species in the US (e.g., Woolley et al. 2012; Hood and Lutes 2017) and Europe (e.g., Fernandes et al. 2008; Catry et al. 2010), and to a lesser degree in Asia (Kwon et al. 2021) and other tropical areas (Rodríguez-Trejo et al. 2007; Pantoja-Campa et al. 2018; Hernández-Correa et al. 2019; Gómez-Mendoza and Rodríguez-Trejo 2021).

2.4.2 Long-Term Fire Effects on Tree Growth and Survival

Growth of trees that survive fire can remain unchanged, increase, or decrease depending on fire-caused injuries, changes to local competition, biophysical and environmental characteristics, and susceptibility to insect attack (Fig. 6). Growth declines due to fire-caused injuries to the crown can reduce photosynthesis while injury to the main stem cambium can limit translocation of photosynthates (Michaletz and Johnson 2007). Several studies have reported a decrease of tree radial growth after fire in different pine species due to injuries (Peterson et al. 1991; De Micco et al. 2013; Battipaglia et al. 2014a; Valor et al. 2015; Seifert et al. 2017; Swann et al. 2022; Wenderott et al. 2022) and with increasing fire intensity (Sparks et al. 2017; Johnston et al. 2019; Swann et al. 2022). Low-intensity fire may not

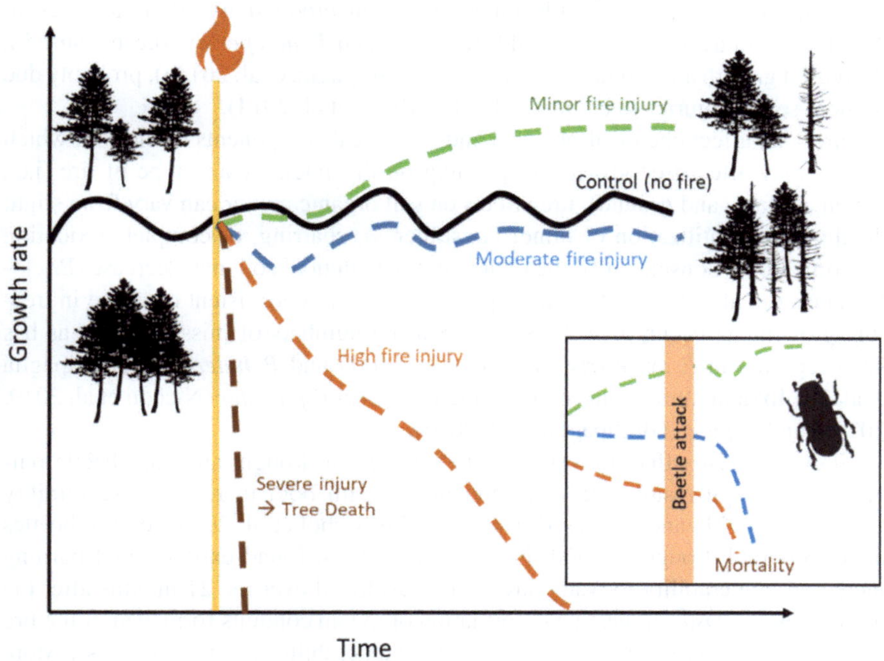

Fig. 6 Hypothetical tree growth responses to fire depending on tree injury, local competition, and bark beetle attack

affect growth if little-to-no tree injury or reduction in forest density occurs (Rozas et al. 2011; Espinosa et al. 2021; Zald et al. 2022). On the other hand, fire-caused reductions in competition can improve post-fire growth of surviving trees if tree injury is low. For example, crown scorch >30% reduced growth in *P. halepensis* relative to unburned trees, while growth increased in pines that had competitive release due to burning coupled with low crown scorch (Valor et al. 2018). Fire reduced local competition and led to an increase in *P. halepensis* growth, but only on the wetter sites, suggesting responses are dependent on environmental factors and water availability (Alfaro-Sánchez et al. 2016). In contrast, high-intensity fire increased growth after an initial reduction in *P. canariensis* on drought-prone sites (Rozas et al. 2011) and via increasing intrinsic water use efficiency (Renninger et al. 2013; Battipaglia et al. 2014b).

With frequent and low-intensity fire, positive long-term growth responses have been reported due to reduced competition with other individuals for soil resources and water was observed (Agee 1993; Stone et al. 1999; Keeling and Sala 2012). Beghin et al. (2011) demonstrated that *P. sylvestris* increased photosynthetic activity during the 5 years following a fire, likely due to trees having less competition for water and nutrients. Prescribed burning enhanced growth of mature *P. halepensis* probably through enhanced short-term water and nutrient availability (Valor et al. 2015), but young pines experienced more intense competition and reduced stomatal conductance, thus showing little improvement in productivity (Battipaglia et al. 2014b). In contrast, successive wildfires in French *P. halepensis* forests caused a decreased growth and stomatal conductance (Battipaglia et al. 2014a), probably due to more severe injuries to carbon sinks (De Micco et al. 2014).

Fires can affect chemical-physical and biological components of the soil, which in turn may affect tree growth. Depending on fire intensity and type of fire (i.e., crown, surface, and ground), fire effects on soil organic matter can vary from slight distillation (volatilization of minor constituents), charring, or complete oxidation (Certini 2005). Consequently, the total nutrient content of soil may decrease (Prieto-Fernández et al. 1993). When those processes occur, a consistent decrease in tree-ring growth can occur, as well as an increasing numbers of missing rings, as has been reported in *P. sylvestris* (Beghin et al. 2011) and *P. halepensis* (Battipaglia et al. 2014b) and intra-annual density fluctuations in *P. pinaster* (Niccoli et al. 2019, 2023b) and *P. pinea* (Battipaglia et al. 2016).

Fires may also affect tree survival to subsequent drought and bark beetle outbreaks, although results are scant and mixed with both increased susceptibility (Steel et al. 2021) and increased resistance (McNichol et al. 2019) to bark beetles under reported drought. Partelli-Feltrin et al. (2020) found experimental burning increased vulnerability to water stress that developed over the 21 months after the fire in *P. ponderosa*, suggesting deformation of xylem conduits formed after the fire can reduce hydraulic function and increase vulnerability to drought stress. More research is needed how fire-caused injury interacts with climate to influence long-term tree survival.

3 Conclusion

This chapter synthesizes knowledge of fire ecology and ecophysiology of *Pinus* species and pine-dominated forests, which is essential to the conservation of species and responsible forest management. In doing so, several knowledge gaps emerged. To provide a comprehensive evolutionary and ecological understanding of how fire affects pines at the individual, stand, and landscape-levels, especially under anthropogenic climate change, more research is needed on the following topics:

- Our understanding of fire-adapted traits is hampered by several factors that stymie synthesis and our ability to quantify advantages conferred to pines and other species. Better quantification of traits beyond binary presence or absence or simple rankings is needed. For some traits (e.g., bark thickness accumulation, litter flammability), robust datasets exist that allow for synthesis and analyses; other traits like crown morphology, self-pruning, and serotiny are relegated to binary possibilities that contrast with reality (and often with the literature), but also lack measurement.
- Literature bias in favor of American, Canadian, and European species largely overlooks the diversity of pines outside of these regions. The high diversity of

pines in Mexico and the aggregations of pines in east Asia lack measurement or reporting in English-language articles such that they may be omitted from syntheses of our global understanding of fire and pines.

- How does fire determine the extant relationships between pines and herbivorous insects, and how would they alter under varying forest management and climate change? Fire and native insects are often both major disturbances in pine forests, with the potential for one to subsequently affect the other. Additional research is needed on how forest management and fire can increase resistance to insects, especially in light of climate change that can alter physiology of both trees and insects.
- More research is needed on the ecophysiological effects of fire on pines and other species. This research is essential to develop process-based models of fire-caused tree mortality under novel climate conditions.

Fire has played a driving role in the speciation and evolution of the genus *Pinus* and continues to be a major influence in many *Pinus* forests around the world (He et al. 2012; Jin et al. 2021). Altered fire regimes due to past management and fire suppression have increased susceptibility of many pine forests to fire through increased fuel loading and forest density. These changes, coupled with hotter and drier climate, can lead to higher intensity fires that make even fire-resistant pine species vulnerable. However, many pine species are well-adapted to warm climates and can survive frequent, low-intensity fire. Applying knowledge of historical fire regimes, trait syndromes, and the effects of fire on tree ecophysiology can guide management to create fire-resilient pine forests.

References

Adams HD, Zeppel MJ, Anderegg WR, Hartmann H, Landhäusser SM, Tissue DT, Huxman TE, Hudson PJ, Franz TE, Allen CD (2017) A multi-species synthesis of physiological mechanisms in drought-induced tree mortality. Nat Ecol Evol 1:1285–1291

Agee JK (1993) Fire ecology of Pacific Northwest forests. Island Press, Washington, DC

Agee JK (1998) Fire and pine ecosystems. In: Richardson DM (ed) Ecology and biogeography of Pinus. Cambridge University Press, pp 193–218

Alfaro-Sánchez R, Camarero JJ, Sánchez-Salguero R, Sangüesa-Barreda G, De Las Heras J (2016) Post-fire Aleppo pine growth, C and N isotope composition depend on site dryness. Trees 30:581–595

Archibald S, Lehmann CER, Belcher CM, Bond WJ, Bradstock RA, Daniau AL, Dexter KG, Forrestel EJ, Greve M, He T, Higgins SI, Hoffmann WA, Lamont BB, McGlinn DJ, Moncrieff GR, Osborne CP, Pausas JG, Price O, Ripley BS, Rogers BM, Schwilk DW, Simon MF, Turetsky MR, Van der Werf GR, Zanne AE (2018) Biological and geophysical feedbacks with fire in the Earth system. Environ Res Lett 13:033003

Badik KJ, Jahner JP, Wilson JS (2018) A biogeographic perspective on the evolution of fire syndromes in pine trees (Pinus: Pinaceae). R Soc Open Sci 5:172412

Bär A, Mayr S (2020) Bark insulation: Ten Central Alpine tree species compared. For Ecol Manag 474:118361

Bär A, Nardini A, Mayr S (2018) Post-fire effects in xylem hydraulics of *Picea abies*, *Pinus sylvestris* and *Fagus sylvatica*. New Phytol 217:1484–1493

Bär A, Michaletz ST, Mayr S (2019) Fire effects on tree physiology. New Phytol 223:1728–1741

Bär A, Schröter DM, Mayr S (2021) When the heat is on: high temperature resistance of buds from European tree species. Plant Cell Environ 44:2593–2603

Barker Plotkin A, Blumstein M, Laflower D, Pasquarella VJ, Chandler JL, Elkinton JS, Thompson JR (2021) Defoliated trees die below a critical threshold of stored carbon. Funct Ecol 35:2156–2167

Battipaglia G, De Micco V, Fournier T, Aronne G, Carcaillet C (2014a) Isotopic and anatomical signals for interpreting fire-related responses in Pinus halepensis. Trees 28:1095–1104

Battipaglia G, Strumia S, Esposito A, Giuditta E, Sirignano C, Altieri S, Rutigliano FA (2014b) The effects of prescribed burning on *Pinus halepensis* Mill. as revealed by dendrochronological and isotopic analyses. For Ecol Manag 334:201–208

Battipaglia G, Savi T, Ascoli D, Castagneri D, Esposito A, Mayr S, Nardini A (2016) Effects of prescribed burning on ecophysiological, anatomical and stem hydraulic properties in *Pinus pinea* L. Tree Physiol 36:1019–1031

Beghin R, Cherubini P, Battipaglia G, Siegwolf R, Saurer M, Bovio G (2011) Tree-ring growth and stable isotopes (13C and 15N) detect effects of wildfires on tree physiological processes in *Pinus sylvestris* L. Trees 25:627–636

Bernal AA, Kane JM, Knapp EE, Zald HSJ (2023) Tree resistance to drought and bark beetle-associated mortality following thinning and prescribed fire treatments. For Ecol Manag 530:120758

Blanche CA, Lorio PL, Sommers RA, Hodges JD, Nebeker TE (1992) Seasonal cambial growth and development of loblolly pine: xylem formation, inner bark chemistry, resin ducts, and resin flow. For Ecol Manag 49:151–165

Blankenship K, Swaty R, Hall KR, Hagen S, Pohl K, Shlisky Hunt A, Patton J, Frid L, Smith J (2021) Vegetation dynamics models: a comprehensive set for natural resource assessment and planning in the United States. Ecosphere 12:e03484

Bond WJ, Keeley JE (2005) Fire as a global 'herbivore': the ecology and evolution of flammable ecosystems. Trends Ecol Evol 20:387–394

Bond WJ, Midgley JJ (1995) Kill thy neighbour: an individulalistic argument for the evolution of flammability. Oikos 73:79

Bond WJ, Scott AC (2010) Fire and the spread of flowering plants in the Cretaceous. New Phytol 188:1137–1150

Börcsök Z, Pásztory Z (2021) The role of lignin in wood working processes using elevated temperatures: an abbreviated literature survey. Eur J Wood Wood Prod 79:511–526

Bradley J, Will R, Stewart J, Nelson C, Guldin J (2016) Post-fire resprouting of shortleaf pine is facilitated by a morphological trait but fire eliminates shortleaf× loblolly pine hybrid seedlings. For Ecol Manag 379:146–152

Businský R (2016) New insight into the morphology of the long shoots of Pinus (Pinaceae). Flora 223:167–190

Cannac M, Barboni T, Ferrat L, Bighelli A, Castola V, Costa J, Trecul D, Morandini F, Pasqualini V (2009) Oleoresin flow and chemical composition of Corsican pine (*Pinus nigra* subsp. *laricio*) in response to prescribed burnings. For Ecol Manag 257:1247–1254

Cansler CA, Hood SM, van Mantgem PJ, Varner JM (2020a) A large database supports the use of simple models of post-fire tree mortality for thick-barked conifers, with less support for other species. Fire Ecol 16:25

Cansler CA, Hood SM, Varner JM, van Mantgem PJ, Agne MC, Andrus RA, Ayres MP, Ayres BD, Bakker JD, Battaglia MA, Bentz BJ, Breece CR, Brown JK, Cluck DR, Coleman TW, Corace RG, Covington WW, Cram DS, Cronan JB, Crouse JE, Das AJ, Davis RS, Dickinson DM, Fitzgerald SA, Fulé PZ, Ganio LM, Grayson LM, Halpern CB, Hanula JL, Harvey BJ, Hiers JK, Huffman DW, Keifer M, Keyser TL, Kobziar LN, Kolb TE, Kolden CA, Kopper KE, Kreitler JR, Kreye JK, Latimer AM, Lerch AP, Lombardero MJ, McDaniel VL, McHugh CW,

McMillin JD, Moghaddas JJ, O'Brien JJ, Perrakis DDB, Peterson DW, Prichard SJ, Progar RA, Raffa KF, Reinhardt ED, Restaino JC, Roccaforte JP, Rogers BM, Ryan KC, Safford HD, Santoro AE, Shearman TM, Shumate AM, Sieg CH, Smith SL, Smith RJ, Stephenson NL, Stuever M, Stevens JT, Stoddard MT, Thies WG, Vaillant NM, Weiss SA, Westlind DJ, Woolley TJ, Wright MC (2020b) The Fire and Tree Mortality Database, for empirical modeling of individual tree mortality after fire. Sci Data 7:194

Catry FX, Rego F, Moreira F, Fernandes PM, Pausas JG (2010) Post-fire tree mortality in mixed forests of central Portugal. For Ecol Manag 260:1184–1192

Certini G (2005) Effects of fire on properties of forest soils: a review. Oecologia 143:1–10

Choat B, Jansen S, Brodribb TJ, Cochard H, Delzon S, Bhaskar R, Bucci SJ, Feild TS, Gleason SM, Hacke UG, Jacobsen AL, Lens F, Maherali H, Martinez-Vilalta J, Mayr S, Mencuccini M, Mitchell PJ, Nardini A, Pittermann J, Pratt RB, Sperry JS, Westoby M, Wright IJ, Zanne AE (2012) Global convergence in the vulnerability of forests to drought. Nature 491:752–755

Climent J, Tapias R, Pardos JA, Gil L (2004) Fire adaptations in the Canary Islands pine (*Pinus canariensis*). Plant Ecol 171:185–196

Davis TS, Jarvis K, Parise K, Hofstetter RW (2011) Oleoresin exudation quantity increases and viscosity declines following a fire event in a ponderosa pine ecosystem. J Ariz Nev Acad Sci 43:6–11

Davis RS, Hood S, Bentz BJ (2012) Fire-injured ponderosa pine provide a pulsed resource for bark beetles. Can J For Res 42:2022–2036

De Micco V, Zalloni E, Balzano A, Battipaglia G (2013) Fire influence on Pinus halepensis: wood responses close and far from the scars. IAWA J 34:446–458

De Micco V, Battipaglia G, Cherubini P, Aronne G (2014) Comparing methods to analyse anatomical features of tree rings with and without intra-annual density fluctuations (IADFs). Dendrochronologia 32:1–6

Della Prasetya C, Syaufina L, Santosa G (2017) The effect of various types of forest fires on pine resin productivity in Gunung Walat University Forest, Sukabumi, Indonesia. Biodiversitas 18:476–482

Dickinson M, Johnson E (2001) Fire effects on trees. In: Johnson E, Miyanishi K (eds) Forest fires: behavior and ecological effects. Academic, New York, pp 477–525

Dickinson MB, Johnson EA (2004) Temperature-dependent rate models of vascular cambium cell mortality. Can J For Res 34:546–559

Dietze MC, Sala A, Carbone MS, Czimczik CI, Mantooth JA, Richardson AD, Vargas R (2014) Nonstructural carbon in woody plants. Annu Rev Plant Biol 65:667–687

Earle CJ (2023) The gymnosperm database: *Pinus devoniana*. https://www.conifers.org/pi/Pinus_devoniana.php

Erbilgin N, Zanganeh L, Klutsch JG, Chen S-h, Zhao S, Ishangulyyeva G, Burr SJ, Gaylord M, Hofstetter R, Keefover-Ring K, Raffa KF, Kolb T (2021) Combined drought and bark beetle attacks deplete non-structural carbohydrates and promote death of mature pine trees. Plant Cell Environ 44:3866–3881

Espinosa J, Martin-Benito D, Rodríguez de Rivera Ó, Hernando C, Guijarro M, Madrigal J (2021) Tree growth response to low-intensity prescribed burning in *Pinus nigra* stands: effects of burn season and fire severity. Appl Sci 11:7462

Fabián-Plesníková I, Sáenz-Romero C, Cruz-De-León J, Martínez-Trujillo M, Sánchez-Vargas NM, Terrazas T (2021) Heritability and characteristics of resin ducts in Pinus oocarpa stems in Michoacán, Mexico. IAWA J 1:1–21

Falk DA, Heyerdahl EK, Brown PM, Farris C, Fulé PZ, McKenzie D, Swetnam TW, Taylor AH, Van Horne ML (2011) Multi-scale controls of historical forest-fire regimes: new insights from fire-scar networks. Front Ecol Environ 9:446–454

Fernandes PM, Vega JA, Jiménez E, Rigolot E (2008) Fire resistance of European pines. For Ecol Manag 256:246–255

Fettig CJ, Runyon JB, Homicz CS, James P, Ulyshen MD (2022) Fire and insect interactions in North American forests. Current For Rep:1–16

Fire Effects Information System (FEIS) Available online at https://www.feis-crs.org/feis/
Fonda RW (2001) Burning characteristics of needles from eight pine species. For Sci 47:390–396
Gagnon PR, Passmore HA, Platt WJ, Myers JA, Paine CT, Harms KE (2010) Does pyrogenicity protect burning plants? Ecology 91:3481–3486
Gauthier S, Bergeron Y, Simon J-P (1996) Effects of fire regime on the serotiny level of jack pine. J Ecol 84:539–548
Gómez-Mendoza FF, Rodríguez-Trejo DA (2021) Fire, mortality and resprouting in forest species at Sierra Norte de Puebla. Madera Bosques 27:e2732148
Greenberg CH, Collins B (2021) Fire ecology and management: past, present, and future of US forested ecosystems. Springer International Publishing
Greene DF, Kane JM, Pounden E, Michaletz ST (2024) Cone allometry and seed protection from fire are similar in serotinous and nonserotinous conifers. New Phytol 242:93–106
Gričar J, Hafner P, Lavrič M, Ferlan M, Ogrinc N, Krajnc B, Eler K, Vodnik D (2020) Post-fire effects on development of leaves and secondary vascular tissues in Quercus pubescens. Tree Physiol 40:796–809
Guyette RP, Stambaugh MC, Dey DC, Muzika R-M (2012) Predicting fire frequency with chemistry and climate. Ecosystems 15:322–335
Hacke UG, Sperry JS (2001) Functional and ecological xylem anatomy. Perspect Plant Ecol Evol Syst 4:97–115
Hammond DH, Varner JM, Kush JS, Fan Z (2015) Contrasting sapling bark allocation of five southeastern USA hardwood tree species in a fire prone ecosystem. Ecosphere 6:1–13
Hansen J, Beck E (1994) Seasonal changes in the utilization and turnover of assimilation products in 8-year-old Scots pine (*Pinus sylvestris* L.) trees. Trees 8:172–182
Hare, R. C. 1961. Heat effects on living plants. Occasional paper 183. USDA, Forest Service, Southern Forest Experiment Station
Hare RC (1965) Contribution of bark to fire resistance. J For 63:248–251
Hartmann H, Trumbore S (2016) Understanding the roles of nonstructural carbohydrates in forest trees–from what we can measure to what we want to know. New Phytol 211:386–403
He T, Pausas JG, Belcher CM, Schwilk DW, Lamont BB (2012) Fire-adapted traits of Pinus arose in the fiery Cretaceous. New Phytol 194:751–759
Hengst GE, Dawson JO (1994) Bark properties and fire resistance of selected tree species from the central hardwood region of North America. Can J For Res 24:688–696
Hernández-Correa R, Rodríguez-Trejo DA, Cruz-Reyes A (2019) Near zero mortality in juvenile Pinus hartwegii Lindl. after a prescribed burn and comparison with mortality after a wildfire. iFor Biogeosci For 12:397
Hillis W, Rozsa A (1985) High temperature and chemical effects on wood stability. Wood Sci Technol 19:57–66
Hoffmann WA, Geiger EL, Gotsch SG, Rossatto DR, Silva LCR, Lau OL, Haridasan M, Franco AC (2012) Ecological thresholds at the savanna-forest boundary: how plant traits, resources and fire govern the distribution of tropical biomes. Ecol Lett 15:759–768
Hoffmann WA, Rodrigues AC, Uncles N, Rossi L (2021) Hydraulic segmentation does not protect stems from acute water loss during fire. Tree Physiol 41:1785
Hoffmann WA, Sherry CDK, Donnelly TM (2024) Stem heating results in hydraulic dysfunction in Symplocos tinctoria: implications for post-fire tree death. Tree Physiol 44:tpae023
Hood SM (2020) Fire and bark beetle interactions. In: Manzello SL (ed) Encyclopedia of wildfires and wildland-urban interface (WUI) fires. Springer International Publishing, pp 1–6
Hood S, Lutes D (2017) Predicting post-fire tree mortality for 12 western US conifers using the First-Order Fire Effects Model (FOFEM). Fire Ecol 13:66–84
Hood S, Sala A (2015) Ponderosa pine resin defenses and growth: metrics matter. Tree Physiol 35:1223–1235
Hood S, Sala A, Heyerdahl EK, Boutin M (2015) Low-severity fire increases tree defense against bark beetle attacks. Ecology 96:1846–1855

Hood S, Baker S, Sala A (2016) Fortifying the forest: thinning and burning increase resistance to a bark beetle outbreak and promote forest resilience. Ecol Appl 26:1984–2000

Hood SM, Varner JM, van Mantgem P, Cansler CA (2018) Fire and tree death: understanding and improving modeling of fire-induced tree mortality. Environ Res Lett 13:113004

Jackson JF, Adams DC, Jackson UB (1999) Allometry of constitutive defense: a model and a comparative test with tree bark and fire regime. Am Nat 153:614–632

Jacquet J-S, Bosc A, O'Grady A, Jactel H (2014) Combined effects of defoliation and water stress on pine growth and non-structural carbohydrates. Tree Physiol 34:367–376

Jin WT, Gernandt DS, Wehenkel C, Xia XM, Wei XX, Wang XQ (2021) Phylogenomic and ecological analyses reveal the spatiotemporal evolution of global pines. Proc Natl Acad Sci 118:e2022302118

Johnson E, Gutsell S (1993) Heat budget and fire behaviour associated with the opening of serotinous cones in two Pinus species. J Veg Sci 4:745–750

Johnston JD, Dunn CJ, Vernon MJ (2019) Tree traits influence response to fire severity in the western Oregon Cascades, USA. For Ecol Manag 433:690–698

Jones MW, Abatzoglou JT, Veraverbeke S, Andela N, Lasslop G, Forkel M, Smith AJP, Burton C, Betts RA, van der Werf GR, Sitch S, Canadell JG, Santín C, Kolden C, Doerr SH, Le Quéré C (2022) Global and regional trends and drivers of fire under climate change. Rev Geophys 60:e2020RG000726

Kavanagh KL, Dickinson MB, Bova AS (2010) A way forward for fire-caused tree mortality prediction: modeling a physiological consequence of fire. Fire Ecol 6:80–94

Keeley J (2012) Ecology and evolution of pine life histories. Ann For Sci 69:445–453

Keeley JE, Pausas JG (2022) Evolutionary ecology of fire. Annu Rev Ecol Evol Syst 53:203–225

Keeley JE, Zedler PH (1998) Evolution of life histories in *Pinus*. In: Richardson DM (ed) Ecology and biogeography of *Pinus*. Cambridge Press, Cambridge, UK, pp 219–251

Keeley JE, Pausas JG, Rundel PW, Bond WJ, Bradstock RA (2011) Fire as an evolutionary pressure shaping plant traits. Trends Plant Sci 16:406–411

Keeling EG, Sala A (2012) Changing growth response to wildfire in old-growth ponderosa pine trees in montane forests of north central Idaho. Glob Chang Biol 18:1117–1126

Kerr B, Schwilk DW, Bergman A, Feldman MW (1999) Rekindling an old flame: a haploid model for the evolution and impact of flammability in resprouting plants. Evol Ecol Res 1:807–833

Kleynhans EJ, Atchley AL, Michaletz ST (2021) Chapter Eleven – Modeling fire effects on plants: from organs to ecosystems. In: Johnson EA, Miyanishi K (eds) Plant disturbance ecology, 2nd edn. Academic, San Diego, pp 383–421

Knapp BO, Pile LS, Walker JL, Wang GG (2018) Fire effects on a fire-adapted species: response of grass stage longleaf pine seedlings to experimental burning. Fire Ecol 14:1–16

Knebel L, Wentworth TR (2007) Influence of fire and southern pine beetle on pine-dominated forests in the Linville Gorge Wilderness, North Carolina. Castanea 72:214–225

Kobziar LN, Hiers JK, Belcher CM, Bond WJ, Enquist CA, Loudermilk EL, Miesel JR, O'Brien JJ, Pausas JG, Hood S, Keane R, Morgan P, Pingree MRA, Riley K, Safford H, Seijo F, Varner JM, Wall T, Watts AC (2024) Principles of fire ecology. Fire Ecol 20:39

Kwon S, Kim S, Kim J, Kang W, Park K-H, Kim C-B, Girona MM (2021) Predicting post-fire tree mortality in a temperate pine forest, Korea. Sustainability 13:569

Lambers H, Chapin FS, Pons TL (2008) Plant physiological ecology. Springer

Lamont BB, Le Maitre D, Cowling R, Enright N (1991) Canopy seed storage in woody plants. Bot Rev 57:277–317

Ledig FT, Fryer JH (1972) A pocket of variability in *Pinus rigida*. Evolution 26:259–266

Lerch AP, Pfammatter JA, Bentz BJ, Raffa KF (2016) Mountain pine beetle dynamics and reproductive success in post-fire lodgepole and ponderosa pine forests in northeastern Utah. PLoS One 11:e0164738

Liang S, Hurteau MD, Westerling AL (2017) Potential decline in carbon carrying capacity under projected climate-wildfire interactions in the Sierra Nevada. Sci Rep 7:2420

Lilly CJ, Will RE, Tauer CG, Guldin JM, Spetich MA (2012) Factors affecting the sprouting of shortleaf pine rootstock following prescribed fire. For Ecol Manag 265:13–19

Little S, Somes HA (1956) Buds enable pitch and shortleaf pines to recover from injury. Station Paper NE-81. US Department of Agriculture, Forest Service, Northeastern Forest Experiment Station, Upper Darby, 14 p

Lodge AG, Dickinson MB, Kavanagh KL (2018) Xylem heating increases vulnerability to cavitation in longleaf pine. Environ Res Lett 13:055007

Lombardero MJ, Ayres MP (2011) Factors influencing bark beetle outbreaks after forest fires on the Iberian Peninsula. Environ Entomol 40:1007–1018

Lombardero MJ, Ayres MP, Ayres BD (2006) Effects of fire and mechanical wounding on *Pinus resinosa* resin defenses, beetle attacks, and pathogens. For Ecol Manag 225:349–358

Losso A, Beikircher B, Dämon B, Kikuta S, Schmid P, Mayr S (2017) Xylem sap surface tension may be crucial for hydraulic safety. Plant Physiol 175:1135–1143

Maloney PE, Smith TF, Jensen CE, Innes J, Rizzo DM, North MP (2008) Initial tree mortality and insect and pathogen response to fire and thinning restoration treatments in an old-growth mixed-conifer forest of the Sierra Nevada, California. Can J For Res 38:3011–3020

Martínez-Vilalta J, Sala A, Asensio D, Galiano L, Hoch G, Palacio S, Piper FI, Lloret F (2016) Dynamics of non-structural carbohydrates in terrestrial plants: a global synthesis. Ecol Monogr 86:495–516

Mayr S, Rothart B, Wolfschwenger M (2006) Temporal and spatial pattern of embolism induced by pressure collar techniques in twigs of *Picea abies*. J Exp Bot 57:3157–3163

McHugh C, Kolb TE, Wilson JL (2003) Bark beetle attacks on ponderosa pine following fire in northern Arizona. Environ Entomol 32:510–522

McMaster GS, Zedler PH (1981) Delayed seed dispersal in *Pinus torreyana* (Torrey pine). Oecologia 51:62–66

McNichol BH, Montes CR, Barnes BF, Nowak JT, Villari C, Gandhi KJK (2019) Interactions between southern Ips bark beetle outbreaks, prescribed fire, and loblolly pine (*Pinus taeda* L.) mortality. For Ecol Manag 446:164–174

Michaletz ST, Johnson EA (2007) How forest fires kill trees: a review of the fundamental biophysical processes. Scand J For Res 22:500–515

Michaletz ST, Johnson EA, Tyree MT (2012) Moving beyond the cambium necrosis hypothesis of post-fire tree mortality: cavitation and deformation of xylem in forest fires. New Phytol 194:254–263

Midgley JJ (2013) Flammability is not selected for, it emerges. Aust J Bot 61:102–106

Midgley JJ, Kruger LM, Skelton R (2011) How do fires kill plants? The hydraulic death hypothesis and Cape Proteaceae “fire-resisters”. S Afr J Bot 77:381–386

Millington W, Chaney W (1973) Shedding of shoots and branches. In: Shedding of plant parts. Academic, pp 149–204

Mirov NT (1967) The genus *Pinus*. The Ronald Press, New York

Moris JV, Reilly MJ, Yang Z, Cohen WB, Motta R, Ascoli D (2022) Using a trait-based approach to asses fire resistance in forest landscapes of the Inland Northwest, USA. Landsc Ecol 37:2149–2164

Moritz MA, Morais ME, Summerell LA, Carlson JM, Doyle J (2005) Wildfires, complexity, and highly optimized tolerance. Proc Natl Acad Sci USA 102:17912–17917

Moya D, Saracino A, Salvatore R, Lovreglio R, de Las Heras J, Leone V (2008) Anatomic basis and insulation of serotinous cones in *Pinus halepensis* Mill. Trees 22:511–519

Muir PS, Lotan JE (1985a) Disturbance history and serotiny of Pinus contorta in western Montana. Ecology 66:1658–1668

Muir PS, Lotan JE (1985b) Serotiny and life history of *Pinus contorta* var. *latifolia*. Can J Bot 63:938–945

Mutch RW (1970) Wildland fires and ecosystems–a hypothesis. Ecology 51:1046–1051

Niccoli F, Esposito A, Altieri S, Battipaglia G (2019) Fire severity influences ecophysiological responses of *Pinus pinaster* Ait. Front Plant Sci 10:539

Niccoli F, Pacheco-Solana A, De Micco V, Battipaglia G (2023a) Fire affects wood formation dynamics and ecophysiology of *Pinus pinaster* Aiton growing in a dry Mediterranean area. Dendrochronologia 77:126044

Niccoli F, Pacheco-Solana A, Delzon S, Kabala JP, Asgharinia S, Castaldi S, Valentini R, Battipaglia G (2023b) Effects of wildfire on growth, transpiration and hydraulic properties of *Pinus pinaster* Aiton forest. Dendrochronologia 79:126086

Nolan RH, Collins L, Leigh A, Ooi MKJ, Curran TJ, Fairman TA, Resco de Dios V, Bradstock R (2021) Limits to post-fire vegetation recovery under climate change. Plant Cell Environ 44:3471–3489

Nolan RH, Reed CC, Hood SM (2024) Mechanisms of fire-caused tree death are far from resolved. Tree Physiol 44:tpae073

Olsson, A.-M., and L. Salmén. 2004. The softening behavior of hemicelluloses related to moisture. ACS Publications

Ormeño E, Céspedes B, Sánchez IA, Velasco-García A, Moreno JM, Fernandez C, Baldy V (2009) The relationship between terpenes and flammability of leaf litter. For Ecol Manag 257:471–482

Pallardy SG (2010) Physiology of woody plants. Academic

Pantoja-Campa V, Rodríguez-Trejo DA, Myers RL, Hernández-Acosta E, González-Santiago MV (2018) Modeling the probability of mortality of *Pinus oocarpa* in sites of prescribed burning or affected by forest fires in Chiapas. Acta Univ 28:1–9

Partelli-Feltrin R, Smith AMS, Adams HD, Kolden CA, Johnson DM (2020) Short- and long-term effects of fire on stem hydraulics in *Pinus ponderosa* saplings. Plant Cell Environ 44:696

Partelli-Feltrin R, Smith AMS, Adams HD, Thompson RA, Kolden CA, Yedinak KM, Johnson DM (2022) Death from hunger or thirst? Phloem death, rather than xylem hydraulic failure, as a driver of fire-induced conifer mortality. New Phytol 237:1154–1163

Paula S, Arianoutsou M, Kazanis D, Tavsanoglu Ç, Lloret F, Buhk C, Ojeda F, Luna B, Moreno JM, Rodrigo A, Espelta JM, Palacio S, Fernández-Santos B, Fernandes PM, Pausas JG (2009) Fire-related traits for plant species of the Mediterranean Basin. Ecology 90:1420–1420

Pausas JG (2015a) Bark thickness and fire regime. Funct Ecol 29:315–327

Pausas JG (2015b) Evolutionary fire ecology: lessons learned from pines. Trends Plant Sci 20:318–324

Pausas JG, Keeley JE (2014) Evolutionary ecology of resprouting and seeding in fire-prone ecosystems. New Phytol 204:55–65

Pausas JG, Su W, Luo C, Shen Z (2021) A shrubby resprouting pine with serotinous cones endemic to Southwest China. Ecology 102:e03282

Pellegrini AFA, Anderegg WRL, Paine CET, Hoffmann WA, Kartzinel T, Rabin SS, Sheil D, Franco AC, Pacala SW (2017) Convergence of bark investment according to fire and climate structures ecosystem vulnerability to future change. Ecol Lett 20:307–316

Perrakis DDB, Agee JK, Eglitis A (2011) Effects of prescribed burning on mortality and resin defenses in old growth ponderosa pine (Crater Lake, Oregon): four years of post-fire monitoring. Nat Areas J 31:14–25

Perry JP (1991) The pines of Mexico and Central America. Timber Press, Inc

Peterson DL, Arbaugh MJ, Pollock GH, Robinson LJ (1991) Postfire growth of *Pseudotsuga menziesii* and *Pinus contorta* in the Northern Rocky Mountains, USA. Int J Wildland Fire 1:63–71

Pile LS, Wang GG, Knapp BO, Liu G, Yu D (2017) Comparing morphology and physiology of southeastern US Pinus seedlings: implications for adaptation to surface fire regimes. Ann For Sci 74:68

Pingree MR, Kobziar LN (2019) The myth of the biological threshold: a review of biological responses to soil heating associated with wildland fire. For Ecol Manag 432:1022–1029

Piper FI, Paula S (2020) The role of nonstructural carbohydrates storage in forest resilience under climate change. Curr For Rep 6:1–13

Plavcová L, Jansen S, Klepsch M, Hacke UG (2013) Nobody's perfect: can irregularities in pit structure influence vulnerability to cavitation? Front Plant Sci 4:453

Powell E, Raffa K (2011) Fire injury reduces inducible defenses of lodgepole pine against mountain pine beetle. J Chem Ecol 37:1184–1192

Powell EN, Townsend PA, Raffa KF (2012) Wildfire provides refuge from local extinction but is an unlikely driver of outbreaks by mountain pine beetle. Ecol Monogr 82:69–84

Prieto-Fernández A, Villar M, Carballas M, Carballas T (1993) Short-term effects of a wildfire on the nitrogen status and its mineralization kinetics in an Atlantic forest soil. Soil Biol Biochem 25:1657–1664

Radeloff VC, Mladenoff DJ, Guries RP, Boyce MS (2004) Spatial patterns of cone serotiny in *Pinus banksiana* in relation to fire disturbance. For Ecol Manag 189:133–141

Raffa K (2014) Terpenes tell different tales at different scales: glimpses into the chemical ecology of conifer – bark beetle – microbial interactions. J Chem Ecol 40:1–20

Reed CC, Hood SM (2024) Nonstructural carbohydrates explain post-fire tree mortality and recovery patterns. Tree Physiol 42:tpad155

Renninger H, Clark K, Skowronski N, Schäfer KR (2013) Effects of a prescribed fire on water use and photosynthetic capacity of pitch pines. Trees 27:1115–1127

Rigling A, Brühlhart H, Bräker OU, Forster T, Schweingruber FH (2003) Effects of irrigation on diameter growth and vertical resin duct production in *Pinus sylvestris* L. on dry sites in the central Alps, Switzerland. For Ecol Manag 175:285–296

Rodman KC, Veblen TT, Andrus RA, Enright NJ, Fontaine JB, Gonzalez AD, Redmond MD, Wion AP (2021) A trait-based approach to assessing resistance and resilience to wildfire in two iconic North American conifers. J Ecol 109:313–326

Rodríguez-García A, Madrigal J, González-Sancho D, Gil L, Guijarro M, Hernando C (2018) Can prescribed burning improve resin yield in a tapped *Pinus pinaster* stand? Ind Crop Prod 124:91–98

Rodríguez-Trejo DA, Fulé PZ (2003) Fire ecology of Mexican pines and a fire management proposal. Int J Wildland Fire 12:23–37

Rodríguez-Trejo DA, Castro-Solis UB, Zepeda-Bautista M, Carr RJ (2007) First year survival of Pinus hartwegii following prescribed burns at different intensities and different seasons in central Mexico. Int J Wildland Fire 16:54–62

Rosell JA (2016) Bark thickness across the angiosperms: more than just fire. New Phytol 211:90–102

Rowell E, Prichard S, Varner JM, Shearman TM (2022) Re-envisioning fire and vegetation feedbacks. In: Speers K, Goodrick S (eds) Wildland fire dynamics: fire effects and behavior from a fluid dynamics perspective. Cambridge University Press, Cambridge, UK

Rozas V, Pérez-de-Lis G, García-González I, Arévalo JR (2011) Contrasting effects of wildfire and climate on radial growth of *Pinus canariensis* on windward and leeward slopes on Tenerife, Canary Islands. Trees 25:895–905

Saladin B (2013) Diversify or specialize: disturbances influence trait evolution in *Pinus*. Master's thesis. University of Zurich, Zurich

Salladay RA, Pittermann J (2023) Using heat plumes to simulate post-fire effects on cambial viability and hydraulic performance in *Sequoia sempervirens* stems. Tree Physiol 43:769–780

Santoro AE, Lombardero MJ, Ayres MP, Ruel JJ (2001) Interactions between fire and bark beetles in an old growth pine forest. For Ecol Manag 144:245–254

Sayer MAS, Tyree MC, Kuehler EA, Jackson JK, Dillaway DN (2020) Physiological mechanisms of foliage recovery after spring or fall crown scorch in young longleaf pine (*Pinus palustris* Mill.). Forests 11:208

Schutz AEN, Bond WJ, Cramer MD (2009) Juggling carbon: allocation patterns of a dominant tree in a fire-prone savanna. Oecologia 160:235–246

Schwilk DW (2003) Flammability is a niche construction trait: canopy architecture affects fire intensity. Am Nat 162:725–733

Schwilk DW, Ackerly DD (2001) Flammability and serotiny as strategies: correlated evolution in pines. Oikos 94:326–336

Schwilk DW, Kerr B (2002) Genetic niche-hiking: an alternative explanation for the evolution of flammability. Oikos 99:431–442

Schwilk DW, Knapp EE, Ferrenberg SM, Keeley JE, Caprio AC (2006) Tree mortality from fire and bark beetles following early and late season prescribed fires in a Sierra Nevada mixed-conifer forest. For Ecol Manag 232:36–45

Seidl R, Spies TA, Peterson DL, Stephens SL, Hicke JA (2016) Searching for resilience: addressing the impacts of changing disturbance regimes on forest ecosystem services. J Appl Ecol 53:120–129

Seifert T, Meincken M, Odhiambo BO (2017) The effect of surface fire on tree ring growth of *Pinus radiata* trees. Ann For Sci 74:34

Shearman T, Wang G, Ma P, Guan S (2018) Patterns of bark growth for juvenile trees of six common hardwood species in the eastern United States and the implications to fire-tolerance. Trees 32:519–524

Six DL, Skov K (2009) Response of bark beetles and their natural enemies to fire and fire surrogate treatments in mixed-conifer forests in western Montana. For Ecol Manag 258:761–772

Slack AW, Zeibig-Kichas NE, Kane JM, Varner JM (2016) Contingent resistance in longleaf pine (*Pinus palustris*) growth and defense 10 years following smoldering fires. For Ecol Manag 364:130–138

Smith CC (1970) The coevolution of pine squirrels (Tamiasciurus) and conifers. Ecol Monogr 40:349–371

Smith KT, Arbellay E, Falk DA, Sutherland EK (2016) Macroanatomy and compartmentalization of recent fire scars in three North American conifers. Can J For Res 46:535–542

Smith AMS, Talhelm AF, Johnson DM, Sparks AM, Kolden CA, Yedinak KM, Apostol KG, Tinkham WT, Abatzoglou JT, Lutz JA, Davis AS, Pregitzer KS, Adams HD, Kremens RL (2017) Effects of fire radiative energy density dose on *Pinus contorta* and *Larix occidentalis* seedling physiology and mortality. Int J Wildland Fire 26:82–94

Snyder JR (1984) The role of fire: Mutch ado about nothing? Oikos 43:404–405

Sparks AM, Smith AM, Talhelm AF, Kolden CA, Yedinak KM, Johnson DM (2017) Impacts of fire radiative flux on mature *Pinus ponderosa* growth and vulnerability to secondary mortality agents. Int J Wildland Fire 26:95–106

Sparks AM, Blanco AS, Wilson DR, Schwilk DW, Johnson DM, Adams HD, Bowman DM, Hardman DD, Smith AM (2023) Fire intensity impacts on physiological performance and mortality in *Pinus monticola* and *Pseudotsuga menziesii* saplings: a dose–response analysis. Tree Physiol 43:tpad051

Sprugel D, Hinckley T, Schaap W (1991) The theory and practice of branch autonomy. Annu Rev Ecol Syst 22:309–334

Steel Z, Goodwin M, Meyer M, Fricker GA, Zald H, Hurteau M, North MP (2021) Do forest fuel reduction treatments confer resistance to beetle infestation and drought mortality? Ecosphere 12:e03344

Stephens SL, Libby WJ (2006) Anthropogenic fire and bark thickness in coastal and island pine populations from Alta and Baja California. J Biogeogr 33:648–652

Stevens JT, Kling MM, Schwilk DW, Varner JM, Kane JM (2020) Biogeography of fire regimes in western U.S. conifer forests: a trait-based approach. Glob Ecol Biogeogr 29:944

Stewart JF, Will RE, Robertson KM, Nelson CD (2015) Frequent fire protects shortleaf pine (*Pinus echinata*) from introgression by loblolly pine (*P. taeda*). Conserv Genet 16:491–495

Stone JE, Kolb TE, Covington WW (1999) Effects of restoration thinning on presettlement *Pinus ponderosa* in northern Arizona. Restor Ecol 7:172–182

Swann DEB, Bellingham PJ, Martin PH (2022) Resilience of a tropical montane pine forest to fire and severe droughts. J Ecol 111:90–109

Swetnam TW, Betancourt JL (1998) Mesoscale disturbance and ecological response to decadal climatic variability in the American Southwest. J Clim 11:3128–3147

Tabacaru CA, Erbilgin N (2015) Competitors and natural enemies may cumulatively mediate *Dendroctonus ponderosae* colonization of burned *Pinus* forests. For Ecol Manag 337:98–109

Tabacaru CA, Park J, Erbilgin N (2016) Prescribed fire does not promote outbreaks of a primary bark beetle at low-density populations. J Appl Ecol 53:222–232

Tapias R, Gil L, Fuentes-Utrilla P, Pardos JA (2001) Canopy seed banks in Mediterranean pines of south-eastern Spain: a comparison between *Pinus halepensis* Mill., *P. pinaster* Ait., *P. nigra* Arn. and *P. pinea* L. J Ecol 89:629–638

Tinker DB, Romme WH, Hargrove WW, Gardner RH, Turner MG (1994) Landscape-scale heterogeneity in lodgepole pine serotiny. Can J For Res 24:897–903

Turner MG (2010) Disturbance and landscape dynamics in a changing world. Ecology 91:2833–2849

Tyree MT, Zimmermann MH (2002) Hydraulic architecture of whole plants and plant performance. In: Zimmerman MH (ed) Xylem structure and the ascent of sap. Springer, pp 175–214

Valor T, González-Olabarria JR, Piqué M (2015) Assessing the impact of prescribed burning on the growth of European pines. For Ecol Manag 343:101–109

Valor T, Casals P, Altieri S, González-Olabarria JR, Piqué M, Battipaglia G (2018) Disentangling the effects of crown scorch and competition release on the physiological and growth response of *Pinus halepensis* Mill. using δ13C and δ18O isotopes. For Ecol Manag 424:276–287

Valor T, Hood SM, Piqué M, Larrañaga A, Casals P (2021) Resin ducts and bark thickness influence pine resistance to bark beetles after prescribed fire. For Ecol Manag 494:119322

Varner JM, Putz FE, O'Brien JJ, Hiers JK, Mitchell RJ, Gordon DR (2009) Post-fire tree stress and growth following smoldering duff fires. For Ecol Manag 258:2467–2474

Varner JM, Hood SM, Aubrey DP, Yedinak K, Hiers JK, Jolly WM, Shearman TM, McDaniel JK, O'Brien JJ, Rowell EM (2021) Tree crown injury from wildland fires: causes, measurement and ecological and physiological consequences. New Phytol 231:1676–1685

Varner JM, Shearman TM, Kane JM, Banwell EM, Jules ES, Stambaugh MC (2022) Understanding flammability and bark thickness in the genus *Pinus* using a phylogenetic approach. Sci Rep 12:7384

Vázquez-González C, Zas R, Erbilgin N, Ferrenberg S, Rozas V, Sampedro L (2020) Resin ducts as resistance traits in conifers: linking dendrochronology and resin-based defenses. Tree Physiol 40:1313–1326

Verrall AF (1938) The probable mechanism of the protective action of resin in fire wounds on red pine. J For 36:1231–1233

Vilà-Vilardell L, Tepley AJ, Sala A, Casals P, and Hood SM (2024) Long-term sensitivity of ponderosa pine axial resin ducts to harvesting and prescribed burning. Forest Ecology and Management 572:122301

Villar-Salvador P, Uscola M, Jacobs DF (2015) The role of stored carbohydrates and nitrogen in the growth and stress tolerance of planted forest trees. New For 46:813–839

Vines R (1968) Heat transfer through bark, and the resistance of trees to fire. Aust J Bot 16:499–514

Wade D, Johansen RW (1986) Effects of fire on southern pines: observations and recommendations. Gen. Tech. Rep. SE-41. U.S. Department of Agriculture, Forest Service, Southeastern Forest Experiment Station

Wallin KF, Kolb TE, Skov KR, Wagner MR (2003) Effects of crown scorch on ponderosa pine resistance to bark beetles in northern Arizona. Environ Entomol 32:652–661

Wallin KF, Kolb TE, Skov KR, Wagner MR (2004) Seven-year results of thinning and burning restoration treatments on old ponderosa pines at the Gus Pearson Natural Area. Restor Ecol 12:239–247

Wenderott Z, van Mantgem PJ, Wright MC, Farris CA, Sherriff RL (2022) Long-term effects of prescribed fire on large tree growth in mixed conifer forests at Lassen Volcanic National Park, California. For Ecol Manag 517:120260

Werner PA, Prior LD (2013) Demography and growth of subadult savanna trees: interactions of life history, size, fire season, and grassy understory. Ecol Monogr 83:67–93

West AG, Nel JA, Bond WJ, Midgley JJ (2016) Experimental evidence for heat plume-induced cavitation and xylem deformation as a mechanism of rapid post-fire tree mortality. New Phytol 211:828–838

Westbrook JW, Walker AR, Neves LG, Munoz P, Resende MF, Neale DB, Wegrzyn JL, Huber DA, Kirst M, Davis JM (2015) Discovering candidate genes that regulate resin canal number in *Pinus taeda* stems by integrating genetic analysis across environments, ages, and populations. New Phytol 205:627–641

Westlind DJ, Kelsey RG (2019) Predicting post-fire attack of red turpentine or western pine beetle on ponderosa pine and its impact on mortality probability in Pacific Northwest forests. For Ecol Manag 434:181–192

Williamson GB, Black EM (1981) High temperature of forest fires under pines as a selective advantage over oaks. Nature 293:643–644

Woolley T, Shaw DC, Ganio LM, Fitzgerald SA (2012) A review of logistic regression models used to predict post-fire tree mortality of western North American conifers. Int J Wildland Fire 21:1–35

Zald HSJ, Callahan CC, Hurteau MD, Goodwin MJ, North MP (2022) Tree growth responses to extreme drought after mechanical thinning and prescribed fire in a Sierra Nevada mixed-conifer forest, USA. For Ecol Manag 510:120107

Zausen GL, Kolb TE, Bailey JD, Wagner MR (2005) Long-term impacts of stand management on ponderosa pine physiology and bark beetle abundance in northern Arizona: a replicated landscape study. For Ecol Manag 218:291–305

Eco-Evolutionary Interactions Between Pines, Fire, and Herbivorous Insects

Kamal J. K. Gandhi and Sharon M. Hood

Abstract Pine (*Pinus* spp.) trees worldwide are subjected to selective evolutionary pressures from both abiotic (fire) and biotic (insects) damage resulting in their extant distribution and abundance, and life-history and morphological traits. These interactions merit serious consideration in long-term pine forest conservation and sustainability plans, and hence, we provide the following elements in this chapter: (1) a synthesis of eco-evolutionary components of these tri-lateral (fire-pine-insect) interactions; (2) a framework for understanding herbivory patterns and processes as based on adaptations of pine trees to different fire regimes and whether they are fire embracers, tolerators, or avoiders; and (3) use of fire as a critical tool to directly and indirectly manage insects while enhancing forest health. Overall, pine traits that confer fire resilience (e.g., thicker bark) and/or promotion (e.g., resin in needles and bark) tend to elicit variable responses by insects. Responses of insects are dependent on fire regimes and resulting host quality (nutrition and defenses) and quantity (e.g., density and dominance). Similarly, insect life-history traits (e.g., whether they are primary or secondary colonizing species, their background density levels, and adaptations to fire) further contribute to herbivory patterns. Prescribed fire can also either indirectly or directly affect herbivorous insect populations depending on burn timing, severity, and proximity. Anthropogenic climate change and fire suppression are leading to significant changes in fire regimes that will continue to alter these ancient trilateral interactions. Future management plans to enhance forest sustainability will benefit from inclusion of these eco-evolutionary interactions in pine-dominated ecosystems.

K. J. K. Gandhi (✉)
D.B. Warnell School of Forestry and Natural Resources, University of Georgia, Athens, GA, USA
e-mail: kjgandhi@uga.edu

S. M. Hood
USDA Forest Service, Rocky Mountain Research Station, Missoula, MT, USA
e-mail: sharon.hood@usda.gov

D. M. Johnson et al. (eds.), *Physiology and Ecology of Pines Worldwide*, Tree Physiology 9, https://doi.org/10.1007/978-3-032-23001-0_4

1 Introduction

Fire is ubiquitous, widespread, and plays a crucial evolutionary and ecological role in pine (*Pinus* spp.)-dominated ecosystems (Pausas 2015). In fact, fire is termed the "global herbivore," as it consumes vegetation in most forested and grassland biomes (Bond and Keeley 2005). Simultaneously, pine trees are also subjected to herbivory (both by invertebrates and vertebrates, but especially forest insects and their associated pathogens) that greatly varies spatially and temporally (Berryman et al. 1987; Kozlov and Zvereva 2017). Along with other natural disturbances (e.g., windstorms, ice storms, flooding, and diseases) (Irland 2000; Gandhi et al. 2007; Cobb and Metz 2017), these selective pressures have resulted in extant abiotic and biotic patterns and processes in pine forests. The interplay between these two directional selection pressures for millions of years has resulted in adaptations and traits of some pine species to thrive in fire-prone regions (Stevens et al. 2020; Jin et al. 2021), while simultaneously coevolving with herbivorous species (de la Mata et al. 2017). We argue that the successional trajectories of pine stands and landscapes are, therefore, best understood under the evolutionary framework created by both fire and herbivorous insect species. Incorporating such relationships into forest management practices may maintain and conserve pine forests for their long-term sustainability, especially under projected climate changes (Seidl et al. 2017).

Reviews of trilateral interactions (fire-pines-insects) have focused on the boreal forest, western U.S., Canada, and/or across North America (McCullough et al. 1998; Parker et al. 2006; Fettig et al. 2022), and have provided valuable frameworks in this field. Here, we synthesize the existing knowledge about direct interactions between fire and associated insects in pine-dominated landscapes around the globe as based on fire regimes. We include both naturally regenerated and planted pine forests in both the northern (the native range of pines) and southern (where pines are imported and managed in monocultures) hemispheres. We focus on Australia, Europe, and North America due to more abundant literature on pine-dominated forests. We recognize the need for more studies elsewhere, such as in Asia and Central America, where native pines represent major centers for speciation (Goldammer and Penafiel 1990). While the studies are more voluminous in some regions (e.g., western U.S.) than others, our goal is to draw general conclusions to provide a framework for future studies under eco-evolutionary patterns and processes and project into the future.

2 Historical and Contemporary Relationship of Pines with Fire

The genus *Pinus* arose ~150 million years ago in the mid-Mesozoic era (Keeley 2012; Singh et al. 2018). Due to competition from rapidly diversifying angiosperms, pines were pushed into sites that either had extreme soil and weather conditions (subgenus *Strobus*) or were fire-prone (subgenus *Pinus*) (Millar 1993; Keeley 2012; Falcon-Lang et al. 2016; Singh et al. 2018). Molecular phylogenetic data indicate that fire-adapted traits appeared at least 126 million years ago when pines showed thicker bark and hence, resistance to low-intensity, surface fires. Pines evolved cone serotiny ~89–96 million years ago, suggesting adaptations to more intense fires (He et al. 2012). One of the oldest known fossil for pines with these fire traits is from ~133 to 140 million years ago (Falcon-Lang et al. 2016). Both of these events were coincident with increased oxygen levels in the mid-Cretaceous in the Mesozoic era (He et al. 2012). Extant pines largely arose during the Miocene, during which speciation was driven in part by fire (Jin et al. 2021). Most species within the subgenus *Pinus* radiated into fire-prone areas, and are categorized as either fire-avoider, fire-tolerator, fire-embracer, or fire-refugee with each having specific morphological and life-history traits (Keeley 2012). For example, pine species such as *P. palustris* P. Mill. (longleaf pine) and *P. pinaster* Ait. (maritime pine) are fire-tolerators that tend to have higher height to crown, thicker bark, and longer needles to allow survival from frequent surface fires. Fire-embracers such as *P. halepensis* Mill. (Aleppo pine) and *P. contorta* Dougl. ex. Loud. (lodgepole pine) retain dead branches as ladder fuels and have short needles and thin bark making them easily killed by fire, but have serotinous cones to allow post-fire regeneration. Pine trees, and their ecosystems, landscapes, and associated fauna and flora are in general, considered adapted to fire regimes—the characterization of fire frequency, intensity, season, size, and type in a region—instead of single fires (Keeley et al. 2011; Keeley 2012). More recently, five global biomes or "pyromes" with similar fire characteristics and global units of fire are recognized of which at least three are prevalent in pine ecosystems (Archibald et al. 2013). (Hood et al. 2026) further provide more details on pine traits associated with fire syndromes and regimes.

Humans have a long evolutionary history with fire, which was considered essential to the stability of communities and faster brain development due to easier absorption of cooked food (Wrangham et al. 1999; Wrangham and Carmody 2010). Indigenous communities traditionally used fire for many purposes, e.g., cooking food, opening up spaces for game wildlife and travel, keeping the fuel loads low and reducing fire hazard, and regeneration of specific plant species (Pyne 1995; Bird et al. 2008; Christianson et al. 2022). Hence, these ecosystems were largely fire-driven because of the influence of Indigenous humans. With the continued loss of Indigenous communities on almost all continents, advent of modern technology and urbanization, and higher human population pressure, use of prescribed fire in forested areas has largely diminished globally (Mistry et al. 2016; Mitsopoulos et al. 2017). In addition, wildfire suppression is common in many pine-dominated regions (Colatskie 2011; Steel et al. 2015; Fernandes et al. 2016). This decline of frequent fire in many pine ecosystems has led to major ecological issues, such as loss of pine regeneration and genetic integrity (Tauer et al. 2012) and disassembly of native communities (Nowacki and Abrams 2008; Diaz-Toribio et al. 2020). With a greater awareness of the importance of fire in pine-dominated ecosystems during the last few decades, there's been a major shift to fire inclusion as a critical tool to maintain ecosystem services along with protecting human lives and structures (Christianson et al. 2022). This paradigm shift represents utilizing historical Indigenous ecological knowledge for restoration of pine-dominated habitats (Robinson et al. 2021) with positive consequences for extant ecological processes.

Climate change is resulting in alterations to wildfire behavior, incidence, and intensity globally. This is primarily due to increases in temperature and drier conditions that lead to greater availability of combustible fuels (Williams et al. 2019; Grünig et al. 2023), as evidenced by the recent extreme forest fires in southeastern Australia (Abram et al. 2021). Coupled with higher fuel loads due to fire suppression, this has resulted in higher fire severity, damage, and extent in many regions (Abatzoglou and Williams 2016; Littell et al. 2016). Similarly, the seasonality of wildfires has changed as the burning season has lengthened in some areas (Fill et al. 2019). Aptly termed" mega-fires," wildfires in many regions are more intense and are burning larger areas, resulting in large economic, social, and ecological costs (Kirchmeier-Young et al. 2019; Boer et al. 2020; Wunder et al. 2021). Global warming has further resulted in significant range expansion of several important pine herbivorous insects (Cullingham et al. 2011; Dodds et al. 2018; Cocos et al. 2023). As these species are coming in contact with both naïve pine hosts and novel fire regimes, their impact on ecosystems may become more pronounced. Climatic changes and ensuing shifts in fire regimes and dynamics of herbivorous insect from the historical norms have serious socio-economic implications (Wunder et al. 2021) to long-term tree health, regeneration, and productivity of pine forests (Stephens et al. 2014).

3 Historical and Contemporary Relationship of Forest Insects with Fire

Forest fires and native herbivorous insects have been interacting with each other for millions of years in pine-dominated ecosystems globally. For example, 90% of extant pine species and 70% of extant *Dendroctonus* species (Coleoptera: Curculionidae) arose in the Miocene, suggesting strong coevolutionary relationships between existing host pines and this insect genus that likely influenced speciation (Ramírez-Reyes et al. 2023). Most pine ecosystems are pyrogenic with various intervals of fire frequency (few to hundreds of years) and intensity (surface to crown fires), which—along with other factors—influences fire severity (Agee 2000; Keeley 2012). Further, amongst gymnosperms, pine trees have a rich diversity of herbivorous insects and their associated pathogens utilizing this valuable resource (Brändle and Brandl 2001). These trilateral interactions are primarily indirect as they are mediated through changes in suitable pine host quality and quantity after a fire. Direct effects of fire are on the survivorship of herbivorous species that have low dispersal abilities at certain life-stages (e.g., they are present either in the litter and duff layer or on bark on the lower bole during a fire) that could be used for management of pest species. Conversely, there may be feedback loops with herbivorous species either contributing to or causing tree mortality which may lead to increases in fuel loads and hence, higher intensity of subsequent fires. While fire is viewed as creating a variety of habitat niches for herbivorous species, there may be opposing directional effects of fire and herbivores allowing microevolution on pines. For example, the continental variation in levels of serotiny (a heritable trait) in *P. contorta* populations is driven both by the population levels of a major seed predator, *Tamiasciurus hudsonicus* (Erxleben) (red squirrel) (positive) (Smith 1970; Talluto and Benkman 2013) and infrequent crown fire regimes (negative) (He et al. 2012; Lamont et al. 2020). Similar eco-evolutionary trends may occur for insect herbivores, but are largely undetermined.

In pine forests where fire is a large influence, the two major fire regimes include either surface or crown fires with varying characteristics and biotic responses. Surface fires, such as occur in the southeastern U.S., are low intensity and low severity (i.e., understory burning) but with high frequency (~2–3 years) (Kirkman et al. 2017). In crown or stand-replacing fires, such as in the boreal pine forests (Rowe and Scotter 1973; Bergeron et al. 2001), fire intensity and fire severity is high, but frequency is low (~80–100 years). Specific attributes of pine tree species occur within different fire regimes and are outlined by Agee (2000) and (Hood et al. 2026). As expected, the dynamics of fire with herbivorous insects vary greatly based on fire regimes, plus anthropogenic alterations, particularly due to climate change and forest management practices including fire suppression.

3.1 *Mechanisms of Interactions Between Forest Fire and Insects*

Oleoresin is a complex mixture of terpenoids (monoterpenes and sesquiterpenes) and resin acids (diterpenes) (Langenheim 1990) that are synthesized and stored in resin ducts (i.e., canals) that occur in needles, cortical tissue, and secondary xylem of all pine species (Wu and Hu 1997). Some components of oleoresin are highly volatile and in addition to other litter attributes, contributes to system flammability (Popović et al. 2021) and also pine survival through compartmentalization of fire-caused wounds (Verrall 1938). Oleoresin is also a highly co-evolved pine tree trait against herbivorous insects and pathogens, and provides formidable physical and chemical defenses (e.g., Raffa 2014; Vázquez-González et al. 2020). There are many important feeding guilds of herbivorous insects such as foliage feeders, sap suckers, bark and woodborers, seed and cone feeders, and root-feeding insects (Baker 1972)—and they invariably come in contact with oleoresin while feeding on plant tissue. Strong selection pressures from both fire and herbivorous species likely resulted in pines having the most sophisticated resin-production duct system among tree genera, and simultaneously to diversify in fire-prone areas (Wu and Hu 1997; Vázquez-González et al. 2020). The oldest known pine fossil (~133–140 million years ago) was preserved as charcoal and had resin ducts, suggesting co-occurring adaptations to both abiotic and biotic disturbances early in their evolutionary history (Falcon-Lang et al. 2016). Consequently, several insect taxa that use pines as a habitat also arose and diversified since that time. For example, bark beetles (Coleoptera: Curculionidae: Scolytinae), of which some species are important pests of pine trees, arose at least 120 million years ago with conifers as the likely basal host (Sequeira et al. 2000; Kirkendall et al. 2015). This was around the same time pines were diversifying in fire-prone areas (He et al. 2012) and angiosperms in other areas (Sequeira et al. 2000). Many co-evolutionary relationships of herbivorous species with pine trees thus, likely developed under various fire regimes that provided a complex mosaic of spatial and temporal habitat conditions.

Fire-injured trees release volatile organic compounds (VOCs), such as many monoterpenes (e.g., alpha-pinene, beta-pinene, myrcene, and 3-carene) to varying degrees (e.g., Perrakis and Agee 2006; Perrakis et al. 2011; Powell and Raffa 2011; Davis et al. 2012). Fire-injured trees also release higher levels of ethanol, which can increase bark beetle attraction (Kelsey and Westlind 2017). As expected, these differences are present and more pronounced after wildfires (Santolamazza-Carbone et al. 2011) than in prescribed fires because of the generally higher intensity of wildfires (Cannac et al. 2009). Resin production and flow, and hence the presence of VOCs increases in recently burned trees and on the charred side of trees in wildfires (Lombardero et al. 2006; Prasetya et al. 2017) and prescribed fires (Feeney et al. 1998; Menges and Deyrup 2001; Perrakis and Agee 2006; Cannac et al. 2009; Perrakis et al. 2011). There's evidence that the resin duct area and production increases post-fire in some frequent-fire pine species (Hood et al. 2015), and there's a positive correlation between resin duct attributes and production of resin (Rodríguez-García et al. 2014; Hood and Sala 2015). Thermal warming of the tree tissue that reduces viscosity of resin along with an induced response to wounding from fire may also increase the flow and production of resin. Resin production tends to increase for a few years in burned forests, indicating elevated responses to fire (Perrakis and Agee 2006; Perrakis et al. 2011), although only immediate responses may occur (Ritger 2023). Based on the relative levels, release rates, types, and persistence of VOCs present in burned forests, insect species show differential responses post-fire.

3.2 *Patterns of Interactions Between Forest Fires and Insects*

Based on fire regime, and as mediated via changes in host quantity and quality, spatial and temporal patterns of insect infestations vary within and across stands. Climate-limited forests tend to support higher prevalence of coniferous species and basal area, with high fuel loads that that support higher fire intensity and severity (Wang 2002; Estes et al. 2017; Lindenmayer et al. 2021; Stephens et al. 2022). Insect attack patterns and density are largely dependent on the degree of fire injury (i.e., percent crown scorch and bark char) on the trees and the amount/depth of burned fuel. As fire injury increases, subcortical beetle on burned trees also increase

(Coleoptera: Buprestidae, Cerambycidae, Curculionidae) (e.g., Furniss 1965; Dixon et al. 1984; Rasmussen et al. 1996; McHugh et al. 2003; Sullivan et al. 2003; Wallin et al. 2003; Breece et al. 2008; Six and Skov 2009; Costello et al. 2011; Santolamazza-Carbone et al. 2011; Davis et al. 2012; Zhan et al. 2022) though they typically decrease on trees with severe fire injury (Powell et al. 2012). This unimodal response is due to reduced host tissue suitability for beetle reproduction in severely burned trees, rendering them as non-hosts beyond a certain threshold. Within high-severity burns, less severely burned snags can allow for persistence of woodboring species that have relatively long life-cycles (Nappi et al. 2010). Smaller pine trees tend to sustain greater damage than larger trees in burned areas (Santolamazza-Carbone et al. 2011; Fettig and McKelvey 2014). However, fires can also cause significant mortality of larger trees in fire-suppressed forests with high fuel loads (Kolb et al. 2007; Maloney et al. 2008; Steel et al. 2021).

In frequent fire ecosystems with fire-tolerant pine species, low-intensity fires may confer increased resistance towards insects. *Pinus ponderosa* subjected to low-intensity wildfires invested more in resin duct-based defenses and had higher survival rates against *Dendroctonus ponderosae* Hopkins (mountain pine beetle) attacks and vice versa when fire was excluded from these sites (Hood et al. 2015). Resin ducts remain functional for several years, and frequent, low-intensity can stimulate persistent induced resistance against bark beetles (Hood et al. 2015). In *P. palustris*, resin ducts were smaller in size up to 10 years after reintroduction of fire, with the greatest reduction in stem and root burns (Slack et al. 2016). However, there were no differences in resin duct size between unburned and burned trees, and resin duct responses were also influenced by climatic factors. In contrast, resin flow that may be positively linked to higher tree defense (Schroeder 1990; Strom et al. 2002), often increasing for weeks to years after fire in numerous pine species (Santoro et al. 2001; Wallin et al. 2003; Lombardero et al. 2006; Knebel and Wentworth 2007; Cannac et al. 2009; Six and Skov 2009; Perrakis et al. 2011; Davis et al. 2012; Ritger 2023). Greater activity of *D. ponderosae* occurred on *P. contorta* than on *P. ponderosa* that required greater burn damage to bole and crown for such attack-levels (Lerch et al. 2016). More detailed studies on fire-tolerant pine species are needed to better assess the presence and various attributes (size and numbers) of resin ducts under different fire regimes.

While there are few studies that directly assessed the influence of post-fire stand characteristics, such as changes in structure (e.g., density, basal area, height, and volume of woody debris) and composition (e.g., percent tree and understory plant species) on insect dynamics, fire plays a significant indirect role. *Dendroctonus frontalis* Zimmerman (southern pine beetle) primarily colonizes *P. taeda* L. (loblolly pine) and rarely *P. palustris* (Friedenberg et al. 2007), which are maintained by frequent, low-intensity prescribed burning (Mitchell et al. 2006). Stand attributes play an important role since prescribed burning creates low basal area and open savannah-like *P. palustris* stands (Gilliam and Platt 1999; Mitchell et al. 2006). Such open stand structure likely dissipates the beetle's aggregation pheromone plume and makes it much more difficult for beetles to locate the source of the attractant. This could make mass aggregation of *D. frontalis* on trees difficult and lessens the likelihood of ensuing herbivory impacts (Thistle et al. 2004, 2011; Martinson et al. 2007). This pattern has also been seen in *P. ponderosa* forests, where stands that had either been thinned or thinned and prescribed burned had much lower mortality during a *D. ponderosae* outbreak (Hood et al. 2016).

Interactions between forest fire and subcortical insects can be broadly categorized by whether insects are primary or secondary colonizers of pine trees (Fig. 1). Most subcortical beetle species are secondary colonizers and only a few are primary colonizers on pine trees. Primary beetles are first colonizers of trees while they are still presumably healthy and can overcome the trees' defenses. Secondary beetles (plus woodwasps, Hymenoptera: Siricidae) typically colonize trees that are stressed, weakened, damaged, or dying due to other abiotic and biotic causes, although in some limited instances (e.g., post-windstorms), secondary insects can become primary colonizers (Gandhi 2005). Overall, subcortical insects function as important phloem and wood degraders (through tunneling activities). This allows other insects, fungi, and bacteria to colonize trees, enhancing forest nutrient cycling and flow. Primary subcortical beetles, mostly in the genus *Dendroctonus*, are important economically in naturally regenerated and planted native pine forests, where they can cause high levels of tree mortality and lost revenue. In some years, economic impacts by subcortical insects cost billions of dollars of lost sawtimber and pulpwood trees, along with costs associated with replanting and reforestation efforts of planted pines (Nowak et al. 2008; Grégoire et al. 2015; Six and Bracewell 2015).

3.2.1 Primary Subcortical Colonizing Insects

Several primary colonizing insects, such as *D. ponderosae* and *D. brevicomis* LeConte (western pine beetle) (Geiszler et al. 1984; Amman and Ryan 1991; Rasmussen et al. 1996; Ryan and Amman 1996; McHugh et al. 2003; Maloney et al. 2008; Six and Skov 2009; Fettig et al. 2010; Davis et al. 2012; Powell et al. 2012; Kulakowski and Jarvis 2013; Steel et al. 2021), *Ips sexdentatus* (Börner) (Valor et al. 2021), and rarely *D. frontalis* (Boyle et al. 2004) are known to attack fire-injured pines, but these insects also commonly attack unburned trees (Fig. 1). These dynamics are likely influenced by fire regime shifts that alter host tree species dominance, and fuel loads and arrangement, and are dependent on insect and host tree species (Meigs et al. 2015; Fettig et al. 2022). Tabacaru et al. (2016) reported that *D. ponderosae* readily colonized fire-damaged *P. contorta* (a fire-embracer species) after a prescribed burn, but these attacks did not lead to outbreaks. This was also found by Powell et al. (2012) after wildfire, even though that fire reduced *P. contorta* resin defenses (Powell and Raffa 2011). In a different system, infestations by *D. frontalis* were higher in young *P. taeda* stands that were prescribed burned than unburned stands (Cameron and Billings 1988). While *P. taeda* is considered fire-tolerant, it has a much lower tolerance to fire (especially in younger age classes) than other prevalent southern pine species (Pile et al. 2017). *Dendroctonus frontalis* infestations were lower in previously prescribed burned but unthinned *P. taeda* stands (Nowak et al. 2015), suggesting that burning confers resilience to trees (even in an over-stocked stand) to subsequent bark beetle outbreaks.

These patterns are insect density-dependent, as at low density levels, *D. ponderosae* did not readily colonize trees that were fire-injured (Elkin and Reid 2004; Powell et al. 2012), and reproductive success was not affected by the level of fire-injury (Elkin and Reid 2004). At higher *D. ponderosae* densities though, trees were successfully attacked irrespective of their fire-caused injury (Elkin and Reid 2004). While burned trees provide habitats for beetle populations at endemic levels, wildfires (Powell et al. 2012) or prescribed fires (Tabacaru et al. 2016) do not typically contribute to beetle outbreaks, unless there are other suitable stand and environmental conditions. Intense competition from secondary beetle species that infest fire-injured trees earlier and have multiple generations per year may also have contributed to these patterns (Powell et al. 2012).

3.2.2 Secondary Subcortical Colonizing Insects

Secondary beetles, in contrast, readily colonize fire-damaged trees, and likely exert strong competitive pressure on primary beetle species, either reducing or excluding them from newly available niches. Examples of secondary beetles associated with burned forests include *Ips* spp. (pine engraver beetles), *Hylastes* spp. (root bark beetles), *Pityogenes* species (twig beetles), *D. valens* LeConte (red turpentine beetle), *D. terebrans* (Olivier) (black turpentine beetle), and many woodboring beetle species (Buprestidae and Cerambycidae: *Monochamus* spp.) (Fig. 1). These taxa are known to increase in numbers on fire-injured pine trees, and even preferably colonize fire-charred parts of the trees (e.g., Geiszler et al. 1984; Amman and Ryan 1991; Rasmussen et al. 1996; Ryan and Amman 1996; McHugh et al. 2003; Sullivan et al. 2003; Lombardero et al. 2006; Costello et al. 2011; Perrakis et al. 2011; Davis et al. 2012; Powell et al. 2012; Ray et al. 2019; Jung et al. 2020). In *P. palustris* and *P. elliotti* Engelm. (slash pine) stands (both fire tolerant), *Ips* and *D. terebrans* populations remained low and those of root-feeding beetles were higher in stands that were moderately and severely burned by a wildfire (Hanula et al. 2002), likely due to root injury. Responses of secondary insects are guild- and species-specific, likely vary across fire regimes, and dependent on the type of host injury, volume and type of available host material, and time since fire. Similar to primary species, there's little evidence that these secondary insects can build up in burned areas and start colonizing healthy or unburned green trees within and surrounding these forests (Hanula et al. 2002; Six and Skov 2009; Santolamazza-Carbone et al. 2011). However, the population spill-over effect of buildups of insects into unburned areas is still a major operational concern in many managed pine stands worldwide.

3.2.3 Variable Feedback Loops Between Insect Outbreaks and Fire

Insect-induced mortality has been hypothesized to contribute to greater probability, severity, and extent of wildfires. In the eastern and western U.S. forests where several bark beetle species outbreaks have resulted in dying and dead trees across millions of hectares (Hicke et al. 2016; McNichol et al. 2019; Audley et al. 2020), there are still numerous questions about the interactions of bark beetle outbreaks with fire. Outstanding research questions in general include (1) do insect outbreaks affect fire behavior via fuel alteration; (2) do insect outbreaks increase in number, severity, and duration with altered fire regimes; and (3) do fuel reduction treatments that are designed to reduce fire hazard also reduce insect outbreak incidence and severity? The empirical evidence for such interactions between bark beetle outbreaks and fires is mixed (Romme et al. 2006; Hicke et al. 2016). Positive interactions were observed by Turner et al. (1999) (fire severity and severe beetle damage) and Lynch et al. (2006) (probability of wildfire), and negative by Turner et al. (1999) (fire severity and intermediate beetle damage), Meigs et al. (2016) (severity), and Simard et al. (2011) (probability of crown fire). A lack of effect of beetle outbreaks on wildfires was observed by Hart et al. (2015) (area burned). Such interactions may be dependent on the conditions of the trees, such as whether standing trees have retained dead needles ("red stage") or fallen needles ("gray stage"), or post-outbreak stage with downed trees (Turner et al. 1999; Fettig et al. 2022). Various site and environmental conditions also likely play a role in these dynamics (Lynch et al. 2006; Black et al. 2013). However, in a rare study on foliage feeding insects about 20-times higher frequency of wildfire occurred in areas of *Dendrolimus sibiricus* (Tschetv.) (Siberian silkmoth) outbreaks that affected multiple conifer species in Siberia (Kharuk and Antamoshkina 2017).

At any given time, pine species are impacted either by stand-replacing wildfire or beetle species. These events may be thought of as operating in near exclusion of one other, although bark beetles are always present in low numbers in these stands. However, fire and/or beetle driven changes in forest composition (tree species and diversity) and structure (age and density) can have feedback loops. Post-burned forests may result in greater regeneration of tree species (e.g., hardwoods, shrubs, or other pine species) that are more tolerant of fire and/or beetles (Retana et al. 2002; Coleman et al. 2008; Perovich and Sibold 2016; Pelz and Smith 2012), particularly at the trailing edge of a tree species range (Renwick et al. 2016). Differential mortality of tree species from wildfires and beetles may negate impacts from the two disturbances. For example, recruitment of *P. pungens* lamb. (Table Mountain pine) was severely limited due to mortality by *D. frontalis* and ice-storms, but promoted by wildfire, as fire killed competitive hardwood species (Lafon and Kutac 2003).

The timing of either disturbance agent may be important for pine recruitment dynamics, both at the individual and population levels. Within a period of 6 years, serotinous *P. contorta* stands killed by *D. ponderosae* lost an estimated 45% more cones than stands unimpacted by beetles (Teste et al. 2011). After a *D. ponderosae* outbreak in *P. contorta* gray-phase forests, heterogeneous burn severities, coupled with areas of unburned fire refugia, fostered subsequent heterogeneity of *P. contorta* regeneration (Talucci et al. 2019). In the gray-phase, unburned patches, canopy seed stored in serotinous cones were reduced via squirrel predation, branch breakage, and cones opening, resulting in reduced regeneration relative to areas that burned at higher intensity. While *P. contorta* cones may germinate from both beetle-killed and live trees (Aoki et al. 2011), germination rates were reduced over time on beetle-killed trees (Rhoades et al. 2022), and short-return interval fires that occurred before serotiny develops can also reduce regeneration (Turner et al. 2019).

Insect outbreaks before prescribed fire can also alter regeneration dynamics of pine stands. In *P. echinata* Mill. (shortleaf pine) stands that were mostly killed by *D. frontalis* and were subsequently prescribed burned, natural regeneration was much higher and more vigorous than in unburned sites (Land and Rieske 2006). Regeneration-attacking insects on younger-aged pines showed a variable response where pine weevils [*Hylobius pales* (Herbst) (Pales weevil) and *Pissodes nemorensis* Germar (pitch-eating weevil)] and *Rhyacionia frustrana* (Comstock) (Nantucket pine tip moth) did not differ between burned and unburned sites (Land and Rieske 2006). However, *Tetralopha robustella* Zeller (pine webworm) infestations were greater on seedlings in burned areas. Conifer sawfly numbers were much lower in burned areas, likely because the fires consumed the duff layer where sawflies overwinter (Land and Rieske 2006). Prescribed fire after *Ips* beetle outbreaks may enhance the health of residual *P. taeda* trees making them more resilient to future beetle colonization activities (McNichol et al. 2019).

3.3 Interactions Between Prescribed Burning and Insects

3.3.1 Indirect Management of Insects Using Prescribed Fire

Prescribed burning is a frequently used tool in managed pine forests for multiple desired outcomes including removing hardwood competition, reducing tree density, improving wildlife habitats dependent on fire, reducing surface and canopy fuel loads, and enhancing tree health in general (Van Lear and Waldrop 1991; Fernandes and Botelho 2004; Ryan et al. 2013; Hunter and Robles 2020; Agbeshie et al. 2022). Low-intensity fire also enhances nutrient cycling and provides a pulse of nutrients that stimulate understory plants and other taxa (Carter and Foster 2004; Agbeshie et al. 2022), and higher intensity fires can cause sufficient injuries to pine trees. Trees stressed during prescribed burning may be attacked by secondary insect species that colonize injured and stressed trees (e.g., Bradley and Tueller 2001; Menges and Deyrup 2001; Sullivan et al. 2003; Campbell et al. 2008; Six and Skov 2009; Davis et al. 2012).

Based on whether the pine species are fire-tolerators [e.g., *P. nigra* Arn. (European black pine), *P. ponderosae*, and *P. palustris*] or fire-embracers (e.g., *P. contorta, P. halepensis*, and *P. pinaster*) (Agee 2000; Tapias et al. 2004), the dynamics of including low-intensity prescribed burning are different and elicit different responses by forest insects. Fire-tolerators have thicker bark and are less prone to fire-caused injury; and may have higher resin duct area and thus, be better defended against insects than fire-embracers (Valor et al. 2021). Insects encounter a more formidable defense system and have a harder time colonizing fire-tolerators than fire-embracers after a prescribed fire. For example, *I. sexdentatus* (Boern) readily colonized *P. sylvestris* L. (Scots pine), which is less tolerant of fire than *P. nigra,* after a prescribed burn (Valor et al. 2021). Most *P. nigra* trees that were attacked were of smaller diameter size that likely experienced greater stem injury (Valor et al. 2021).

Fire-excluded pine stands are more vulnerable to basal injury from prescribed burning due to long-term smoldering of basal duff accumulations (Hood 2010), as reported for *P. ponderosa* (e.g., Perrakis and Agee 2006; Fule et al. 2007; Six and Skov 2009; Perrakis et al. 2011), *P. palustris* (O'Brien et al. 2010), and mixed pine forests (Maloney et al. 2008), particularly under drier conditions (Varner et al. 2007; Ritger et al. 2023). After reintroduction of fire to *P. palustris*, trees with stem and root burns had smaller resin ducts which were correlated with lower precipitation levels, suggesting that climate change is an important consideration in restoration activities (Slack et al. 2016). Fire-suppressed (~15 years) *P. palustris* had higher trap catches of *Ips* beetles and higher resin flows on xeric than mesic sites (Ritger et al. 2023). However, exactly how climate change will affect these dynamics is unknown. In fire-suppressed forests, deep litter and duff can accumulate at the bases of trees, which if burned can lead to long-duration, low-intensity smoldering of the duff and basal cambial kill. In some pine species (e.g., *P. palustris*), when stands are not regularly burned, fine roots grow into developing forest floor layers and may be killed during burning, causing tree stress and mortality (Swezy and Agee 1991; O'Brien et al. 2010). In these cases, tree mortality may occur a few years post-fire, indicating a lag phase (Sullivan et al. 2003; Fettig and McKelvey 2014) where interactions with climate (e.g., drought) can further impact tree health and cause additional mortality (Steel et al. 2021). Hence, longer-term monitoring of stands where prescribed fire is reintroduced is recommended. Slowly reintroducing fire through a series of low-intensity prescribed fires to gradually reduce duff accumulations after long periods of fire exclusion or thinning of smaller diameter trees will reduce tree injury and mortality (Pollet and Omi 2002; Perrakis et al. 2011). This may also increase their resilience to drought (Vilà-Vilardell et al. 2023) and reduce activity of herbivorous insects (Ritger et al. 2023). If gradual consumption of deep duff layers is difficult to achieve due to duff moisture, other mitigation treatments such as raking of basal duff may be warranted (Hood 2010).

Cessation of frequent prescribed burning cycles may have implications for bark beetle outbreaks. During 2016–2017, many states in the southeastern U.S. experienced higher numbers of, and damage by, *Ips* beetles, presumably under water deficit conditions (Georgia Forestry Commission 2017). Deferred prescribed burning was recommended for *P. taeda* stands (under 2–3 year fire frequency cycle) that had *Ips* beetle outbreaks until the drought alleviated. However, these fire-deferred pine stands had greater infestation of, and higher mortality caused by *Ips* beetles than stands kept on their regular burn cycle (McNichol et al. 2019). Further, in some sites, these *Ips* beetle outbreaks were immediately followed by *D. frontalis* outbreaks in unburned pine stands (*personal observations*) as fire suppression altered relationships between pine trees and subcortical beetles.

Pine tree mortality and associated insect activity vary based on the season of prescribed burning. Fall burning of *P. ponderosa* stands may result in greater tree injury than spring burning due to generally wetter and cooler conditions in the spring (Perrakis and Agee 2006; Perrakis et al. 2011). In other studies, no differences in tree mortality or beetle infestations were observed after prescribed burning in different seasons (Ganz et al. 2003; Schwilk et al. 2006; Fettig et al. 2010), or the highest tree mortality occurred in spring burns and lowest in fall burns (McHugh et al. 2003). In *P. ponderosa* stands burned at 5- and 15-year intervals in spring and fall, the lowest mortality of pines and the highest activity of *Neophasia menapia* Felder (pine butterfly) and *D. brevicomis* were found in stands burned in the fall at 5-year intervals (Westlind and Kerns 2021). Phenological responses of trees and insects to fire may also be attributed to stand, fuel, weather conditions, and ignition patterns in prescribed burned forests, and may explain this variation in herbivore responses across studies.

Trapping studies report increased numbers of bark beetle species (mostly secondary colonizers) in recently prescribed burned areas (e.g., Santoro et al. 2001; Sullivan et al. 2003; Ritger et al. 2023; Palmer et al. 2024), suggesting immigration of insects from surrounding areas. However, after low-intensity prescribed fires in fire-tolerator ecosystems, little insect activity on trees was observed, as there were fewer trees that were sufficiently fire-injured (Cannac et al. 2009; Ritger et al. 2023). Burned areas may provide beetles with a short-term refuge from their natural enemies that are not responding positively to fire, and that have similar population sizes in burned and unburned areas (Six and Skov 2009; Powell et al. 2012). If trees are well or even better defended (Hood et al. 2015), then insects would have used their valuable energy reserves to find suitable hosts (Kinn et al. 1994; Evenden et al. 2014), particularly during the spring-time when both the burning and first insect dispersal activity occurs. Fire can influence constitutive and induced defenses, and post-fire, bark beetle attack success may be more influenced by induced defenses of pine trees (Howe et al. 2020). Stands that are well maintained by prescribed fire likely have stronger constitutive and induced defenses post-fire, and thus may be acting as an ecological trap for these insects. In landscapes where multiple stands are regularly burned on a rotating basis in spring and fall of each year (e.g., in many *P. palustris* stands in Florida and Georgia), subsequent generations of beetles may be dispersing from one burn to the next in attempts to find suitable hosts, yet prescribed burned sites may still yield a greater number of suitable hosts than unburned sites, where stressed and damaged hosts are a rare and ephemeral resource at the landscape-level. Trees that have lower crown vigor, growth rates, and smaller diameter, along with greater degree of bark charring, provide more suitable hosts for colonizing insects in burned (Menges and Deyrup 2001; Wallin et al. 2003; Perrakis et al. 2011; Valor et al. 2021) than in unburned areas.

Prescribed burning after catastrophic storms (wind and ice) has been used as a fuel reduction treatment to reduce woody debris on the landscape (Emery et al. 2020), prevent insect outbreaks and accelerate pine regeneration (Gilmore et al.

2002; Gandhi et al. 2009). Burning these forests also presumably counteracts the effects of fire suppression in fire embracer species, although that remains untested. Subcortical beetles and their associated predators can increase in numbers after windstorms and in post-storm-prescribed burned *P. banksiana* Lamb. (jack pine) stands (Gandhi et al. 2009). Their populations decrease after a few years as natural enemies build up and host resources degrade.

3.3.2 Direct Management of Insects Using Prescribed Fire

While the concept of "thermosanitation" is decades-old, the alternate use of chemicals and the logistical and legal challenges of conducting prescribed burning have taken precedence in modern silviculture (Miller 1979; Schmid and Parker 1990). It is well recognized that prescribed fire can be used as a tool to directly kill and reduce populations of several herbivorous insects in pine stands, although this management tactic is rarely applied in many pine ecosystems (McCullough et al. 1998; Fettig et al. 2022). Insects that have specific life-stages that overwinter on the forest floor and in fallen woody material are especially appropriate targets for prescribed burning, as they are in a vulnerable stage (McCullough et al. 1998). Rarely is fire used to control foliage-feeding insects (Parker et al. 2006), although those utilizing younger age classes of trees are more vulnerable. Prescribed fire has been recommended to reduce populations of seed predators such as *Conophthorus resinosae* Hopkins (red pine cone beetle), and *Eucosma monitorana* (Heinrich) (red pinecone borer moth) in *P. resinosa* Ait. (red pine) seed production areas (and when cones are on the forest floor) (Miller 1978), *Coloradia pandora* Blake (Pandora moth), in *P. ponderosa* stands (when moths are pupating in the soil) (Miller and Wagner 1984; Gerson and Kelsey 1997), and sawflies in *P. echinata* stands (on smaller trees) (Land and Rieske 2006). Burning of slash piles from logging and thinning is also conducted to reduce infestations by subcortical beetles (Dahl 1971; Schmid and Parker 1990; DeGomez et al. 2008) and likelihood of beetles colonizing trees near slash piles. Interestingly, both the load of non-native pinewood nematodes [*Bursaphelenchus xylophilus* (Steiner & Buhrer)] that causes pine wilt disease and its vector the Japanese pine sawyer (*Monochamus alternatus* Hope) were reduced greatly in *Pinus massoniana* Lambert (Masson pine) stands after prescribed burning (Chen and Liu 2020). This was attributed to greater mortality and/or disruption of life-cycle of cerambycid beetles and nematodes in the lower part of tree boles due to fire-injury (Chen and Liu 2020).

Prescribed fire may be used as a preventative management tool to reduce the probability of insect outbreaks (Miller 1979), but this has rarely been implemented. *Dendroctonus ponderosae* brood density was reduced by 50% in prescribed burned than unburned *P. contorta* forests (Safranyik et al. 2001). On the burned trees, the incidence of attacks, average attacks, and brood production per tree were lower than on unburned trees (Safranyik et al. 2001). Beetles preferentially attacked uncharred areas on burned trees. The following year, *D. ponderosae* tended to attack trees in unburned rather than burned sites, as previously reported by Elkin and Reid (2004).

Similar studies need to be conducted on other primary subcortical insect species under various fire regimes, especially some of them have expanded their distribution range into novel habitats due to climate change.

3.4 Adaptations of Insects to Fire

Like many other animal species, herbivorous insects have developed behavioral, morphological, or physiological adaptations to fire in pine ecosystems (Pausas and Parr 2018). These adaptations are advantageous to insects in that fire creates empty niches for a host of insects and other organisms to colonize—and the first colonizers experience the lowest competitive and predation pressures. Only a few forest herbivorous insects on pines are known to be attracted to post-burned pine stands, although other pyrophilous genera and species of insects are attracted to burned trees (e.g., the carabid beetle, *Sericoda* species) (Wikars 1997; Koivula et al. 2006) in pine-dominated and pine-hardwood ecosystems. A few species (e.g., *Calosoma frigidum* Kirby) have behavioral adaptations where they construct underground burrows, and after low-intensity fire, start foraging aboveground *en masse* (Jacobs et al. 2011). Pyrophilous species have been well studied in Canadian and European pine forests, and several of them are threatened with extinction due to fire-suppression and habitat changes in European forests (e.g., Gongalsky et al. 2003; Süda et al. 2009; Milberg et al. 2015; Heikkala et al. 2017). These insects have even been discovered in fossil records of burned wetland pine forests in Europe (Whitehouse 2000) indicating their long evolutionary history with fire. Smoke can also be both an attractant and deterrent to other species in pine forests that are not necessarily using pine trees directly as a resource (Liu et al. 2022). For example, *Exyra* spp. (pitcher plant moth) lives in pitcher plants (*Sarracenia* spp.) in frequently burned southern U.S. pine forests, and when exposed only to smoke, exits the plant (Lee et al. 2016), indicating strong behavioral adaptations to fire. Many other vertebrate and invertebrate species in pine forests also escape from fire either by running or burrowing deeper into the soil (Certini et al. 2021).

Several insect species cue into forest fires rapidly. Beetles in the genus *Melanophila* (Buprestidae) have sensors on their thorax to detect infrared radiation (Evans 1966) and their antennae can also detect smoke (Schütz et al. 1999). They were reported to find forest fires >95 km away from the nearest conifers in California and can even pick up anthropogenic fire sources (e.g., sugar refineries, smelting plants, and trash fires) (Hart 1998). Many buprestid beetles in the genera *Dicerca* and *Melanophila* have colors and reflective patterns resembling burnt wood which allow them to blend well when they land on burnt trees (*personal observations*). The cerambycid beetle, *Monochamus galloprovincialis* (Olivier), also has antennal sensillae that detect smoke-related compounds and is frequently found on burnt pine wood (Álvarez et al. 2015). Interestingly, *C. pandora* (Pandora moth) is attracted to night-burning prescribed fires, causing ~2%–17% mortality of its local populations; therefore, night burning may be a better management tool for that pest species

(Gerson and Kelsey 1997). It's quite possible that other pine herbivorous insects have various adaptations to fire, but they have been vastly under documented.

4 Fire and Insects in Non-native Planted Pines

Pine species [e.g., *P. caribaea* Morelet (Caribbean pine), *P. elliottii*, *P. patula* Schldl. et Cham. (Patula pine), *P. pinaster*, *P. radiata* D. Don (Monterey pine), and *P. taeda*] planted in the Southern Hemisphere are all non-native and imported from Europe, Mexico, and North America (Pryor 1991; Richardson 1998; Richardson and Nsikani 2021). There is a positive feedback loop where fire promotes invasion of these pine species into novel habitats, and pines also alter stand attributes to promote persistence of fire in these ecosystems (Zalba et al. 2008; Franzese and Raffaele 2017; Taylor et al. 2017; Davis et al. 2019). This has led to major changes in invaded habitats with adverse consequences for their native biodiversity and ecological processes (García et al. 2018; Cifuentes-Croquevielle et al. 2020).

Many insect species are introduced to these non-native planted pines and can cause damage irrespective of their colonizing status in their native range (Neumann 1987; Zhang et al. 1993; Yan et al. 2005; Carnegie et al. 2018; Estay 2020; Graziosi et al. 2020). This is due to planting off-site that stresses trees, and release from host defenses, natural enemies, and inter-species competitive pressures. Similar to their native ranges, bark and woodboring insects positively interact with wildfires due to highly available scorched and burnt host material in the non-native ranges (Bradbury 1998; Wylie et al. 1999; Zhan et al. 2022). *Dendroctonus valens* was introduced to China and also readily attacks pine trees after burning (Zhan et al. 2022). Females of *Arhopalus tristi* (F.) (burnt pine longhorn beetle), a cerambycid beetle introduced to planted pines in New Zealand, orient positively towards odors from burnt pine on which they lay eggs (Suckling et al. 2001), and colonize trees within days after a wildfire (Bradbury 1998). Wildfires can interfere with integrated pest management strategies in non-native pine stands. For example, mega-wildfires have resulted in catastrophic pine timber losses in southeastern Australia (Keeves and Douglas 1983; Bousfield et al. 2023). This led to increases in populations of *I. grandicollis* Eichhoff, a North American bark beetle, and interference with biocontrol of *Sirex noctilio* F., a Eurasian woodwasp, due to competitive pressure on trap logs (Gitau et al. 2013; Hayes and Nahrung 2022) (Fig. 1).

5 Conclusions

The long-evolved, complex, and co-occurring nature of fire and herbivorous insects as natural disturbance agents from individual pine trees to stand and landscape levels contributes to extant forest patterns and processes. Fire may positively or negatively affect pine tree and stand health, herbivorous insects, and their interactions as

Fig. 1 Examples of activity (as evidenced by resin pitch tubes) of herbivorous insects on pine trees after fires: (**a**) *Dendroctonus valens* on *Pinus contorta* after a low intensity prescribe fire; (**b**) *Dendroctonus valens* on *Pinus ponderosa* after a wildfire; (**c**) *Dendroctonus ponderosae* on *Pinus ponderosa* after a wildfire

based on its frequency and intensity, and hence severity (Figs. 2 and 3). Hence, an understanding of historical fire regimes is critical to restoring ecosystem services in pinelands. A mechanistic model of these trilateral interactions indicates that attributes of pine trees and stands that are adapted to fire provide a positive feedback loop to fire's persistence in the ecosystem (Fig. 3). Similarly, fires affect insect herbivores either indirectly through alterations in host quality/quantity or directly through mortality. There is mixed evidence for feedback loops from insects to pine tree-insect interactions to persistence of fires, and these patterns remain unknown for many outbreaking species (Fig. 3). Pine trees and herbivorous insects may positively or negatively affect each other. From an ecological perspective, insects act as a disturbance agent retriggering or augmenting forest succession and thus, are an integral positive component of forested ecosystems. From an economic perspective, herbivorous insects are capturing the mortality of pine trees instead of foresters, hence causing economic damage particularly in planted pine stands. Interestingly, fire and insects as abiotic and biotic disturbance agents, respectively, contribute to similar altered pine dynamics such as stand composition and structure, and tree nutrition, defenses, morphology, and various life history strategies including reproduction (Fig. 3). This lends greater support towards simultaneously including their evolutionary history in forest pine management plans with a focus on their differential impacts.

In managed pine stands, interventions against herbivorous insect pests may result in variable forestry responses depending on population levels and responses to fire (Fig. 4). In pre-fire conditions and low insect population levels, prevention is the

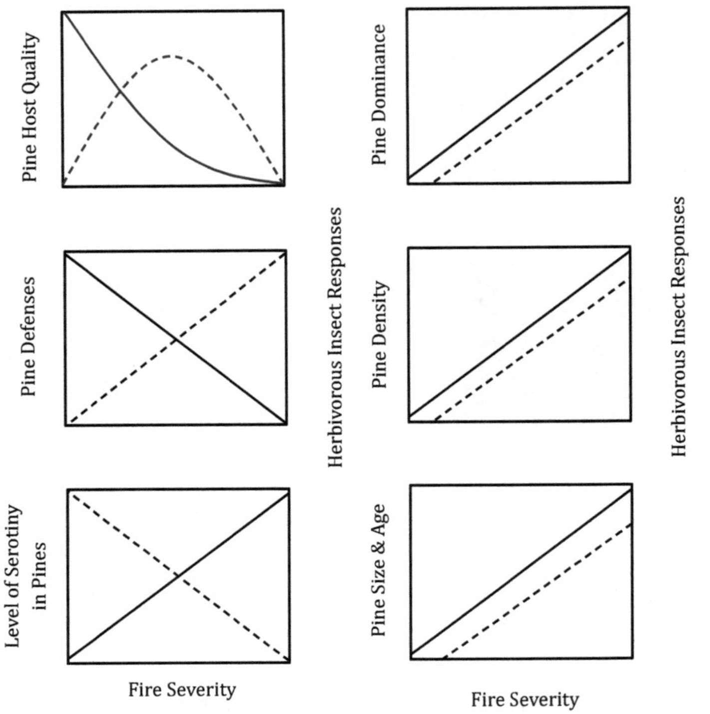

Fig. 2 Examples of potential effects of fire severity on pine trees and stand attributes (solid line) and resulting herbivorous pine insect responses (dashed line) levels

best option and may include reduction in fuel loads (e.g., inclusion of appropriate fire regimes and optimal silvicultural practices such as thinning) to avoid high-severity fires. At higher insect population levels (i.e., outbreak phase) pre-fire, management techniques such as removal and treatment of infested trees to reduce insect levels are warranted. Post-fire, increased insect activity (higher population levels) may trigger management activities such as pest reduction (e.g., removal of infested trees) and pest prevention (e.g., thinning) practices. Moderate responses by insects post-fire may be managed via prevention activities, while little to no responses from insects to fire won't trigger any management interventions. Interestingly, these strategies may be used in pine stands irrespective of their fire regimes, and can be refined based on insect and tree attributes.

We recognize several gaps and areas for future research on these trilateral interactions in pine-dominated ecosystems: (1) little is known about these interactions in Mesoamerica, which is the epicenter for pine diversity, and in Asia, where many species of pines occur at high elevation and latitude sites; (2) a better understanding of morphological, physiological, and other adaptations of herbivorous insects to fire regimes is warranted; (3) a greater focus on insect guilds other than bark beetles,

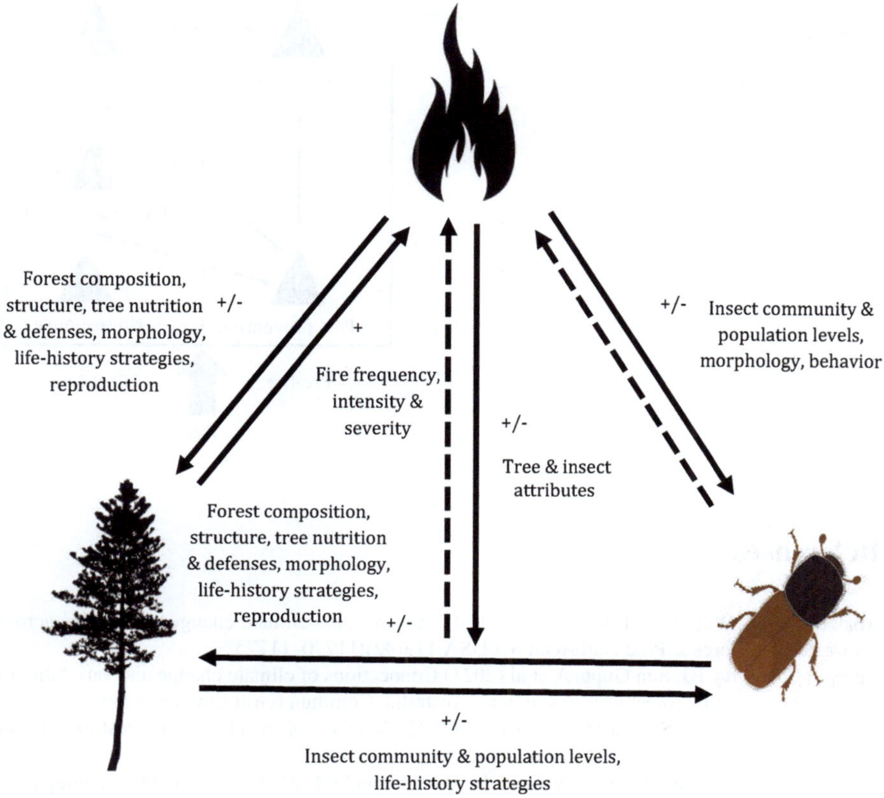

Fig. 3 Proposed interactions [positive (+) or negative (−)] between fire, pine trees, and herbivorous insects. Solid lines indicate established interactions and dashed lines indicate mixed or hypothesized evidence for interactions

insect-associated pathogens, and vertebrates will be useful; (4) many feedback loops from pine tree-insect interactions and those of insects to fire regimes are largely unknown (Fig. 3); and (5) a thorough underpinning of the relative selection pressures by the abiotic and biotic agents on pine tree and stand attributes are critical to restoring pinelands under continuing fire regime and climatic changes worldwide.

Acknowledgements The authors thank D.B. Warnell School of Forestry and Natural Resources, University of Georgia and the USDA Forest Service for their support. Brittany Barnes (D.B. Warnell School of Forestry and Natural Resources, University of Georgia) assisted with formatting of the paper. A special thanks to Daniel Johnson (D.B. Warnell School of Forestry and Natural Resources, University of Georgia) for providing this tremendous opportunity to work on this chapter. We also thank Kier Klepzig (Jones Center at Ichauway) and Ken Raffa (University of Wisconsin) for providing thoughtful comments on the chapter.

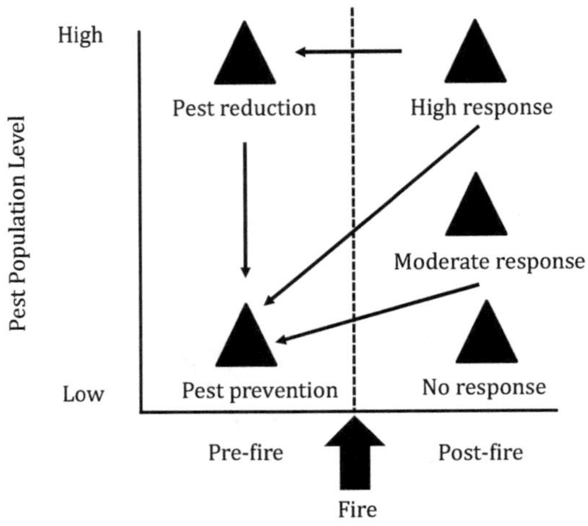

Fig. 4 Various management interventions for herbivorous insects (pests) as based on their population levels and responses after pre- and post-fire

References

Abatzoglou JT, Williams AP (2016) Impact of anthropogenic climate change on wildfire across western US forests. Proc Natl Acad Sci USA 113(42):11770–11775

Abram NJ, Henley BJ, Sen Gupta A et al (2021) Connections of climate change and variability to large and extreme forest fires in southeast Australia. Commun Earth Environ 2(1):8

Agbeshie AA, Abugre S, Atta-Darkwa T et al (2022) A review of the effects of forest fire on soil properties. J For Res 33:1419–1441

Agee JK (2000) Fire and pine ecosystems. In: Richardson DM (ed) Ecology and biogeography of *Pinus*. Cambridge University Press

Álvarez G, Ammagarahalli B, Hall DR et al (2015) Smoke, pheromone and kairomone olfactory receptor neurons in males and females of the pine sawyer *Monochamus galloprovincialis* (Olivier) (Coleoptera: Cerambycidae). J Insect Physiol 82:46–55

Amman GD, Ryan KC (1991) Insect infestation of fire-injured trees in the Greater Yellowstone Area. USDA Forest Service, Intermountain Research Station, Research Note INT-398, p 9

Aoki CF, Romme WH, Rocca ME (2011) Lodgepole pine seed germination following tree death from mountain pine beetle attack in Colorado, USA. Am Midl Nat 165(2):446–451

Archibald S, Lehmann CE, Gómez-Dans JL et al (2013) Defining pyromes and global syndromes of fire regimes. Proc Natl Acad Sci 110(16):6442–6447

Audley JP, Fettig CJ, Munson AS et al (2020) Impacts of mountain pine beetle outbreaks on lodgepole pine forests in the Intermountain West, US, 2004–2019. For Ecol Manag 475:118403

Baker WL (1972) Eastern forest insects, vol 1166. US Department of Agriculture, Forest Service

Bergeron Y, Gauthier S, Kafka V et al (2001) Natural fire frequency for the eastern Canadian boreal forest: consequences for sustainable forestry. Can J For Res 31(3):384–391

Berryman AA, Stenseth NC, Isaev AS (1987) Natural regulation of herbivorous forest insect populations. Oecologia 71:174–184

Bird RB, Bird DW, Codding BF et al (2008) The "fire stick farming". Hypothesis: Australian Aboriginal foraging strategies, biodiversity, and anthropogenic fire mosaics. Proc Natl Acad Sci 105(39):14796–14,801

Black SH, Kulakowski D, Noon BR et al (2013) Do bark beetle outbreaks increase wildfire risks in the central US Rocky Mountains? Implications from recent research. N Areas J 33(1):59–65

Boer MM, Resco de Dios V, Bradstock R (2020) Unprecedented burn area of Australian mega forest fires. Nat Clim Change 10:171–172

Bond WJ, Keeley JE (2005) Fire as a global 'herbivore': the ecology and evolution of flammable ecosystems. Trends Ecol Evol 20:387–394

Bousfield CG, Lindenmayer DB, Edwards DP (2023) Substantial and increasing global losses of timber-producing forest due to wildfires. Nat Geosci 16(12):1145–1150

Boyle MF, Hedden RL, Waldrop TA (2004) Impact of prescribed fire and thinning on host resistance to the southern pine beetle: preliminary results of the National Fire and Fire Surrogate Study. In: Proceedings, 12th biennial southern silvicultural research conference. General Technical Report SRS-71. US Department of Agriculture, Forest Service, Southern Research Station, Asheville, pp 60–64

Bradbury PM (1998) The effects of the burnt pine longhorn beetle and wood-staining fungi on fire damaged *Pinus radiata* in Canterbury. N Z For 43:28–31

Bradley T, Tueller P (2001) Effects of fire on bark beetle presence on Jeffrey pine in the Lake Tahoe Basin. For Ecol Manag 142(1–3):205–214

Brändle M, Brandl R (2001) Species richness of insects and mites on trees: expanding Southwood. J Anim Ecol 70(3):491–450

Breece CR, Kolb TE, Dickson BG et al (2008) Prescribed fire effects on bark beetle activity and tree mortality in southwestern ponderosa pine forests. For Ecol Manag 255(1):119–128

Cameron RS, Billings RF (1988) Southern pine beetle: factors associated with spot occurrence and spread in young plantations. South J Appl For 12(3):208–214

Campbell JW, Hanula JL, Waldrop TA (2008) Effects of prescribed fire and fire surrogates on saproxylic Coleoptera in the southern Appalachians of North Carolina. J Entomol Sci 43:57–75

Cannac M, Barboni T, Ferrat L et al (2009) Oleoresin flow and chemical composition of Corsican pine (*Pinus nigra* subsp. *laricio*) in response to prescribed burnings. For Ecol Manag 257(4):1247–1254

Carnegie AJ, Venn T, Lawson S et al (2018) An analysis of pest risk and potential economic impact of pine wilt disease to *Pinus* plantations in Australia. Aust For 81(1):24–36

Carter MC, Foster CD (2004) Prescribed burning and productivity in southern pine forests: a review. For Ecol Manag 191(1–3):93–109

Certini G, Moya D, Lucas-Borja ME et al (2021) The impact of fire on soil-dwelling biota: a review. For Ecol Mang 488:118989

Chen XW, Liu FL (2020) Effects of prescribed burning on pine wood nematode (*Bursaphelenchus xylophilus*). Appl Ecol Environ Res 18(5):1–14

Christianson AC, Sutherland CR, Moola F et al (2022) Centering Indigenous voices: the role of fire in the boreal forest of North America. Curr For Rep 8(3):257–276

Cifuentes-Croquevielle C, Stanton DE, Armesto JJ (2020) Soil invertebrate diversity loss and functional changes in temperate forest soils replaced by exotic pine plantations. Sci Rep 10(1):7762

Cobb RC, Metz MR (2017) Tree diseases as a cause and consequence of interacting forest disturbances. Forests 8(5):147

Cocos D, Klapwijk MJ, Schroeder M (2023) Tree species preference and impact on native species community by the bark beetle *Ips amitinus* in a recently invaded region. NeoBiota 84:349–367

Colatskie R (2011) Fuel management considerations in Caribbean pine forests of the Mountain Pine Ridge of Belize, Central America. Dissertation, University of Missouri-Columbia

Coleman TW, Clarke SR, Meeker JR et al (2008) Forest composition following overstory mortality from southern pine beetle and associated treatments. Can J For Res 38(6):1406–1418

Costello SL, Negrón JF, Jacobi WR (2011) Wood-boring insect abundance in fire-injured ponderosa pine. Agric For Entomol 13(4):373–381

Cullingham CI, Cooke JE, Dang S et al (2011) Mountain pine beetle host-range expansion threatens the boreal forest. Mol Ecol 20(10):2157–2171

Dahl BM (1971) Mortality of *Monochamus* larvae in slash fires. Environ Can For Ser Bimon Res Notes 27(2):12

Davis RS, Hood S, Bentz BJ et al (2012) Fire-injured ponderosa pine provide a pulsed resource for bark beetles. Can J For Res 42(12):2022–2036

Davis KT, Maxwell BD, Caplat P et al (2019) Simulation model suggests that fire promotes lodgepole pine (*Pinus contorta*) invasion in Patagonia. Biol Invasions 21:2287–2300

de la Mata R, Hood S, Sala A (2017) Insect outbreak shifts the direction of selection from fast to slow growth rates in the long-lived conifer *Pinus ponderosa*. Proc Natl Acad Sci 114(28):7391–7396

DeGomez T, Fettig CJ, McMillin JD et al (2008) Managing slash to minimize colonization of residual leave trees by *Ips* and other bark beetle species following thinning in southwestern ponderosa pine. University of Arizona, College of Agriculture and Life Sciences Bulletin, AZ1448, Tucson, pp 1–12

Diaz-Toribio MH, Carr S, Putz FE (2020) Pine savanna plant community disassembly after fire suppression. J Veg Sci 31(2):245–254

Dixon WN, Corneil JA, Wilkinson RC et al (1984) Using stem char to predict mortality and insect infestation of fire-damaged slash pines. South J Appl For 8(2):85–88

Dodds KJ, Aoki CF, Arango-Velez A et al (2018) Expansion of southern pine beetle into northeastern forests: management and impact of a primary bark beetle in a new region. J For 116(2):178–191

Elkin CM, Reid ML (2004) Attack and reproductive success of mountain pine beetles (Coleoptera: Scolytidae) in fire-damaged lodgepole pines. Environ Entomol 33(4):1070–1080

Emery RK, Kleinman JS, Goode JD (2020) Effects of catastrophic wind disturbance, salvage logging, and prescribed fire on fuel loading and composition in a *Pinus palustris* woodland. For Ecol Manag 478:118515

Estay SA (2020) Insect pests affecting exotic trees in Chile and their management. In: Estay S (ed) Forest Pest and Disease Management in Latin America. Springer, Cham, pp 185–195

Estes BL, Knapp EE, Skinner CN et al (2017) Factors influencing fire severity under moderate burning conditions in the Klamath Mountains, northern California, USA. Ecosphere 8(5):e01794

Evans WG (1966) Perception of infrared radiation from forest fires by *Melanophila acuminata* De Geer (Buprestidae, Coleoptera). Ecology 47(6):1061–1065

Evenden ML, Whitehouse CM, Sykes J (2014) Factors influencing flight capacity of the mountain pine beetle (Coleoptera: Curculionidae: Scolytinae). Environ Entomol 43(1):187–196

Falcon-Lang HJ, Mages V, Collinson M (2016) The oldest *Pinus* and its preservation by fire. Geology 44(4):303–306

Feeney SR, Kolb TE, Covington WW (1998) Influence of thinning and burning restoration treatments on presettlement ponderosa pines at the Gus Pearson Natural Area. Can J For Res 28(9):1295–1306

Fernandes P, Botelho H (2004) Analysis of the prescribed burning practice in the pine forest of northwestern Portugal. J Environ Manage 70(1):15–26

Fernandes PM, Pacheco AP, Almeida R et al (2016) The role of fire-suppression force in limiting the spread of extremely large forest fires in Portugal. Eur J For Res 135:253–262

Fettig CJ, McKelvey SR (2014) Resiliency of an interior ponderosa pine forest to bark beetle infestations following fuel-reduction and forest-restoration treatments. Forests 5(1):153–176

Fettig CJ, McKelvey SR, Cluck DR et al (2010) Effects of prescribed fire and season of burn on direct and indirect levels of tree mortality in ponderosa and Jeffrey pine forests in California, USA. For Ecol Manag 260(2):207–218

Fettig CJ, Runyon JB, Homicz CS et al (2022) Fire and insect interactions in North American forests. Curr For Rep 8(4):301–316

Fill JM, Davis CN, Crandall RM (2019) Climate change lengthens southeastern USA lightning-ignited fire seasons. Glob Chang Biol 25(10):3562–3356

Franzese J, Raffaele E (2017) Fire as a driver of pine invasions in the Southern Hemisphere: a review. Biol Invasions 19:2237–2246

Friedenberg NA, Whited BM, Slone DH (2007) Differential impacts of the southern pine beetle, *Dendroctonus frontalis*, on *Pinus palustris* and *Pinus taeda*. Can J For Res 37(8):1427–1437

Fule PZ, Roccaforte JP, Covington WW (2007) Posttreatment tree mortality after forest ecological restoration, Arizona, United States. Environ Manage 40:623–634

Furniss MM (1965) Susceptibility of fire-injured Douglas-fir to bark beetle attack in southern Idaho. J For 63(1):8–11

Gandhi KJK (2005) The responses of sub-boreal forest insects to a catastrophic wind-disturbance event and subsequent fuel-reduction practices in northeastern Minnesota. Dissertation, University of Minnesota

Gandhi KJK, Gilmore DW, Katovich SA et al (2007) Physical effects of weather events on the abundance and diversity of insects in North American forests. Environ Rev 15:113–152

Gandhi KJK, Gilmore DW, Haack RA et al (2009) Application of semiochemicals to assess the biodiversity of subcortical insects following an ecosystem disturbance in a sub-boreal forest. J Chem Ecol 35:1384–1410

Ganz DJ, Dahlsten DL, Shea PJ (2003) The post-burning response of bark beetles to prescribed burning treatments. In: Fire, fuel treatments, and ecological restoration. US Forest Service, p 143

García RA, Franzese J, Policelli N et al (2018) Non-native pines are homogenizing the ecosystems of South America. In: Biocultural homogenization to biocultural conservation. Springer, pp 245–263

Geiszler DR, Gara RI, Littke WR (1984) Bark beetle infestations of Lodgepole pine following a fire in South Central Oregon. Z Angew Entomol 98(1–5):389–394

Georgia Forestry Commission (2017) *Ips* engraver beetle outbreak update, 30 January 2017. Georgia Forestry Commission, Forest Health Program

Gerson EA, Kelsey RG (1997) Attraction and direct mortality of pandora moths, *Coloradia pandora* (Lepidoptera: Saturniidae), by nocturnal fire. For Ecol Manag 98(1):71–75

Gilliam FS, Platt WJ (1999) Effects of long-term fire exclusion on tree species composition and stand structure in an old-growth *Pinus palustris* (longleaf pine) forest. Plant Ecol 140:15–26

Gilmore DW, Seybold SJ, Zasada JC et al (2002) Cumulative effects of a severe windstorm and subsequent silvicultural treatments on plant and arthropod diversity in the Gunflint Corridor of the Superior National Forest in northern Minnesota: project design. In: Forestry at the Great Divide, Proceedings: Society of American Foresters 2001 National Convention. Denver, CO, pp 364–337

Gitau CW, Carnegie AJ, Nicol HI et al (2013) Incidence of *Ips grandicollis* (Coleoptera: Scolytinae) in trap trees prepared for biological control of *Sirex noctilio* (Hymenoptera: Siricidae) in Australia: influence of environment and silviculture. For Ecol Manag 310:865–874

Goldammer JG, Penafiel SR (1990) Fire in the pine-grassland biomes of tropical and subtropical Asia. In: Fire in the tropical biota: ecosystem processes and global challenges. Springer, Berlin/Heidelberg, pp 45–62

Gongalsky KB, Wikars LO, Persson T (2003) Dynamics of pyrophilous carabids in a burned pine forest in Central Sweden. Balt J Coleopterol 3(2):107–111

Graziosi I, Tembo M, Kuate J et al (2020) Pests and diseases of trees in Africa: a growing continental emergency. Plants People Planet 2(1):14–28

Grégoire JC, Raffa KF, Lindgren BS (2015) Economics and politics of bark beetles. In: Bark beetles. Academic Press, pp 585–613

Grünig M, Seidl R, Senf C (2023) Increasing aridity causes larger and more severe forest fires across Europe. Glob Chang Biol 29(6):1648–1659

Hanula JL, Meeker JR, Miller DR et al (2002) Association of wildfire with tree health and numbers of pine bark beetles, reproduction weevils and their associates in Florida. For Ecol Manag 170(1–3):233–247

Hart S (1998) Beetle mania: an attraction to fire. BioScience 48(1):3–5

Hart SJ, Schoennagel T, Veblen TT et al (2015) Area burned in the western United States is unaffected by recent mountain pine beetle outbreaks. Proc Natl Acad Sci USA 112(14):4375–4380

Hayes A, Nahrung H (2022) An a-luring solution for bark beetles. The Forester 2022:44–45

He T, Pausas JG, Belcher CM et al (2012) Fire-adapted traits of *Pinus* arose in the fiery Cretaceous. New Phytol 194(3):751–759

Heikkala O, Martikainen P, Kouki J (2017) Prescribed burning is an effective and quick method to conserve rare pyrophilous forest-dwelling flat bugs. Insect Conserv Divers 10(1):32–41
Hicke JA, Meddens AJ, Kolden CA (2016) Recent tree mortality in the western United States from bark beetles and forest fires. For Sci 62(2):141–115
Hood SM (2010) Mitigating old tree mortality in long-unburned, fire-dependent forests: a synthesis. RMRS-GTR-238. U.S. Department of Agriculture, Forest Service, Rocky Mountain Research Station, Fort Collins
Hood S, Sala A (2015) Ponderosa pine resin defenses and growth: metrics matter. Tree Physiol 35(11):1223–1235
Hood S, Sala A, Heyerdahl EK et al (2015) Low-severity fire increases tree defense against bark beetle attacks. Ecology 96(7):1846–1855
Hood SM, Baker S, Sala A (2016) Fortifying the forest: thinning and burning increase resistance to a bark beetle outbreak and promote forest resilience. Ecol Appl 26(7):1984–2000
Howe M, Mason CJ, Gratton C et al (2020) Relationships between conifer constitutive and inducible defenses against bark beetles change across levels of biological and ecological scale. Oikos 129(7):1093–1107
Hunter ME, Robles MD (2020) Tamm review: The effects of prescribed fire on wildfire regimes and impacts: a framework for comparison. For Ecol Manag 475:118435
Irland LC (2000) Ice storms and forest impacts. Sci Total Environ 262(3):231–242
Jacobs JM, Bergeron JC, Work TT et al (2011) Low intensity surface fire instigates movement by adults of *Calosoma frigidum* (Coleoptera, Carabidae). ZooKeys 147:641
Jin WT, Gernandt DS, Wehenkel C et al (2021) Phylogenomic and ecological analyses reveal the spatiotemporal evolution of global pines. Proc Natl Acad Sci USA 118(20):e2022302118
Jung JK, Kim M, Nam Y et al (2020) Changes in spatial and temporal distributions of *Monochamus* beetles along the fire severity in burned *Pinus densiflora* forests. J Asia Pac Entomol 23(2):404–410
Keeley JE (2012) Ecology and evolution of pine life histories. Ann For Sci 69(4):445–453
Keeley JE, Pausas JG, Rundel PW (2011) Fire as an evolutionary pressure shaping plant traits. Trends Plant Sci 16(8):406–411
Keeves A, Douglas DR (1983) Forest fires in South Australia on 16 February 1983 and consequent future forest management aims. Aust For 46(3):148–162
Kelsey RG, Westlind DJ (2017) Ethanol and primary attraction of red turpentine beetle in fire stressed ponderosa pine. For Ecol Manage 396:44–54
Kharuk VI, Antamoshkina OA (2017) Impact of silkmoth outbreak on taiga wildfires. Contemp Probl Ecol 10:556–562
Kinn DN, Perry TJ, Guinn FH et al (1994) Energy reserves of individual southern pine beetles (Coleoptera: Scolytidae) as determined by a modified phosphovanillin spectrophotometry method. J Entomol Sci 29(1):152–163
Kirchmeier-Young MC, Gillett NP et al (2019) Attribution of the influence of human-induced climate change on an extreme fire season. Earth's Future 7(1):2–10
Kirkendall LR, Biedermann P, Jordal BH (2015) Evolution and diversity of bark and Ambrosia beetles. In: Vega FE, Hofstetter RW (eds) Bark Beetles: biology and ecology of native and invasive species. Elsevier, pp 85–156
Kirkman LK, Jack SB, McIntyre RK (2017) The fire forest of the past and present. In: Ecological Restoration and Management of Longleaf Pine Forests. CRC Press, pp 3–16
Knebel L, Wentworth TR (2007) Influence of fire and southern pine beetle on pine-dominated forests in the Linville Gorge Wilderness, North Carolina. Castanea 72(4):214–225
Koivula M, Cobb T, Déchêne A et al (2006) Responses of two *Sericoda* Kirby, 1837 (Coleoptera: Carabidae) species to forest harvesting, wildfire, and burn severity. Entomol Fenn 17(3):315–324
Kolb TE, Agee JK, Fule PZ et al (2007) Perpetuating old ponderosa pine. For Ecol Manag 249(3):141–157
Kozlov MV, Zvereva EL (2017) Background insect herbivory: impacts, patterns and methodology. Prog Bot 79:313–355

Kulakowski D, Jarvis D (2013) Low-severity fires increase susceptibility of lodgepole pine to mountain pine beetle outbreaks in Colorado. For Ecol Manag 289:544–550

Lafon CW, Kutac MJ (2003) Effects of ice storms, southern pine beetle infestation, and fire on Table Mountain pine forests of southwestern Virginia. Phys Geogr 24(6):502–519

Lamont, B. B., P. J. G., H. Tianhua, W. E. T. F., and M. E. and Hanley. 2020. Fire as a selective agent for both serotiny and nonserotiny over space and time. Crit Rev Plant Sci 39:140–172.

Land AD, Rieske LK (2006) Interactions among prescribed fire, herbivore pressure and shortleaf pine (*Pinus echinata*) regeneration following southern pine beetle (*Dendroctonus frontalis*) mortality. For Ecol Manage 235(1–3):260–269

Langenheim JH (1990) Plant resins. Am Sci 78(1):16–24

Lee J, Brumley J, Ryckeley M et al (2016) Pitcher plant moths (*Exyra*) fly from pitchers in response to smoke. J Lepid Soc 70(4):268–270

Lerch AP, Pfammatter JA, Bentz BJ et al (2016) Mountain pine beetle dynamics and reproductive success in post-fire lodgepole and ponderosa pine forests in northeastern Utah. PLoS One 11(10):e0164738

Lindenmayer D, Taylor C, Blanchard W (2021) Empirical analyses of the factors influencing fire severity in southeastern Australia. Ecosphere 12(8):e03721

Littell JS, Peterson DL, Riley KL et al (2016) A review of the relationships between drought and forest fire in the United States. Glob Chang Biol 22:2353–2369

Liu Y, Francis RA, Wooster MJ et al (2022) Systematic mapping and review of Landscape Fire Smoke (LFS) exposure impacts on insects. Environ Entomol 51(5):871–884

Lombardero MJ, Ayres MP, Ayres BD (2006) Effects of fire and mechanical wounding on *Pinus resinosa* resin defenses, beetle attacks, and pathogens. For Ecol Manag 225(1–3):349–358

Lynch HJ, Renkin RA, Crabtree RL et al (2006) The influence of previous mountain pine beetle (*Dendroctonus ponderosae*) activity on the 1988 Yellowstone fires. Ecosystems 9:1318–1327

Maloney PE, Smith TF, Jensen CE et al (2008) Initial tree mortality and insect and pathogen response to fire and thinning restoration treatments in an old-growth mixed-conifer forest of the Sierra Nevada, California. Can J For Res 38(12):3011–3020

Martinson S, Hofstetter RW, Ayres MP (2007) Why does longleaf pine have low susceptibility to southern pine beetle? Can J For Res 37(10):1966–1977

McCullough DG, Werner RA, Neumann D (1998) Fire and insects in northern and boreal forest ecosystems of North America. Annu Rev Entomol 43(1):107–127

McHugh CW, Kolb TE, Wilson JL (2003) Bark beetle attacks on ponderosa pine following fire in northern Arizona. Environ Entomol 32(3):510–522

McNichol BH, Montes CR, Barnes BF et al (2019) Interactions between southern *Ips* bark beetle outbreaks, prescribed fire, and loblolly pine (*Pinus taeda* L.) mortality. For Ecol Manag 446:164–174

Meigs GW, Campbell JL, Zald HSJ et al (2015) Does wildfire likelihood increase following insect outbreaks in conifer forests? Ecosphere 6(7):art118

Meigs GW, Zald HS, Campbell JL et al (2016) Do insect outbreaks reduce the severity of subsequent forest fires? Environ Res Lett 11(4):045008

Menges ES, Deyrup MA (2001) Postfire survival in south Florida slash pine: interacting effects of fire intensity, fire season, vegetation, burn size, and bark beetles. Int J Wildland Fire 10(1):53–63

Milberg P, Bergman KO, Norman H et al (2015) A burning desire for smoke? Sampling insects favoured by forest fire in the absence of fire. J Insect Conserv 19:55–65

Millar CI (1993) Impact of the Eocene on the evolution of *Pinus* L. Ann Mo Bot Gard 80:471–498

Miller WE (1978) Use of prescribed burning in seed production areas to control red pine cone beetle. Environ Entomol 7(5):698–702

Miller WE (1979) Fire as an insect management tool. Bull ESA 25(2):137–140

Miller KK, Wagner MR (1984) Factors influencing pupal distribution of the pandora moth (Lepidoptera: Saturniidae) and their relationship to prescribed burning. Environ Entomol 13(2):430–431

Mistry J, Bilbao BA, Berardi A (2016) Community owned solutions for fire management in tropical ecosystems: case studies from Indigenous communities of South America. Philos Trans R Soc B Biol Sci 371(1696):20150174

Mitchell RJ, Hiers JK, O'Brien JJ (2006) Silviculture that sustains: the nexus between silviculture, frequent prescribed fire, and conservation of biodiversity in longleaf pine forests of the southeastern United States. Can J For Res 36(11):2724–2736

Mitsopoulos I, Mallinis G, Zibtsev S et al (2017) An integrated approach for mapping fire suppression difficulty in three different ecosystems of Eastern Europe. J Spat Sci 62(1):139–155

Nappi A, Drapeau P, Saint-Germain M et al (2010) Effect of fire severity on long-term occupancy of burned boreal conifer forests by saproxylic insects and wood-foraging birds. Int J Wildland Fire 19(4):500–511

Neumann FG (1987) Introduced bark beetles on exotic trees in Australia with special reference to infestations of *Ips grandicollis* in pine plantations. Aust For 50(3):166–178

Nowacki GJ, Abrams MD (2008) The demise of fire and "mesophication" of forests in the eastern United States. BioScience 58(2):123–138

Nowak J, Asaro C, Klepzig K et al (2008) The southern pine beetle prevention initiative: working for healthier forests. J For 106(5):261–267

Nowak JT, Meeker JR, Coyle DR et al (2015) Southern pine beetle infestations in relation to forest stand conditions, previous thinning, and prescribed burning: evaluation of the southern pine beetle prevention program. J For 113(5):454–462

O'Brien JJ, Kevin Hiers J, Mitchell RJ et al (2010) Acute physiological stress and mortality following fire in a long-unburned longleaf pine ecosystem. Fire Ecol 6(2):1–12

Palmer JF, Coyle DR, Heaton WC et al (2024) Response of *Ips* bark beetles (Coleoptera: Curculionidae: Scolytinae) to prescribed fire in South Carolina. For Ecol Manag 561:121870

Parker TJ, Clancy KM, Mathiasen RL (2006) Interactions among fire, insects and pathogens in coniferous forests of the interior western United States and Canada. Agric For Entomol 8(3):167–189

Pausas JG (2015) Evolutionary fire ecology: lessons learned from pines. Trends Plant Sci 20(5):318–324

Pausas JG, Parr CL (2018) Towards an understanding of the evolutionary role of fire in animals. Evol Ecol 32(2–3):113–125

Pelz KA, Smith FW (2012) Thirty year change in lodgepole and lodgepole/mixed conifer forest structure following 1980s mountain pine beetle outbreak in western Colorado, USA. For Ecol Manag 280:93–102

Perovich C, Sibold JS (2016) Forest composition change after a mountain pine beetle outbreak, Rocky Mountain National Park, CO, USA. For Ecol Manag 366:184–192

Perrakis DD, Agee JK (2006) Seasonal fire effects on mixed-conifer forest structure and ponderosa pine resin properties. Can J For Res 36(1):238–254

Perrakis DD, Agee JK, Eglitis A (2011) Effects of prescribed burning on mortality and resin defenses in old growth ponderosa pine (Crater Lake, Oregon): four years of post-fire monitoring. Nat Areas J 31(1):14–25

Pile LS, Wang GG, Knapp BO et al (2017) Comparing morphology and physiology of southeastern US *Pinus* seedlings: implications for adaptation to surface fire regimes. Ann For Sci 74:1–12

Pollet J, Omi PN (2002) Effect of thinning and prescribed burning on crown fire severity in ponderosa pine forests. Int J Wildland Fire 11(1):1–10

Popović Z, Bojović S, Marković M et al (2021) Tree species flammability based on plant traits: a synthesis. Sci Total Environ 800:149625

Powell EN, Raffa KF (2011) Fire injury reduces inducible defenses of lodgepole pine against mountain pine beetle. J Chem Ecol 37:1184–1192

Powell EN, Townsend PA, Raffa KF (2012) Wildfire provides refuge from local extinction but is an unlikely driver of outbreaks by mountain pine beetle. Ecol Monogr 82(1):69–84

Prasetya CD, Syaufina L, Santosa G (2017) The effect of various types of forest fires on pine resin productivity in Gunung Walat University Forest, Sukabumi, Indonesia. Biodiversitas 18(1):476–482

Pryor LD (1991) Forest plantations and invasions in the Mediterranean zones of Australia and South Africa. In: Biogeography of Mediterranean Invasions. Cambridge University Press, pp 405–412

Pyne SJ (1995) World fire: the culture of fire on earth. Henry Holt

Raffa KF (2014) Terpenes tell different tales at different scales: glimpses into the chemical ecology of conifer-bark beetle-microbial interactions. J Chem Ecol 40:1–20

Ramírez-Reyes T, Armendáriz-Toledano F, Rodríguez LGC (2023) Rearranging and completing the puzzle: phylogenomic analysis of bark beetles *Dendroctonus* reveals new hypotheses about genus diversification. Mol Phylogenet Evol 187:107885

Rasmussen LA, Amman G, Vandygriff JC et al (1996) Bark beetle and wood borer infestation in the Greater Yellowstone Area during four postfire years. USDA Forest Service, Intermountain Research Station, Research Paper INT-RP-487, p 9

Ray C, Cluck DR, Wilkerson RL et al (2019) Patterns of woodboring beetle activity following fires and bark beetle outbreaks in montane forests of California, USA. Fire Ecol 15:21

Renwick KM, Rocce ME, Stohlgren TJ (2016) Biotic disturbance facilitates range shift at the trailing but not the leading edge of lodgepole pine's altitudinal distribution. J Veg Sci 27(4):780–788

Retana J, Maria Espelta J, Habrouk A et al (2002) Regeneration patterns of three Mediterranean pines and forest changes after a large wildfire in northeastern Spain. Ecoscience 9(1):89–97

Rhoades CC, Fegel TS, Hubbard RM et al (2022) Limited seed viability in long-dead serotinous lodgepole pine trees in the Southern Rockies, USA. For Ecol Manag 526:120565

Richardson DM (1998) Forestry trees as invasive aliens. Conserv Biol 12(1):18–26

Richardson DM, Nsikani MM (2021) Mediterranean pines as invasive species in the Southern Hemisphere. In: Pines and their mixed forest ecosystems in the Mediterranean Basin. Springer, pp 83–99

Ritger HMW (2023) Forest Health implications of restoring longleaf pine (*Pinus palustris* Mill.) ecosystems with prescribed fire. Ph.D. Dissertation, University of Georgia

Ritger HW, Brantley ST, Klepzig KD et al (2023) Differential effects of fire regime and site conditions on bark beetle (Coleoptera: Curculionidae) activity in longleaf pine (*Pinus palustris* Mill.) ecosystem. For Ecol Manag 549:121488

Robinson JM, Gellie N, MacCarthy D et al (2021) Traditional ecological knowledge in restoration ecology: a call to listen deeply, to engage with, and respect Indigenous voices. Rest Ecol 29(4):e13381

Rodríguez-García A, López R, Martín JA et al (2014) Resin yield in *Pinus pinaster* is related to tree dendrometry, stand density and tapping-induced systemic changes in xylem anatomy. For Ecol Manag 313:47–54

Romme WH, Clement J, Hicke J et al (2006) Recent forest insect outbreaks and fire risk in Colorado forests: a brief synthesis of relevant research. Colorado Forest Restoration Institute, p 24

Rowe JS, Scotter GW (1973) Fire in the boreal forest. Quat Res 3(3):444–464

Ryan KC, Amman GD (1996) Bark beetle activity and delayed tree mortality in the Greater Yellowstone Area following the 1988 fires. In: The Bark Beetles, fuels, and fire bibliography, vol 15. Utah State University

Ryan KC, Knapp EE, Varner JM (2013) Prescribed fire in North American forests and woodlands: history, current practice, and challenges. Front Ecol Environ 11(s1):e15–e24

Safranyik L, Linton DA, Shore TL et al (2001) The effects of prescribed burning on mountain pine beetle in lodgepole pine. Pacific Forestry Center, Information Report BC-X-391

Santolamazza-Carbone S, Pestana M, Vega JA (2011) Post-fire attractiveness of maritime pines (*Pinus pinaster* Ait.) to xylophagous insects. J Pest Sci 84(3):343

Santoro AE, Lombardero MJ, Ayres MP, Ruel JJ (2001) Interactions between fire and bark beetles in an old growth pine forest. For Ecol Manag 144(1–3):245–254

Schmid JM, Parker DL (1990) Fire and forest insect pests. Paper presented at Effect of fire management of southwestern natural resources, Tucson, AZ, Nov 15–17
Schroeder LM (1990) Duct resin flow in Scots pine in relation to the attack of the bark beetle *Tomicus piniperda* (L.) (Col., Scolytidae). J Appl Entomol 109(1–5):105–112
Schütz S, Weissbecker B, Hummel HE et al (1999) Insect antenna as a smoke detector. Nature 398(6725):298–299
Schwilk DW, Knapp EE, Ferrenberg SM et al (2006) Tree mortality from fire and bark beetles following early and late season prescribed fires in a Sierra Nevada mixed-conifer forest. For Ecol Manag 232(1–3):36–45
Seidl R, Thom D, Kautz M et al (2017) Forest disturbances under climate change. Nat Clim Change 7(6):395–402
Sequeira AS, Normark BB, Farrell BD (2000) Evolutionary assembly of the conifer fauna: distinguishing ancient from recent associations in bark beetles. Proc R Soc Lond Ser B: Biol Sci 267(1460):2359–2366
Simard M, Romme WH, Griffin JM et al (2011) Do mountain pine beetle outbreaks change the probability of active crown fire in lodgepole pine forests? Ecol Monogr 81(1):3–24
Singh SP, Inderjit Singh JS, Majumdar S et al (2018) Insights on the persistence of pines (*Pinus* species) in the Late Cretaceous and their increasing dominance in the Anthropocene. Ecol Evol 8(20):10345–10359
Six DL, Bracewell R (2015) *Dendroctonus*. In: Bark beetles. Academic Press, pp 305–350
Six DL, Skov K (2009) Response of bark beetles and their natural enemies to fire and fire surrogate treatments in mixed-conifer forests in western Montana. For Ecol Manag 258(5):761–772
Slack AW, Zeibig-Kichas NE, Kane JM et al (2016) Contingent resistance in longleaf pine (*Pinus palustris*) growth and defense 10 years following smoldering fires. For Ecol Manag 364:130–138
Smith CC (1970) The coevolution of pine squirrels (*Tamiasciurus*) and conifers. Ecol Monogr 40(3):349–371
Steel ZL, Safford HD, Viers JH (2015) The fire frequency-severity relationship and the legacy of fire suppression in California forests. Ecosphere 6(1):1–23
Steel ZL, Goodwin MJ, Meyer MD et al (2021) Do forest fuel reduction treatments confer resistance to beetle infestation and drought mortality? Ecosphere 12(1):e03344
Stephens SL, Burrows N, Buyantuyev A et al (2014) Temperate and boreal forest mega-fires: characteristics and challenges. Front Ecol Environ 12(2):115–122
Stephens SL, Bernal AA, Collins BM et al (2022) Mass fire behavior created by extensive tree mortality and high tree density not predicted by operational fire behavior models in the southern Sierra Nevada. For Ecol Manag 518:120258
Stevens JT, Kling MM, Schwilk DW et al (2020) Biogeography of fire regimes in western US conifer forests: a trait-based approach. Glob Ecol Biogeogr 29(5):944–955
Strom BL, Goyer RA, Ingram LL Jr et al (2002) Oleoresin characteristics of progeny of loblolly pines that escaped attack by the southern pine beetle. For Ecol Manag 158(1–3):169–178
Suckling DM, Gibb AR, Daly JM et al (2001) Behavioral and electrophysiological responses of *Arhopalus tristis* to burnt pine and other stimuli. J Chem Ecol 27:1091–1104
Süda I, Voolma K, Õunap H (2009) Short-term monitoring of fire-adapted Coleoptera in burnt pine forest of northern Estonia. Acta Biol Univ Daugavpiliensis 9(1):43–48
Sullivan BT, Fettig CJ, Otrosina WJ et al (2003) Association between severity of prescribed burns and subsequent activity of conifer-infesting beetles in stands of longleaf pine. For Ecol Manag 185(3):327–340
Swezy DM, Agee JK (1991) Prescribed-fire effects on fine-root and tree mortality in old-growth ponderosa pine. Can J For Res 21(5):626–634
Tabacaru CA, Park J, Erbilgin N (2016) Prescribed fire does not promote outbreaks of a primary bark beetle at low-density populations. J Appl Ecol 53(1):222–232
Talluto MV, Benkman CW (2013) Landscape-scale eco-evolutionary dynamics: selection by seed predators and fire determine a major reproductive strategy. Ecol 94(6):1307–1316

Talucci AC, Lertzman KP, Krawchuk MA (2019) Drivers of lodgepole pine recruitment across a gradient of bark beetle outbreak and wildfire in British Columbia. For Ecol Manag 451:117500
Tapias R, Climent J, Pardos JA et al (2004) Life histories of Mediterranean pines. Plant Ecol 171:53–68
Tauer CG, Stewart JF, Will RE et al (2012) Hybridization leads to loss of genetic integrity in shortleaf pine: unexpected consequences of pine management and fire suppression. J For 110(4):216–224
Taylor KT, Maxwell BD, McWethy DB et al (2017) *Pinus contorta* invasions increase wildfire fuel loads and may create a positive feedback with fire. Ecology 98(3):678–687
Teste FP, Lieffers VJ, Landhäusser SM (2011) Seed release in serotinous lodgepole pine forests after mountain pine beetle outbreak. Ecol Appl 21(1):150–162
Thistle HW, Peterson H, Allwine G et al (2004) Surrogate pheromone plumes in three forest trunk spaces: composite statistics and case studies. For Sci 50(5):610–625
Thistle HW, Strom B, Strand T (2011) Atmospheric dispersion from a point source in four southern pine thinning scenarios: basic relationships and case studies. Trans ASABE 54(4):1219–1236
Turner MG, Romme WH, Gardner RH (1999) Prefire heterogeneity, fire severity, and early post-fire plant reestablishment in subalpine forests of Yellowstone National Park, Wyoming. Int J Wildland Fire 9(1):21–36
Turner MG, Braziunas KH, Hansen WD et al (2019) Short-interval severe fire erodes the resilience of subalpine lodgepole pine forests. Proc Natl Acad Sci 116(23):11319–11,328
Valor T, Hood SM, Piqué M et al (2021) Resin ducts and bark thickness influence pine resistance to bark beetles after prescribed fire. For Ecol Manag 494:119322
Van Lear DH, Waldrop TA (1991) Prescribed burning for regeneration. In: Forest regeneration manual. Springer, pp 235–250
Varner JM, Hiers JK, Ottmar RD et al (2007) Overstory tree mortality resulting from reintroducing fire to long-unburned longleaf pine forests: the importance of duff moisture. Can J For Res 37(8):1349–1358
Vázquez-González C, Zas R, Erbilgin N et al (2020) Resin ducts as resistance traits in conifers: linking dendrochronology and resin-based defenses. Tree Physiol 40(10):1313–1326
Verrall AF (1938) The probable mechanism of the protective action of resin in fire wounds on red pine. J For 36(12):1231–1233
Vilà-Vilardell L, De Cáceres M, Piqué M et al (2023) Prescribed fire after thinning increased resistance of sub-Mediterranean pine forests to drought events and wildfires. For Ecol Manag 527:120602
Wallin KF, Kolb TE, Skov KR et al (2003) Effects of crown scorch on ponderosa pine resistance to bark beetles in northern Arizona. Environ Entomol 32(3):652–661
Wang GG (2002) Fire severity in relation to canopy composition within burned boreal mixedwood stands. For Ecol Manag 163(1–3):85–92
Westlind DJ, Kerns BK (2021) Repeated fall prescribed fire in previously thinned *Pinus ponderosa* increases growth and resistance to other disturbances. For Ecol Manag 480:118645
Whitehouse NJ (2000) Forest fires and insects: palaeoentomological research from a subfossil burnt forest. Palaeogeogr Palaeoclimatol Palaeoecol 164(1–4):231–246
Wikars LO (1997) Effects of forest fire and the ecology of fire-adapted insects, vol 272. Acta University Uppsala Comprehensive Summaries of Uppsala Dissertations from the Faculty of Science and Technology, p 35
Williams AP, Abatzoglou JT, Gershunov A et al (2019) Observed impacts of anthropogenic climate change on wildfire in California. Earth's Future 7(8):892–910
Wrangham RW, Carmody RN (2010) Human adaptation to the control of fire. Evol Anthropol 19(5):187–199
Wrangham RW, Jones JH, Laden G et al (1999) The raw and the stolen: cooking and the ecology of human origins. Curr Anthropol 40(5):567–594
Wu H, Hu ZH (1997) Comparative anatomy of resin ducts of the Pinaceae. Trees 11:135–143

Wunder S, Calkin DE, Charlton V et al (2021) Resilient landscapes to prevent catastrophic forest fires: Socioeconomic insights towards a new paradigm. For Policy Econ 128:102458

Wylie FR, Peters B, DeBaar M et al (1999) Managing attack by bark and ambrosia beetles (Coleoptera: Scolytidae) in fire-damaged *Pinus* plantations and salvaged logs in Queensland, Australia. Aust For 62(2):148–153

Yan Z, Sun J, Don O et al (2005) The red turpentine beetle, *Dendroctonus valens* LeConte (Scolytidae): an exotic invasive pest of pine in China. Biodivers Conserv 14:1735–1760

Zalba SM, Cuevas YA, Boó RM (2008) Invasion of *Pinus halepensis* Mill. following a wildfire in an Argentine grassland nature reserve. J Environ Manag 88(3):539–546

Zhan Z, Yu L, Ren L (2022) Infestation patterns of incipient red turpentine beetle populations in fire-affected, logged and undisturbed forest stands of northern China. For Ecol Manage 521:120424

Zhang QH, Byers JA, Zhang XD (1993) Influence of bark thickness, trunk diameter and height on reproduction of the longhorned beetle, *Monochamus sutor* (Col., Cerambycidae) in burned larch and pine. J Appl Entomol 115(1–5):145–154

Shoot Architecture: A Crown Attribute Regulating Biosphere-Atmosphere Exchanges Via Coordinated Light Absorption and Water Delivery

Pantana Tor-Ngern, Sari Palmroth, Hyungwoo Lim, and Ram Oren

Abstract This chapter examines the interplay between shoot structure and canopy function in pine forests in three parts. The first two synthesize research on shoot hydraulics and architecture in relation to light absorption, establishing the foundation for understanding resource delivery within crowns. Using a modeling approach, the last part of the chapter explores the combined effects of crown structure and active-season solar angles on canopy light absorption (as a proxy for potential photosynthetic production) of virtual stands of three pine species with differing shoot characteristics. The simulation results suggest potential limits to the migration of different pine species beyond their current latitudinal ranges and the possible consequences of such migration on species composition of pine ecosystems.

P. Tor-Ngern
Department of Environmental Science, Faculty of Science, Chulalongkorn University, Bangkok, Thailand

Center of Excellence in Water Science and Technology for Sustainable Environment, Chulalongkorn University, Bangkok, Thailand

S. Palmroth (✉)
Nicholas School of the Environment, Duke University, Durham, NC, USA

Department of Forest Sciences, Institute for Atmospheric and Earth System Research (INAR), University of Helsinki, Helsinki, Finland
e-mail: sari.palmroth@duke.edu

H. Lim
Department of Forest Ecology and Management, Swedish University of Agricultural Sciences (SLU), Umeå, Sweden

R. Oren
Nicholas School of the Environment, Duke University, Durham, NC, USA

Department of Forest Sciences, Institute for Atmospheric and Earth System Research (INAR), University of Helsinki, Helsinki, Finland

Pratt School of Engineering, Department of Civil and Environmental Engineering, Duke University, Durham, NC, USA

D. M. Johnson et al. (eds.), *Physiology and Ecology of Pines Worldwide*, Tree Physiology 9, https://doi.org/10.1007/978-3-032-23001-0_5

1 Background

Shoots represent the fundamental structural units in the crowns of conifers (Sprugel et al. 1991), where a "shoot" refers to a flush of growth (of twig and needles; once or several times per growing season) from a terminal or lateral bud at the end of the main stem or a branch. In pine shoots, the density, length, flexibility, and orientation of needles or fascicles relative to the shoot axis (needle angle) collectively determine mutual shading patterns within and among shoots. These architectural features control the interception of light, or photosynthetic photon flux density (PPFD) by leaf surfaces throughout tree crowns and stand canopies under varying solar angles and atmospheric conditions (Leverenz and Hinckley 1990; Niinemets et al. 2002a; Smolander et al. 1994; Stenberg et al. 2001; Thérézien et al. 2007). Shoot hydraulic properties—including vulnerability to decreasing water potential, leaf area supported per unit xylem cross-sectional area, and xylem conductivity—determine the water delivery capacity to leaves under different soil and atmospheric moisture conditions. Therefore, a given investment in vascular architecture and transport tissue should be expected to maximize hydraulic safety vs hydraulic efficiency of the structure (McCulloh et al. 2003). These hydraulic characteristics influence both stomatal opening and the maintenance of moist mesophyll surfaces for carbon dioxide (CO_2) absorption. Through these combined mechanisms, shoot structure regulates the availability of both solar energy and leaf-internal CO_2 concentration, the two primary resources for photosynthesis. The substantial variation in shoot and needle dimensions within species (responding to climate) and across species has important implications for both light interception and hydraulic efficiency (Mäkelä et al. 2019; Niinemets et al. 2006; Thérézien et al. 2007; Wang et al. 2019).

The effects of shoots on photosynthetic resource availability must be considered in the context of entire crown architecture, where the shoot position relative to others determines light availability, and location along the water flow path affects water delivery to foliage. Research demonstrates that shoot traits are coordinated within the crown to enhance resource utilization under varying conditions. Shoots in low light environments orient more horizontally with needles placed less densely along the shoot axis compared to shoots at high light, which orient more vertically, thereby minimizing mutual shading and maximizing light absorption (Niinemets et al. 2002a, 2006; Stenberg et al. 2001). Needles produced in lower light environments exhibit reduced leaf mass per area, reflecting decreased photosynthetic machinery and nutrient content per unit leaf area (Oren et al. 1986), which imposes greater biochemical limitations to photosynthesis relative to shoots developed under higher light intensities. Additionally, shoot axes in shaded positions demonstrate reduced water delivery capacity per unit leaf area (Grönlund et al. 2016), supporting lower stomatal conductance and imposing greater biophysical limitations to photosynthesis than in sunnier locations. However, the reduced physicochemical capacity of shaded needles represents an adaptive advantage rather than a lost opportunity. Since the prevailing light environment, rather than leaf photosynthetic characteristics, ultimately limits photosynthesis in these conditions, the lower respiration costs associated with producing and maintaining shade-adapted shoots provide a net benefit.

Under genetic constraints that vary among species and genotypes to accommodate environmental controls, the vertical pattern of shoot traits within the canopy is expected to distribute available carbon efficiently for new carbon fixation (Niinemets 2010). This efficiency extends to dynamic processes, as even temporary shading of adjacent branches during diurnal solar angle changes may reduce their transpiration, thereby increasing water delivery to (and thus stomatal conductance of) unshaded shoots (Ewers et al. 2007). Throughout the growing season, the same shoot, crown, and canopy architecture will generate varying amounts of light absorption and carbohydrates reflecting the dynamics of solar angles over the active period. These architectural adaptations have been shown empirically to influence not only individual shoot performance but scale up to affect canopy gross primary productivity. However, evaluating the effect of vertical trends in shoot structure in the context of intact canopies on crown- and canopy-scale carbon acquisition can be accomplished only through modeling experiments (Kim et al. 2011).

2 Shoot Hydraulic Properties and Water Supply to Leaves

Data on hydraulic traits related to water transport efficiency of shoots are relatively rare; therefore, to increase the scope, this chapter includes studies on the hydraulics of pine needles and of branches, typically composed of several shoots. Hydraulic investment is often aligned with needle morphology and shoot structure. Leaf hydraulic conductance has been shown to increase with needle length across pine species (Wang et al. 2019) and shoot xylem hydraulic conductivity with shoot size (twig length and diameter within a species; Mäkelä et al. 2019). Long needles of *Pinus palustris* show greater investment in xylem tissue compared to the short needles of *Pinus echinata*, demonstrating a strong correlation between hydraulic supply and needle size (Wang et al. 2019, 2025; Zwieniecki et al. 2006). The sapwood area supporting leaf area (or sapwood-to-leaf area ratio, the Huber value, Hv) is expected to vary with the envionment such that, for a given species, both shoot and whole-tree Hv likely increase with evaporative demand of the environment (DeLucia et al. 2000; Grönlund et al. 2016; Mencuccini and Grace 1995; Waring et al. 1982). However, variation in Hv along geographical gradients may reflect responses to other environmental factors or be accompanied by adjustments in other hydraulic traits (Mencuccini et al. 2019).

Hydraulic architecture of pine shoots exhibits remarkable plasticity in response to canopy position and environmental conditions, with important implications for whole-tree water transport and gas exchange. In *Pinus sylvestris*, leaf-area-specific conductivity of current-year shoots increases with light availability and decreases with parent branch age, demonstrating that hydraulic properties respond to both photosynthetic demand and position within the crown—two variables that are typically confounded (Grönlund et al. 2016). Recent comparative studies across five southeastern US pine species (*Pinus virginiana*, *Pinus taeda*, *Pinus elliottii*, *P. echinata*, and *P. palustris*) revealed that shorter-needle species exhibit greater drought tolerance, with root vulnerability to embolism (P50, the xylem water potential corresponding to a 50% loss of hydraulic conductivity) systematically varying with needle morphology and canopy conductance responding primarily to vapor pressure deficit rather than soil water availability (Wang et al. 2025; Zhang et al. 2025). Shoot hydraulic architecture in *P. taeda* shows considerable developmental plasticity, with fertilization treatments altering leaf-specific hydraulic conductance by up to 50% through changes in leaf-to-sapwood area ratios, highlighting the coordinated adjustment of hydraulic supply and demand at the shoot level (Ewers et al. 2000). These findings emphasize that pine shoot hydraulics represent a dynamic interface between structural constraints and physiological demands, with species-specific strategies for balancing water transport efficiency and safety across heterogeneous canopy environments.

When vertical variation in light availability is excluded as a factor in shoot structure and function, such as in a study of open-grown *Pinus ponderosa* trees, vertical trends in shoot hydraulic properties primarily reflect the effects of path length and tissues developing under different water potentials (Hubbard et al. 2002). The study demonstrated that sap flux was lower at 25 m compared to 10 m height, consistent with the increased path length from soil to leaf at greater heights. More importantly, branches at the higher position had developed a proportionally lower leaf-to-sapwood area ratio (the inverse of Hv; 0.17 versus 0.27 m^2 cm^{-2}), which compensated for the increased hydraulic resistance and resulted in similar leaf-specific hydraulic conductance from soil to leaf. This hydraulic compensation mechanism maintained comparable gas-exchange rates between canopy positions despite the height difference. These findings demonstrate that the lower shoot hydraulic conductance commonly observed in old versus young trees cannot be explained by height differences alone.

Soil texture affects xylem growth and anatomical properties, which in turn, influences branch hydraulic properties, but only when the soil textural differences are substantial. For example, P50 values of *P. taeda* branches were similar between soils with modest textural differences (Domec et al. 2015) but became more negative (more resistant) when comparing trees growing on loamy versus sandy soils with markedly different properties (Hacke et al. 2000). It is noteworthy that nitrogen fertilization decreased leaf hydraulic conductance of *P. taeda* despite no discernible effect on P50, suggesting compensatingchanges in other traits (Domec et al. 2009, Domec et al. 2016). In contrast, elevated CO_2 (eCO_2) conditions did not affect leaf hydraulic conductance in this species (Domec et al. 2009). The ecological significance of branch hydraulic variation may be limited because branch and stem xylem comprise a relatively small fraction of whole-tree hydraulic resistance in pines (Domec et al. 2010). This pattern holds across the five most southern pine species, where branch hydraulic traits showed no relationship to climate of origin (Wang et al. 2025). More fundamentally, according to the hydraulic vulnerability segmentation hypothesis, branches consistently demonstrate greater resistance to embolism than any other plant organs, including stems—a pattern observed across multiple pine species including *P. taeda*, *P. virginiana*, and *Pinus pinea* (Johnson et al. 2016). Because more vulnerable distal organs (leaves and roots) become hydraulically dysfunctional at higher (less negative) water potentials than shoots, the latter maintaining wider safety margins, whole-tree vulnerability to drought is relatively independent of branch hydraulic properties.

Rainwater and nutrients can be directly absorbed through surfaces of shoot stems (Katz et al. 1989), a process recently confirmed in experiments on three pine species (*Pinus cembra*, *Pinus mugo*, *P. sylvestris*; Chin et al. 2025). Surface water absorption involves a surface capillary phase spreading droplets across the stem surface, some of which stored for many hours in extra-vascular macropores overlaying dead epidermal cells—a phenomenon observed across nearly the entire surface of *P. sylvestris* shoots. Water movement from the surface follows both apoplastic and symplastic pathways, driven by water potential gradients generated by the xylem. Intracellular airspaces in leaf bases, cortex, and pith serve as hydraulic capacitors that facilitate water storage and redistribution (Chin et al. 2025). While the vascular system can transport water among adjacent locations, a substantial portion of absorbed surface water remains localized rather than being immediately transported. This retained water may become available for embolism repair as the gradual movement of surface water toward the pith increases local xylem water potential. This mechanism suggests that higher water storage capacity in shoots serves dual function: maintaining elevated stomatal conductance during water stress (McCulloh et al. 2019) and possibly facilitating embolism repair to recover shoot hydraulic conductance (Chin et al. 2025).

3 Shoot Structure and Light Transfer

Leaf photosynthetic responses are scaled to canopies based in part on estimates of PPFD intercepted and absorbed by leaf surfaces. Because of strong coupling between photosynthetic rates and stomatal conductance, PPFD received by leaf surfaces can also be used in estimating transpiration and informing energy balance calculations. The light regime on leaf surfaces cannot be routinely measured and is thus commonly estimated based on radiative transfer models, describing how incoming solar radiation is intercepted by canopy elements (Pearcy 1989). The same principles are used to invert optically measured canopy transmittance or reflectance for estimating leaf area index (L; Knyazikhin et al. 1998; Nilson and Ross 1996). Because estimation of canopy photosynthesis is often based on both indirect estimates of L and modeled distributions of PPFD, it relies on the ability of models to simulate realistic canopies.

Crown architecture influences canopy photosynthesis by controlling the amount and distribution of canopy light absorption. Absorption of light and light-use efficiency (carbohydrate production per unit light absorbed) may vary 50%–80% among forests of different species composition, growing on sites of different fertility and subjected to different silvicultural treatments (Binkley et al. 2013).

Species- and site-specific foliage and crown aggregation (to shoots, branches and crowns) influence the amount and spatiotemporal distribution of canopy light interception (Chen et al. 1994; Kellomäki and Oker-Blom 1983; Oker-Blom and Kellomäki 1983; Sampson and Smith 1993; Stenberg 1998). The spatial organization and morphology of pine shoots appear to have evolved to facilitate light capture across complex canopy environments. Shade shoots in lower canopy positions tend to be horizontally inclined and flattened to maximize light interception efficiency, while sun shoots in upper canopy positions show more vertical orientations and clumped distributions (Kellomäki and Oker-Blom 1983; Stenberg 1996a; Stenberg et al. 2001).

The light incident on crowns and canopies is affected by stand conditions and latitude, respectively. Thus, the crown architecture of a species reflects the light environment prescribed by both factors (Kuuluvainen 1992). Leaf orientation plays a critical role in facilitating light interception and photosynthesis within plant canopies, as modeling studies show that vertical leaf arrangements can enhance photosynthetic performance compared to horizontal orientations, particularly at lower latitudes during summer (Chen et al. 1994). Furthermore, differential performance across latitudes may drive morphological divergence. For example, long-needled southern pines like *P. palustris* (needle length >300 mm) may optimize for high sun angles, while northern ecotypes such as *Pinus contorta* has compact shoots with short needles (needle length <70 mm), may be better suited to low sun angles and diffuse light (Critchfield 1980; Wang et al. 2019).

Needle clustering in a cylindrical arrangement generates within-shoot self shading, thus reducing interception of light where it is ample, but allowing radiation to penetrate deeper into the canopy, enhancing canopy-scale photosynthesis (Niinemets et al. 2002b; Stenberg 1998). Moreover, the three-dimensional structure of shoots leads to multiple scattering of short-wave radiation between the needles thus decreasing canopy reflectance compared to flat leaves (Knyazikhin et al. 1998; Shabanov et al. 2000; Smolander and Stenberg 2003, 2005).

Shoot architecture also plays a role in determining the magnitude of partial shading or the penumbral effect (Miller and Norman 1971; Smith et al. 1989). A leaf (or part of the leaf) in a canopy can lie in a penumbra if the fundamental canopy element (leaf or shoot) is narrow enough such that a shading element positioned between the sun and the leaf does not fully cover the solar disk. The penumbral effect redistributes the direct light, reducing the fraction of leaf surfaces receiving peak values and that in full shade (Stenberg 1998). The magnitude of the effect depends on the depth of the canopy, the distribution of the shoots and their "transparency," i.e., the dimensions of the needles and their organization in shoots. Stenberg (1998) quantified this transparency with a transmission coefficient (*c*), the gap area within the projected shoot envelope area. The spherically averaged value of *c* is related to the average shoot silhouette (i.e., projected area of a shoot) to total leaf area ratio ($\overline{STAR}$; Oker-Blom and Smolander 1988; Thérézien et al. 2007), another parameter used in radiative transfer models. Thus defined, shading of one shoot by another reduces the fraction of visible sun by a factor *c*. Penumbra leads to a more even distribution of direct light on needle surfaces at a given canopy depth without affecting the mean fraction of visible sun (Palmroth et al. 1999). In *P. sylvestris*, the effect redistributes up to 35% of incoming light (Stenberg 1998). The transmission coefficient varies among species and along crowns; *c* of *P. taeda* shade shoots was higher (~40%) than those of sun shoots (Thérézien et al. 2007). Despite the potentially significant effects on canopy photosynthesis, penumbra is seldom considered in models (cf. Bertin et al. 2011; Kim et al. 2011; Ryel et al. 2001; Schäfer et al. 2003; Stenberg 1998). Including penumbral effects, estimates of photosynthesis in lower canopy layers, where indirect light dominates, improved by 12%–18% (Ryel et al. 2001), which is critical to sustaining lower branches.

In modeling canopy absorbed photosynthetically active radiation (APAR) and photosynthesis, shoot structure can be incorporated by multiplying the mean projection of unit flat area (*G*; Nilson 1971; Stenberg 2006) by a clumping factor that reduces the light extinction efficiency of the canopy foliage elements. The mean shoot silhouette to total leaf area ratio is analogous to the shoot-level clumping factor ($\text{CF} = 4\overline{\text{STAR}}$) as it quantifies the amount of effective, light intercepting leaf area with respect to the investment in total leaf surface area (Oker-Blom and Smolander 1988) and can be used for, e.g., estimating *L* from ground-based transmittance measurements (Chen et al. 1997; Stenberg 1996b). It is also related to the mean probability of escape (i.e., probability that a photon emitted from needle surface of a shoot will not hit another needle; $1 - p_{sh} = 4\overline{\text{STAR}}$; Smolander and Stenberg 2003), which is an important parameter in physically-based radiative transfer models that are applied in monitoring of vegetation using remote sensing data (Smolander and Stenberg 2005; Stenberg et al. 2016;).

Within and across pine species, the estimates of STAR not only vary with needle area density, but also with twig and needle length, such that, all else being equal, $\overline{\text{STAR}}$ first increases and the saturates with increasing needle length (Wang et al. 2019). However, as needle area density in the shoot volume declines with (or without) increasing needle length across or within species, e.g., in response to site quality, STAR and thus efficiency of light interception may increase (Niinemets et al. 2002b, 2006; Smolander et al. 1990). STAR also increases with decreasing light availability (Niinemets et al. 2002b, 2006; Stenberg et al. 2001; Thérézien et al. 2007). For example, shade shoots in *P. taeda* display a $\overline{\text{STAR}}$ of 0.22, about 40% higher than sun-exposed shoots (0.16), allowing more efficient light interception under low irradiance (Thérézien et al. 2007). Although STAR and related parameters are fundamental to characterizing radiation regimes in coniferous canopies, they are rarely estimated accurately for each application. However, several approaches facilitate rapid generation of these radiative transfer parameters for applications requiring angle-dependent estimates (Fig. 1 after Thérézien et al. 2007; Pisek et al. 2023).

Considering canopy light absorption as setting the potential for growth rate (Monteith 1977), it is possible to evaluate the likely impact of crown architecture (i.e., foliage organization on shoots within the crown) and solar elevation (i.e., site latitude) on growth of the canopy as well as associated ecological consequences to the understory. Translating shoot structure and leaf physiology into canopy photosynthesis estimates requires models that incorporate species-specific structural

Fig. 1 Comparison between actual shoots (upper panels) and their corresponding models (lower panel) for *P. echinata* (needle length = ~40 mm), *P. taeda* (~140 mm) and *P. palustris* (~290 mm; left to right; all reference disks have the same diameter). Shoots are shown at different orientations, ranging from the side (left) to from above. (Taken from Thérézien et al. (2007))

traits that affect light distribution in the canopy. The considerable variability apparent in the shoot structure of coniferous species is reflected in the variation of the values of the shoot-structure related parameters such as STAR (Thérézien et al. 2007). Furthermore, shoot structure varies greatly within the canopy because of shading (Stenberg et al. 2001), meaning that structural parameters obtained from one species or light environment should not be employed to another species or throughout the canopy.

4 Shoot Growth Effects on Light Absorption by and Water Delivery to Needles

The regular crown architecture characteristic of most pines reflects efficient resource allocation, with shoots developing where they best enhance individual tree fitness. This architectural coordination is evident in the re-establishment of typical crown structure following damage (Långström et al. 1992; Honkanen et al. 1999). However, pine crown architecture exhibits considerable variation that reflects both environmental conditions (light, wind, snow, competition) and inherent flexibility of the developmental system. This variation occurs at multiple scales—among locations, between branches, and across whorls—and is driven by a degree of autonomous nature of sub-crown units. This autonomy enables phenotypic plasticity that allows spatial coordination within the crown and temporal adjustments as conditions change. Environmental factors influence shoot development in predictable ways. Shoots tend to develop longer axes during years with, e.g., non-limiting moisture or temperature, while within-year variation among shoots reflects local light conditions. Notably, shoots that develop long axes at initiation tend to maintain this pattern in subsequent years. Crown connectivity also affects shoot development, as demonstrated in *P. sylvestris* where shaded branches showed suppressed shoot production (reduced number and length) when connected to unshaded branches compared to when most branches experienced shade (Stoll and Schmid 1998). This resource-sharing effect has been documented across multiple tree species (Sprugel 2002), indicating a general pattern of within-crown resource coordination. Despite this variation and plasticity, pines maintain strong overall crown symmetry through regulatory mechanisms that balance local resource use with architectural integrity.

Shoot structure responds not only to interannual variation in growing conditions and vertical light gradients within the crown, but also to the availability of growth resources such as CO_2 and nutrients. For example, when *P. taeda* trees were exposed to eCO_2 and balanced nutrient application, both treatments increased foliage retention and produced deeper crowns (Maier et al. 2002). Elevated CO_2 produced distinct structural changes in *P. taeda* shoots compared to fertilization treatments. Under eCO_2 (but not fertilization), shoots developed approximately 40% longer axes, potentially increasing hydraulic path-length resistance proportionally. However, the effects on leaf area differed markedly between treatments: eCO_2 increased shoot leaf area by only ~10%, while fertilization increased it by ~40%. This difference reflected fertilization's enhancement of fascicle number in the second flush and an increase in needle length, whereas eCO_2 decreased needle length despite promoting shoot elongation. The structural changes under eCO_2 had important implications for light capture and photosynthesis. Mutual shading within shoots decreased substantially, as evidenced by ~20% lower foliage area density per unit shoot length. This reduced self-shading, combined with higher photosynthetic rate under eCO_2, suggests that light availability did not become more limiting to photosynthesis despite the development of deeper crowns with higher total leaf area (Maier et al. 2002). These findings point to a potential shift in photosynthetic limitations under eCO_2. While light limitations were reduced, the increased path-length resistance from longer shoots may have imposed greater hydraulically driven stomatal constraints on photosynthesis. However, this potential hydraulic limitation was largely offset by compensatory changes in shoot anatomy—specifically, a near doubling of Hv associated with increased basal diameter in eCO_2-grown shoots.

Fertilization produced a contrasting pattern of shoot development. Unlike eCO_2 treatments, fertilization increased leaf area without increasing shoot length, thus foliage density increased by ~40% (Maier et al. 2002). The increase in shoot basal diameter did not match the increase in leaf area, resulting in ~20% lower Hv. Despite this apparent hydraulic constraint, fertilized shoots likely did not experience increased stomatal limitation to photosynthesis. The much higher leaf density substantially reduced light interception per unit leaf area of the shoot, which would correspondingly reduce CO_2 demand and thus the need for high water transport rates. This suggests a functional balance between resource supply and demand even under altered shoot architecture. A scaling between Hv (water transport capacity per unit leaf area) and STAR (light interception efficiency inversely related to the packing of foliage) could be expected but remains to be empirically demonstrated.

These contrasting responses to eCO_2 and fertilization illustrate an important principle: light availability at external leaf surfaces and water delivery to internal leaf surfaces appear to be coordinated such that both factors co-limit photosynthesis rather than operating independently. This coordination implies that quantifying limitations imposed by one factor provides insight into limitations imposed by the other. Given this apparent coordination between light and water limitations at the shoot level, the following section explores how shoot structure affects light absorption patterns throughout entire pine canopies across different geographical locations.

5 Estimating Light Absorption of the Case Study Stands

Estimation of the light absorption profile for two case studies (described below) was based on the radiation transfer module of the Canopy Conductance Constrained Carbon Assimilation model (4C-A; Kim et al. 2011; Schäfer et al. 2003). Although 4C-A was designed to estimate diurnal patterns of canopy-integrated ratio between leaf internal and ambient CO_2 concentration and photosynthesis when coupled with other modules, the following analyses focus on the light module only. All subsequent among-canopies comparisons were therefore made based solely on vertical profiles of APAR down the canopy. A brief description of the model and its parameterization are provided below; a complete description can be found elsewhere (4C-A model; Kim et al. 2011; Schäfer et al. 2003).

The light module of 4C-A is a one-dimensional, multi-layer radiative transfer scheme (Kim et al. 2008; Schäfer et al. 2003). Integrating essential elements of three models: Campbell and Norman (1998), Nilson (1999), and Stenberg (1998), 4C-A computes the proportion of sunlit area on a horizontal surface for each sun angle and at different levels in the canopy. The sunlit fraction, at a given direction of the sun, is computed based on the spatial distribution of the canopy elements (leaves, branches and stems). For foliage, the leaf angle distribution is ellipsoidal (Campbell and Norman 1998), and foliage area is clumped into shoots ($CF = 4\overline{STAR}$) and crowns (Nilson 1999), and the cumulative leaf (and branch and stem) area density in each layer determines the probability of transmission of direct, diffuse, and scattered radiation through the canopy (Campbell and Norman 1998). The incoming PPFD is quantified above the canopy and its transfer (by components) is estimated sequentially along 1-m layers down the canopy. The light absorbed by each layer is determined by taking the difference (between the mean PPFD on horizontal surface) of two adjacent layers and integrating over all layers to estimate total canopy APAR. 4C-A also simulates the distribution of PPFD on leaf surfaces as well as the penumbral effect (Stenberg 1998); however, this redistribution of direct radiation has no effect on the estimates of APAR.

5.1 Species and Locations

Canopy light absorption, a proxy for photosynthetic production, was quantified across species and locations (Table 1). First, to separate the effect of latitude from crown architecture on canopy APAR, a "hypothetical" *P. taeda* (*Pt*) stand was virtually established in central Sweden (SE) and a *P. sylvestris* (*Ps*) stand in North Carolina, USA, (NC; *Case 1*). Second, a hypothetical *P. contorta* (*Pc*) stand was placed in SE to determine whether differences in estimated canopy APAR could explain the faster growth observed relative to *P. sylvestris* stands (Elfving and Norgren 1993; Elfving et al. 2001; Gundale et al. 2014; Norgren 1995; *Case 2*). The performance of these "exotic" stands of *Pt*, *Ps* and *Pc* were compared to that in their "native" habitats in NC, SE, and British Columbia, Canada (BC), respectively. In both case studies, the impact of canopy APAR on light availability for understory and competing vegetation was also assessed.

All simulations were done using species-specific shoot and crown characteristics. However, many factors, such as differences in stand density and *L*, may confound the effect of crown architecture when comparing APAR at different latitudes.

Table 1 Site and stand information

	Actual stand		Hypothetical stand		
Variables	*P. taeda* (NC)	*P. sylvestris* (SE)	*P.taeda*	*P. contorta*	*P. sylvestris*
Location	36°20′N 79°28′W	64°10′N 19°45′E	36°N 79°W	58°N 125°W	64°N 19°E
Elevation (m asl)	212	145	–		
Age (year)	38	~90			
H (m)	23	19	19	19	20
qDBH (cm)	36	18	29		
SD (trees ha^{-1})	308	1017	660		
RSD	0.68	0.69	1		
L (m^2 m^{-2})	1.60	2.02	2.00		
S (m^2 m^{-2})	2.90	2.30	2.80	2.70	2.40
Native GS period	–		Mar 16–Dec 6	Apr 20–Oct 11	May 1–Sep 30
Number of GS days			266	175	153

Values for the hypothetical stand were calculated based on allometric relationships, assuming mean tree size as indicated on the self-thinning line (see Fig. 2a; star). *H* mean canopy height, *qDBH* quadratic mean diameter at 1.3 m, *SD* stand density, *RSD* relative stand density, *L* mean growing season leaf area index, *S* mean growing season plant surface area index, *GS* growing season

To simplify interpretation, this analysis focused on stands (i.e., hypothetical stands) with uniform tree arrangements, identical mean tree sizes and stand densities. Data showed that all three species shared a common maximum size-density relationship (Fig. 2a, solid line; Farnden 1996; Harrington and Edwards 1996; Lim et al. 2015; Reineke 1933; Scott 1997) and the structure of the hypothetical stand was set between that of the actual stands in NC and SE (Fig. 2a; star). Furthermore, several additional, simplifying assumptions were made. First, light absorbed by live branches at the bottom of the crowns in the native habitat were assumed to represent the minimum "tolerable" daily average light absorbed over the growing season by a branch of that species based on the branch autonomy rule (Sprugel et al. 1991). Second, changes in growing season length were assumed not to affect the balance between photosynthesis and respiration. Third, leaf area dynamics of a species in an exotic environment were assumed to remain the same as in the native environment over the simulation year. Further details on the parameters, conditions, and input variables used in the 4C-A simulations of APAR are detailed in Appendix.

5.2 *Simulated Impact of Shoot Architecture on Stand Performance at Two Distant Latitudes*

The consequences of virtual introduction of a pine species into an exotic latitude with different growing season solar angle distribution (Fig. 2b) were evaluated considering *only* its crown structural attributes through their effects on APAR. The vertical profiles of the surface area for aboveground components were similar among all three species (Fig. 3a, b). Consequently, light absorption was primarily influenced by shoot-scale characteristics, particularly the grouping of foliage (clumping factor, $CF = 4\overline{STAR}$) and needle inclination (needle angle distribution, *NAD*; Niinemets and Kull 1995; Stenberg 1998; Thérézien et al. 2007). Clumping of foliage decreases APAR at fixed *L*, but high clumping at the canopy top (i.e., small CF values) allows greater light penetration to lower canopy layers, thereby increasing APAR for less clumped lower foliage (Oker-Blom et al. 1989; Stenberg et al. 2001). While this adjustment of shoot structure was evident in all three species studied, the two northern species exhibited higher clumping (Fig. 3c) and more vertically angled foliage compared to *Pt* (Fig. 3d), possibly reflecting necessary adaptations to the low solar incidence angles that dominate northern latitudes (Kuuluvainen 1992). In contrast, *Pt* stand featured less clumped leaves and more horizontally oriented needles, causing rapid light extinction below the canopy top.

While canopy photosynthetic production was not simulated here, only APAR, the differences in shoot structure within and among species imply that higher clumping at the top of the canopy likely help preventing excessive energy loads that cause photoinhibition (Björkman et al. 1988; Gamon and Pearcy 1989) while simultaneously extending favorable light environment at the canopy base (Stenberg 1998). Additionally, greater shoot-level clumping reduces the fraction of gap area

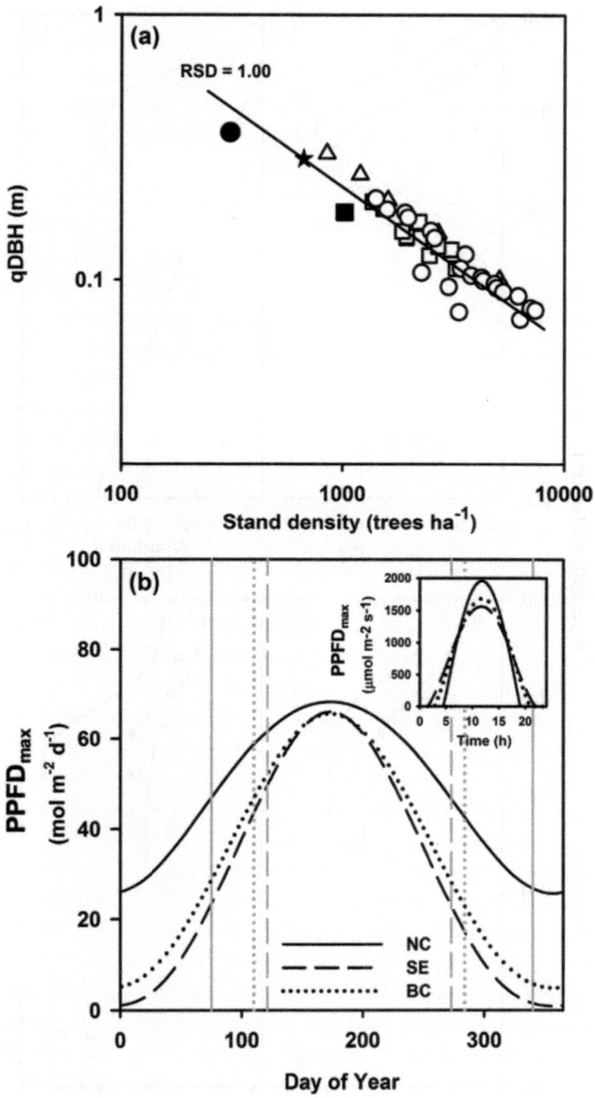

Fig. 2 (**a**) Size-density line. The relationship between average tree size, expressed as quadratic mean diameter at breast height (qDBH), and stand density, with the maximum size-density line shown (relative stand density, RSD = 1.00). Open symbols show published or measured data of *P. taeda* in Piedmont region in USA (circles), *P. sylvestris* near Rosinedal, Sweden (squares) and *P. contorta* in British Columbia, Canada (triangles) used to generate the relation. Filled symbols refer to the study sites of *P. taeda* in North Carolina (filled circle) and *P. sylvestris* in Rosinedal (filled square). The hypothetical stand (star) used in all analyses of this study is also shown. (**b**) Theoretical maximum photosynthetic photon flux density ($PPFD_{max}$). Daily patterns of $PPFD_{max}$ used in the analyses at the three locations. Inset shows diurnal pattern of $PPFD_{max}$ on the summer solstice (DOY 173)

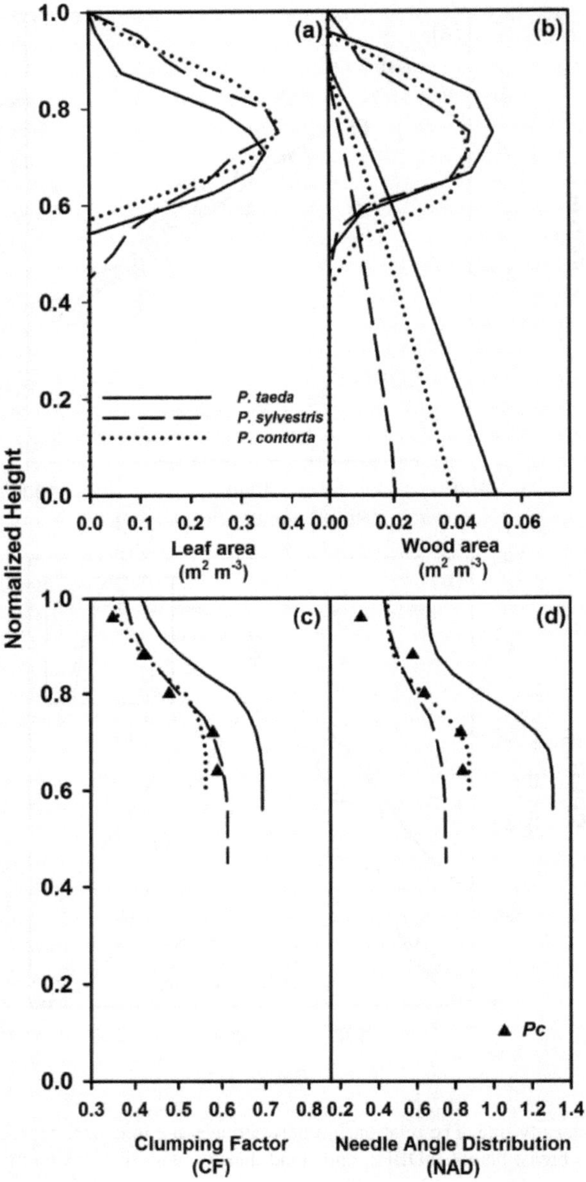

Fig. 3 Canopy structures. Vertical profiles of canopy attributes as input parameters for the radiative transfer model. The structural properties were generated from field measurements. (**a, b**) Plant surface area profile (the leaf area profile is during the summer solstice). Wood surface area includes that of branches and stem. (**c, d**) Shoot characteristic profiles. Measured shoot parameters for *P. contorta* (*Pc*) are shown as triangles

within the shoot envelope (i.e., decreasing the shoot transmission coefficient), which limits light transfer through shoots and reduces the frequency of penumbra (Stenberg 1998; Thérézien et al. 2007). *Pt* and *Ps* shared a similar vertical pattern of light transmission coefficient (not shown), indicating increasing penumbral effects with canopy depth. However, this phenomenon was more pronounced throughout the *Pt* canopy, indicating a more even distribution of direct light on leaf surfaces and resulting in higher potential canopy photosynthesis per unit *L* (Horn 1971; Miller and Norman 1971; Palmroth et al. 1999). Taken together, *Ps* crown architecture facilitates a more even APAR distribution down the canopy, while *Pt* crown architecture may achieve higher APAR and photosynthtic rates for a given amount of above-canopy PPFD.

The average daily pattern during the growing season (Fig. 4) can be used to demonstrate the effect of latitudinal placement on APAR. When *Pt* performance was simulated in Sweden, the bottom branches received ~10%–35% lower APAR than in their native NC environment (Fig. 4c). Conversely, *Ps* shows the opposite response upon southward transfer. When averaged across all sky conditions, the bottom of *Ps* canopies received more light in NC than in their native SE environment (Fig. 4i). These contrasting responses highlight a fundamental mismatch: the distinct shoot characteristics that have evolved in each species may become maladaptive when exposed to the solar angle regimes of different latitudes. Each species' architecture appears optimized for its native light environment rather than for conditions at other latitudes.

While *Ps* would absorb relatively less light at the canopy top when moved to NC compared to its native latitude (Fig. 5a, dashed line), APAR levels throughout its entire canopy would remain sufficiently high to support crown function (Fig. 5b, dashed line). In contrast, a substantial portion of *Pt* crown would receive lower mean daily APAR over the growing season compared to conditions in its native NC environment (Fig. 5a, solid line). Note, however, that the greater penumbral effect of *Pt* shoots would likely reduce to some degree the negative effect of reduced APAR in lower branches. Nevertheless, this large reduction in light availability presents a significant physiological constraint that could compromise lower canopy branch survival (considering branch autonomy; Sprugel et al. 1991). Compensating for the reduction would require adjusting leaf area to levels that would either force the species below the maximum size-density line or push allometric relationships well beyond typical species bounds (Fig. 5c, d). These findings demonstrate that when stand and tree characteristics are otherwise similar, the inherent crown characteristics of both species are poorly suited for the exotic latitudes tested in this simulation. The shoot-scale differences that appear to optimize performance in native environments may become constraints when species encounter different solar angle regimes.

Understory vegetation plays a crucial role in boreal forest ecosystems, influencing tree regeneration, decomposition, nutrient cycling, and overall forest productivity (Nilsson and Wardle 2005; Turetsky 2003; Wardle et al. 1997). Therefore, crown architectures that limit light penetration to lower branches would also affect light availability for subcanopy species when forests are established in new

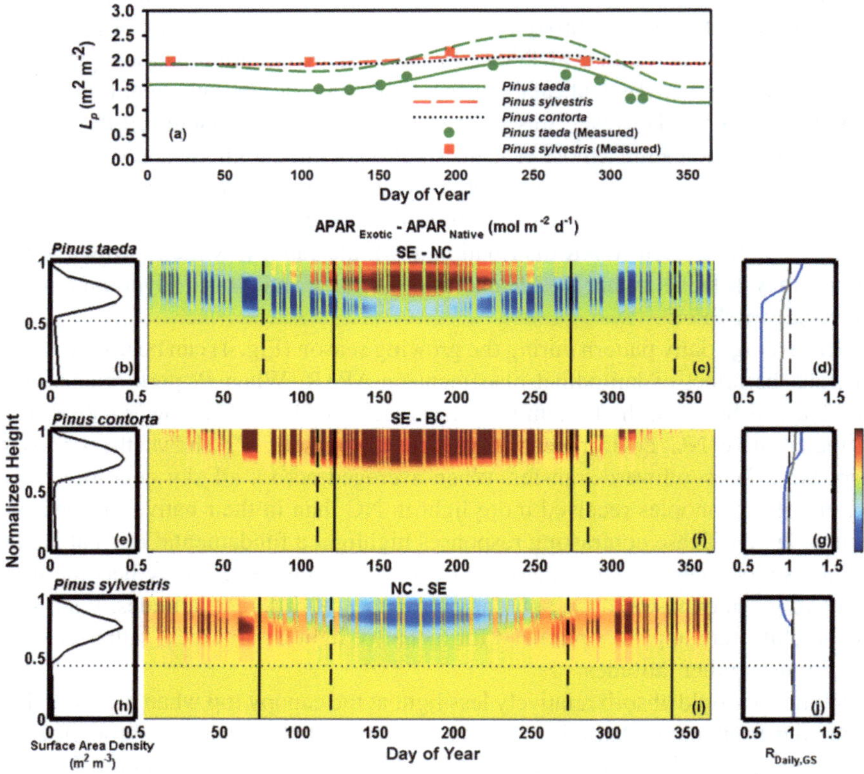

Fig. 4 Daily growing season average absorbed photosynthetically active radiation (APAR) profiles. (**a**) Projected leaf area index (L_p) dynamics of each species during the year. Symbols represent field measurements from the actual sites of *P. taeda* (*Pt*) and *P. sylvestris* (*Ps*). The L_p dynamic of *P. contorta* (*Pc*) was assumed to be similar to that of *Ps*, adjusted for the different growing season length in the native site. Green dashed line represents L_p of the hypothetical stand of *Pt* in the analyses. (**b**), (**e**), and (**h**) Vertical profiles of plant (foliage+branch+stem) surface area density in the three pine canopies. (**c**), (**f**), and (**i**) Absolute difference of daily APAR between exotic and native latitude, of each species. Solid lines represent the start and the end of growing season in the exotic location while those of the native location are denoted by dashed lines. (**d**), (**g**), and (**j**) Relative changes of daily growing season average APAR ($R_{\text{Daily,GS}}$) calculated as $R_{\text{Daily,GS}} = \frac{\sum_{\text{GS,Exotic}} \text{APAR}_{\text{Exotic}} / N_{\text{Exotic}}}{\sum_{\text{GS,Native}} \text{APAR}_{\text{Native}} / N_{\text{Native}}}$ where N represents the number of days during the growing season (GS) of either exotic or native sites. The $R_{\text{Daily,GS}}$ values are shown separately for clear (blue lines) and cloudy (gray lines) days. Horizontal dotted lines delineate the base of the canopy

environments. To illustrate these cascading effects, how exotic species replacements would alter both within-canopy APAR and non-intercepted PPFD reaching the understory during the growing season was assessed (Fig. 6). These results reveal dramatic differences between the two scenarios. In SE, *Pt* absorbed 26% more light than the native *Ps*, reducing understory light availability by nearly half (Fig. 6d), which would significantly alter understory species composition and ecosystem

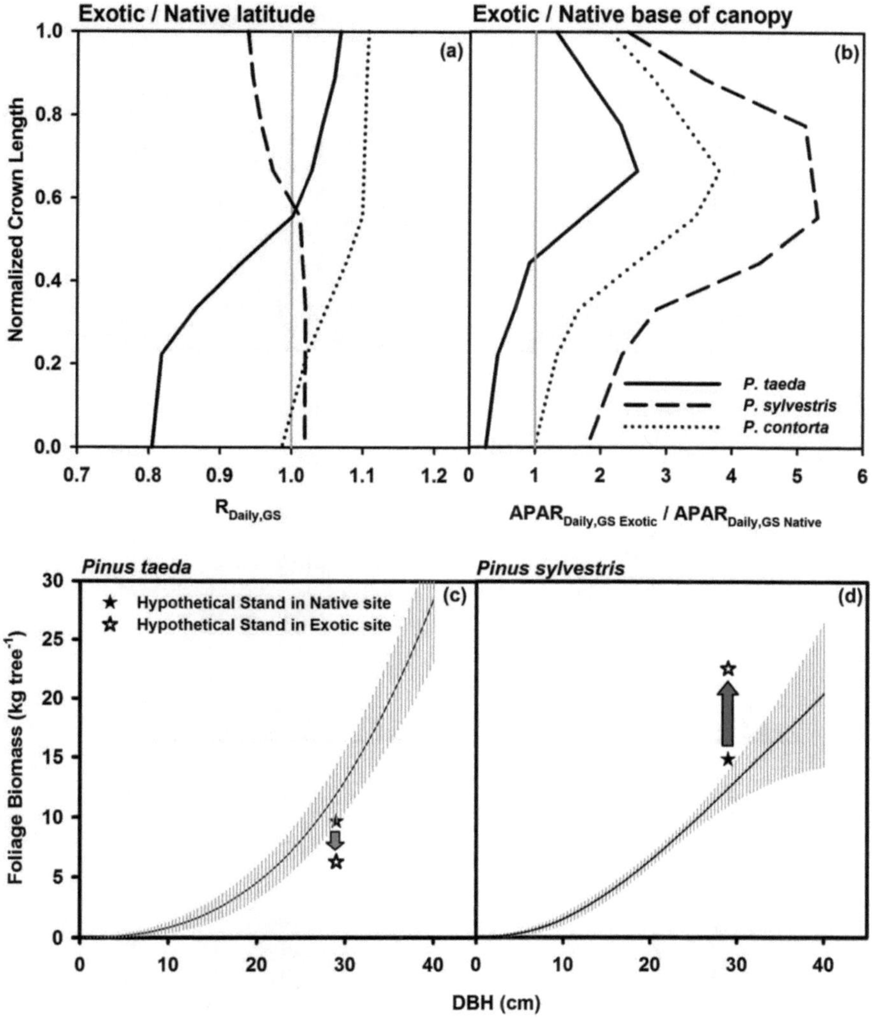

Fig. 5 Comparisons of vertical distribution of absorbed photosynthetically active radiation (APAR) within crown. (**a**) Vertical distribution of daily growing season average APAR for each species in an exotic relative to its native latitude, defined as $R_{\text{Daily,GS}}$. The ratio was calculated as $R_{\text{Daily,GS}} = \dfrac{\sum_{\text{GS,Exotic}} \text{APAR}_{\text{Exotic}} / N_{\text{Exotic}}}{\sum_{\text{GS,Native}} \text{APAR}_{\text{Native}} / N_{\text{Native}}}$ where N represents the number of days during the growing season (GS) period of either exotic or native sites. The results are based on representative PAR derived from theoretical PAR with varying sky conditions. Gray lines indicate no change of the daily growing season average APAR in the exotic compared to its native latitude. (**b**) Vertical distribution of APAR of each species in its exotic location relative to the minimum APAR required by the bottom branches at the native site. (**c, d**) Literature-based allometric relations between foliage biomass and diameter at 1.3 m from the ground or breast height (DBH), the position of the average tree of the hypothetical *P. taeda* and *P. sylvestris* stands in their native latitude, and the requisite change of foliage biomass necessary to generate a similar APAR for bottom branches in the exotic latitude as in the native latitude

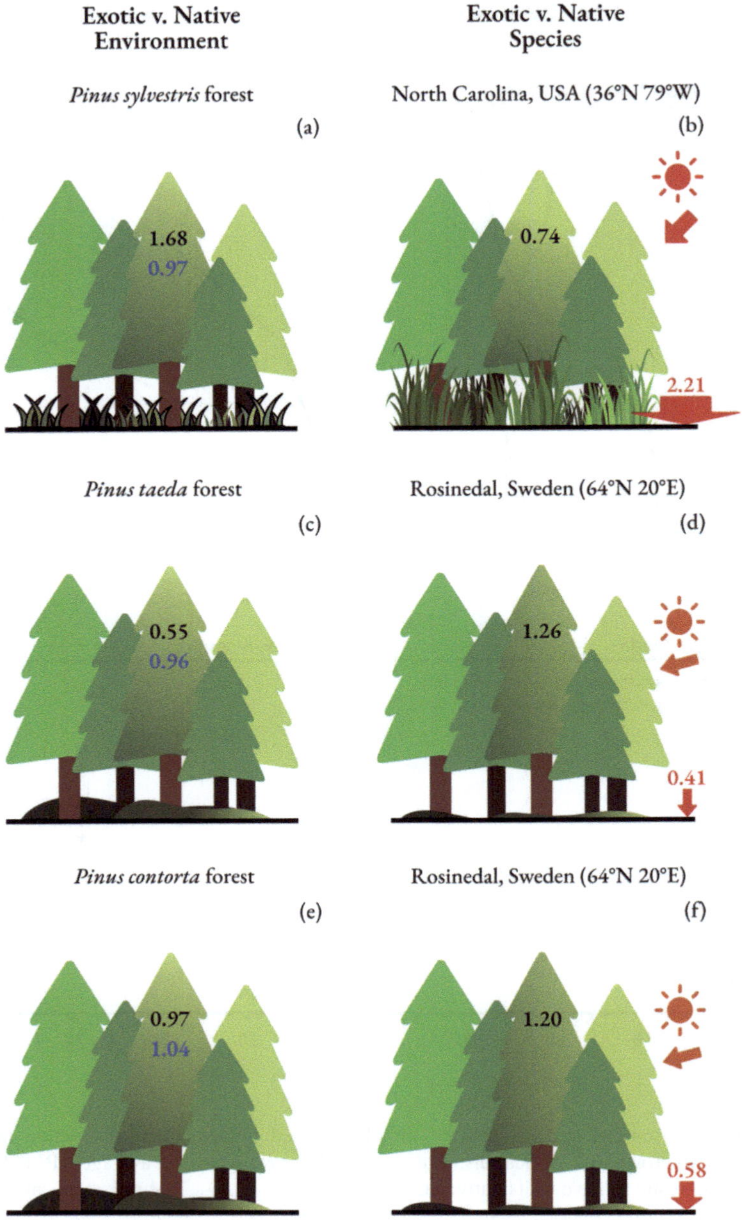

Fig. 6 Summary of findings. *Case I* (**a**, **c**) Comparison of total canopy absorption by each species in the exotic versus native latitude. Ratios are the amount of absorbed photosynthetically active radiation (APAR) in the canopy at the exotic latitude normalized by that at the native latitude. Ratios in black are calculated from the respective growing season days in each location (i.e., taking into account the difference in growing season length between the two locations). Ratios in blue are computed from the growing season days of the exotic site. (**b**, **d**) Comparison of total canopy absorption in each latitude of exotic versus native species. Ratios are the amount of APAR in the canopy of the exotic species normalized by that at the native species. Ratios in red are for non-intercepted PAR that will be available for the understory in stands of the exotic versus native species. *Case II*: comparisons of APAR of *P. contorta*, originally from British Columbia, growing in northern Sweden relative to its native environment (**e**) or the native *P. sylvestris* (**f**)

function. Conversely, introducing *Ps* to NC would create a more open light environment. *Ps* absorbed only 74% of the light captured by native *Pt*, more than doubling the light supply reaching the forest floor (Fig. 6b).

The extra light reaching the understory would stimulate growth and intensify competition for soil resources with the canopy species (Canham 1988; McGuire et al. 2001). Real-world evidence supports these simulation predictions when *P. echinata* is considered as an analog for *Ps*, given their similar needle and shoot characteristics (Lawson 1990; Thérézien et al. 2007). Plantations of *P. echinata* consistently require intensive understory management to maintain acceptable growth rates (Grano 1970; Shelton and Murphy 1991), suggesting that this species allows sufficient light penetration to promote vigorous understory growth. This is in line with predictions that *Ps*-type crown architecture would create more open understory conditions in NC. Furthermore, in areas where the ranges of *Pt* and *P. echinata* overlap, *Pt* consistently outperforms the local *Ps*-analog (Chandler et al. 1943; Meade 1951; Williston 1972). This superior performance matches the simulations showing higher light absorption by *Pt* compared to *Ps*. The competitive advantage is particularly evident during stand establishment, as *Pt* quickly overtop *P. echinata* when growing in mixed stands (Cain 1990). These field observations validate the modeling predictions and demonstrate how crown architecture differences translate into measurable competitive advantages and management challenges in real forest ecosystems.

5.3 Simulated Impact of Shoot Architecture on Stand Performance at Two Proximal Latitudes

It has long been observed that exotic species can maintain and even increase production relative to that at their original latitudes, and relative to native species, yet mechanistic explanations for these observations remain elusive (Gurevitch et al. 2011; Hierro et al. 2005). Several mechanisms have been proposed to explain higher stem wood production in species' exotic relative to native environment and in exotic versus native species. These include reduced pressure from nematodes, insects and diseases (Keane and Crawley 2002), more advantageous associations with soil microbiota (Gundale et al. 2014), or physiological factors affecting canopy photosynthesis and carbon allocation patterns between above- and belowground compartments (Jansons et al. 2013; Wilsey and Polley 2006). Additionally, intensive genetic selection and genotype-environment matching in exotic plantation systems can create substantial productivity advantages through optimized clonal deployment and silvicultural practices that may not be achievable in native forest contexts (Binkley et al. 2017).

Increased APAR by exotic canopies is one potential explanation for their success. Daily increases in growing season APAR for the exotic *Pc* stand in SE were persistent throughout the entire crown, with maximum increases of 11% at the top and minimal reductions of less than 1% at the bottom (Figs. 5 and 6). Moreover, if each canopy layer in an exotic location is assumed to require at least the same mean growing season daily APAR as the bottom crown layer receives in the native site, then the entire crown of *Pc* in SE exceeded the minimum APAR threshold established by its native BC environment (Fig. 5b, dotted line).

Because growing season length was similar in BC and SE, the total canopy APAR of *Pc* was comparable at both locations (Fig. 6e). However, when compared to the simulated canopy APAR of native *Ps* in SE, *Pc* showed 20% greater light interception (Fig. 6f), an advantage that was insufficient to explain the observed 36% growth superiority of the introduced species (Elfving and Norgren 1993). This initial comparison, however, was based on hypothetical *Pc* stands situated on the maximum size-density line and did not represent a typical structure of a managed stand. A more realistic assessment emerges when evaluating *Pc* performance under standard silvicultural regimes with relative stand densities ranging from 0.35 to 0.4. Under these conditions, *Pc* demonstrated a clear structural advantage: it intercepted approximately 50% more light than its *Ps* counterpart in SE (not shown)—a difference sufficient to explain the observed growth superiority. This enhanced light interception by *Pc* carries ecological consequences beyond tree growth. The ~40% reduction in PAR reaching the understory in *Pc* stands (Fig. 6f; or −50% if scaled to a typical management density) would significantly alter forest floor conditions, resulting in understory communities with reduced species diversity and fewer shade-intolerant species (Engelmark et al. 2001; Nilsson et al. 2008).

6 Summary

From the tips of fine roots to the margins of leaves, the structure of tree organs is designed to coordinate, under given environmental conditions (short-term) and constraints (long-term), the supply to the mesophyll of two resources used in carbohydrate production: Carbon dioxide and the products of light energy absorbed by leaves. The former flux is largely determined by the degree of stomatal opening, which depends on the structure and function of the soil-plant-atmosphere pathway for water flow. The latter is largely determined by the capacity of shoots to intercept light efficiently, which depends on the architectural properties of shoots and how these vary spatially in the canopy. To determine at which scales carbohydrate production may be optimized, more research is needed to quantify spatially (e.g., at different distances from the outer crown envelope or upper canopy surface) and temporally (e.g., diurnally, over drying cycles, over growing seasons) the coordination between hydraulic and light absorption properties. The chapter began with a review of the relationships between shoot structure and hydraulic properties. Building on the coordination of traits responsible for light interception and water delivery, the chapter then focused on the variation in properties affecting light absorption through the canopy. Crown and canopy architecture governs total canopy light absorption at any given stand density and tree size, thereby placing an upper limit on forest productivity. By integrating observed crown and stand attributes, modeling APAR permits evaluation of how solar elevation—representing both time and latitude—influences total light capture and its vertical distribution in theoretical pine forests across the northern hemisphere. To explore this from a comparative perspective, APAR was simulated in stands of two wide-ranging species with similar tree size-density relationships: *P. taeda* in southeastern USA (North Carolina) and *P. sylvestris* in central Sweden, both found and established on relatively nutrient-poor sites. The simulation results suggest that neither species (or a species with similar shoot/crown structure) would perform well at the other's latitude, albeit for different reasons. In central Sweden, *P. taeda* would experience a 15% reduction in lower-crown light absorption, necessitating a 31% decrease in leaf biomass— a reduction that exceeds the species' known allometric plasticity. Conversely, the higher solar elevation in North Carolina would paradoxically cause *P. sylvestris* to absorb less total canopy light than in its native environment, while simultaneously allowing 2.2 times more light to penetrate through the canopy. To match the APAR performance of *P. taeda* in North Carolina, *P. sylvestris* requires a 52% increase in canopy leaf biomass—again exceeding its allometric constraints.

The same modeling framework was applied to a practical case study examining the performance of exotic *P. contorta* versus native *P. sylvestris* in Sweden. The superior growth rate of *P. contorta* was directly attributable to its more efficient crown architecture, which enabled greater canopy light absorption than the native species. This enhanced light capture by *P. contorta*, however, came with ecological trade-offs: the substantially reduced light penetration to the understory corresponded with observed declines in both diversity and abundance of shade-intolerant species. Critically, *P. contorta* achieved this competitive advantage while maintaining sufficient light availability for its lower branches—a key factor enabling its successful establishment and higher production rates in Swedish forests.

Theoretical analyses reveal how crown architecture influences species distribution patterns across latitudinal gradients. The crown characteristics such as those of *P. sylvestris*, which allow deeper light penetration and support moderate understory development in northern latitudes, would transmit substantially more light in temperate regions. This increased light transmission would facilitate the establishment of highly competitive understory layers. Conversely, crown traits that are typical of fast-growing temperate species such as *P. taeda* create a different set of constraints. These species promote efficient canopy light absorption, yet their architecture restricts light availability to their own lower branches, especially in latitudes where solar angle is low during the growing season. Such structural limitations may prevent the poleward migration of temperate species, even as climate warming creates increasingly favorable physiological conditions for their growth (Hughes 2000; McCarty 2001; Woodward 1987).

The case of exotic *P. contorta* in Sweden exemplifies how crown architecture can drive productivity differences between species. Its canopy shows more efficient light absorption than native *P. sylvestris*, likely directly contributing to greater fiber production that has been demonstrated earlier (Elfving and Norgren 1993; Elfving et al. 2001; Gundale et al. 2014; Norgren 1995). This highlights a broader principle:

crown traits vary not only between species but also within species, creating opportunities to optimize canopy architecture for specific latitudes and thereby enhance APAR and overall growth (Emhart et al. 2007; Laforest-Lapointe et al. 2014). Previous theoretical studies have demonstrated that crown shape adapts for efficient light utilization and varies predictably with latitude (Kuuluvainen 1992). However, other structural factors, such as shoot-level clumping, also significantly influence light absorption and may differ greatly even within a single latitude (Thérézien et al. 2007). Despite recognition of these factors, few studies have examined how foliage arrangement within the crown interacts with seasonal changes in solar elevation to affect light interception and the vertical distribution of APAR across latitudinal gradients. Such analyses are needed for quantifying species productivity potential, which depends on both total APAR and its distribution within the canopy, as well as the species' capacity to modify crown attributes within allometric constraints.

7 Challenges

During the synthesis of information for this chapter, several topics emerged as insufficiently studied to allow generalization.

1. The relationship between shoot hydraulic properties and structural properties that affect light interception remains poorly understood. Exploring this link across multiple species along vertical gradients of both light availability and water potential—two key factors that jointly influence shoot structure, hydraulics, and photosynthetic capacity—would provide valuable empirical data on trait coordination. Such research would also inform the development of more realistic canopy models by elucidating how these interconnected physiological and structural parameters vary within forest canopies.
2. The rate at which newly produced shoots acclimate to environmental changes—whether to gradual changes such as climate shifts or to abrupt changes such as sudden canopy opening—requires investigation. Specifically, research is needed on how quickly shoots adjust key stuctural traits including STAR, Hv, and other properties that regulate both water supply to and light interception by needles.
3. Genetic variation in crown dimensions is already considered in silviculture operations to enhance forest production (Cortés et al. 2020; Lai et al. 2017; Mazanec and Mason 1993). Similarly, genetic variation in shoot structure within species may prove valuable for matching genotypes to specific planting environments, particularly in assisted migration programs as species ranges shift poleward with climate warming. However, the simplified simulations presented above are insufficient to justify such management interventions. More comprehensive analyses incorporating empirical data on genetic variation in shoot structure are needed before these approaches can be confidently implemented in forest management practices.

Appendix

Simulating Absorbed Photosynthetically Active Radiation (APAR)

The simulations were designed to compare light interception patterns across three coniferous species (*P. taeda*, *P. sylvestris*, and *P. contorta*) in different geographic locations by modeling APAR during the active growing season. The analyses employed theoretical maximum photosynthetically active radiation calculations based on solar geometry and atmospheric conditions, with direct and diffuse radiation components estimated for both clear and overcast days. Stand structure was parameterized using empirical data on tree dimensions, crown characteristics, and surface areas, while shoot-level optical properties including clumping factors, transmission coefficients, and needle angle distributions were incorporated from field measurements. To ensure valid cross-species comparisons, stands with similar density and mean tree size were selected despite the species' different native environments, allowing the simulations to isolate the effects of species-specific structural characteristics on canopy light absorption and vertical distribution patterns.

Atmospheric Variables

Absorbed photosynthetically active radiation was estimated for the active growing season delineated by the first day after daily mean temperature exceeded +5 °C for five consecutive days until it dropped below +5 °C for five consecutive days (Table 1; Mäkelä et al. 2006). For the comparative analyses of light interception, theoretical maximum photosynthetic photo flux density ($PPFD_{max}$) was used as the incident radiation. Half-hourly $PPFD_{max}$ was calculated from solar zenith angle and atmospheric transmittance of 0.7 (Campbell and Norman 1998). The direct and diffuse components of incoming radiation were estimated according to Liu and Jordan (1960). APAR was then estimated for direct and diffuse PPFD separately for clear days, while all incoming radiation was assumed to be diffuse on overcast days. To anchor the theoretical results to reality, daily sky conditions over an average year were estimated based on 8 years (2002–2009) of measured PPFD data from representative sites in both NC and SE. The number of cloudy days was similar at the two sites ($p > 0.1$ in each of five non-overlapping intra-annual periods). Therefore, all analyses in this study were based on a single annual sequence derived from the statistics of clear and overcast days during the 8 years, with 48% of days designated as clear. The same approach to calculate maximum PPFD ($PPFD_{max}$) was employed using the site coordinates of BC (Table 1) with growing season length determined based on daily temperature data obtained from Wang et al. (2016). The profile $PPFD_{max}$ is illustrated in Fig. 2b.

Parameterization of Canopy and Shoot Structure

Tree height, crown length and DBH were available for 90 *Pt* trees in NC and 307 *Ps* trees in SE and were used to generate branch and stem surface areas and crown dimensions (i.e., crown width assuming ellipsoidal shape). Branch surface area was estimated from a logarithmic function of DBH and crown ratio for *Pt* (Baldwin et al. 1997; McCarthy et al. unpublished). For *Ps*, branch surface area was estimated from a relation of branch dry weight to DBH (Lim et al. 2015) and branch surface area versus branch dry weight (Halldin 1985; Ilvesniemi and Liu 2001). The vertical profile of branch surface area of *Pt* was taken from a nearby site (Kim et al. 2011), and that of *Ps* from a similar stand (Halldin 1985). Stem surface area was determined from segmented polynomial tapering equations (Sharma and Burkhart 2003) using diameter measurements along stems from the nearby site in NC (Kim et al. 2011) and harvested trees in SE (Lim et al. 2015). Crown radii were derived from equations presented in Baldwin et al. (1997) for *Pt* and Rautiainen and Stenberg (2005) for *Ps*. Hemi-surface L of *Pt* was estimated from monthly measurements using an optical sensor, converted to projected L (L_p) using a factor of 1/2 (Stenberg 1996b). A correction factor of 0.59 (Thérézien et al. 2007) was applied to L_p to account for foliage clumping (Stenberg 1996b). For *Ps*, annual L_p values estimated from allometric functions and annual production and litterfall in SE were used (Lim et al. 2015), matched to seasonal dynamics of L_p of similar-age *Ps* stand on a similar site (Rautiainen et al. 2012). The vertical distribution of L_p was obtained from the harvested individuals (Lim et al. 2015). Structural attributes for *Pc* such as height, crown length, crown width, wood surface areas, foliar vertical distribution, and biomass components were generated from published data (Gary 1978; Lambert et al. 2005; Pearson et al. 1984; Sampson and Smith 1993; Sattler and LeMay 2011; Wonn 2001).

The model requires specification of shoot characteristics including shoot clumping factor (CF), shoot transmission coefficient (c) and needle angle distribution (*NAD*). The CF value ranges from 0 (highly clumped) to 1 (no clumping). *NAD* refers to the average ratio of projected areas of canopy elements on horizontal and vertical surfaces for ellipsoidal needle angle distribution. In more vertically oriented leaves, *NAD* is less than 1, while NAD is greater than 1 for more horizontally oriented ones (Barclay 2001; Thomas and Winner 2000; Wirth et al. 2001). c is the mean gap area within the projected shoot envelope area (Stenberg 1998). The CF and *NAD* parameters of *Pt* were assumed to be similar to those measured at a nearby site (Kim et al. 2011; Thérézien et al. 2007). CF and *NAD* of *Ps* were obtained from measurements of shoot structure in a *Ps* stand in central Finland (Stenberg et al. 2001). Shoot structural parameters of *Pc* trees were estimated from needle angle, fascicles per shoot, and projected leaf area measurements taken along the crowns in SE near the source of *Ps* data. CF and c were calculated using the approach outlined in Thérézien et al. (2007). These estimates may not represent a continuous *Pc* canopy, but they should provide a rough estimate of the differences in shoot characteristics between *Pc* and *Ps*. The final canopy structure of the three species is presented in Fig. 3.

References

Baldwin VC, Peterson KD, Burkhatt HE, Amateis RL, Dougherty PM (1997) Equations for estimating loblolly pine branch and foliage weight and surface area distributions. Can J For Res 27:918–927

Barclay H (2001) Distribution of leaf orientations in six conifer species. Botany 79:389–397

Bertin S, Palmroth S, Kim H-S, Perks M, Mencuccini M, Oren R (2011) Modelling understorey light for seedling regeneration in continuous cover forestry canopies. Forestry 84:397–409

Binkley D, Campoe OC, Gspalti M, Forrester DI (2013) Light absorption and use efficiency in forests: why patterns differ for trees and stands. For Ecol Manag 288:5–13

Binkley D, Campoe OC, Alvares C, Carneiro RL, Cegatta I, Stape JL (2017) The interactions of climate, spacing and genetics on clonal eucalyptus plantations across Brazil and Uruguay. For Ecol Manag 405:271–283

Björkman O, Demmig B, Andrew TJ (1988) Mangrove photosynthesis: response to high-irradiance stress. Aust J Plant Physiol 15:43–61

Cain MD (1990) Incidental observations on the growth and survival of loblolly and shortleaf pines in an even-aged natural stand. South J Appl For 14:81–84

Campbell GS, Norman JM (1998) An introduction to environmental biophysics. Springer-Verlag

Canham CD (1988) Growth and canopy architecture of shade-tolerant trees: response to canopy gaps. Ecology 69(3):786–795

Chandler RF Jr, Schoen PW, Anderson DA (1943) Relation between soil types and the growth of loblolly pine and shortleaf pine in East Texas. J For 41:505–506

Chen SG, Shao BY, Impens I, Ceulemans R (1994) Effects of plant canopy structure on light interception and photosynthesis. J Quant Spectrosc Radiat Transf 52:115–123

Chen JM, Rich PM, Gower ST, Norman JM, Plummer S (1997) Leaf area index of boreal forests: theory, techniques, and measurements. J Geophys Res 102(D24):29429–29443

Chin ARO, Gessler A, Guzmán-Delgado P, Manzanedo RD, Saurer M, Ris Lambers JH (2025) Rainwater uptake in conifer twigs: five experiments tell a story of absorption, storage, and transport. J Exp Bot 76(12):3515–3526. https://doi.org/10.1093/jxb/eraf087

Cortés AJ, Restrepo-Montoya M, Bedoya-Canas LE (2020) Modern strategies to assess and breed forest tree adaptation to changing climate. Front Plant Sci 11:583323

Critchfield WB (1980) Genetics of lodgepole pine. USDA Forest Service Research Paper WO-37

DeLucia EH, Maherali H, Carey EV (2000) Climate-driven changes in biomass allocation in pines. Glob Chang Biol 6:587–593

Domec J-C, Noormets A, King JS, Sun G, McNulty SG, Gavazzi M, Boggs JL, Treasure EA (2009) Decoupling the influence of leaf and root hydraulic conductances on stomatal conductance and its sensitivity to vapor pressure deficit as soil dries in a drained loblolly pine plantation. Plant Cell Environ 32:980–991

Domec J-C, Schafer K, Oren R, Kim H, McCarthy H (2010) Variable conductivity and embolism in roots and branches of four contrasting tree species and their impacts on whole-plant hydraulic performance under future atmospheric CO2 concentration. Tree Physiol 30:1001–1015

Domec J-C, King JS, Ward E, Oishi AC, Palmroth S, Radecki A, Bell DM, Miao G, Gavazzi M, Johnson DM, McNulty SG, Sun G, Noormets A (2015) Conversion of natural forests to managed forest plantations decreases tree resistance to prolonged droughts. For Ecol Manag 355:58–71

Domec J.-C, Palmroth S, Oren R (2016) Effects of Pinus taeda leaf anatomy on vascular and extravascular leaf hydraulic conductance as influenced by N-fertilization and elevated CO2 .Journal of Plant Hydraulics 3e007

Elfving B, Norgren O (1993) Volume yield superiority of lodgepole pine compared to scots pine in Sweden. In: Lindgren D (ed) Pinus contorta—from untamed forest to domesticated crop. Proceedings of a meeting with IUFRO WP 2.02.06 and Frans Kempe symposium. Department of Forest Genetics and Plant Physiology, Swedish University of Agricultural Sciences, Report 11, pp 69–80

Elfving B, Ericsson T, Rosvall O (2001) The introduction of lodgepole pine for wood production in Sweden—a review. For Ecol Manag 141:15–29

Emhart VI, Martin TA, White TL, Huber DA (2007) Clonal variation in crown structure, absorbed photosynthetically active radiation and growth of loblolly pine and slash pine. Tree Physiol 27(3):421–430

Engelmark O, Sjöberg K, Andersson B, Rosvall O, Ågren GI, Baker WL, Barklund P, Björkman C, Despain DG, Elfving B, Ennos RA, Karlman L, Knecht MF, Knight DH, Ledgard NJ, Lindelöw Å, Nilsson C, Peterken GF, Sörlin S, Sykes MT (2001) Ecological effects and management aspects of an exotic tree species: the case of lodgepole pine in Sweden. For Ecol Manag 141:3–13

Ewers BE, Oren R, Sperry S (2000) Influence of nutrient versus water supply on hydraulic architecture and water balance in *Pinus taeda*. Plant Cell Environ 23(10):1055–1066

Ewers BE, Oren R, Kim H-S, Bohrer G, Lai C-T (2007) Effects of hydraulic architecture and spatial variability in light on mean stomatal conductance of tree branches and crowns. Plant Cell Environ 30:483–496

Farnden C (1996) Stand density management diagrams for lodgepole pine, white spruce and interior Douglas-fir. Information Report BC-X-360, Canadian Forest Service

Gamon JA, Pearcy RW (1989) Leaf movement stress avoidance and photosynthesis in *Vitis californica*. Oecologia 79:475–481

Gary HL (1978) The vertical distribution of needles and branchwood in thinned and unthinned 80-year-old lodgepole pine. Northwest Sci 52:303–309

Grano CX (1970) Small hardwoods reduce growth of pine overstory. USDA Forest Service Southern Research Paper SO-55

Grönlund L, Hölttä T, Mäkelä A (2016) Branch age and light conditions determine leaf-area-specific conductivity in current shoots of Scots pine. Tree Physiol 36:994–1006

Gundale MJ, Kardol P, Nilsson MC, Nilsson U, Lucas RW, Wardle DA (2014) Interactions with soil biota shift from negative to positive when a tree species is moved outside its native range. New Phytol 204:364–378. https://doi.org/10.1111/nph.12699

Gurevitch J, Fox GA, Wardle GM, Inderjit, Taub D (2011) Emergent insights from the synthesis of conceptual frameworks for biological invasions. Ecol Lett 14:407–418

Hacke UG, Sperry JS, Ewers BE, Schäfer KVR, Oren R, Ellsworth DS (2000) Influence of soil porosity on water use in *Pinus taeda*. Oecologia 124:495–505

Halldin S (1985) Leaf and bark area distribution in a pine forest. In: Hutchinson BA, Hicks BB (eds) The forest atmosphere interaction. Springer Netherlands, pp 39–58

Harrington TB, Edwards MB (1996) Structure of mixed pine and hardwood stands 12 years after various methods and intensities of site preparation in the Georgia Piedmont. Can J For Res 26:1490–1500

Hierro JL, Maron JL, Callaway RM (2005) A biogeographical approach to plant invasions: the importance of studying exotics in their introduced and native range. J Ecol 93:5–15

Honkanen T, Haukioja E, Kitunen V (1999) Responses of Pinus sylvestris branches to simulated herbivory are modified by tree sink/source dynamics and by external resources. Funct Ecol 13:126–140

Horn HS (1971) The adaptive geometry of trees. Princeton University Press

Hubbard RM, Bond BJ, Senock RS, Ryan MG (2002) Effects of branch height on leaf gas exchange, branch hydraulic conductance and branch sap flux in open-grown ponderosa pine. Tree Physiol 22:575–581. https://doi.org/10.1093/treephys/22.8.575

Hughes L (2000) Biological consequences of global warming: is the signal already apparent? Trends Ecol Evol 15:56–61

Ilvesniemi H, Liu C (2001) Biomass distribution in a young scots pine stand. Boreal Environ Res 6:3–8

Jansons A, Sisenis L, Neimane U, Rieksts-Riekstins J (2013) Biomass production of young lodgepole pine (*Pinus contorta* var. *latifolia*) stands in Latvia. iForest 6:10–14

Johnson DM, Wortemann R, McCulloh KA, Jordan-Meille L, Ward E, Warren JM, Palmroth S, Domec J-C (2016) A test of the hydraulic vulnerability segmentation hypothesis in angiosperm and conifer tree species. Tree Physiol 36:983–993

Katz C, Oren R, Schulze E-D, Milburn JA (1989) Uptake of water and solutes through twigs of *Picea abies* (L.) Karst. Trees 3(3):3–37

Keane RM, Crawley MJ (2002) Exotic plant invasions and the enemy release hypothesis. Trends Ecol Evol 17(4):164–170

Kellomäki S, Oker-Blom P (1983) Canopy structure and light climate in a young Scots pine stand. Silva Fenn 17(1983):1–21

Kim H-S, Oren R, Hinckley TM (2008) Actual and potential transpiration and carbon assimilation in an irrigated poplar plantation. Tree Physiol 28:559–577

Kim H-S, Palmroth S, Thérézien M, Stenberg P, Oren R (2011) Analysis of the sensitivity of absorbed light and incident light profile to various canopy architecture and stand conditions. Tree Physiol 30:31–47

Knyazikhin Y, Martonchik JV, Myneni RB, Diner DJ, Running SW (1998) Synergistic algorithm for estimating vegetation canopy leaf area index and fraction of absorbed photosynthetically active radiation from MODIS and MISR data. J Geophys Res 103(D24):32257–32275. https://doi.org/10.1029/98JD02462

Kuuluvainen T (1992) Tree architectures adapted to efficient light utilization: is there a basis for latitudinal gradients? Oikos 65:275–284

Laforest-Lapointe I, Martínez-Vilalta J, Retana J (2014) Intraspecific variability in functional traits matters: case study of scots pine. Oecologia 175:1337–1348

Lai M, Dong L, Yi M, Sun S, Zhang Y, Fu L, Xu Z, Lei L, Leng C, Zhang L (2017) Genetic variation, heritability and genotype × environment interactions of resin yield, growth traits and morphologic traits for *Pinus elliottii* at three progeny trials. Forests 8(11):409

Lambert M-C, Ung C-H, Raulier F (2005) Canadian national tree aboveground biomass equations. Can J For Res 35:1996–2018

Långström B, Ericsson CA, Gref R (1992) Induced defence reaction in scots pine following stem attacks by *Tomicus piniperda*. Ecography 15:318–327

Lawson ER (1990) Shortleaf pine. In: Burns RM, Honkala BH (eds) Silvics of North America, vol 1. US Department of Agriculture, Forest Service, Agriculture Handbook 654

Leverenz JW, Hinckley TM (1990) Shoot structure, leaf area index and productivity of evergreen conifer stands. Tree Physiol 6:135–149

Lim H, Oren R, Palmroth S, Tor-ngern P, Mörling T, Näsholm T, Lundmark T, Helmisaari H, Leppälammi-Kujansuu J, Linder S (2015) Inter-annual variability of precipitation constrains the production response of boreal *Pinus sylvestris* to nitrogen fertilization. For Ecol Manag 348:31–45

Liu BY, Jordan RC (1960) The interrelationship and characteristic distribution of direct, diffuse and total solar radiation. Sol Energy 4:1–19

Maier CA, Johnson KH, Butnor J, Kress LW, Anderson PH (2002) Branch growth and gas exchange in 13-year-old loblolly pine (*Pinus taeda*) trees in response to elevated carbon dioxide concentration and fertilization. Tree Physiol 22:1093–1106

Mäkelä A, Kolari P, Karimäki J, Nikinmaa E, Perämäki M, Hari P (2006) Modelling five years of weather-driven variation of GPP in a boreal forest. Agric For Meteorol 139:382–398

Mäkelä A, Grönlund L, Schiestl-Aalto P, Kalliokoski T, Hölttä T (2019) Current-year shoot hydraulic structure in two boreal conifers—implications of growth habit on water potential. Tree Physiol 39(12):1995–2007

Mazanec RA, Mason ML (1993) Genetic variation in *Eucalyptus diversicolor* F. Muell. in western Australia and potential gains from selection. For Ecol Manag 62(1–4):285–301

McCarty JP (2001) Ecological consequences of recent climate change. Conserv Biol 15:320–331

McCulloh K, Sperry JS, Adler FR (2003) Water transport in plants obeys Murray's law. Nature 421:939–942

McCulloh KA, Domec JC, Johnson DM, Smith DD, Meinzer FC (2019) A dynamic yet vulnerable pipeline: integration and coordination of hydraulic traits across whole plants. Plant Cell Environ 42:2789–2807

McGuire JP, Mitchell RJ, Moser EB, Pecot SD, Gjerstad DH, Hedman CW (2001) Gaps in a gappy forest: plant resources, longleaf pine regeneration, and understory response to tree removal in longleaf pine savannas. Can J For Res 31(5):765–778

Meade FM (1951) Forest plantations in Arkansas—plantations in the Arkansas Ozarks and experimental planting on a Coastal Plain site in Southwest Arkansas. Arkansas Agricultural Experiment Station Bulletin 512

Mencuccini M, Grace J (1995) Climate influences the leaf area/sapwood area ratio in scots pine. Tree Physiol 15:1–10

Mencuccini M, Rosas T, Rowland L, Choat B, Cornelissen H, Jansen S, Kramer K, Lapenis A, Manzoni S, Niinemets Ü, Reich P, Schrodt F, Soudzilovskaia N, Wright IJ, Martínez-Vilalta J (2019) Leaf economics and plant hydraulics drive leaf: wood area ratios. New Phytol 224(4):1544–1556

Miller EE, Norman JM (1971) A sunfleck theory for plant canopies. I. Lengths of sunlit segments along a transect. Agron J 63:735–738

Monteith JL (1977) Climate and efficiency of crop production in Britain. Philos Trans R Soc Ser B 281:277–294

Niinemets Ü (2010) A review of light interception in plant stands from leaf to canopy in different plant functional types and in species with varying shade tolerance. Ecol Res 25:693–714

Niinemets Ü, Kull O (1995) Effects of light availability and tree size on the architecture of assimilative surface in the canopy of *Picea abies*: variation in shoot structure. Tree Physiol 15:791–798

Niinemets Ü, Ellsworth DS, Lukjanova A, Tobias M (2002a) Dependence of needle architecture and chemical composition on canopy light availability in three North American *Pinus* species with contrasting needle length. Tree Physiol 22(11):747–761

Niinemets Ü, Cescatti A, Lukjanova A, Tobias M, Truus L (2002b) Modification of light-acclimation of *Pinus sylvestris* shoot architecture by site fertility. Agric For Meteorol 115(1–2):83–100

Niinemets Ü, Tobias M, Cescatti A, Sparrow A (2006) Size-dependent variation in shoot light-harvesting efficiency in shade-intolerant conifers. Int J Plant Sci 167(1):19–32

Nilson T (1971) Theoretical analysis of frequency of gaps in plant stands. Agric Meteorol 8(1):25–38

Nilson T (1999) Inversion of gap frequency data in forest stands. Agric For Meteorol 98-99:437–448

Nilson T, Ross J (1996) Modeling radiative transfer through forest canopies: implications for canopy photosynthesis and remote sensing. In: Gholz HL, Nakane K, Shimoda H (eds) The use of remote sensing in the modeling of forest productivity. Kluwer Academic Publishers, pp 23–60

Nilsson MC, Wardle DA (2005) Understory vegetation as a forest ecosystem driver: evidence from the northern Swedish boreal forest. Front Ecol Environ 3:421–428

Nilsson C, Engelmark O, Cory J, Forsslund A, Carlborg E (2008) Differences in litter cover and understory flora between stands of introduced lodgepole pine and native Scots pine in Sweden. For Ecol Manag 255:1900–1905

Norgren O (1995) Growth differences between Pinus sylvestris and Pinus contorta. Doctoral dissertation, Department of Silviculture, Swedish University of Agricultural Sciences

Oker-Blom P, Kellomäki S (1983) Effect of grouping of foliage on the within-stand and within-crown light regime: comparison of random and grouping canopy models. Agric Meteorol 28:143–155

Oker-Blom P, Smolander H (1988) Shoot silhouette to needle area ratio in scots pine. For Sci 34:894–896

Oker-Blom P, Pukkala T, Kuuluvainen T (1989) Relationship between radiation interception and photosynthesis in forest canopies: effect of stand structure and latitude. Ecol Model 49:73–87

Oren R, Schulze E-D, Matyssek R, Zimmermann R (1986) Estimating photosynthetic rate and annual carbon gain in conifers from specific leaf weight and leaf biomass. Oecologia 70:187–193

Palmroth S, Palva L, Stenberg P, Kotisaari A (1999) Fine scale measurement and simulation of penumbral radiation formed by a pine shoot. Agric For Meteorol 95:15–25
Pearcy RW (1989) Radiation and light measurements. In: Pearcy RW, Ehleringer JR, Mooney HA, Rundel PW (eds) Plant physiological ecology. Springer, Dordrecht
Pearson JA, Fahey TJ, Knight DH (1984) Biomass and leaf area in contrasting lodgepole pine forests. Can J For Res 14(2):259–265
Pisek J, Borysenko O, Janoutová R, Homolová L (2023) Estimation of coniferous shoot structure by high precision blue light 3D photogrammetry scanning. Remote Sens Environ 291:113568
Rautiainen M, Stenberg P (2005) Simplified tree crown model using stand forest mensuration data for Scots pine. Agric For Meteorol 128:123–129
Rautiainen M, Heiskanen J, Korhonen L (2012) Seasonal changes in canopy leaf area index and MODIS vegetation products for a boreal forest site in Central Finland. Boreal Environ Res 17:72–84
Reineke LH (1933) Perfecting a stand density index for even-aged forests. J Agric Res 46:627–638
Ryel R, Falge E, Joss U, Geyer R, Tenhunen J, Köstner B et al (2001) Penumbral and foliage distribution effects on *Pinus sylvestris* canopy gas exchange. Theor Appl Climatol 68:109–124
Sampson DA, Smith FW (1993) Influence of canopy architecture on light penetration in lodgepole pine forests. Agric For Meteorol 64:63–79
Sattler DF, LeMay V (2011) A system of nonlinear simultaneous equations for crown length and crown radius for the forest dynamics model SORTIE-ND. Can J For Res 41(8):1567–1576
Schäfer KVR, Oren R, Ellsworth DS, Lai C-T, Herrick JD, Finzi AC, Richter DD, Katul GG (2003) Exposure to an enriched CO2 atmosphere alters carbon assimilation and allocation in a pine forest ecosystem. Glob Chang Biol 9:1378–1400
Scott JS (1997) An examination of crown characteristics, growth efficiency, and bole wood maturation in loblolly pine of various stand densities grown in the absence of interspecific competition. Master's thesis, School of Forest Resources, University of Georgia
Shabanov N, Knyaziknin Y, Frederic B, Myneni RB (2000) Stochastic modeling of radiation regime in discontinuous vegetation canopies. Remote Sens Environ 74(1):125–144
Sharma M, Burkhart HE (2003) Selecting a level of conditioning for the segmented polynomial taper equation. For Sci 49:324–330
Shelton MG, Murphy PA (1991) Age and size structure of a shortleaf pine-oak stand in the Ouachita Mountains—implications for uneven aged management. In: Coleman SS, Neary DG (eds) Proceedings 6th biennial southern silvicultural research conference, vol 2. General Technical Report SE-70, US Department of Agriculture, Forest Service, Southeastern Forest Experiment Station, pp 852–857
Smith WK, Knapp AK, Reiners WA (1989) Penumbral effects on sunlight penetration in plant communities. Ecology 70(6):1603–1609
Smolander S, Stenberg P (2003) A method to account for shoot scale clumping in coniferous canopy reflectance models. Remote Sens Environ 88:363–373
Smolander S, Stenberg P (2005) Simple parameterizations of the radiation budget of uniform broadleaved and coniferous canopies. Remote Sens Environ 94(3):355–363
Smolander H, Kellomäki S, Oker-Blom P (1990) The effect of nitrogen concentration on needle photosynthesis and within shoot shading in scots pine. Silva Fenn 24(1):54–14
Smolander H, Stenberg P, Linder S (1994) Dependence of light interception efficiency of Scots pine shoots on structural parameters. Tree Physiol 14(7–9):971–980
Sprugel DG (2002) When branch autonomy fails: Milton's Law of resource availability and allocation. Tree Physiol 22:1119–1124
Sprugel DG, Hinckley TM, Schaap W (1991) The theory of branch autonomy. Annu Rev Ecol Syst 22:309–334
Stenberg P (1996a) Simulations of the effects of shoot structure and orientation on vertical gradients in intercepted light by conifer canopies. Tree Physiol 16(1–2):99–108
Stenberg P (1996b) Correcting LAI-2000 estimates for the clumping of needles in shoots of conifers. Agric For Meteorol 79:1–8

Stenberg P (1998) Implications of shoot structure on the rate of photosynthesis at different levels in a coniferous canopy using a model incorporating grouping and penumbra. Funct Ecol 12:82–91
Stenberg P (2006) A note on the G-function for needle leaf canopies. Agric For Meteorol 136:76–79
Stenberg P, Palmroth S, Bond BJ, Sprugel DG, Smolander H (2001) Shoot structure and photosynthetic efficiency along the light gradient in a Scots pine canopy. Tree Physiol 21:805–814
Stenberg P, Mõttus M, Rautiainen M (2016) Photon recollision probability in modelling the radiation regime of canopies — a review. Remote Sens Environ 183:98–108
Stoll P, Schmid B (1998) Architecture, biomass allocation, and bud demography of *Pinus sylvestris* branches in heterogeneous light environments. J Ecol 86:934–944
Thérézien M, Palmroth S, Brady R, Oren R (2007) Estimation of light interception properties of conifer shoots by an improved photographic method and a 3D model of shoot structure. Tree Physiol 27(10):1375–1387
Thomas SC, Winner WE (2000) A rotated ellipsoidal angle density function improves estimation of foliage inclination distributions in forest canopies. Agric For Meteorol 100:19–24
Turetsky MR (2003) The role of bryophytes in carbon and nitrogen cycling. Bryologist 106:395–409
Wang T, Hamann A, Spittlehouse D. L, Carroll C (2016) Locally downscaled and spatially customizable climate data for historical and future periods for North America. PLOS ONE, 11(6), e0156720
Wang N, Palmroth S, Maier CA, Domec JC, Oren R (2019) Anatomical changes with needle length are correlated with leaf structural and physiological traits across five *Pinus* species. Plant Cell Environ 42(5):1690–1704
Wang N, Domec JC, Palmroth S, Maier CA, Xie L, Yin C, Chen YJ, Oren R (2025) Xylem hydraulic properties of five *Pinus* species grown in common environment vary from needles to roots with needle length and native-range climate. Plant Cell Environ 48:7225. (early view)
Wardle DA, Zackrisson O, Hörnberg G, Gallet C (1997) Influence of Island area on ecosystem properties. Science 277:1296–1299
Waring RH, Schroeder PE, Oren R (1982) Pipe model theory and canopy leaf area. Can J For Res 12:556–560
Williston HL (1972) Shortleaf and loblolly pine growth in the mid-South. J For 70:290–291
Wilsey BJ, Polley HW (2006) Aboveground productivity and root-shoot allocation differ between native and introduced grass species. Oecologia 150:300–309
Wirth R, Weber B, Ryel RJ (2001) Spatial and temporal variability of canopy structure in a tropical moist forest. Acta Oecol 22:235–244
Wonn HT (2001) Height:Diameter ratios and stability relationships for four northern Rocky mountain tree species. West J Appl For 16:87–94
Woodward FI (1987) Climate and plant distribution. Cambridge University Press
Zhang H, Domec JC, Maier CA, Boroski CA, Wang N, Palmroth S, Oren R (2025) Similar response of canopy conductance to increasing vapor pressure deficit and decreasing soil conductivity with drought among five morphologically contrasting but co-occurring pine species. Agric For Meteorol 366:110479
Zwieniecki MA, Stone HA, Leigh A, Boyce CK, Holbrook NM (2006) Hydraulic design of pine needles: one-dimensional optimization for single-vein leaves. Plant Cell Environ 29(5):803–809

Pine Photosynthesis: Needle Longevity, Structural and Physiological Controls, Needle and Tree Age Dependent Changes and Plasticity

Ülo Niinemets and Ziyan Zhang

Abstract Pines colonize a wide range of habitats from subarctics to tropics, but the comparative information of global variation in pine needle photosynthesis and associated traits has been limited. In this chapter, we first introduce the key functional traits determining needle photosynthesis rate and discuss the ways of expressing needle photosynthesis rates in pines. We use novel global datasets of needle longevity (Λ) and structural, chemical, and physiological traits to assess the overall trait variation across the pines and compare the pine trait variation with global trait variation. We further analyze differences in needle photosynthetic characteristics (photosynthetic capacity per unit area, A_{area}, and dry mass, A_{mass}, photosynthetic nitrogen use efficiency, PNUE) in relation to needle structural characteristics (leaf dry mass per unit area, LMA), nitrogen content per dry mass (N_{mass}) and area (N_{area}), and Λ, and develop the pine needle economics spectrum. We use the quantitative limitation analysis to characterize the degree to which needle photosynthetic rates are constrained by stomatal and mesophyll diffusion and biochemical limitations, and how the limitations scale with needle functional traits. We conclude by reviewing alterations in needle photosynthetic rates as driven by needle age, plant age and size, and acclimation to environment. The analysis demonstrates about 15-fold variation in key needle traits A_{area}, A_{mass}, PNUE, N_{area} and LMA, with a smaller, 4.5-fold variation

Ü. Niinemets (✉)
Institute of Agricultural and Environmental Sciences, Estonian University of Life Sciences, Tartu, Estonia

Estonian Academy of Sciences, Tallinn, Estonia
e-mail: ylo.niinemets@emu.ee

Z. Zhang
Institute of Agricultural and Environmental Sciences, Estonian University of Life Sciences, Tartu, Estonia

School of Ecology, Northeast Forestry University, Harbin, China

Key Laboratory of Sustainable Forest Ecosystem Management-Ministry of Education, Northeast Forestry University, Harbin, China

Northeast Asia Biodiversity Research Center, Northeast Forestry University, Harbin, China

D. M. Johnson et al. (eds.), *Physiology and Ecology of Pines Worldwide*, Tree Physiology 9, https://doi.org/10.1007/978-3-032-23001-0_6

in N_{mass}. Needle average longevity varied almost 30-fold across pines, albeit most pines are characterized by moderate needle longevity of 2–4 yr. Classic leaf economics spectrum relationships were uncovered across pines, characterizing the trade-offs between investments in leaf structural and photosynthetic tissues as driven by variation in needle longevity. In addition to classic economics spectrum traits, available data of different photosynthetic constraints indicated that mesophyll diffusion limitations are greater in pines with more robust needles. Apart from the broad trait relationships, we highlight major impacts of needle ontogeny (juvenile vs adult), age of adult needles, and tree size and environmental acclimation (plasticity) on needle functional traits. Overall, we demonstrate remarkable variation in pine needle photosynthetic capacity across species and during needle and plant lifetime, broadly consistent with their conservative N use strategy and position in the lower end of the leaf economics spectrum.

1 Introduction

Pines are traditionally considered as shade intolerant and drought tolerant species thriving in open habitats (Teskey et al. 1994; Barbéro et al. 1998; Richardson and Rundel 1998; Niinemets and Valladares 2006; Reich et al. 2014; Schmiege et al. 2021a). However, pines colonize habitats from subarctic to tropical biomes and there is a remarkable diversity in pine life strategies (Richardson and Rundel 1998; Álvarez-Lopeztello and del Castillo 2023; Wasowicz et al. 2025). Their wide range is associated with an extensive variation in needle longevity, anatomy, and content of limiting nutrients, leading to concomitant changes in needle physiological activity (Teskey et al. 1994; Reich et al. 1995a, b; Ross and Loik 2021; Schmiege et al. 2021a; Song et al. 2022). However, there is little comparative information of co-variation of needle functional traits across pine species and overall position of pines in global leaf trait space.

Leaf economics spectrum (LES) is an ecological concept that characterizes the coordinated variation in the suites of leaf functional traits across plants with differing photosynthetic activity, from highly active high turnover/high return species to species with robust low-activity long-living leaves (Wright et al. 2004b; Onoda et al. 2017; Pan et al. 2020; Joswig et al. 2022). The fundamental compromises underlying the leaf economics spectrum (LES) are the trade-off between the investments in assimilative and support structures and consequent trade-off between the functional activity and leaf longevity. In essence, these trade-offs reflect the circumstance that it is impossible to maximize simultaneously both the within-leaf biomass investments in assimilative and support structures. Making leaves more robust to increase longevity inevitably reduces the share of physiologically highly active mesophyll. Consequently, plants can either allocate more biomass in photosynthetic tissues that can be used for a short period of time, or they can allocate more biomass in support structures to construct firm, resistant long-living leaves, but with lower investment in photosynthetic tissues (Reich et al. 1991, 1992; Wright et al. 2004b).

The key relationships in LES are the positive scaling of leaf photosynthetic capacity per dry mass (A_{mass}) with nitrogen content per dry mass (N_{mass}), and negative scaling of A_{mass} with leaf dry mass per unit area (LMA) and leaf longevity (Wright et al. 2004b). LES relationships have been tested across species, but there is also evidence of presence of such relationships within different plant phylogenetic groups such as ferns (Tosens et al. 2016; Jin et al. 2020; Zhao et al. 2024a) or mosses (Carriquí et al. 2019; Grau-Andrés et al. 2022) or even within individual species with wide geographic range (Niinemets 2015; Hayes et al. 2018; Sporbert et al. 2021). Nevertheless, there is a certain variability in leaf trait relationships within the broad leaf economics spectrum as driven, for instance, plant ecological strategy (Hallik et al. 2009; Ren et al. 2022). However, uncovering group-specific relationships among leaf structural and physiological traits depends on trait range (Wright et al. 2004a). In this chapter, we address the questions of how large is the variability in key leaf functional traits across pines, whether the trait relationships analogous to global leaf economics spectrum can be found across pines, and where are pines positioned in the global LES?

The position of any species in A_{mass}-LMA-N_{mass} axes is driven by the variation in the photosynthetic nitrogen use efficiency (PNUE). PNUE depends on the investment of nitrogen in rate-limiting proteins that affects the leaf biochemical photosynthesis potentials at given N content, and by the variation in the efficiency with which the leaf photosynthetic machinery operates, i.e., how much a given investment of N in rate-limiting photosynthetic proteins increases the rate of photosynthesis. This efficiency is driven by diffusion limitations that affect the CO_2 concentration at Rubisco in the chloroplasts. The CO_2 diffusion limitations of net assimilation rate, A, are driven by both non-finite stomatal conductance for CO_2 (g_s) and mesophyll conductance (g_m):

$$A = g_s \left(C_a - C_i\right) = g_m \left(C_i - C_c\right), \tag{1}$$

where C_a is the ambient CO_2 concentration, C_i the CO_2 concentration in the intercellular air space and C_c the CO_2 concentration in the chloroplasts. From Eq. (1),

$$C_i = C_a - A / g_s, \tag{2}$$

and

$$C_c = C_i - A / g_m. \tag{3}$$

The ratio A/g_s is the CO_2 drawdown due to limited stomatal conductance and A/g_m the CO_2 drawdown due to limited mesophyll conductance. There are strong positive correlations among foliage photosynthetic potentials and maximum values of g_s and g_m (Evans and Loreto 2000; Niinemets et al. 2009; Evans 2021). Despite these broad correlations, the degree to which photosynthesis limited by CO_2 diffusion limitations varies across species with different suites of leaf structural traits and nitrogen content (Onoda et al. 2017; Knauer et al. 2022). In particular, g_m is expected to be a stronger limitation of photosynthesis in species with more robust leaves at the lower end of LES than in species with less robust leaves at the higher return end of LES (Niinemets et al. 2011; Onoda et al. 2017; Knauer et al. 2022). In the current chapter, we go beyond the traditional LES by analyzing the limitations of pine foliage photosynthetic rate due to differences in needle biochemical photosynthesis potentials, and stomatal and mesophyll conductance. We expect that in pines with structurally robust leaves, g_m is the most important component of the three limitations.

There are further major juvenile-to-adult and tree height and age-dependent changes in leaf structure and function, reflected in dimorphism in leaf traits in juvenile and adult leaves, called heteroblasty (Zotz et al. 2011). The juvenile-to-adult transition leads to particularly large modifications in pine needle photosynthesis, and such ontogenetic differences in foliage physiological activity play a major role in seedling establishment (Boddi et al. 2002; Kuusk et al. 2018a, b; Fernández-Marín et al. 2021). Changes in pine needle photosynthesis also occur in adult needles with increasing tree size and with needle aging. The tree size-dependent modifications reflect the phenotypic response of needle functional traits to increased water potential gradients in stems of larger trees (Niinemets 2002; Koch et al. 2004; Han 2011; Steppe et al. 2011), but can also be partly caused by epigenetic alterations during whole tree juvenile-to-adult transition (Greenwood 1995; Mencuccini et al. 2007).

Needle age-dependent changes in non-senescent needles result from multiple processes, including secondary needle growth, thickening of cell walls and concomitant alterations in the share of structural and photosynthetic biomass and needle water transport capacity as well as reacclimation of high-light formed needles to shading by newly formed foliage (Op de Beeck et al. 2010; Deligöz et al. 2018; Kuusk et al. 2018a; Roth-Nebelsick et al. 2023). Consideration of these leaf age effects on photosynthesis is of paramount significance for accurate estimation of whole canopy photosynthetic production (Beyschlag et al. 1994; Wilson et al. 2001; Crous et al. 2021).

Leaf economics spectrum considers traits of non-stressed high light exposed leaves, yet, even in relatively open canopies typical for pines, there are strong within-canopy light gradients (Bond et al. 1999; Niinemets et al. 2001; Stenberg et al. 2001; Niinemets 2023; Zhao et al. 2024b). Despite their low shade tolerance, pines can colonize habitats with moderately broad variation in understory light availability (Messier and Puttonen 1995; Coates 2002; de Chantal et al. 2003). Pines exhibit pronounced phenotypic plasticity in response to these gradients, especially in key anatomical and photosynthetic traits (Bond et al. 1999; Niinemets et al. 2001; Stenberg et al. 2001; Niinemets 2023; Zhao et al. 2024b). In addition, pines also exhibit plasticity in functional traits in response to variation in other environmental characteristics including water and nutrient availabilities and temperature environment, and these sources of trait variation can be of similar magnitude as interspecific trait variability (Fernandez et al. 1999; Teskey and Will 1999; Niinemets et al. 2001; Tjoelker et al. 2008; Wieser et al. 2010; Tarvainen et al. 2018; Gauthey et al. 2024).

In this analysis, we characterize the overall range in key functional traits and develop LES relationships for pines worldwide. In addition, we review the substantial variation in needle photosynthetic activity as driven by needle and whole plant age and briefly consider modifications in needle photosynthetic activity as driven by acclimation to environmental changes. We demonstrate that pines generally have a conservative resource use strategy, but nevertheless exhibit a remarkable interspecific variability in key photosynthetic traits consistent with their worldwide broad range. We emphasize that pines exhibit a high phenotypic variability in photosynthetic characteristics that needs consideration when scaling up photosynthesis from leaves to ecosystems.

For this analysis, we compiled a global dataset of needle longevity for 87 *Pinus* taxa based on published data and authors' original assessments (Table 1). We further constructed a database of needle structural, chemical and photosynthetic traits based on 59 studies providing data of net assimilation rate and associated suite of structural and chemical traits for 51 *Pinus* taxa (available at DOI: https://doi.org/10.15159/eds.dt.26.02, Niinemets and Zhang 2026, Appendix I for species and data sources). Through the analysis, we define needle photosynthetic capacity as the light-saturated net assimilation rate per unit projected needle area standardized to a leaf temperature of 25 °C and an intercellular CO_2 concentration of 300 μmol mol^{-1} (Text Box 1 for details of data standardization).

2 Variation in Needle Longevity Across Pines Worldwide

Needle longevity is a key trait in LES that directly links foliage photosynthetic activity to the economics of construction of photosynthesizing biomass (short vs. long duration photosynthetic organs) and ecological strategy (slow return vs. fast return) (Wright et al. 2004b). Average needle longevity (Λ_a) is typically defined as the oldest needle-age class with half of the needles remaining (Wright et al. 2004b; Mediavilla et al. 2014), while the maximum needle longevity (Λ_m) refers to the oldest needles present. Needle longevity is a trait that has both genotypic and phenotypic components. In particular, within a species, a classic understanding is that needle longevity increases as site conditions become harsher. Thus, pine needle longevity increases with increasing site altitude (Schoettle 1990; Nebel and Matile 1992; Schoettle and Fahey 1994; Reich et al. 1996; Schoettle and Rochelle 2000; Boratyński et al. 2009), and decreasing nutrient availability (Niinemets and Lukjanova 2003a, b). In addition, needle longevity differs for juvenile and adult needles (Table 1, Sect. 5.2). Here we first introduce a global pine needle longevity dataset and discuss hurdles in estimating pine needle longevity and possible problems in previous longevity assessments. Based on the new compilation we further analyze the variation in needle longevity across pines worldwide.

Table 1 Average needle longevity, Λ_a (average ± SE (*n*)), and maximum needle longevity, Λ_m, in 87 *Pinus* species

Species[a]	Λ_a (yr)	Λ_m (yr)[b]	Data sources
P. albicaulis	7.0 ± 0.3 (4)	8	Sargent (1897), Sudworth (1908), Abrams (1923), Ewers and Schmid (1981), Richardson and Rundel (1998)
P. aristata	12.4 ± 1.1 (6)	15	Sargent (1897), Bailey (1970), Mastrogiuseppe and Mastrogiuseppe (1980), Ewers and Schmid (1981), Schoettle (1994), Richardson and Rundel (1998), Smith et al. (2019)
P. arizonica	4 (1)		Smith et al. (2019)
P. armandii	2.5 ± 0.3 (3)	3	Richardson and Rundel (1998), Luo et al. (2002), Smith et al. (2019)
P. attenuata	4.5 ± 0.0 (3)	5	Sudworth (1908), Abrams (1923), Ewers and Schmid (1981), Richardson and Rundel (1998)
P. ayacahuite	2.5 ± 0.5 (2)		Richardson and Rundel (1998), Smith et al. (2019)
P. balfouriana subsp. austriana	11.6 (1)	18	Bailey (1970), Mastrogiuseppe and Mastrogiuseppe (1980), Ewers and Schmid (1981)
P. balfouriana subsp. balfouriana	13.4 ± 2.2 (4)	15	Sargent (1897), Sudworth (1908), Bailey (1970), Mastrogiuseppe and Mastrogiuseppe (1980), Ewers and Schmid (1981), Richardson and Rundel (1998)
P. banksiana	2.7 ± 0.2 (5)	4	Sargent (1897), Ewers and Schmid (1981), Reich et al. (1998, 1999), Richardson and Rundel (1998), Wright et al. (2004b), Hallik et al. (2009), Niinemets (2010), Smith et al. (2019)
P. bhutanica	2.5 (1)	3	Richardson and Rundel (1998)
P. brutia	2.2 ± 0.2 (2)	3	This study/images[c]; This study/Ü. Niinemets[d]
P. bungeana	4.4 ± 0.9 (2)		Richardson and Rundel (1998), Smith et al. (2019)
P. canariensis, mature	2.3 ± 0.3 (2)	3	Richardson and Rundel (1998); This study/Ü. Niinemets[d]
P. canariensis, juvenile	1.5 (1)		This study/Ü. Niinemets[d]
P. caribaea	1.8 ± 0.3 (2)		Richardson and Rundel (1998); This study/images[c]
P. cembra	5.3 ± 1.3 (3)	12	Nebel and Matile (1992), Smith et al. (2019)
P. cembroides	3.5 ± 0.0 (2)	4	Sargent (1897), Ewers and Schmid (1981), Richardson and Rundel (1998)
P. clausa	3.0 ± 0.5 (2)	4	Sargent (1897), Ewers and Schmid (1981), Richardson and Rundel (1998)
P. contorta	5.8 ± 0.7 (5)	8	Sargent (1897), Sudworth (1908), Abrams (1923), Critchfield (1957), Ewers and Schmid (1981), Richardson and Rundel (1998), Hessl et al. (2004), Niinemets (2010), Smith et al. (2019)

(continued)

Table 1 (continued)

Species[a]	Λ_a (yr)	Λ_m (yr)[b]	Data sources
P. contorta subsp. contorta	4.1 ± 0.4 (4)	6	Critchfield (1957), Ewers and Schmid (1981), Richardson and Rundel (1998), Hallik et al. (2009), Niinemets (2010)
P. contorta subsp. bolanderi	3.3 ± 0.6 (3)	6	Critchfield (1957), Ewers and Schmid (1981), Richardson and Rundel (1998)
P. contorta subsp. murrayana	5.8 ± 0.8 (2)	11	Critchfield (1957), Ewers and Schmid (1981)
P. contorta subsp. latifolia	10.1 ± 1 (2)	13	Critchfield (1957), Ewers and Schmid (1981), Schoettle (1989, 1994), Richardson and Rundel (1998)
P. coulteri	3.5 ± 0.0 (3)	4	Sudworth (1908), Abrams (1923), Ewers and Schmid (1981), Richardson and Rundel (1998)
P. dalatensis	1.8 ± 0.3 (2)		This study/images[c]
P. densata	2.4 ± 0.2 (4)		Richardson and Rundel (1998), Luo et al. (2002); This study/images[c]
P. densiflora	2.3 ± 0.2 (3)	3	Richardson and Rundel (1998), Kume et al. (2000), Smith et al. (2019)
P. durangensis	2 (1)		This study/images[c]
P. echinata	3.3 ± 0.4 (3)	5	Sargent (1897), Ewers and Schmid (1981), Richardson and Rundel (1998), Smith et al. (2019)
P. edulis	4.9 ± 0.7 (4)	9	Sargent (1897), Ewers and Schmid (1981), Reich et al. (1998, 1999), Richardson and Rundel (1998), Wright et al. (2004b), Hallik et al. (2009), Niinemets (2010)
P. elliottii	1.9 ± 0.1 (3)		Sargent (1922), Preston (1976), Ewers and Schmid (1981), Richardson and Rundel (1998), Holder (2000)
P. engelmannii	2 (1)		Preston (1976), Ewers and Schmid (1981)
P. fenzeliana	1.5 (1)		This study/images[c]
P. flexilis	4.8 ± 0.5 (7)	10	Sargent (1897), Sudworth (1908), Abrams (1923), Ewers and Schmid (1981), Reich et al. (1998, 1999), Richardson and Rundel (1998), Schoettle and Rochelle (2000), Wright et al. (2004b), Hallik et al. (2009), Niinemets (2010), Smith et al. (2019)
P. gerardiana	2.5 (1)		Smith et al. (2019)
P. glabra	2.5 ± 0.0 (2)	3	Sargent (1897), Ewers and Schmid (1981), Richardson and Rundel (1998)
P. greggii	2.5 (1)	3	Richardson and Rundel (1998)
P. halepensis, mature	2.6 ± 0.2 (7)	4	Richardson and Rundel (1998), Villar Salvador (2000), Mediavilla and Escudero (2003a, b), Mediavilla et al. (2008), Hallik et al. (2009), Niinemets (2010), Kuusk et al. (2018a, b)
P. halepensis, juvenile	1.2 (1)		Kuusk et al. (2018a, b)
P. henryi	3 (1)		Smith et al. (2019)

(continued)

Table 1 (continued)

Species[a]	Λ_a (yr)	Λ_m (yr)[b]	Data sources
P. jeffreyi	5.2 ± 0.9 (6)	8	Sargent (1897), Sudworth (1908), Abrams (1923), Ewers and Schmid (1981), Richardson and Rundel (1998), Hallik et al. (2009), Niinemets (2010), Smith et al. (2019)
P. kesiya	1.1 ± 0.4 (3)		Kumar Das and Ramakrishnan (1986), Richardson and Rundel (1998), Luo et al. (2002)
P. koraiensis	3.5 ± 0.5 (4)		Richardson and Rundel (1998), Luo et al. (2002), Smith et al. (2019), Liu et al. (2020)
P. krempfii	2.5 (1)		This study/images[c]
P. lambertiana	2.8 ± 0.3 (7)	4	Sargent (1897), Sudworth (1908), Abrams (1923), Ewers and Schmid (1981), Richardson and Rundel (1998), Hallik et al. (2009), Niinemets (2010), Smith et al. (2019)
P. leiophylla	2 (1)		Sargent (1897), Ewers and Schmid (1981), Richardson and Rundel (1998)
P. leucodermis	3.9 ± 0.7 (3)		Richardson and Rundel (1998), Smith et al. (2019)
P. longaeva	19.6 ± 3.8 (3)	33	Bailey (1970), Mastrogiuseppe and Mastrogiuseppe (1980), Ewers and Schmid (1981), Richardson and Rundel (1998)
P. massoniana	1.5 (1)		Luo et al. (2002)
P. maximartinezii	5 (1)		Richardson and Rundel (1998)
P. merkusii	1.4 ± 0.2 (3)	2	Richardson and Rundel (1998), Pumijumnong and Palakit (2020); This study/images[c]
P. monophylla	8.3 ± 0.1 (6)	12	Sargent (1897), Sudworth (1908), Ewers and Schmid (1981), Richardson and Rundel (1998), Hallik et al. (2009), Niinemets (2010)
P. monteuzumae	3 (1)		Richardson and Rundel (1998)
P. monticola	3.0 ± 0.2 (5)	4	Sargent (1897), Abrams (1923), Cronquist et al. (1972), Ewers and Schmid (1981), Richardson and Rundel (1998), Smith et al. (2019)
P. mugo	4.3 ± 0.8 (2)		Smith et al. (2019)
P. muricata	2.6 ± 0.3 (5)	5	Sargent (1897), Sudworth (1908), Abrams (1923), Ewers and Schmid (1981), Richardson and Rundel (1998)
P. nigra	3.5 ± 0.3 (4)		Richardson and Rundel (1998), Reich et al. (1999), Wright et al. (2004b), Hallik et al. (2009), Niinemets (2010), Smith et al. (2019)
P. nigra subsp. salzmannii, juvenile	1.2 (1)		Kuusk et al. (2018a, b)
P. nigra subsp. salzmannii, mature	3.8 (1)		Kuusk et al. (2018a, b)

(continued)

Table 1 (continued)

Species[a]	Λ_a (yr)	Λ_m (yr)[b]	Data sources
P. palustris	2.3 ± 0.2 (4)		Reich et al. (1998, 1999), Hallik et al. (2009), Niinemets (2010), Sargent (1897), Ewers and Schmid (1981), Richardson and Rundel (1998), Wright et al. (2004b)
P. parviflora	2.8 ± 0.8 (2)	4	Richardson and Rundel (1998), Smith et al. (2019)
P. patula	2.0 ± 0.0 (2)	4	Richardson and Rundel (1998); This study/Ü. Niinemets[d]
P. peuce	2.3 ± 0.3 (2)		Smith et al. (2019); This study/images[c]
P. pinaster	3.2 ± 0.3 (7)	5	Maugé (1987), Richardson and Rundel (1998), Mediavilla and Escudero (2003a, b), Mediavilla et al. (2008), Eimil-Fraga et al. (2015b), Smith et al. (2019)
P. pinea, mature	3.1 ± 0.3 (5)	4	Richardson and Rundel (1998), Mediavilla and Escudero (2003a, b), Mediavilla et al. (2008), Kuusk et al. (2018a, b)
P. pinea, juvenile	1.2 (1)		Kuusk et al. (2018a, b)
P. ponderosa	3.3 ± 0.4 (6)	9	Sargent (1897), Abrams (1923), Weidman (1939), Ewers and Schmid (1981), Richardson and Rundel (1998), Hessl et al. (2004), Hallik et al. (2009), Niinemets (2010), Smith et al. (2019)
P. pumila	4.3 ± 0.8 (2)		Richardson and Rundel (1998), Smith et al. (2019)
P. pungens	2.8 ± 0.3 (2)		Sargent (1897), Ewers and Schmid (1981), Richardson and Rundel (1998)
P. quadrifolia	3.5 (1)	4	Sargent (1897), Ewers and Schmid (1981)
P. radiata	3.0 ± 0.2 (5)	4	Sargent (1897), Sudworth (1908), Abrams (1923), Ewers and Schmid (1981), Gower et al. (1993), Richardson and Rundel (1998), Hallik et al. (2009), Niinemets (2010)
P. resinosa	3.8 ± 0.3 (6)	5	Sargent (1897), Britton and Shafer (1908), Ewers and Schmid (1981), Reich et al. (1998, 1999), Richardson and Rundel (1998), Wright et al. (2004b), Hallik et al. (2009), Niinemets (2010)
P. rigida	2.6 ± 0.1 (6)	3	Sargent (1897), Britton and Shafer (1908), Ewers and Schmid (1981), Reich et al. (1998, 1999), Richardson and Rundel (1998), Wright et al. (2004b), Hallik et al. (2009), Niinemets (2010), Smith et al. (2019)
P. roxburghii	1.5 ± 0.5 (2)	3	Singh et al. (1994), Richardson and Rundel (1998)
P. sabiniana	3.5 ± 0.0 (3)	4	Sargent (1897), Sudworth (1908), Ewers and Schmid (1981), Richardson and Rundel (1998)
P. serotina	2.8 ± 0.3 (5)	4	Sargent (1897), Abrams (1923), Ewers and Schmid (1981), Reich et al. (1998, 1999), Richardson and Rundel (1998), Wright et al. (2004b), Hallik et al. (2009), Niinemets (2010)

(continued)

Table 1 (continued)

Species[a]	Λ_a (yr)	Λ_m (yr)[b]	Data sources
P. sibirica	4 (1)		Smith et al. (2019)
P. strobiformis	3.8 ± 0.3 (2)		Sargent (1897), Ewers and Schmid (1981), Smith et al. (2019)
P. strobus	2.3 ± 0.1 (8)	3	Sargent (1897), Small (1972), Ewers and Schmid (1981), Whitney (1982), Gower et al. (1993), Reich et al. (1998, 1999), Richardson and Rundel (1998), Wright et al. (2004b), Hallik et al. (2009), Niinemets (2010), Smith et al. (2019)
P. sylvestris	3.3 ± 0.3 (9)	8	Reich et al. (1998, 1999), Richardson and Rundel (1998), Villar Salvador (2000), Mediavilla and Escudero (2003a, b), Wright et al. (2004b), Mediavilla et al. (2008), Hallik et al. (2009), Niinemets (2010), Smith et al. (2019)
P. tabuliformis	2.8 ± 0.7 (3)		Luo et al. (2002), Xiao (2003), Smith et al. (2019)
P. taeda	1.4 ± 0.2 (6)		Sargent (1897), Britton and Shafer (1908), Ewers and Schmid (1981), Richardson and Rundel (1998), Hallik et al. (2009), Albaugh et al. (2010), Niinemets (2010)
P. teocote	3 (1)		Richardson and Rundel (1998)
P. thunbergii	3.8 ± 0.3 (2)		Richardson and Rundel (1998), Smith et al. (2019)
P. torreyana	3.5 ± 0.0 (3)	4	Sudworth (1908), Abrams (1923), Ewers and Schmid (1981), Richardson and Rundel (1998)
P. tropicalis	1.3 (1)		Alvarez et al. (2001); This study/images[c]
P. uncinata	3 (1)		Smith et al. (2019)
P. virginiana	3.3 ± 0.2 (3)	4	Sargent (1897), Ewers and Schmid (1981), Richardson and Rundel (1998), Smith et al. (2019)
P. wallichiana	3.2 ± 0.4 (2)	4	Richardson and Rundel (1998), Smith et al. (2019)
P. washoensis	5 (1)	6	Richardson and Rundel (1998)
P. yunnanensis	1.5 ± 0.0 (2)		Luo et al. (2002), He et al. (2021); This study/images[c]

Average needle longevity is the oldest needle age class with half of the needles remaining, and maximum needle longevity refers to the oldest living needles present. In the literature, especially in older data compilations, it is not always clear whether average, Λ_a, or maximum, Λ_m, needle longevity has been reported. When a study provided a range, i.e., the minimum and maximum longevity, Λ_a was calculated as the average of these two values for the given species in the given study. Direct estimates of Λ_a and those calculated from the range of reported values were strongly correlated ($r^2 = 0.95$, $P < 0.001$). In studies providing only one value for a given species, it was assumed to be the average longevity. Ultimately, an average value of Λ_a was calculated across the studies for every species. As explained in the main text, the variation in Λ_a within any given species can be large, primarily reflecting the acclimation of needle longevity to site conditions

(continued)

Table 1 (continued)

The maximum needle longevity in this table refers to the reported typical maximum longevity across the studies not including exceptional values. Exceptional maximum needle longevities observed in high altitude species, e.g., in *P. aristata* and *P. longaeva*, are reported in the main text. In several cases, the difference between Λ_m and average Λ_a is relatively small, less than one year. This might reflect lack of data of maximum Λ_m at species environmental limits. In addition, it might also indicate a certain overestimation of average Λ_a due to difficulties in separating data of Λ_m and Λ_a reported in the literature. Future studies are needed to estimate full ranges of pine needle average and maximum longevities, focusing on species with limited number of observations and small differences between Λ_m and Λ_a

[a]For four species, data are provided for both juvenile and mature needles (Sect. 5.2). Juvenile (primary) needles are the first needles formed in seedlings, and they are gradually replaced by adult (secondary) needles. Compared with mature needles, juvenile needles are softer with a lower leaf dry mass per unit area (LMA), and physiologically more active with a higher photosynthetic capacity per dry mass (A_{mass}), and they also have different cross-sectional geometry. In most pines, juvenile needles are carried for a short period of time. The transition from juvenile to adult needles typically occurs in plants less than one year old. In some pine species, e.g., Mediterranean pines, juvenile needles can still be formed in 3–4-year-old plants. Mature trees of some species such as *P. canariensis* can form juvenile needles after damage and resprouting (Sect. 5.2)

[b]Maximum longevities were rounded to an integer when non-integer values were reported in original studies

[c]This study. Leaf longevity was estimated from shoot images in public domain by counting the number of needle age classes (needle flushes) present and considering the information of shoot flushing (single vs. multiple shoot flushes formed in the given year as explained in the main text)

[d]This study. Ü. Niinemets personal observations: *P. brutia* (Cyprus); *P. canariensis* (Tenerife); *P. patula* (New Zealand)

2.1 Construction of a Worldwide Pine Needle Longevity Dataset

For the assessment of leaf economics spectrum relationships, we created a dataset of needle average and maximum longevities for adult needles of 87 and juvenile needles of four *Pinus* species. Data were compiled from multiple literature sources, and included several novel estimations based on authors' own observations (Table 1). The compiled dataset was cleaned by removing erroneous estimates that did not consider the circumstance that some pine species form multiple needle flushes in a given year. In particular, when compiling the data of needle longevity, we noticed that there is a certain controversy in literature over needle longevities in subtropical and tropical pines that have multiple leaf flushes during a growing season. In species forming only one annual shoot flush, the maximum needle longevity can be conveniently estimated by counting the number of annual shoot flushes that still carry needles, whereas the average needle longevity is typically taken as the oldest shoot flush with half of the needles remaining. In species forming multiple shoot flushes per year, it is also necessary to estimate the number of shoot flushes formed during the year (n_{sf}), and thus, the needle longevity is given as the number of shoot flushes carrying needles divided by n_{sf}. In early studies, the needle longevity of subtropical and tropical pines has been apparently taken as the total number of needled shoot flushes irrespective of n_{sf}. These overestimated values of needle

Text Box 1: Outline of Extraction and Standardization of Published Foliage Photosynthetic Traits in *Pinus* Species

Foliage light-saturated net assimilation rate, photosynthetic capacity (A_{max}) varies in dependence on CO_2 concentration in the leaf intercellular air space (C_i) and leaf temperature. C_i varies with stomatal conductance that in turn depends on environmental characteristics and plant physiological conditions. In this analysis, A_{max} was standardized to C_i = 300 μmol mol^{-1} and leaf temperature of 25 °C to remove the environmental sources of variation due to differences in C_i and leaf temperature.

Extraction of photosynthetic traits and details of measurement conditions, support information and preliminary data clearing

From each study (Appendix I for data sources and DOI: https://doi.org/10.15159/eds.dt.26.02) we attempted to extract a comprehensive set of variables with corresponding support information characterizing the measurement conditions and plant material measured.

The following categories of information were extracted:

- Photosynthetic characteristics: light-saturated net assimilation rate (A_{max}) corresponding stomatal conductance and intercellular CO_2 concentration (C_i), mesophyll conductance and chloroplastic CO_2 concentration (C_c), dark respiration rate (R_d), non-photorespiratory respiration rate in light (R_{nd}), maximum carboxylase activity of Rubisco (V_{cmax}) and capacity for photosynthetic electron transport (J_{max});
- Rubisco kinetic data used to estimate V_{cmax} and J_{max};
- Information whether the stomatal conductance was for water vapor or for CO_2;
- Environmental conditions during measurements: chamber/leaf temperature, light intensity, ambient CO_2 concentration;
- Whether the "ambient CO_2 concentration" refers to the CO_2 concentration outside the chamber or inside the chamber (C_a);
- Details of measurement system: instrument make and model, measurement chamber type and size (e.g., clip-on chamber, whole leaf/conifer chamber), flow rate used;
- Expression basis of photosynthetic characteristics: projected, total, half-of-the-total, or silhouette area and information of how the area was estimated;
- Support information of needle material, plant and site: type of needles measured (juvenile vs. mature), needle age, tree age and size, date and year of measurement, growth conditions (field vs. artificial, e.g., growth chamber, greenhouse), site geographic coordinates and altitude for field sites, used treatment if any.

Standardization of photosynthetic data to a common temperature and intercellular CO_2 concentration

(continued)

Text Box 1 (continued)

Assuming Rubisco-limited conditions, for calculation of light-saturated net assimilation rate at the standard conditions of $C_i = 300$ μmol mol^{-1} and leaf temperature of 25 °C, one needs an estimate of maximum Rubisco carboxylase activity (V_{cmax}), non-photorespiratory respiration rate in light (R_{nd}) or dark respiration rate (R_d), and corresponding Rubisco kinetic traits, all at 25 °C. Once this information is available, A_{max} can be calculated for standardized conditions by Farquhar et al. (1980) photosynthesis model as:

$$A_{max} = V_{cmax}\left(\frac{C_i - \Gamma_*}{C_i + K_m}\right) - R_{nd}, \tag{4}$$

where Γ_* is the hypothetical CO_2 compensation point in the absence of R_{nd}, and K_m is the effective Michaelis-Menten constant (at 21% O_2) and C_i here is the standard value of 300 μmol mol^{-1}. Γ_* and K_m are the inherent characteristics of Rubisco that can vary among species (Galmés et al. 2016), but the data of Rubisco kinetics are very limited for pines. Thus, we used the Rubisco kinetic traits of Jordan and Ogren (1984) as fitted by Niinemets and Tenhunen (1997) frequently used in the photosynthetic community. Use of these kinetic traits yields a Γ_* value of 45.4 μmol mol^{-1} and a K_m value of 414.9 μmol mol^{-1} at 25 °C. When R_d rather than R_{nd} was available, R_{nd} was taken as $0.5R_d$ (Brooks and Farquhar 1985; Niinemets et al. 2015).

We encountered the following situations with available data: (1) both V_{cmax} and R_{nd} or R_d were available; (2) V_{cmax} was available, but neither R_{nd} nor R_d were available; (3) non-standardized A_{max} (i.e., A_{max} at an arbitrary value of C_i and leaf temperature; $A_{max,ns}$), C_i and R_d were available; (4) $A_{max,ns}$ and C_i were available and R_d was not available; (5) as (3), but instead of C_i, stomatal conductance and C_a were available; (6) as (5), but R_d was not available; (7) as (5), but C_a was not available (C_a referred to as "ambient"); (8) as (7), but R_d was not available; (9) only $A_{max,ns}$ and R_d were available (C_a referred to as "ambient"); (10) as (9), but R_d was not available. We followed Niinemets et al. (2015) approach to deal with the data gaps with some modifications as outline here.

First, in all cases when an estimate of respiration rate was available, it was standardized to 25 °C using a Q_{10} value of 2.3 (Niinemets et al. 2015). When an estimate of respiration rate was missing, i.e., the cases (2), (4), (6), (8) and (10), R_{nd} was found as $0.025V_{cmax}$ (Niinemets et al. 2015). For data reporting $A_{max,ns}$ rather than V_{cmax}, i.e., cases (4), (6), (8) and (10), R_{nd} and V_{cmax} were calculated simultaneously (Niinemets and Zhang 2026).

In (1) and (2), V_{cmax} was first standardized to 25 °C using the Rubisco kinetic traits used in the original studies. Thereafter, the values of V_{cmax} were standardized to the common set of Rubisco kinetic traits of Jordan and Ogren (1984) according to Niinemets et al. (2015), and A_{max} was calculated by Eq. (4). In (3)–(10), V_{cmax} was calculated by inverting Eq. (4) and then standardized to 25 °C according to Niinemets and Tenhunen (1997). In (3) and

(continued)

Text Box 1 (continued)

(4), V_{cmax} could be directly calculated using C_i, $A_{max,ns}$ and the available estimate of R_{nd}. In (5) and (6), C_i was calculated from C_a, $A_{max,ns}$ and stomatal conductance. In (7)–(10), atmospheric CO_2 concentration (C_{at}) was estimated using the Mauna Loa CO_2 concentration time series using the information of measurement date and year. Thereafter, chamber CO_2 concentration, C_a, was estimated based on air flow rate, chamber size and $A_{max,ns}$. The values C_a/C_{at} obtained ranged between 0.85 and 0.96. When not enough information was available to estimate C_a/C_{at}, a C_a/C_{at} value of 0.96 was used. In (9) and (10), C_i was estimated using a C_i/C_a ratio of 0.734 obtained as the average across the studies with available C_i/C_a data.

In the final database (available at DOI: https://doi.org/10.15159/eds.dt.26.02, Niinemets and Zhang 2026), the expression basis for all the area-based trait values is the projected needle area. Whenever the expression basis was not projected area, and the pertinent conversion factor was reported or could be derived based on the information provided in the study, the data were converted to projected needle area basis. We encountered multiple situations when the basis of expression was not reported or cases when it was reported, but the numeric values of data were seemingly unrealistic, suggesting that either the reported conversion factor or statement of the expression basis used was incorrect. In such cases, we contacted the authors of the original studies, and converted the data based on the information provided. However, not all authors responded and not always the information provided was helpful and added to what was already reported in the original publication. When it was impossible to identify the expression basis of a specific trait, the data were not included in the analysis.

longevity have propagated through compilations and are still occasionally used. For example, the subtropical pine *P. taeda* has up to five shoot flushes per year and average needle longevity of 1.1–1.8 yr (Hallik et al. 2009; Albaugh et al. 2010; Niinemets 2010), yet early studies have suggested a needle longevity of 3–3.5 yr for this species (Sargent 1897; Britton and Shafer 1908), and these values have been repeatedly used in compilations of pine needle longevity (e.g., Richardson and Rundel 1998). To the best of our knowledge, we have left out such overestimations when calculating species averages. However, the flushing pattern can differ depending on site climate and tree growth conditions (Harrington 1991; Albaugh et al. 2010; Sung et al. 2013), implying that retrospective identification of the number of yearly flushes, e.g., by using plant photographs, is not always straightforward for pine species with a large bioclimatic range. We argue that estimations of pine needle longevity should carefully scrutinize species- and environment-dependent flushing behavior.

Although there are some studies that have looked at needle longevity in species grown in a common garden (Reich et al. 1996; Smith et al. 2019), pine ranges extend from subarctic to tropical climates and we emphasize that in such a global synthesis, the longevity data inevitably include both genotypic and phenotypic sources of variation. Nevertheless, within the global dataset constructed, within-species variation in average Λ_a was smaller than among-species variation (Table 1), suggesting that the database provides an adequate summary of among-species variation in Λ_a.

2.2 *Global Variation in Pine Needle Longevity*

In the global synthesis, Λ_a of adult needles varied almost 30-fold, from 1.13 ± 0.36 yr in the tropical pine *Pinus kesiya* to 19.6 ± 3.8 yr in the subalpine to alpine pine *P. longaeva* (Table 1), whereas the global average across pine species was 3.75 ± 0.30 yr (Fig. 1a). The distribution of needle longevity is right-skewed, indicating that there are fewer species with Λ_a higher than the average than species with Λ_a smaller than the average (Fig. 1a). The highest maximum average longevities (Table 1) were observed in the high altitude species *P. contorta* subsp. *latifolia* (13 yr, in exceptional cases 22 yr), *P. aristata* (15 yr, exceptional 20 yr), *Pinus balfouriana* (15–18 yr, exceptional 30 yr), and *P. longaeva* (33 yr, exceptional 45 yr) (Sargent 1897; Sudworth 1908; Critchfield 1957; Bailey 1970; Mastrogiuseppe and Mastrogiuseppe 1980; Ewers and Schmid 1981; Schoettle 1989, 1994; Richardson and Rundel 1998). The exceptionally high values of longevity are particularly

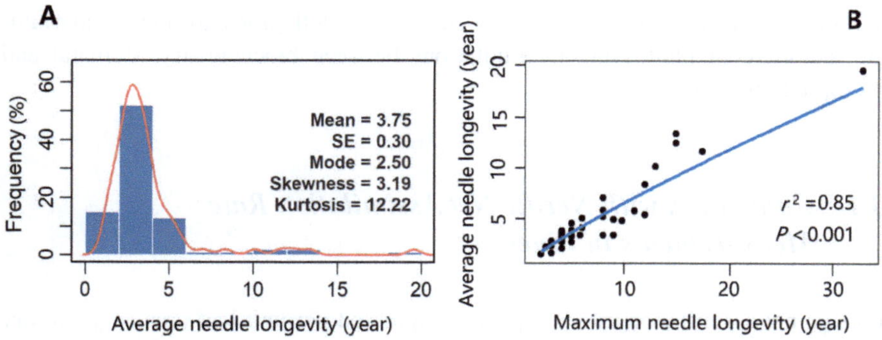

Fig. 1 Distribution of average needle longevity (Λ_a, (**a**)) and the correlation between Λ_a and maximum needle longevity (Λ_m, (**b**)) for 87 *Pinus* species. Λ_a is defined as the oldest needle age class with half of the needles remaining and Λ_m the age of the oldest needles present. Table 1 provides data and data sources (for (**a**): n = 87 for adult needles and n = 4 for juvenile needles, for (**b**): n = 54). Section 5.2 provides a discussion of juvenile to adult needle transition. The distribution of average needle longevity in (**a**) was fitted using kernel density estimation (KDE) and the relationship in (**b**) was fitted by the power function $y = ax^b$ using the *stats* package in R 4.3.3 (R Development Core Team 2024)

interesting as they are close to the maximum limits of leaf longevity in the plant kingdom (Wright et al. 2004b). Although the average and maximum needle longevity are strongly correlated (Fig. 1b), surprisingly few studies have looked at the upper limit of pine needle longevity. In fact, the upper limit of Λ_m might be underestimated for many species where no targeted surveys have been conducted to find the oldest living needles (Table 1), whereas the reported exceptional values largely rest on anecdotal evidence, calling for future studies to characterize the limits of pine needle longevity.

3 Variation Patterns in Needle Photosynthesis and Associated Traits in Pines: Structural, Chemical, and Physiological Constraints

We constructed a database of key leaf economics spectrum (LES) traits LMA, N_{mass}, N_{area}, A_{mass}, A_{area} (all area-based traits expressed per unit needle projected area) and photosynthetic nitrogen use efficiency (PNUE = $A_{mass}/N_{mass} = A_{area}/N_{area}$), and stomatal ($g_s$) and mesophyll diffusion conductance (g_m) as explained in the Text Box 1 (Appendix I for data sources). Furthermore, we standardized the light-saturated net assimilation rates to a common leaf temperature of 25 °C and intercellular CO_2 concentration (C_i) of 300 μmol mol^{-1} allowing to look at leaf trait relationships without interfering environmental sources of variation in assimilation rate (Text Box 1 for detailed explanations). The value of C_i used corresponds to a C_i to ambient (C_a) CO_2 concentration ratio of about 0.7, characteristic to non-water-stressed leaves. In the following we highlight the key hurdles in evaluating reported pine photosynthesis data, characterize the extent of variation in the functional traits across pine species, compare the pine trait averages with global averages and evaluate the share of photosynthetic limitations between biochemistry, stomatal and mesophyll diffusion.

3.1 Challenges with Needle Net Assimilation Rate Measurements in Pines

Foliage net assimilation rate is typically expressed per unit leaf area (A_{area}) or dry mass (A_{mass}), whereas A_{area} is a combined trait, a product of A_{mass} and leaf dry mass per unit area (LMA). Likewise, needle nitrogen content per area (N_{area}) is the product of needle N content per dry mass (N_{mass}) and LMA. In broad-leaved species, leaf thickness is small compared with leaf width, but in needle-leaved species like pines, needle thickness and width are of similar magnitude (Fig. 2). Only one pine species, *Pinus krempfii*, a rare tropical pine, the only species in the *Pinus* subsection *Krempfianeae*, has relatively broad flat leaves (Wang et al. 2000; Brodribb and Feild

2008; Schmiege et al. 2021a; Zhang et al. 2021). Thus, the needle three-dimensional structure importantly alters the calculation of the rate of needle photosynthesis per unit needle area, and therefore, the numeric values of A_{area} differ in dependence on the basis of expression. In the literature, one can find pine photosynthesis rates and chemical element contents expressed per unit projected (S_P) or total (S_T) needle area, or sometimes per unit half of the total needle area ($0.5S_T$) or whole shoot silhouette area (S_S). Furthermore, the numeric values of LMA also depend on the type of area used.

S_P is the easiest to measure; it is typically derived from images of needles removed from the shoot axis and laid flat on a light table using commercial imaging systems or using optical planimeters or estimated as the product of average needle length (L_n) and width (W). Estimation of S_T is more cumbersome; S_T can be estimated by measuring needle length and perimeter from needle cross-sections (Fig. 2, Kuusk et al. 2018a) or from measurements of L_n, W and thickness (T) using idealized geometric approximations of needle shape (e.g., Beets 1977; Benecke 1979; Johnson 1984; Shelton and Switzer 1984; Niinemets et al. 2001, 2002b). Methods involving estimation of needle volume and derivation of S_T using geometric approximations (Johnson 1984) or covering needle surface with a monolayer of small balls with known area/mass ratio and estimating the surface area from the mass difference (Davies and Benecke 1980; Koppel and Frey 1984) have also been used. Once the conversion factor S_T/S_P is known, net assimilation rates can be converted between projected and total area basis.

As demonstrated in Fig. 2, the conversion factor, S_T/S_P varies with the number of needles in the fascicle, emphasizing that the basis of expression of area-based functional traits is by no means a trivial issue. For three-needled pines, a conversion

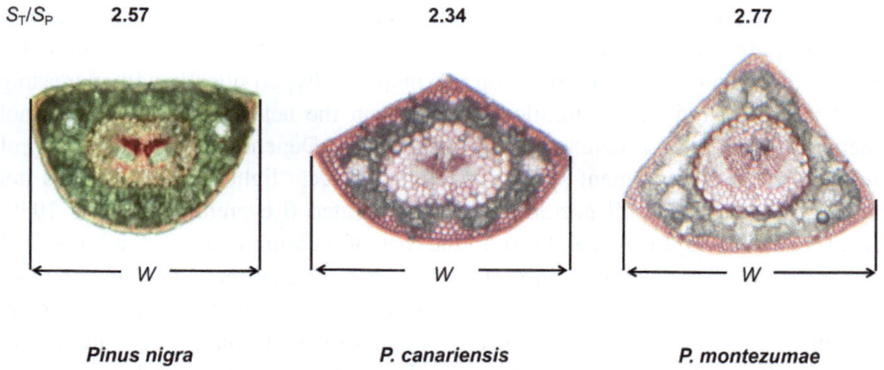

Fig. 2 Representative free-hand needle cross-sections for *Pinus* species with two (*Pinus nigra* subsp. *salzmannii*), three (*P. canariensis*) and five (*P. montezumae*) needles in a fascicle. Needle total (S_T) to projected (S_P) area ratio (S_T/S_P) is equal to needle circumference to needle width (W). The values of S_T/S_P provided in the figure are calculated assuming that the needle cross-section can be approximated by half of the circle (*P. nigra*) or sector of a circle (*P. canariensis* and *P. montezumae*). The needle cross-sections were stained by phloroglucinol that renders the lignified tissues reddish. (The cross-section of *P. nigra*: modified from Kuusk et al. (2018a), the other two cross-sections: Ü. Niinemets unpublished data)

factor of 2.36 is obtained assuming the needle cross-sectional shape is a sector of a circle. This conversion factor has often been used to convert the measurements of projected needle area by an optical planimeter to total area for these pines (Johnson 1984; Ginn et al. 1991; Law et al. 2001). For five-needled pines, the same assumption yields a conversion factor of 2.77 (Fig. 2). Assuming that the needles in two-needled species can be approximated by half of a circle, yields a conversion factor of 2.57 (Niinemets et al. 2001). However, comparisons of geometric approximations with measurements of needle circumference from needle cross-sections demonstrate that needles can significantly deviate from the idealized shape (Niinemets et al. 2001, 2002b; Kuusk et al. 2018a). A further difficulty can result from needle twisting around its axis such that the estimates of S_P obtained by optical planimeters might not correspond to the estimates based on L_n and W, resulting in different estimates of S_T/S_P depending on the method used for S_P determination. When no direct measurements of S_T/S_P are available, the values reported in Fig. 2 for pines with different needle number per fascicle provide a good approximation for the area conversion for non-twisting needles. For twisting needles, additional empirical measurements might be needed to assess the impact of twisting on S_P derived from optical measurements.

To further complicate matters, sometimes pine needle photosynthesis rates are expressed per unit shoot silhouette area (S_S) or for small plants, per unit whole plant projected area (Smith et al. 1991; Maher et al. 2005). S_S to total shoot needle area ratio, ($S_S/(nS_T)$, where n is the needle number per shoot, depends on needle inclination angle distribution and spatial clumping of needles and reflects the degree of self-shading within the shoot (Stenberg et al. 1995; Niinemets et al. 2002a, 2006). Net assimilation rate per S_S can be converted to total area basis when $S_S/(nS_T)$ is known or to projected area basis when $S_S/(nS_P)$ is known.

Apart from the basis of expression, the numeric values of photosynthetic capacity, "the light-saturated net assimilation rate," depend on how the measurements have been conducted. Net assimilation rate increases hyperbolically with increasing light intensity, and light saturation occurs when the net assimilation rate is not increasing with further increases in light availability. Depending on the geometry of foliage in the measurement system and light source, "light saturation" does not necessarily mean that all needles are light saturated (Leverenz and Jarvis 1979; Brunes et al. 1980; Leverenz 1995). In the case of measurements with a broad-leaf clamp-on cuvette as common in portable gas exchange systems, needles are positioned side-by-side, and thus, there is no self-shading among the needles. However, the inherent curvature of needle surface might still imply that light-saturation of photosynthesis cannot be achieved for a unidirectional irradiance (Jordan and Smith 1993). When the entire shoot is enclosed in the measurement chamber, e.g., in a custom-made or commercial conifer chamber, there is typically a certain degree of self-shading among foliage elements, especially for a unidirectional illumination, with the degree of shading depending on shoot architecture. Pine shoots characteristically have a uniform foliage angular distribution with random dispersion, implying that self-shading among the needles within a shoot is relatively large (Stenberg et al. 2001; Niinemets et al. 2002a). This is different from *Abies* and *Picea* spp., where the shoots can be almost flat, especially in low light-acclimated shoots, resembling a flat leaf in broad-leaved species (Stenberg et al. 1998, 1999, 2001). Due to high self-shading, shoot-level measurements do not result in light-saturation of all foliage area enclosed in the measurement chamber, resulting in underestimation of foliage net assimilation rate expressed per unit needle area. Classical studies have demonstrated dramatic differences in photosynthetic rates in conifer shoots exposed to diffuse (e.g., enclosing the foliage in an integrating sphere) or unidirectional irradiance (Żelawski et al. 1973; Szaniawski and Wierzbicki 1978; Brunes et al. 1980), emphasizing that single needle and whole shoot measurements are not directly comparable (Niinemets 2012 for a review).

This discussion emphasizes that when comparing area-based foliage traits across studies, researchers should pay meticulous attention to the basis of expression. When preparing this review, we primarily considered pine studies that included measurements of photosynthesis and also extracted associated structural and chemical traits from these studies. We observed an overall lack of standard of expression of needle photosynthesis rates across studies with all possible expression bases, projected, half-of-the-total, total, silhouette surface area used. In this analysis, all area-based traits are expressed per unit projected needle area as explained in the Text Box 1. We acknowledge that the selection of the expression basis is often practical, but we emphasize that what type of an area has been used must always be stated and whenever conversion factors have been used to convert from one basis of expression to the other, the value of the conversion factor must be reported too.

3.2 *Overall Variation in Key LES Traits*

A large variation, more than ten-fold, was uncovered in most of leaf structural, chemical, and photosynthetic traits in the compiled global pine dataset (Fig. 3 for sample numbers for different traits) with A_{area} varying 14-fold (2.54–35.0 μmol m^{-2} s^{-1}), A_{mass} 18-fold (from 7.9 to 142 nmol g^{-1} s^{-1}), N_{area} 11-fold (0.72–7.90 g m^{-2}), PNUE 15-fold (0.75–11.2 μmol g^{-1} s^{-1}), and LMA 15-fold (51–748 g m^{-2}). The variation in N_{mass} was smaller, but still substantial, 4.5-fold (0.59–2.64%) (Fig. 3). The worldwide variation across vascular plants in these key traits is somewhat larger, on the order of 30–40-fold for most of these traits, with larger variation, 100–150-fold for LMA and A_{mass} (Wright et al. 2004b; Kattge et al. 2011, 2020). Compared with global trait averages (Fig. 3 for pine vs. global average for vascular plants in Kattge et al. (2011)), A_{mass} (48 vs. 115 nmol g^{-1} s^{-1}) and PNUE (4.0 vs. 10.8 μmol g^{-1} s^{-1}) were more than twice lower in pines than the global average, N_{mass} was moderately lower (1.2 vs. 1.74%), A_{area} was similar (10.3 vs. 11.2 μmol m^{-2} s^{-1}), whereas N_{area} was more than two-fold (3.39 vs. 1.59 g m^{-2}) and LMA more than four-fold greater in pines than the global average (267 vs. 60 g m^{-2}). A similar average value of A_{area} and greater N_{area} in pines compared with global averages despite lower mass-based trait values underscore the importance of stacking foliage photosynthetic biomass per unit area. This results from high LMA and high needle thickness.

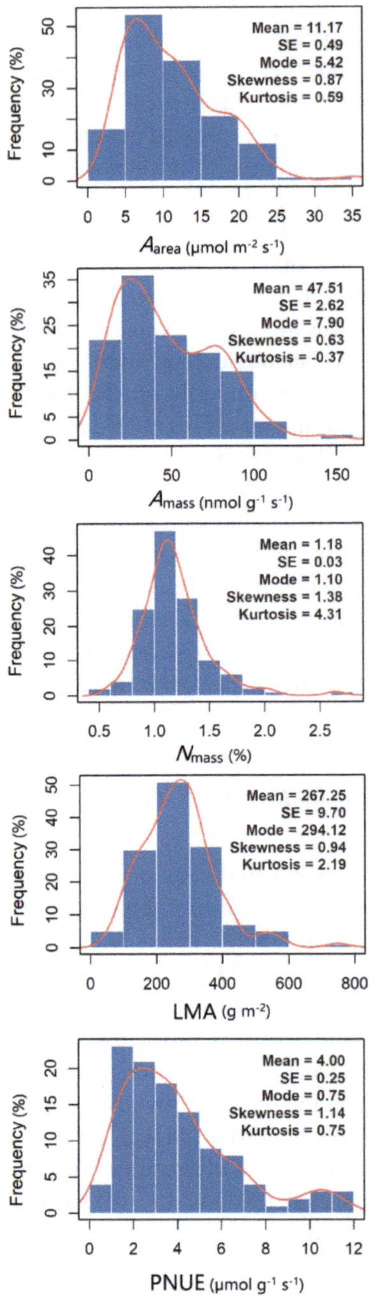

Fig. 3 Distribution of light-saturated needle net assimilation rate per unit needle projected area (A_{area}, $n = 141$), and dry mass (A_{mass}, $n = 120$), nitrogen content per dry mass (N_{mass}, $n = 130$), leaf dry mass per unit projected area (LMA, $n = 126$) and photosynthetic nitrogen use efficiency (PNUE, $A_{mass}/N_{mass} = A_{area}/N_{area}$, $n = 130$) in pines (*Pinus* spp.) across the world (Appendix I for data

In general, the distributions of photosynthetic traits, A_{area}, A_{mass} and PNUE are right-skewed (Fig. 3), indicating that the frequency of values lower than the average is greater than expected for a normal distribution. This evidence together with lower A_{mass} and PNUE compared with the worldwide averages suggests that, globally, a large fraction of pines are positioned at the relatively low end of the leaf economics spectrum. A much lower PNUE in pines compared with the global average can be indicative of a lower investment of N in non-photosynthetic functions as well as greater mesophyll diffusion limitations of photosynthesis as commonly observed for species with structurally robust leaves with high LMA (Niinemets et al. 2009; Knauer et al. 2022) as analyzed in the next section.

3.3 Share of Photosynthetic Limitations Among Biochemistry, and Diffusion Through Stomata and Mesophyll

Maximum stomatal conductance to water vapor ($g_{s,max}$) is a trait that is generally difficult to measure because it sensitively responds to ambient air humidity as well as to soil water limitations. For needle-leaved species, a further complication is accurate estimation of needle temperature that is needed for calculation of stomatal conductance from the transpiration measurements. To analyze the variation in maximum conductance values, we used only the data with intercellular (C_i) to ambient (C_a) ratio (C_i/C_a) > 0.6. Across this truncated pine dataset ($n = 70$), stomatal conductance to water vapor per unit needle projected area varied 6.6-fold (average ± SE = 0.189 ± 0.010 mol m^{-2} s^{-1}), and the corresponding average C_i/C_a ratio was 0.745 ± 0.010.

For mesophyll conductance, g_m, we identified nine studies providing information for ten pine species (Fig. 7 for data sources), indicating that the data of mesophyll conductance are overall limited for pines. Across all the data (n = 32), g_m per unit needle projected area, $g_{m,area}$, varied nine-fold (average ± SE = 0.145 ± 0.012 mol m^{-2} s^{-1}), and g_m per dry mass, $g_{m,mass}$, varied 22-fold (0.69 ± 0.08 mmol g^{-1} s^{-1}).

Fig. 3 (continued) sources). Needle projected area-based trait values are given as $A_{area} = A_{mass}$LMA, and $N_{area} = N_{mass}$LMA. The dataset only includes data of high-light-grown, non-water-stressed plants at peak season (available at DOI: 10.15159/eds.dt.26.02, Niinemets and Zhang 2026, Appendix I for data sources). For several species, multiple observations were included. The sources of variation in trait values within species could be due to plant age (juvenile vs adult needles; seedlings, saplings vs mature trees), needle age and site geography (altitudinal, latitudinal variation). These sources of trait variation are expected to affect needle robustness and thus, the changes in needle traits are anticipated to broadly follow the leaf economics spectrum (Sect. 5.1 for impacts of needle age and Sect. 5.2 for the impacts of plant age and size on the trait values). The number of species available for different functional traits varied between 28 (PNUE) to 36 (A_{area}). The values of net assimilation rate were standardized to a leaf temperature of 25 °C and an intercellular CO_2 concentration of 300 µmol mol^{-1} allowing to look at the trait relationships without interfering physiological sources of variation as explained in the Text Box 1. The distributions were fitted by kernel density estimation (KDE) using the 'density()' function from the *stats* package in R 4.3.3 (R Development Core Team 2024)

This variation in g_m was associated with 1.7-fold variation in chloroplastic (C_c) to intercellular (C_i) CO_2 concentration ratio (C_c/C_i, calculated for a $C_i = 300$ μmol mol^{-1}) with an average of 0.733 ± 0.016. The average g_m value obtained for pines here is more than two-fold greater than the global average for evergreen gymnosperms (Knauer et al. 2022). Given the limited data, we argue that g_m needs to be measured in more pine species to confirm the higher average g_m in pines.

The share of photosynthesis limitations due to biochemistry (l_b), stomata (l_s) and mesophyll diffusion (l_m, $l_b + l_s + l_m = 1$) were calculated according to Tomás et al. (2013) for a C_i range of 200–300 μmol mol^{-1}. Not all studies reporting g_m provided estimates of g_s, so photosynthetic limitations could be calculated only for seven species ($n = 20$ for different site and treatment combinations): *P. contorta* (Guo et al. 2022), *P. densata* (Zhang 2014), *P. densiflora* (Han 2011), *P. kesiya* (Zhang 2014), *P. radiata* (Bown et al. 2009), *P. sylvestris* (Stangl et al. 2022), and *P. yunnanensis* (Zhang 2014). In this dataset, l_b varied 2.2-fold (0.32–0.71; average ± SE = 0.498 ± 0.027), l_s 3.6-fold (0.16–0.60; 0.326 ± 0.031) and l_m 4.1-fold (0.08–0.31; 0.176 ± 0.016). Unexpectedly, the lowest g_m-limitations were observed in the study of Guo et al. (2022) for a subalpine population of *P. contorta*. This study was outlying also because of higher-than-average stomatal limitations of photosynthesis (0.35–0.57).

The relatively low contribution of g_m to photosynthesis limitations in pines is surprising and suggests that the existing data might not be representative. In evergreen gymnosperms analyzed globally, g_m-limitation is larger, 0.15–0.35 (Knauer et al. 2022). Overall, the measurements of g_m are challenging (Pons et al. 2009; Flexas et al. 2013; Bunce 2018; Evans 2021), especially for needle-leaved conifers due to small to moderate gas-exchange fluxes, difficulties in estimating the light intensity on leaf surface, accurate estimation of needle temperature and possible diffusion leaks. Thus, the relatively large variability in g_m-limitations of photosynthesis among pines uncovered in this analysis might partly be associated with study-to-study differences in the uncertainty of g_m estimates. We argue that more research is needed to gain conclusive insight into the quantitative importance of g_m in pines.

4 Pine Needle Economics Spectrum: Scaling Structure and Function

Sections 2.2, 3.2, and 3.3 demonstrated that there is a large variation in foliage structural, chemical and photosynthetic traits and needle longevity among pine species worldwide. Although positioned at the lower end of the global leaf economics spectrum, we demonstrate here that the variation uncovered in the needle traits results in a remarkably broad continuum of trait combinations across pines. In the following, we show the relationships with all the data combined (Fig. 3 for the sources of variation within species included in the relationships) as in classical studies (e.g., Wright et al. 2004b) and with species averages. The relationships are qualitatively similar for both sets of data, although the data ranges are somewhat narrower for species average data.

4.1 Relationships Among Foliage Structural and Functional Traits

Across the pines, we observed typical LES relationships, including the negative scaling of A_{mass} with LMA (Fig. 4a, b), positive scaling of A_{mass} with N_{mass} (Fig. 4c, d), negative scaling of N_{mass} and LMA (Fig. 4e, f) and negative scaling of PNUE with LMA (Fig. 4g, h). For the area-based relationships, the correlations of A_{area} with N_{area} (Fig. 5a, b) and with LMA (for power functions, $r^2 = 0.05$, $P < 0.05$ for all data, and $r^2 = 0.08$, $P > 0.05$ for species averages) were much weaker. However, there was a strong positive relationship among N_{area} and LMA (Fig. 5c, d). Such a positive correlation also exists within the global LES, albeit it is weaker in the global LES (e.g., $r^2 = 0.23$, $P < 0.001$ in Wright et al. 2004b). The strong positive relationship between LMA and N_{area} and the weak scaling of A_{area} with N_{area} indicates diminishing returns, i.e., the greater N_{area} the smaller the impact of any further increase of nitrogen content on assimilation rate. Indeed, the relationship of diminishing returns is reflected in a negative scaling of PNUE with LMA (Fig. 4g, h).

As the global pine data demonstrate, N_{mass} can be potentially relatively high, more than 2% especially in fertilized pines (Fig. 3) (Makoto and Koike 2007; Tarvainen et al. 2016; Wu et al. 2021), but high N investments in leaves will be inefficiently used in constructing photosynthetic machinery. This evidence of efficient use of low N and inefficient use of high N is in agreement with widely perceived low nutrient tolerance of pines and their success in colonizing infertile habitats.

The low PNUE in leaves with greater LMA has been typically associated with enhanced mesophyll diffusion limitations (Niinemets 1999; Onoda et al. 2017). However, as discussed in Sect. 3.3, the data available suggest that mesophyll diffusion might be less limiting in pines than in other evergreen gymnosperms. Nevertheless, there was a significant variation in chloroplastic to intercellular CO_2 concentration ratio (C_c/C_i), and the variation in C_c/C_i was driven by changes in $g_{m,mass}$ (Fig. 6a). Furthermore, $g_{m,mass}$ scaled negatively with LMA for ten pine species with available data (Fig. 6b). This evidence suggests that mesophyll conductance might play an important role in LMA-dependent reductions in PNUE, again underscoring the need for a systematic survey of mesophyll diffusion limitations across pines with contrasting structure and physiological activity.

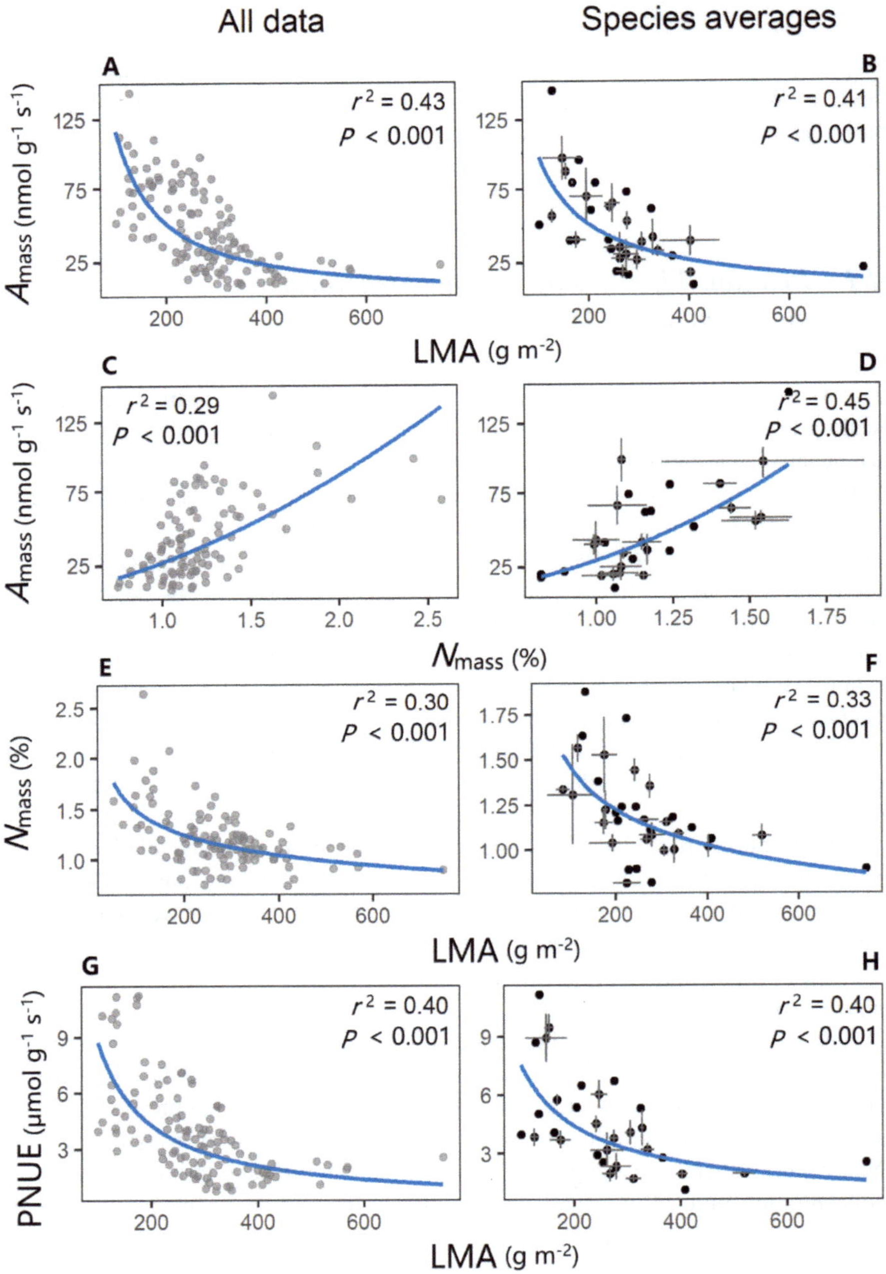

Fig. 4 Relationships of pine photosynthetic capacity per dry mass (A_{mass}) with leaf dry mass per unit projected area (LMA; **a**, **b**) and nitrogen content per dry mass (N_{mass}, **c**, **d**), N_{mass} with LMA (**e**, **f**) and photosynthetic nitrogen use efficiency (PNUE) with LMA (**g**, **h**) across the global pine dataset (**a**, **c**, **e**, **g**) and with species averages (**b**, **d**, **f**, **h**). Sample sizes and details of data as reported in Fig. 3 (Appendix I for data sources). A_{mass} data were standardized to a leaf temperature of 25 °C and intercellular CO_2 concentration of 300 µmol mol^{-1} (Text Box 1). In plots with species averages, error bars show ± SE. Data were fitted by the power function $y = ax^b$ using the *stats* package in R 4.3.3 (R Development Core Team 2024). r^2 values of the nonlinear fits (proportion of variance explained) are calculated as 1-RSS/TSS, where RSS is the residual sum of squares and TSS is the total sum of squares

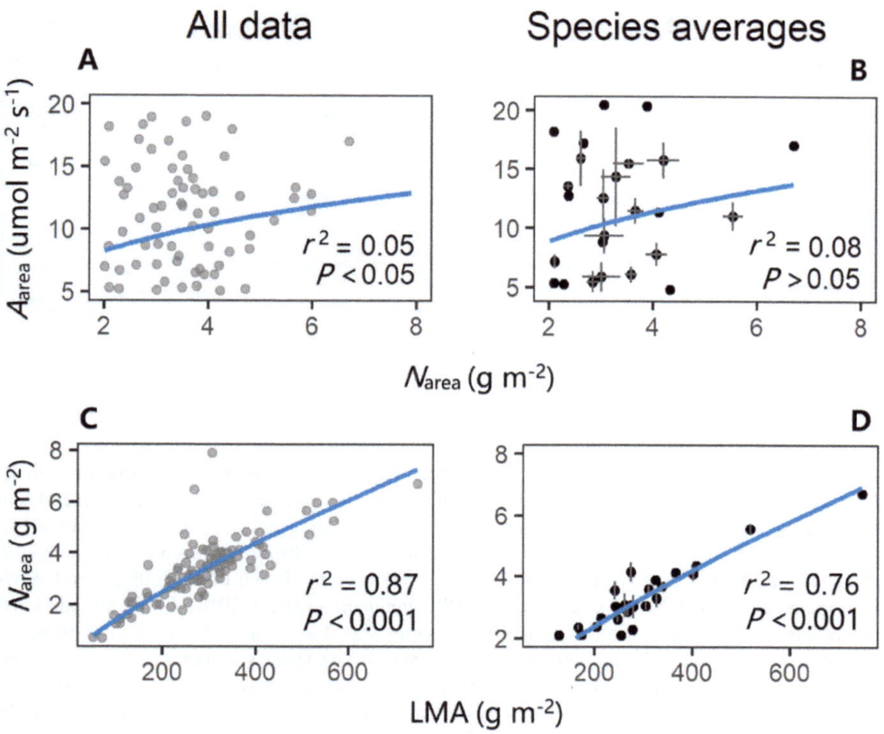

Fig. 5 Relationships of pine photosynthetic capacity per unit projected needle area (A_{area}) with needle nitrogen content per unit projected area (N_{area}) across all data (**a**) and for species averages (**b**), N_{area} with LMA across all data (**c**) and for species averages (**d**). Data and data presentation as in Fig. 4. Data were fitted by the power function $y = ax^b$ using the *stats* package in R 4.3.3 (R Development Core Team 2024). r^2 values as in Fig. 4

4.2 LES Traits in Relation to Needle Longevity

The premise of scaling leaf structural, chemical, and physiological traits with leaf longevity is that the increase in leaf longevity requires increased investment is structural and chemical defenses and as the result, the investment in photosynthetic machinery inevitably decreases (Wright et al. 2004b; Mason and Donovan 2015; Onoda et al. 2017; Alonso-Forn et al. 2020). In addition, enhanced structural investments can be associated with increased diffusive limitations of photosynthesis (Sect. 4.1, Fig. 6b). Given that needle traits also change with increasing needle age (Sect. 5.1), the correlations of needle structural, chemical, and photosynthetic traits with average needle longevity were examined for young needles (current year and one-year-old). Across the pines, A_{mass} (Fig. 7a, b) and N_{mass} (Fig. 7c, d) scaled negatively, LMA (Fig. 7e, f) positively, and PNUE (Fig. 7g, h) negatively with average needle longevity (Λ_a). These patterns are in agreement with worldwide LES (Wright et al. 2004b), and highlight the major trait differentiation across pines with varying

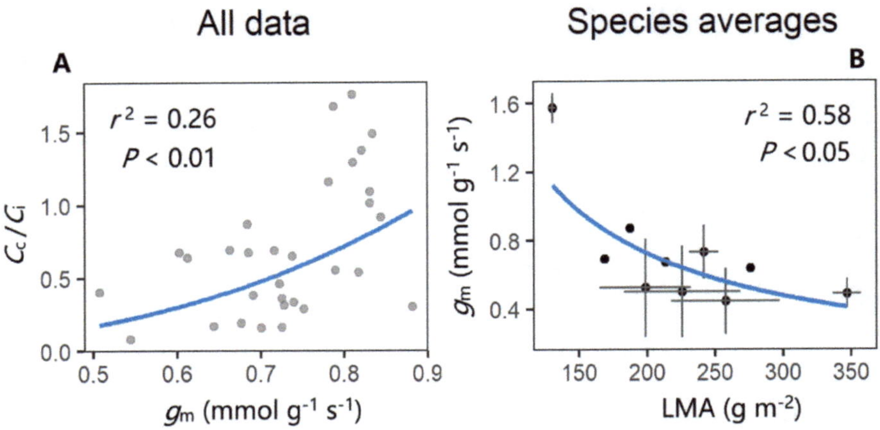

Fig. 6 Relationships of (**a**) the ratio of CO_2 concentration in chloroplasts (C_c) to that in intercellular air space (C_i, C_c/C_i) and (**b**) mesophyll conductance per needle dry mass (g_m) with LMA in ten *Pinus* species: *P. contorta* (Guo et al. 2022), *P. densata* (Zhang 2014), *P. densiflora* (Han 2011), *P. halepensis* (Kuusk et al. 2018a, b), *P. kesiya* (Zhang 2014), *P. nigra* subsp. *salzmannii* (Kuusk et al. 2018a, b), *P. pinea* (Kuusk et al. 2018a, b), *P. radiata* (Bown et al. 2009), *P. sylvestris* (Veromann et al. 2017; Stangl et al. 2022), and *P. yunnanensis* (Zhang 2014). In (**a**), all individual datapoints are shown and the variation within species was due to site location, needle ontogeny (juvenile vs. adult), tree size and fertilization treatment. In (**b**), species averages (± SE) are shown, and data for both juvenile and adult needles are reported for *P. halepensis*, *P. nigra* subsp. *salzmannii* and *P. pinea* ($n = 13$). Data were fitted by the power function $y = ax^b$ using the *stats* package in R 4.3.3 (R Development Core Team 2024). r^2 values as in Fig. 4

leaf turnover from tropics to subarctics and from lowlands to treeline. Due to lack of information of leaf structural, chemical and physiological traits for pines at the extremes of Λ_a in this combined trait dataset, however, the range of Λ_a was smaller than observed in pines globally (Table 1 for the full Λ_a range). Structural, chemical and especially photosynthetic data are limited for pines with both exceptionally high and low needle longevities, calling for studies of pine photosynthetic and structural differentiation at the limits of Λ_a.

Needle longevity vs. photosynthetic, chemical and structural trait relationships were stronger for species averages than for all data (e.g., Fig. 7a, b for A_{mass} vs. Λ_a for all vs. species averages). This likely highlights the circumstance that Λ_a as used in constructing these relationships is a species average value (Table 1). Λ_a itself is a plastic trait (Sect. 2.1), and thus, it is also expected that plastic alterations in Λ_a in response to site conditions, e.g., increase in Λ_a with altitude (Sect. 2.1), also lead to concomitant modifications in the suite of needle structural, chemical and photosynthetic traits. For example, an increase of LMA with increasing altitude has been observed in several pine species (Hultine and Marshall 2000; Wieser et al. 2010) and this is further associated with reduced photosynthetic activity (Wieser et al. 2010) as predicted by LES. Although the available evidence is broadly consistent with the predictions based on LES, the data of concurrent modifications in Λ_a and associated suite of LES traits are surprisingly limited for pines. In addition, apart

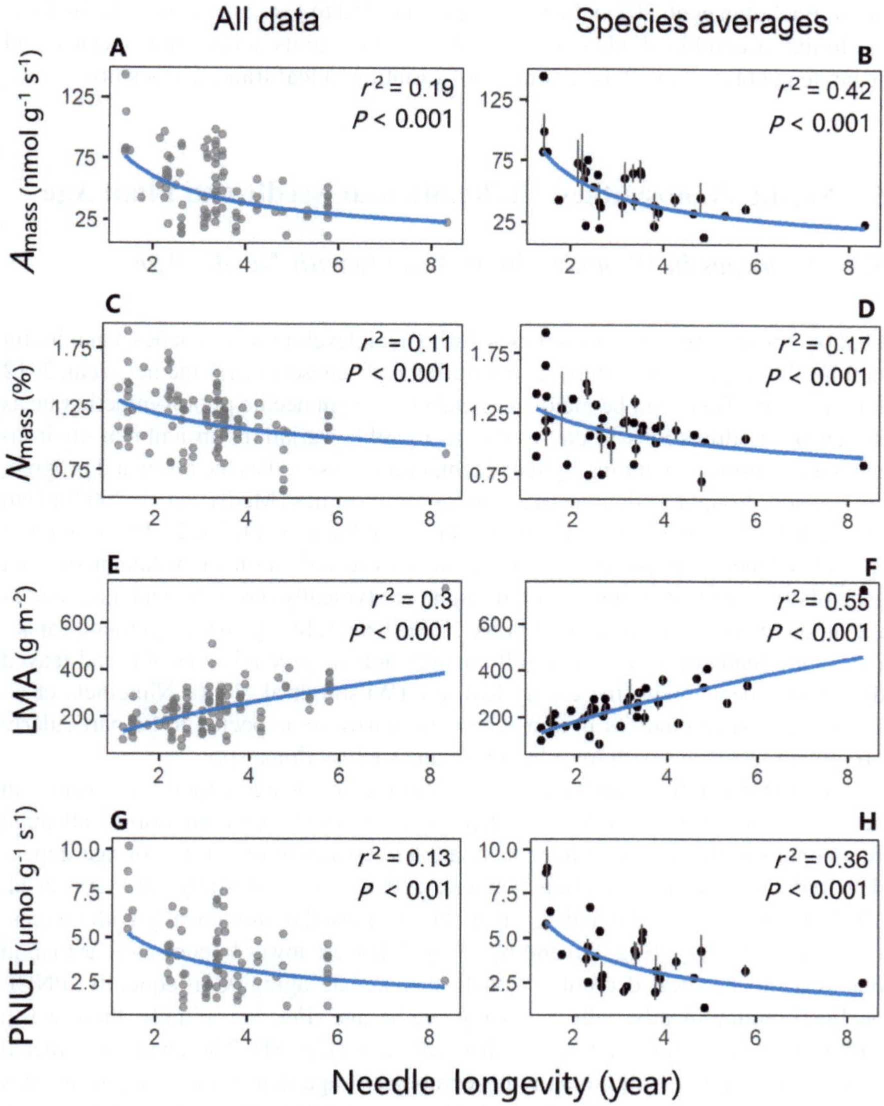

Fig. 7 Dependencies of A_{mass} (**a**, **b**), N_{mass} (**c**, **d**), LMA (**e**, **f**) and PNUE (**g**, **h**) on average needle longevity (Λ_a) for all pine data (**a**, **c**, **e**, **g**) and for species averages (**b**, **d**, **f**, **h**). In this analysis, trait values for only current year and one-year-old needles were used. Data of leaf photosynthetic, structural and chemical traits as in Fig. 3. Data of average and maximum needle longevity are reported in Table 1. Data presentation and fitting as in Fig. 4

from the phenotypic modifications, there is also genetic variation in pines along altitudinal gradients (Sáenz-Romero et al. 2011, 2012; Huang et al. 2015; Leal-Saenz et al. 2020), but only a few studies have attempted to resolve plastic and ecotypic sources of trait variation in pines (Huang et al. 2015; Leal-Saenz et al.

2020; Bachofen et al. 2021). Further work is needed to gain insight into the biological limits of combined adjustments of Λ_a and LES traits across pine species, and separate genotypic and plastic sources of variation in leaf trait relationships.

5 Needle Photosynthesis in Relation to Needle and Plant Age

5.1 Changes in Photosynthetic Activity with Needle Age

Leaf photosynthetic activity increases during leaf development, reaches a maximum in fully-developed leaves and declines during leaf senescence (Niinemets et al. 2012 for a review). There can be major seasonal changes in needle photosynthetic rate as driven by modifications in day-to-day to monthly variations in ambient environmental conditions, including light and temperature, as well as the onset and progression of soil drought as demonstrated in multiple studies (Medlyn et al. 2002b; Han et al. 2004; Peters et al. 2008; Deligöz et al. 2018; Stangl et al. 2022). These changes in realized net assimilation rate, e.g., the classic reduction in instantaneous net assimilation rate in response to soil drought, are typically physiological, i.e., reversible upon return of previous conditions. Nevertheless, foliage photosynthetic capacity can gradually decline even in fully mature non-senescent leaves of broad-leaved deciduous species with increasing leaf age (Wilson et al. 2000; Niinemets et al. 2004a), and such changes in non-senescent leaves are expected to be particularly prominent in species with long-living foliage such as *Pinus* spp.

As a representative example, we highlight the needle age-dependent changes in key LES traits in *P. leucodermis* (Oleksyn et al. 1997). Examination of all main functional needle traits simultaneously allows separation of sources of age-dependent reductions in needle photosynthetic activity. In their study (Oleksyn et al. 1997), both A_{mass} (Fig. 8a) and A_{area} (Fig. 8b) decreased monotonically with increasing needle age. N_{mass} (Fig. 8c) and N_{area} (Fig. 8d) were lower in current-year than in one-year-old needles, declining with further needle aging. Consequently, PNUE declined monotonically with increasing needle age (Fig. 8e). Finally, there was a strong positive scaling of LMA with needle age (Fig. 8f). The changes depicted broadly correspond with literature observations of age-dependent changes in pines (Warren 2006; Peters et al. 2008; Op de Beeck et al. 2010; Wei et al. 2011; Yan et al. 2012; Eimil-Fraga et al. 2015a; Wyka et al. 2016; Kuusk et al. 2018b; Ji et al. 2021; Wu et al. 2021).

The needle age-dependent decline in photosynthetic activity in conifers can reflect thickening of cell walls, and enhanced mesophyll diffusion limitations with concomitant reductions in PNUE as has been observed in evergreen broad-leaved species (Niinemets et al. 2005). However, despite reductions in mesophyll conductance, the CO_2 drawdown was similar in different-aged needles of *P. halepensis*, *P. nigra* subsp. *salzmannii*, and *P. pinea* (Kuusk et al. 2018b) and in *P. pinaster* (Warren 2006). This indicates that the reductions in mesophyll conductance and

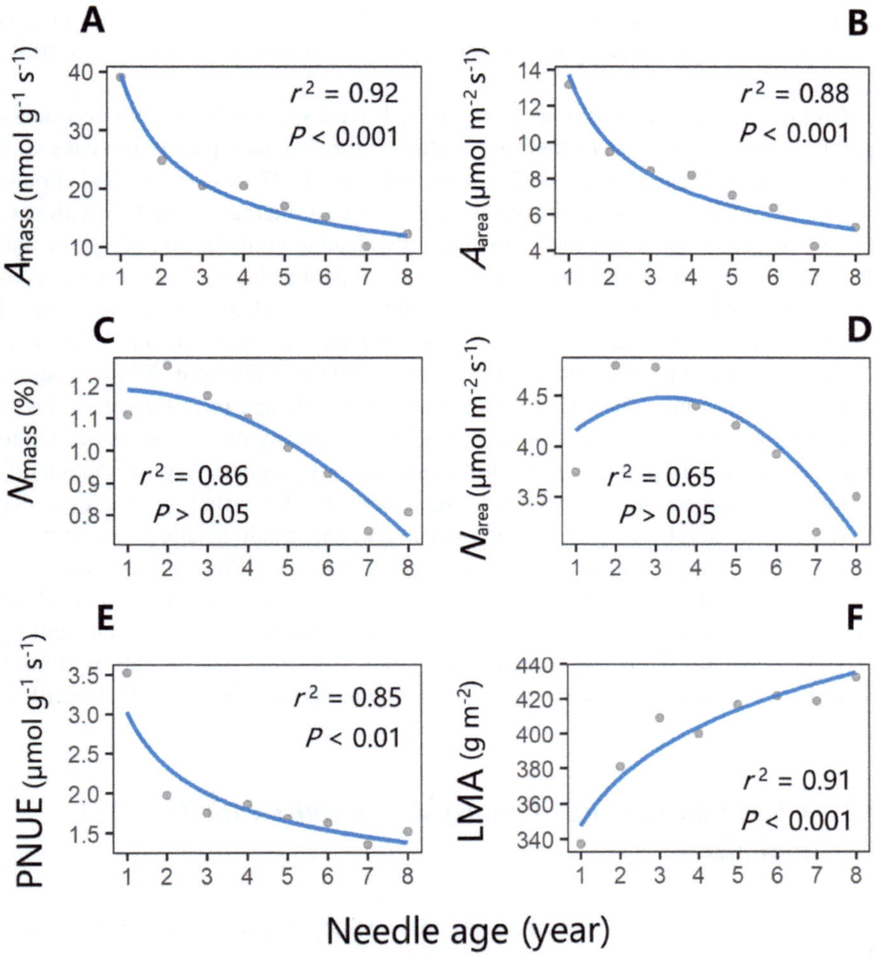

Fig. 8 Sample relationships of A_{mass} (**a**), A_{area} (**b**), N_{mass} (**c**), N_{area} (**d**), PNUE (**e**), and LMA (**f**) vs. needle age in *Pinus leucodermis*. Data for A_{mass}, PNUE, N_{area} and A_{area} were fitted by power relationships in the form of $y = ax^b$, whereas second order polynomial regressions were used for N_{mass} and N_{area}. Data modified from Oleksyn et al. (1997). For the sake of data fitting with power relationships, needle age of 1 yr in this figure corresponds to current year needles. r^2 values as in Fig. 4

photosynthetic capacity were of similar magnitude in these species. In *P. pinaster*, the decrease of PNUE with needle aging was associated with deactivation of Rubisco (Warren 2006), suggesting a coordination between increased diffusion limitations and concomitant reductions in Rubisco activity. Deactivation of Rubisco in older needles might be related to increased shading of older foliage by younger needles as has been observed in conifers (Brooks et al. 1994, 1996). The question is, however, why excess Rubisco is not dismantled to invest N in other rate-limiting processes, e.g., in enhanced light harvesting, or is not used to support new foliage development. Such "luxurious" N partitioning pattern may further underscore the inefficiency of N use when N availability is high, but the share of photosynthesis limitations through needle aging has not been studied across sites of varying fertility in pines.

Conifer needles are different from broad-leaved species by having secondary needle growth, such that needle cross-sectional dimensions typically increase with increasing needle age (Ewers 1982; Ewers and Aloni 1987; Niinemets 1997; Fouda 2004). Secondary growth can be partly responsible for increases in LMA with needle age (Fig. 8f), but secondary growth can also lead to compression of xylem, and disruption of xylem water transport with concomitant reductions of stomatal conductance in older needles. The data of needle age-dependent changes in stomatal conductance in pines are limited and controversial. For example, in *P. pinaster*, stomatal conductance was little affected by needle age (Warren 2006), whereas stomatal conductance decreased with increasing needle age in *P. sylvestris* (Wyka et al. 2016) and *P. canariensis* (Peters et al. 2008). Clearly more work is needed to separate the share of photosynthetic limitations among biochemistry, mesophyll diffusion and stomata in different-aged pine needles. Nevertheless, the evidence reviewed collectively suggests that age-dependent changes in needle photosynthetic activity play a major role in whole canopy photosynthesis. Pine species that maintain a high proportion of old needles are expected to have a much smaller whole canopy photosynthetic activity than predicted on the basis of physiological activity of young needles. Thus, we suggest that studies on pine needle photosynthesis should not only measure the youngest fully mature needles, but also older needles.

5.2 Changes in Foliage Photosynthetic Rates with Tree Age and Size

Many plant species are characterized by tree age-dependent leaf dimorphism with distinct differences among juvenile and adult leaves, a phenomenon called heteroblasty (Zotz et al. 2011). In pines, juvenile needles are solitary and have often a circular or rhomboidal cross-sectional shape (Kuusk et al. 2018a; Fernández-Marín et al. 2021), while the adult needles are in fascicles (short shoots) with the characteristic half-circular or sectorial cross-sectional shape depending on the number of needles in the fascicle (Fig. 2). Juvenile needles are softer and thinner, with lower LMA and greater photosynthetic capacity and higher mesophyll conductance (Kuusk et al. 2018b). Lower needle robustness of juvenile needles is associated with their smaller average needle longevity (average ± SE = 1.28 ± 0.07 yr for juvenile and 2.95 ± 0.30 yr for adult needles for the four species in Table 1; $P < 0.03$ for the comparison of averages with a paired samples *t*-test). Although heteroblasty occurs in all pines, the duration of formation of juvenile needles differs among species. For example, in Mediterranean pines, juvenile needles can be found in seedlings of 2–3 years old, whereas Eurasian boreal and temperate pines typically start forming adult needles already in the first or after the first growing season (Climent et al. 2006,

2011). Furthermore, in some pine species, e.g., in *P. canariensis*, a phenomenon of rejuvenation, i.e., formation of juvenile needles in mature trees can be observed; this species forms juvenile needles on epicormic resprouts, e.g., after stem mechanical damage or after fire (Pausas and Keeley 2017; Fernández-Marín et al. 2021).

Apart from alterations in foliage functional traits during juvenile-to-adult transition, there is a continuous change in foliage chemical, structural, and photosynthetic traits of adult needles with increasing tree size and age, a change sometimes called ontogenetic drift (Zotz et al. 2011). Typically, tree height increases with increasing tree age, and this is associated with greater water limitations in larger trees due to increased resistance of longer water conducting elements and gravitational component of water potential (Mullin et al. 2009; Steppe et al. 2011; Williams et al. 2017). In pines, LMA strongly increases with increasing tree height (Fig. 9a, Niinemets 2002; Han 2011), and this is associated with reduced A_{mass} (Fig. 9b, Niinemets 2002; Han 2011). The size-dependent reduction in photosynthetic activity has been associated with reduced stomatal conductance due to greater water stress in larger trees (Yoder et al. 1994; Han 2011), but the reduction in A_{mass} with increasing tree size is also evident when A_{mass} values are examined at a constant intercellular CO_2 concentration (Fig. 9b, Niinemets 2002; Han 2011). The reduction of A_{mass} at a constant C_i suggests that PNUE and/or mesophyll conductance might decrease with increasing tree size. Indeed, PNUE decreased with increased tree size in *P. sylvestris* (Niinemets 2002) and in *P. densiflora* (Han 2011). The study of Han (2011) further demonstrated that mesophyll conductance decreased with increasing tree size in *P. densiflora* and this was associated with reduced chloroplastic CO_2 concentration, indicating that mesophyll conductance limited photosynthesis more in larger trees.

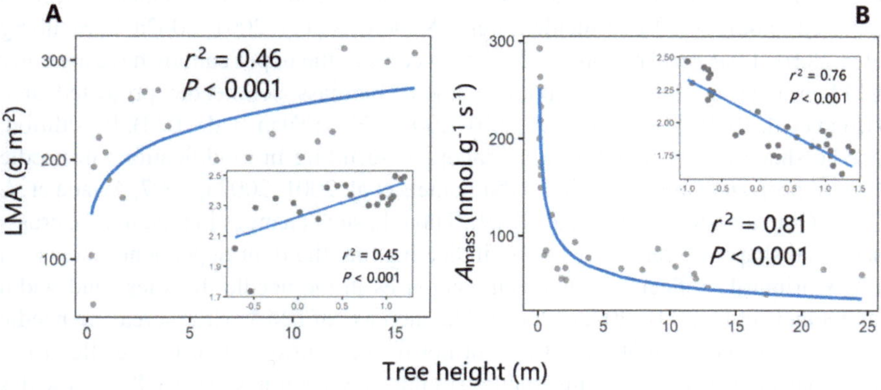

Fig. 9 Dependencies of LMA (**a**) and A_{mass} (**b**) on tree size in *Pinus sylvestris*. The insets show the data in log-log axes. This figure includes all the data from the meta-analysis of Niinemets (2002) ($n = 30$) and additional data from four additional studies ($n = 9$) (Mediavilla and Escudero 2003a; Wright et al. 2004b; Tarvainen et al. 2016; Wyka et al. 2016). Data of only current and one-year-old needles were included. All A_{mass} data are converted to 25 °C and intercellular CO_2 concentration of 300 μmol mol^{-1} (Text Box 1). Both the main panel (nonlinear power function, $y = ax^b$) and the inset (log-log linear regression, $\log_{10}y = \log_{10}a + b\log_{10}x$) were fitted using the *stats* package in R 4.3.3 (R Core Team 2024). r^2 values as in Fig. 4

This suggests that the increased diffusive limitations are an inevitable consequence of tree size-dependent needle structural changes, including increases in LMA.

The size-dependent changes in needle structure and function are in agreement with enhanced water stress in large trees, but we argue that it is difficult to tease apart age- and height-dependent changes in needle structure and function as age and size are correlated and therefore, the increases in LMA and reductions in A_{mass} correlate with both tree height and age (Niinemets 2002). Grafting experiments with various-aged rootstocks and scions (shoots) do indicate a certain impacts of tree age on needle traits, e.g., on needle size (Greenwood 1987, 1995; Mencuccini et al. 2007). Such age-dependent changes have been associated with alterations in DNA methylation profiles (Fraga et al. 2002; Gardner et al. 2023). Nevertheless, in grafted *P. sylvestris* plants, scion age had no effect on foliage photosynthetic activity and the photosynthetic modifications were associated with tree size (Vanderklein et al. 2007). So far, there is very limited information of the velocity, extent and interspecific variability of age-dependent modifications in foliage characteristics. Thus, we conclude that the trait alterations in adult needle characteristics as the trees increase in size might include an age-dependent component, but more work is needed with different *Pinus* species. Especially interesting might be to look at these alterations in species that have a large capacity for rejuvenation after damage.

6 Plasticity of Pine Needle Photosynthetic Characteristics

Although pine canopies are often relatively open, there are nevertheless strong light availability gradients from canopy top to bottom, with lower canopy needles exposed to only 5–10% of incident light (Niinemets et al. 2001, 2002a, b; Stenberg et al. 2001; Leal and Thomas 2003). The needles in the upper canopy have a greater LMA, reflecting stacking of photosynthetic biomass per needle projected area (Bond et al. 1999; Niinemets et al. 2001, 2002a, 2007; Yuan et al. 2023). In addition, needle shape can vary within pine canopies, resulting in modifications in needle total to projected area ratio (S_T/S_P) (Niinemets et al. 2001, 2002a, 2007; Awada et al. 2003; González-Rodríguez et al. 2019). Though, such changes in pine needle cross-sectional shape are relatively minor, indicating that the light-dependent increase in LMA primarily reflects simultaneous increases in the needle thickness and width (Niinemets et al. 2001, 2002a, 2007). The increase of LMA with increasing needle width might seem counterintuitive, but not if one considers that the needle cross-sectional shape is relatively invariable (similar S_T/S_P) in pines. As needle area scales linearly with thickness and width, while volume and mass scale with the product of thickness and width, it follows that both increases in thickness and width increase LMA in pines (Niinemets et al. 2001 for representative calculations; Niinemets et al. 2002a). These anatomical changes are accompanied by increases in N_{area} and Rubisco content per needle area, leading to a higher A_{area} in needles at high light (Bond et al. 1999; Niinemets et al. 2007; Tarvainen et al. 2016; Yuan et al. 2023).

Differently from A_{area}, A_{mass} is invariable or changes much less within the canopy light gradients in pines, further underscoring the importance of LMA in light acclimation of A_{area} (Malkina 1982; Bond et al. 1999; Niinemets et al. 2007; Niinemets 2023). In addition to light per se, high light at the top of the canopy is often associated with limited water availability and greater mechanical stress (Niinemets et al. 2004b), and this can lead to greater investments in structural tissues and thicker cell walls (Niinemets et al. 2007; Yuan et al. 2023). These structural changes can result in a stronger internal CO_2 diffusion limitation, reducing mesophyll conductance and altering the balance between stomatal and mesophyll limitations (Guo et al. 2023). In addition, as mentioned in Sect. 5.1, older needles become gradually shaded as canopy expands and new needles are formed, but the extent of structural reacclimation of the older foliage to lower light conditions is inherently limited (Brooks et al. 1996; Leal and Thomas 2003). In particular, needle anatomy can change little in mature needles, and secondary needle growth (Sect. 5.1) is expected to render the older shaded needles even less efficient under new shadier conditions. Nevertheless, available evidence for other conifers suggests that older shaded needles can increase light harvesting efficiency by increasing chlorophyll content (Brooks et al. 1994, 1996).

So far, the data of within-canopy variation in the share of photosynthetic limitations are scarce for pines, but in other conifers, within-canopy variation in mesophyll conductance varies within the canopy (Peguero-Pina et al. 2016). Future studies should examine the reacclimation of pine needle photosynthetic traits to altered light climate as well as the variation in the share of photosynthesis limitations within the canopy.

The vertical heterogeneity in leaf traits is further influenced by environmental resource availability, particularly soil nutrients, which modulate the degree of photosynthetic acclimation across canopy layers. In nutrient-rich environments, pine needles exhibit greater morphological and physiological responsiveness to light gradients, including simultaneous increases in needle thickness and width, contributing to increased LMA, N_{area} and A_{area} (Niinemets et al. 2001; Maier et al. 2008; Zhao et al. 2024b). In contrast, trees growing in nutrient-poor soils display reduced photosynthetic plasticity, with diminished capacity for structural and biochemical adjustments in leaves exposed to different light environments (Niinemets et al. 2001, 2002a). Such a lower plasticity under nutrient limitation is associated with overall lower light range due to more open canopies under strong nutrient limitation as well as with reduced carbon availability for needle growth, constraining changes in LMA and A_{area} over the given light range (Niinemets et al. 2001; Niinemets 2020). In summary, we argue that the within-canopy variation in LES traits and the impact of belowground resource availability on photosynthetic acclimation should be included in scaling up photosynthesis from needles to canopies.

In addition to light and nutrient availability, temperature and water availability are critical drivers that shape the plastic responses of conifer needles across both spatial and temporal gradients, and are largely responsible for seasonal changes in needle photosynthetic activity. Seasonal variability in pine needle photosynthesis has been examined in numerous studies (e.g., Anderson et al. 1997; Irvine et al.

1998; Medlyn et al. 2002b; Han et al. 2004; Deligöz et al. 2018; Stangl et al. 2022; Liu et al. 2023). Seasonal changes to temperature and water availability include both modification in foliage photosynthetic capacity as driven by chemical and structural traits as well as reversible physiological changes. In addition, temperature acclimation involves changes in the shape of the temperature response curve of photosynthesis as well as the onset of winter dormancy and breakage of dormancy. As this chapter focuses mainly on photosynthetic capacity and interspecific and intraspecific variability in the photosynthetic capacity, the instantaneous physiological changes in photosynthetic activity and plastic modifications in the shape of response curves are beyond the scope of this review. There is a large body of literature of alterations of the shape of the temperature response of photosynthesis upon acclimation to changed temperature conditions (Medlyn et al. 2002a, b; Han et al. 2004; Tjoelker et al. 2008; Wieser et al. 2010; Tarvainen et al. 2018), and frost responses of subarctic to temperate conifers (Buchner and Neuner 2011; Pardos et al. 2014; Menzies and Holden 1981; Greer and Warrington 1982; Greer and Stanley 1985; Savitch et al. 2002; Strand and Öquist 1988; Sutinen et al. 2000), and we refer to the original studies for in depth consideration. Taken together, pines are characterized by strong photosynthetic plasticity that needs consideration in predicting pine photosynthetic performance in the field.

7 Conclusions and Outlook

We constructed a global database of pine structural, chemical, and photosynthetic traits and needle longevity, and demonstrated a major coordinated variation in these traits—pine needle economics spectrum. The economics spectrum interconnects key needle functional traits that drive the variation in needle light-saturated net assimilation rate (photosynthetic capacity) across pines worldwide. Although pines are traditionally perceived to possess a similar ecological strategy, characteristically colonizing open high light habitats, and typically be shade intolerant and drought tolerant, pine economics spectrum indicates a remarkable variation in the way pines allocate carbon and nitrogen for efficient carbon assimilation. Pines as long-living evergreen trees are also characterized by distinct transition from carrying juvenile to adult needles, by changes in needle photosynthetic activity with increasing tree age and size, and by large needle age-dependent changes in needle functional traits and significant phenotypic modifications in response to variations in environmental drivers, including within-canopy variations in light. Our analysis demonstrates that there is relatively good information of interspecific variation in needle photosynthetic capacity, albeit the data are still limited for several widespread pine species, especially for multiple widely planted tropical pines.

This chapter indicates that little research has been conducted to characterize the share of photosynthetic limitations between mesophyll diffusion, stomata, and biochemistry among pine species. Quantitative characterization of the degree of photosynthetic limitations is highly important in predicting how different species respond to the global increase in atmospheric [CO_2]. Likewise, more comparative studies are needed to quantify the interspecific variability in needle and tree age-dependent changes in needle photosynthetic rates and interspecific variation in photosynthetic acclimation to environmental changes.

We emphasize that in analyses of trait relationships in species with complex three-dimensional photosynthetic surface like pines, the expression basis of needle traits requires special attention. Studies in pine photosynthesis have reported data on projected, total, half-of-the-total, shoot silhouette or mass basis. For comparative analyses, all data included must use the same expression basis. We feel that photosynthesis community has not paid due attention to relevant details in reporting gas exchange data. We have observed that many studies did not report the expression basis of photosynthesis rate and other area-based traits, or the expression basis reported did not match the trait numeric values shown. In addition, photosynthetic measurements of thick foliage elements with clamp-on cuvettes of portable photosynthesis systems are challenging due to possible diffusion leaks, and problems with needle temperature and achievement of light saturation. Unrealistic values of photosynthetic capacity and other needle traits have been reported in several studies, and we suggest that when using published data of pine needle functional traits, extreme scrutiny is needed, including careful examination of the measurement methods and calculation procedures. Datasets not providing sufficient information of measurement methods should not be used in broad-scale analyses. A protocol for reporting gas-exchange data has been recently proposed (Ely et al. 2021), and we suggested that future studies on foliage gas exchange should adhere to this protocol.

We underscore that LES provides an envelope of maximum photosynthetic rate and associated biochemical and diffusional limitations as determined by structural and chemical constraints. The instantaneous value of net assimilation rate is highly variable; it depends on short-term variations in environmental drivers, light, temperature and humidity. Environmental response curves of pine needle photosynthesis, e.g., light, humidity or temperature response curves, have been measured in multiple studies (Sandford and Jarvis 1986; Leverenz 1988; Sheriff and Mattay 1995; Dang et al. 1998; Cheng et al. 2012; Zhu et al. 2014; García-Quintana et al. 2020; Schmiege et al. 2021a), but the environmental responses, especially for humidity, are only available for a small number of pine species. This is a crucial gap of knowledge for process-based models of pine forest photosynthesis, and we call for more comparative studies of environmental responsiveness of foliage photosynthetic rate and stomatal and mesophyll conductances across pine species.

Acknowledgements Preparation of this chapter has been supported by funding from the Estonian Ministry of Science and Education (Center of Excellence AgroCropFuture, TK200) and the Estonian Research Council (PRG2207).

Appendix I: Data Sources for Photosynthetic, Chemical and Structural Traits of *Pinus* Species

Pinus albicaulis (Maher et al. 2005), *P. aristata* (Trueba et al. 2022), *P. armandii* (Du et al. 2020), *P. banksiana* (Reich et al. 1998, 1999; Wright et al. 2004b; Trueba et al. 2022), *P. brutia* (Awada et al. 2003; Cherif et al. 2020; Fkiri et al. 2020; Houminer et al. 2022), *P. canariensis* (Peters et al. 2003; González-Rodríguez et al. 2019; Cherif et al. 2020; Fernández-Marín et al. 2021), *P. caribaea* var. *caribaea* (García-Quintana et al. 2020), *P. cembra* (Wieser et al. 2010), *P. contorta* (Guo et al. 2022, 2023; Trueba et al. 2022), *P. dalatensis* (Schmiege et al. 2021a, b), *P. densata* (Zhang 2014), *P. densiflora* (Han 2011; Trueba et al. 2022), *P. echinata* (Wang et al. 2019), *P. edulis* (Reich et al. 1998, 1999; Wright et al. 2004b), *P. elliottii* (Wang et al. 2019; Trueba et al. 2022), *P. fenzeliana* (Trueba et al. 2022), *P. flexilis* (Reich et al. 1998, 1999; Wright et al. 2004b; Trueba et al. 2022), *P. glabra* (Trueba et al. 2022), *P. halepensis* (Gulías et al. 2003; Mediavilla and Escudero 2003a; Wright et al. 2004b; Baquedano et al. 2008; Kuusk et al. 2018a, b; Cherif et al. 2020; Fkiri et al. 2020; Houminer et al. 2022; Trueba et al. 2022), *P. halepensis* × *P. brutia* (Houminer et al. 2022), *P. jeffreyi* (Schlesinger et al. 1989; DeLucia and Schlesinger 1991; Wright et al. 2004b; Grulke 2010), *P. kesiya* (Rundel et al. 2001; Zhang 2014; Schmiege et al. 2021a, b), *P. kesiya* var. *langbianensis* (Liu et al. 2009), *P. koraiensis* (Choi et al. 2006; Makoto and Koike 2007; Cheng et al. 2012; Zhu et al. 2014; Sun et al. 2016; Fan et al. 2019; Wu et al. 2021), *P. krempfii* (Du et al. 2020; Schmiege et al. 2021a, b), *P. lambertiana* (Grulke 2010), *P. leucodermis* (Oleksyn et al. 1997), *P. merkusii* (Rundel et al. 2001), *P. monophylla* (Schlesinger et al. 1989; DeLucia and Schlesinger 1991; Wright et al. 2004b), *P. monticola* (Trueba et al. 2022), *P. nigra* (Wright et al. 2004b; Trueba et al. 2022), *P. nigra* subsp. *salzmannii* (Kuusk et al. 2018a, b), *P. palustris* (Reich et al. 1998, 1999; Wright et al. 2004b; Wang et al. 2019; Trueba et al. 2022), *P. patula* (Niinemets et al. 2007), *P. pinaster* (Awada et al. 2003; Mediavilla and Escudero 2003a; de Miguel et al. 2012, 2014; Feinard-Duranceau et al. 2018; Trueba et al. 2022), *P. pinea* (Awada et al. 2003; Mediavilla and Escudero 2003a; Kuusk et al. 2018a, b; Fkiri et al. 2020; Trueba et al. 2022), *P. ponderosa* (Schlesinger et al. 1989; DeLucia and Schlesinger 1991; Wright et al. 2004b; Grulke 2010; Trueba et al. 2022), *P. pumila* (Nagano et al. 2018), *P. pungens* (Trueba et al. 2022), *P. radiata* (Bown et al. 2007, 2009; Niinemets et al. 2007; Trueba et al. 2022), *P. resinosa* (Gower et al. 1993; Reich et al. 1998, 1999; Wright et al. 2004b), *P. rigida* (Reich et al. 1998, 1999; Wright et al. 2004b; Choi et al. 2006; Trueba et al. 2022), *P. sabiniana* (Trueba et al. 2022), *P. serotina* (Reich et al. 1998, 1999; Wright et al. 2004b; Trueba et al. 2022), *P. strobus* (Small 1972; Gower et al. 1993; Reich et al. 1998, 1999; Leal and Thomas 2003; Wright et al. 2004b; Trueba et al. 2022), *P. sylvestris* (Reich et al. 1998, 1999; Zagirova 2001; Mediavilla and Escudero 2003a; Wright et al. 2004b; Tarvainen et al. 2016, 2018; Wyka et al. 2016; Veromann et al. 2017; Stangl et al. 2022; Trueba et al. 2022), *P. tabuliformis* (Du et al. 2020), *P. taeda* (Myers et al. 1999; Ingwers et al. 2016; Yáñez et al. 2017; Wang et al. 2019), *P. thunbergii* (Trueba et al. 2022),

P. tropicalis (García-Quintana et al. 2020), *P. virginiana* (Wang et al. 2019; Trueba et al. 2022), *P. yunnanensis* (Zhang 2014).

For the whole dataset: (Niinemets and Zhang 2026 available at DOI: https://doi.org/10.15159/eds.dt.26.02)

References

Abrams L (1923) An illustrated flora of the Pacific states. Washington, Oregon and California. Vol. I. Ophigolossaceae to Aristolochaceae. Stanford University Press, Stanford

Albaugh TJ, Allen HL, Stape JL, Fox TR, Rubilar RA, Carlson CA, Pezzutti R (2010) Leaf area duration in natural range and exotic *Pinus taeda*. Can J For Res 40:224–234

Alonso-Forn D, Sancho-Knapik D, Ferrio JP, Peguero-Pina JJ, Bueno A, Onoda Y et al (2020) Revisiting the functional basis of sclerophylly within the leaf economics spectrum of oaks: different roads to Rome. Curr For Rep 6:260–281

Alvarez A, Suárez JT, Hechavarría O, Diago I (2001) *Pinus tropicalis* Morelet: its characteristics and genetic resource status. In: Palmberg-Lerche C, Iversen PA, Sigaud P (eds) Forest genetic resources. No. 29. Food and Agriculture Organization of the United Nations, Rome. https://www.fao.org/4/Y2316E/y2316e2304.htm

Álvarez-Lopeztello J, del Castillo RF (2023) Pine plantations in savanna areas can alter vegetation dynamics facilitating tropical rainforest regeneration at the expense of savannas: a case study in Southeast Mexico. Plant Ecol 224:943–952

Anderson PD, Houpis JLJ, Helms JA, Momen B (1997) Seasonal variation of gas exchange and pigmentation in branches of three grafted clones of mature ponderosa pine exposed to ozone and acid rain. Environ Pollut 97:253–263

Awada T, Radoglou K, Fotelli MN, Constantinidou HI (2003) Ecophysiology of seedlings of three Mediterranean pine species in contrasting light regimes. Tree Physiol 23:33–41

Bachofen C, Perret-Gentil A, Wohlgemuth T, Vollenweider P, Moser B (2021) Phenotypic plasticity versus ecotypic differentiation under recurrent summer drought in two drought-tolerant pine species. J Ecol 109:3861–3876

Bailey DK (1970) Phytogeography and taxonomy of *Pinus* subsection Balfourianae. Ann Mo Bot Gard 57:545–554

Baquedano FJ, Valladares F, Castillo FJ (2008) Phenotypic plasticity blurs ecotypic divergence in the response of *Quercus coccifera* and *Pinus halepensis* to water stress. Eur J For Res 127:495–506

Barbéro M, Loisel R, Quézel P, Richardson DM, Romane F (1998) Pines of the Mediterranean basin. In: Richardson DM (ed) Ecology and biogeography of *Pinus*. Cambridge University Press, Cambridge, pp 153–170

Beets P (1977) Determination of the fascicle surface area for *Pinus radiata*. N Z J For Sci 7:397–407

Benecke U (1979) Surface area of needles in *Pinus radiata* – variation with respect to age and crown position. N Z J For Sci 9:267–271

Beyschlag W, Ryel RJ, Dietsch C (1994) Shedding of older needle age classes does not necessarily reduce photosynthetic primary production of Norway spruce. Analysis with a 3-dimensional canopy photosynthesis model. Trees Struct Funct 9:51–59

Boddi S, Bonzi LM, Calamassi R (2002) Structure and ultrastructure of *Pinus halepensis* primary needles. Flora 197:10–23

Bond BJ, Farnsworth BT, Coulombe RA, Winner WE (1999) Foliage physiology and biochemistry in response to light gradients in conifers with varying shade tolerance. Oecologia 120:183–192

Boratyński A, Jasińska AK, Boratyńska K, Iszkuło G, Piórkowska M (2009) Life span of needles of *Pinus mugo* Turra: effect of altitude and species origin. Pol J Ecol 57:546–550

Bown HE, Watt MS, Clinton PW, Mason EG, Richardson B (2007) Partitioning concurrent influences of nitrogen and phosphorus supply on photosynthetic model parameters of *Pinus radiata*. Tree Physiol 27:335–344

Bown HE, Watt MS, Mason EG, Clinton PW, Whitehead D (2009) The influence of nitrogen and phosphorus supply and genotype on mesophyll conductance limitations to photosynthesis in *Pinus radiata*. Tree Physiol 29:1143–1151

Britton NL, Shafer JA (1908) North American trees : being descriptions and illustrations of the trees growing independently of cultivation in North America, north of Mexico and the West Indies. H. Holt and Co, New York

Brodribb TJ, Feild TS (2008) Evolutionary significance of a flat-leaved *Pinus* in Vietnamese rainforest. New Phytol 178:201–209

Brooks A, Farquhar GD (1985) Effects of temperature on the CO_2/O_2 specificity of ribulose-1,5-bisphosphate carboxylase/oxygenase and the rate of respiration in the light. Estimates from gas-exchange measurements on spinach. Planta 165:397–406

Brooks JR, Hinckley TM, Sprugel DG (1994) Acclimation responses of mature *Abies amabilis* sun foliage to shading. Oecologia 100:316–324

Brooks JR, Sprugel DG, Hinckley TM (1996) The effects of light acclimation during and after foliage expansion on photosynthesis of *Abies amabilis* foliage within the canopy. Oecologia 107:21–32

Brunes L, Öquist G, Eliasson L (1980) On the reason for the different photosynthetic rates of seedlings of *Pinus sylvestris* and *Betula verrucosa*. Plant Physiol 66:940–944

Buchner O, Neuner G (2011) Winter frost resistance of *Pinus cembra* measured in situ at the alpine timberline as affected by temperature conditions. Tree Physiol 31:1217–1227

Bunce J (2018) Three methods of estimating mesophyll conductance agree regarding its CO_2 sensitivity in the Rubisco-limited C_i range. Plants (Basel) 7:62

Carriquí M, Roig-Oliver M, Brodribb TJ, Coopman R, Gil W, Mark K et al (2019) Anatomical constraints to nonstomatal diffusion conductance and photosynthesis in lycophytes and bryophytes. New Phytol 222:1256–1270

Cheng XB, Wu J, Han SJ, Zhou YM, Wang XX, Wang CG et al (2012) Photosynthesis, leaf morphology and chemistry of *Pinus koraiensis* and *Quercus mongolica* in broadleaved Korean pine mixed forest. Photosynthetica 50:56–66

Cherif S, Ezzine O, Khouja ML, Nasr Z (2020) A comparison of the physiological responses of three pine species in different bioclimatic zones in Tunisia. Appl Ecol Environ Res 18:1–13

Choi DS, Kayama M, Jin HO, Lee CH, Izuta T, Koike T (2006) Growth and photosynthetic responses of two pine species (*Pinus koraiensis* and *Pinus rigida*) in a polluted industrial region in Korea. Environ Pollut 139:421–432

Climent J, Chambel MR, López R, Mutke S, Alía R, Gil L (2006) Population divergence for heteroblasty in the Canary Island pine (*Pinus canariensis*, Pinaceae). Am J Bot 93:840–848

Climent J, San-Martín R, Chambel MR, Mutke S (2011) Ontogenetic differentiation between Mediterranean and Eurasian pines (sect. *Pinus*) at the seedling stage. Trees Struct Funct 25:175–186

Coates KD (2002) Tree recruitment in gaps of various size, clearcuts and undisturbed mixed forest of interior British Columbia, Canada. For Ecol Manag 155:387–398

Critchfield WB (1957) Geographic variation in *Pinus contorta*. Harvard University, Cambridge

Cronquist A, Holmgren AH, Holmgren NH, Reveal JL (1972) Intermountain flora: vascular plants of the intermountain West, U. S. A., volume one: Geological and botanical history of the region, its plant geography and a glossary. Vascular cryptogams and the gymnosperms. New York Botanical Garden, New York

Crous KY, Campany C, Lopez R, Cano FJ, Ellsworth DS (2021) Canopy position affects photosynthesis and anatomy in mature *Eucalyptus* trees in elevated CO_2. Tree Physiol 41:206–222

Dang Q-L, Margolis HA, Collatz GJ (1998) Parameterization and testing of a coupled photosynthesis-stomatal conductance model for boreal trees. Tree Physiol 18:141–153

Davies CE, Benecke V (1980) Fluidized bed coating of conifer needles with glass beads for determination of leaf surface area. For Sci 26:29–32

de Chantal M, Leinonen K, Kuuluvainen T, Cescatti A (2003) Early response of *Pinus sylvestris* and *Picea abies* seedlings to an experimental canopy gap in a boreal spruce forest. For Ecol Manag 176:321–336

de Miguel M, Sanchez-Gomez D, Cervera MT, Aranda I (2012) Functional and genetic characterization of gas exchange and intrinsic water use efficiency in a full-sib family of *Pinus pinaster* Ait. in response to drought. Tree Physiol 32:94–103

de Miguel M, Cabezas JA, de Maria N, Sanchez-Gomez D, Guevara MA, Velez MD et al (2014) Genetic control of functional traits related to photosynthesis and water use efficiency in *Pinus pinaster* Ait. drought response: integration of genome annotation, allele association and QTL detection for candidate gene identification. BMC Genomics 15:464

Deligöz A, Bayar E, Genç M, Karatepe Y, Kirdar E, Cankara FG (2018) Seasonal and needle age-related variations in the biochemical characteristics of *Pinus nigra* subsp. *pallasiana* (Lamb.) Holmboe. J For Sci 64:379–386

DeLucia EH, Schlesinger WH (1991) Resource-use efficiency and drought tolerance in adjacent Great Basin and Sierran plants. Ecology 72:51–58

Du H, Ran JH, Feng YY, Wang XQ (2020) The flattened and needlelike leaves of the pine family (Pinaceae) share a conserved genetic network for adaxial-abaxial polarity but have diverged for photosynthetic adaptation. BMC Evol Biol 20:131

Eimil-Fraga C, Sánchez-Rodríguez F, Álvarez-Rodríguez E, Rodríguez-Soalleiro R (2015a) Relationships between needle traits, needle age and site and stand parameters in *Pinus pinaster*. Trees Struct Funct 29:1103–1113

Eimil-Fraga C, Sánchez-Rodríguez F, Álvarez-Rodríguez E, Rodríguez-Soalleiro R (2015b) Variability in needle lifespan and foliar biomass along a gradient of soil fertility in maritime pine plantations on acid soils rich in organic matter. For Ecol Manag 343:34–41

Ely KS, Rogers A, Agarwal DA, Ainsworth EA, Albert LP, Ali A et al (2021) A reporting format for leaf-level gas exchange data and metadata. Eco Inform 61:101232

Evans JR (2021) Mesophyll conductance: walls, membranes and spatial complexity. New Phytol 229:1864–1876

Evans JR, Loreto F (2000) Acquisition and diffusion of CO_2 in higher plant leaves. In: Leegood RC, Sharkey TD, von Caemmerer S (eds) Photosynthesis: physiology and metabolism. Kluwer Academic Publishers, Dordrecht, pp 321–351

Ewers FW (1982) Secondary growth in needle leaves of *Pinus longaeva* (bristlecone pine) and other conifers: quantitative data. Am J Bot 69:1552–1559

Ewers FW, Aloni R (1987) Seasonal secondary growth in needle leaves of *Pinus strobus* and *Pinus brutia*. Am J Bot 74:13–22

Ewers FW, Schmid R (1981) Longevity of needle fascicles of *Pinus longaeva* (bristlecone pine) and other North American pines. Oecologia 51:107–115

Fan Y, Moser WK, Cheng Y (2019) Growth and needle properties of young *Pinus koraiensis* Sieb. et Zucc. trees across an elevational gradient. Forests 10:54

Farquhar GD, von Caemmerer S, Berry JA (1980) A biochemical model of photosynthetic CO_2 assimilation in leaves of C_3 species. Planta 149:78–90

Feinard-Duranceau M, Berthier A, Vincent-Barbaroux C, Marin S, Lario F-J, Rozenberg P (2018) Plastic response of four maritime pine (*Pinus pinaster* Aiton) families to controlled soil water deficit. Ann For Sci 75:47

Fernandez M, Gil L, Pardos JA (1999) Response of *Pinus pinaster* Ait. provenances at early age to water supply. I. Water relation parameters. Ann For Sci 56:179–187

Fernández-Marín B, Ruiz-Medina MA, Miranda JC, González-Rodríguez ÁM (2021) Coexistent heteroblastic needles of adult *Pinus canariensis* C. Sm. ex DC. in Buch trees differ structurally and physiologically. Forests 12:341

Fkiri S, Rzigui T, Elkhorchani A, Hassine AB, Khaldi A, Khouja ML, Nasr Z (2020) Tree growth and leaf gas exchange variability of three Mediterranean *Pinus* spp. growing in a common garden in Northeastern Tunisia. Euro-Mediterr J Environ Integr 5:2

Flexas J, Scoffoni C, Gago J, Sack L (2013) Leaf mesophyll conductance and leaf hydraulic conductance: an introduction to their measurement and coordination. J Exp Bot 64:3965–3981

Fouda RA (2004) Some botanical variations in the needles of two *Pinus* species affected by age and parts of needle, shading and position of the needle on the tree. J Agric Sci Mansoura Univ 29:6993–7010

Fraga MF, Rodríguez R, Cañal MJ (2002) Genomic DNA methylation-demethylation during aging and reinvigoration of *Pinus radiata*. Tree Physiol 22:813–816

Galmés J, Hermida-Carrera C, Laanisto L, Niinemets Ü (2016) A compendium of temperature responses of Rubisco kinetic traits: variability among and within photosynthetic groups and impacts on photosynthesis modeling. J Exp Bot 67:5067–5091

García-Quintana Y, Arteaga-Crespo Y, de Decker M, Vega-Rosete S, Geada-López G (2020) The comparative photosynthetic responses of *Pinus caribaea* var. *caribaea* and *Pinus tropicalis* from Western Cuba. Floresta Ambiente 27:e20170386

Gardner ST, Bertucci EM, Sutton R, Horcher A, Aubrey D, Parrott BB (2023) Development of DNA methylation-based epigenetic age predictors in loblolly pine (*Pinus taeda*). Mol Ecol Resour 23:131–144

Gauthey A, Bachofen C, Chin A, Cochard H, Gisler J, Mas E et al (2024) Twenty years of irrigation acclimation is driven by denser canopies and not by plasticity in twig- and needle-level hydraulics in a *Pinus sylvestris* forest. J Exp Bot 75:3141–3152

Ginn SE, Seiler JR, Cazell BH, Kreh RE (1991) Physiological and growth responses of eight-year-old loblolly pine stands to thinning. For Sci 37:1030–1040

González-Rodríguez ÁM, Brito P, Lorenzo JR, Jiménez MS (2019) Photosynthetic performance in *Pinus canariensis* at semiarid treeline: phenotype variability to cope with stressful environment. Forests 10:845

Gower ST, Reich PB, Son Y (1993) Canopy dynamics and aboveground production of five tree species with different leaf longevities. Tree Physiol 12:327–345

Grau-Andrés R, Kardol P, Gundale MJ (2022) Trait coordination in boreal mosses reveals a bryophyte economics spectrum. J Ecol 110:2493–2506

Greenwood MS (1987) Rejuvenation of forest trees. Plant Growth Regul 6:1–12

Greenwood MS (1995) Juvenility and maturation in conifers: current concepts. Tree Physiol 15:433–438

Greer DH, Stanley CJ (1985) Regulation of the loss of frost hardiness in *Pinus radiata* by photoperiod and temperature. Plant Cell Environ 8:111–116

Greer DH, Warrington IJ (1982) Effect of photoperiod, night temperature, and frost incidence on on development of frost hardiness in *Pinus radiata*. Aust J Plant Physiol 9:333–342

Grulke NE (2010) Plasticity in physiological traits in conifers: implications for response to climate change in the western U.S. Environ Pollut 158:2032–2042

Gulías J, Flexas J, Mus M, Cifre J, Lefi E, Medrano H (2003) Relationship between maximum leaf photosynthesis, nitrogen content and specific leaf area in Balearic endemic and non-endemic Mediterranean species. Ann Bot 92:215–222

Guo J, Beverly DP, Mercer JJ, Cook CS, Ewers BE, Williams DG (2022) Topographic controls on stomatal and mesophyll limitations to photosynthesis in two subalpine conifers. Int J Plant Sci 183:205–219

Guo J, Beverly DP, Ewers BE, Williams DG (2023) Stomatal, mesophyll and biochemical limitations to photosynthesis and their relationship with leaf structure over an elevation gradient in two conifers. Photosynth Res 157:85–101

Hallik L, Niinemets Ü, Wright IJ (2009) Are species shade and drought tolerance reflected in leaf-level structural and functional differentiation in Northern Hemisphere temperate woody flora? New Phytol 184:257–274

Han Q (2011) Height-related decreases in mesophyll conductance, leaf photosynthesis and compensating adjustments associated with leaf nitrogen concentrations in *Pinus densiflora*. Tree Physiol 31:976–984

Han Q, Kawasaki T, Nakano T, Chiba Y (2004) Spatial and seasonal variability of temperature responses of biochemical photosynthesis parameters and leaf nitrogen content within a *Pinus densiflora* crown. Tree Physiol 25:737–744

Harrington CA (1991) Retrospective shoot growth analysis for three seed sources of loblolly pine. Can J For Res 21:306–317

Hayes FJ, Buchanan SW, Coleman B, Gordon AM, Reich PB, Thevathasan NV et al (2018) Intraspecific variation in soy across the leaf economics spectrum. Ann Bot 123:107–120

He XF, Wang SW, Sun H, Körner C, Yang Y (2021) Water relations of "trailing-edge" evergreen oaks in the semi-arid upper Yangtze region, SE Himalaya. J Syst Evol 59:1256–1265

Hessl AE, Milesi C, White MA, Peterson DL, Keane RE (2004) Ecophysiological parameters for Pacific Northwest trees. US Department of Agriculture, Forest Service, Pacific Northwest Research Station, Portland

Holder CD (2000) Geography of *Pinus elliottii* Engelm. and *Pinus palustris* Mill. leaf life-spans in the southeastern U.S.A. J Biogeogr 27:311–318

Houminer N, Riov J, Moshelion M, Osem Y, David-Schwartz R (2022) Comparison of morphological and physiological traits between *Pinus brutia*, *Pinus halepensis*, and their vigorous F1 hybrids. Forests 13:1477

Huang Y, Mao J, Chen Z, Meng J, Xu Y, Duan A, Li Y (2015) Genetic structure of needle morphological and anatomical traits of *Pinus yunnanensis*. J For Res 27:13–25

Hultine KR, Marshall JD (2000) Altitude trends in conifer leaf morphology and stable carbon isotope composition. Oecologia 123:32–40

Ingwers MW, Urban J, McGuire MA, Bhuiyan RA, Teskey RO (2016) Physiological attributes of three- and four-needle fascicles of loblolly pine (*Pinus taeda* L.). Trees Struct Funct 30:1923–1933

Irvine J, Perks MP, Magnani F, Grace J (1998) The response of *Pinus sylvestris* to drought: stomatal control of transpiration and hydraulic conductance. Tree Physiol 18:393–402

Ji M, Jin G, Liu Z (2021) Effects of ontogenetic stage and leaf age on leaf functional traits and the relationships between traits in *Pinus koraiensis*. J For Res 32:2459–2471

Jin DM, Zhou XL, Schneider H, Wei HJ, Wei HY, Yan YH (2020) Functional traits: adaption of ferns in forest. J Syst Evol 59:1040–1050

Johnson JD (1984) A rapid technique for estimating total surface area of pine needles. For Sci 30:913–921

Jordan DB, Ogren WL (1984) The CO_2/O_2 specificity of ribulose 1,5-bisphosphate carboxylase/oxygenase. Dependence on ribulosebisphosphate concentration, pH and temperature. Planta 161:308–313

Jordan DN, Smith WK (1993) Simulated influence of leaf geometry on sunlight interception and photosynthesis in conifer needles. Tree Physiol 13:29–39

Joswig JS, Wirth C, Schuman MC, Kattge J, Reu B, Wright IJ et al (2022) Climatic and soil factors explain the two-dimensional spectrum of global plant trait variation. Nat Ecol Evol 6:36–50

Kattge J, Díaz S, Lavorel S, Prentice C, Leadley P, Bönisch G et al (2011) TRY – a global database of plant traits. Glob Chang Biol 17:2905–2935

Kattge J, Bönisch G, Díaz S, Lavorel S, Prentice IC, Leadley P et al (2020) TRY plant trait database – enhanced coverage and open access. Glob Chang Biol 26:119–188

Knauer J, Cuntz M, Evans JR, Niinemets Ü, Tosens T, Veromann-Jürgenson L-L et al (2022) Contrasting anatomical and biochemical controls on mesophyll conductance across plant functional types. New Phytol 236:357–368

Koch GW, Sillett SC, Jennings GM, Davis SD (2004) The limits to tree height. Nature 428:851–854

Koppel AT, Frey JM (1984) Bystroye opredeleniye poverhnosti pobega hvoinyh. Fiziol Rast 31:1000–1003

Kumar Das A, Ramakrishnan PS (1986) Adaptive growth strategy of Khasi pine (*Pinus kesiya* Royle ex Gordon). Proc Indian Acad Sci Plant Sci 96:25–36

Kume A, Tsuboi N, Satomura T, Suzuki M, Chiwa M, Nakane K et al (2000) Physiological characteristics of Japanese red pine, *Pinus densiflora* Sieb. et Zucc., in declined forests at Mt. Gokurakuji in Hiroshima Prefecture, Japan. Trees Struct Funct 14:305–311

Kuusk V, Niinemets Ü, Valladares F (2018a) A major trade-off between structural and photosynthetic investments operative across plant and needle ages in three Mediterranean pines. Tree Physiol 38:543–557

Kuusk V, Niinemets Ü, Valladares F (2018b) Structural controls on photosynthetic capacity through juvenile-to-adult transition and needle ageing in Mediterranean pines. Funct Ecol 32:1479–1491

Law BE, Van Tuyl S, Cescatti A, Baldocchi DD (2001) Estimation of leaf area index in open-canopy ponderosa pine forests at different successional stages and management treatments in Oregon. Agric For Meteorol 108:1–14

Leal DB, Thomas SC (2003) Vertical gradients and tree-to-tree variation in shoot morphology and foliar nitrogen in an oldgrowth *Pinus strobus* stand. Can J For Res 33:1304–1314

Leal-Saenz A, Waring KM, Menon M, Cushman SA, Eckert A, Flores-Renteria L et al (2020) Morphological differences in *Pinus strobiformis* across latitudinal and elevational gradients. Front Plant Sci 11:559697

Leverenz JW (1988) The effects of illumination sequence, CO_2 concentration, temperature and acclimation on the convexity of the photosynthetic light response curve. Physiol Plant 74:332–341

Leverenz JW (1995) Shade shoot structure of conifers and the photosynthetic response to light at two CO_2 partial pressures. Funct Ecol 9:413–421

Leverenz JW, Jarvis PG (1979) Photosynthesis in Sitka spruce. VIII. The effects of light flux density and direction on the rate of net photosynthesis and the stomatal conductance of needles. J Appl Ecol 16:919–932

Liu J, Liang J-S, Wang J-M, Chen X-M, Yang Z-X, Chen H (2009) Diurnal change of photosynthetic characteristics and response to light intensity of *Pinus kesiya* Royle ex Gordon var. *langbianensis*. For Res 22:677–682

Liu Z, Hikosaka K, Li F, Jin G, Ostertag R (2020) Variations in leaf economics spectrum traits for an evergreen coniferous species: tree size dominates over environment factors. Funct Ecol 34:458–467

Liu Y, Wu J, Wu D, Li S, Wang L (2023) Seasonal variation in delta^{13}C of *Pinus yunnanensis* and *Pinus armandii* at different stand ages. Sci Rep 13:7938

Luo T, Neilson RP, Tian H, Vörösmarty CJ, Zhu H, Liu S (2002) A model for seasonality and distribution of leaf area index of forests and its application to China. J Veg Sci 13:817–830

Maher EL, Germino MJ, Hasselquist NJ (2005) Interactive effects of tree and herb cover on survivorship, physiology, and microclimate of conifer seedlings at the alpine tree-line ecotone. Can J For Res 35:567–574

Maier CA, Palmroth S, Ward E (2008) Short-term effects of fertilization on photosynthesis and leaf morphology of field-grown loblolly pine following long-term exposure to elevated CO_2 concentration. Tree Physiol 28:597–606

Makoto K, Koike T (2007) Effects of nitrogen supply on photosynthetic and anatomical changes in current-year needles of *Pinus koraiensis* seedlings grown under two irradiances. Photosynthetica 45:99–104

Malkina IS (1982) Vliyaniye osveshtshonnosti i vozrasta dereva na assimilyatsionnuyu sposobnost hvoi sosny obyknovennoi. (Effect of illumination and tree age on CO_2 assimilation in pine needles). Fiziol Rast 29:465–470

Mason CM, Donovan LA (2015) Does investment in leaf defenses drive changes in leaf economic strategy? A focus on whole-plant ontogeny. Oecologia 177:1053–1066

Mastrogiuseppe RJ, Mastrogiuseppe JD (1980) A study of *Pinus balfouriana* Grev. & Balf. (Pinaceae). Syst Bot 5:545–554

Maugé JP (1987) Le pin maritime premier résineux de France. Centre de Productivité et d'Action Forestiére d'Aquitaine – Institut pour le Développment Forestier, Paris

Mediavilla S, Escudero A (2003a) Photosynthetic capacity, integrated over the lifetime of a leaf, is predicted to be independent of leaf longevity in some tree species. New Phytol 159:203–211

Mediavilla S, Escudero A (2003b) Relative growth rate of leaf biomass and leaf nitrogen content in several Mediterranean woody species. Plant Ecol 168:321–332

Mediavilla S, Garcia-Ciudad A, Garcia-Criado B, Escudero A (2008) Testing the correlations between leaf life span and leaf structural reinforcement in 13 species of European Mediterranean woody plants. Funct Ecol 22:787–793

Mediavilla S, Herranz M, González-Zurdo P, Escudero A (2014) Ontogenetic transition in leaf traits: a new cost associated with the increase in leaf longevity. J Plant Ecol 7:567–575

Medlyn BE, Dreyer E, Ellsworth D, Forstreuter M, Harley PC, Kirschbaum MUF et al (2002a) Temperature response of parameters of a biochemically based model of photosynthesis. II. A review of experimental data. Plant Cell Environ 25:1167–1179

Medlyn BE, Loustau D, Delzon S (2002b) Temperature response of parameters of a biochemically based model of photosynthesis. I. Seasonal changes in mature maritime pine (*Pinus pinaster* Ait.). Plant Cell Environ 25:1155–1165

Mencuccini M, Martínez-Vilalta J, Hamid HA, Korakaki E, Vanderklein D (2007) Evidence for age- and size-mediated controls of tree growth from grafting studies. Tree Physiol 27:463–473

Menzies MI, Holden DG (1981) Seasonal frost resistance of *of Pinus radiata*, *Pinus muricata*, and *Pseudotsuga menziesii*. N Z J For Sci 11:92–99

Messier C, Puttonen P (1995) Spatial and temporal variation in the light environment of developing Scots pine stands: the basis for a quick and efficient method of characterizing light. Can J For Res 25:343–354

Mullin LP, Sillett SC, Koch GW, Tu KP, Antoine ME (2009) Physiological consequences of height-related morphological variation in *Sequoia sempervirens* foliage. Tree Physiol 29:999–1010

Myers DA, Thomas RB, DeLucia EH (1999) Photosynthetic capacity of loblolly pine (*Pinus taeda* L.) trees during the first year of carbon dioxide enrichment in a forest ecosystem. Plant Cell Environ 22:473–481

Nagano S, Nakano T, Hikosaka K, Maruta E (2018) *Pinus pumila* photosynthesis is suppressed by water stress in a wind-exposed mountain site. Arct Antarct Alp Res 45:229–237

Nebel B, Matile P (1992) Longevity and senescence of needles in *Pinus cembra* L. Trees Struct Funct 6:156–161

Niinemets Ü (1997) Acclimation to low irradiance in *Picea abies*: influences of past and present light climate on foliage structure and function. Tree Physiol 17:723–732

Niinemets Ü (1999) Research review. Components of leaf dry mass per area – thickness and density – alter leaf photosynthetic capacity in reverse directions in woody plants. New Phytol 144:35–47

Niinemets Ü (2002) Stomatal conductance alone does not explain the decline in foliar photosynthetic rates with increasing tree age and size in *Picea abies* and *Pinus sylvestris*. Tree Physiol 22:515–535

Niinemets Ü (2010) A review of light interception in plant stands from leaf to canopy in different plant functional types and in species with varying shade tolerance. Ecol Res 25:693–714

Niinemets Ü (2012) Whole plant photosynthesis. In: Flexas J, Loreto F, Medrano H (eds) Terrestrial photosynthesis in a changing environment. A molecular, physiological and ecological approach. Cambridge University Press, Cambridge, pp 399–423

Niinemets Ü (2015) Is there a species spectrum within the world-wide leaf economics spectrum? Major variations in leaf functional traits in the Mediterranean sclerophyll *Quercus ilex*. New Phytol 205:79–96

Niinemets Ü (2020) Leaf trait plasticity and evolution in different plant functional types. Annu Plant Rev 3:473–522

Niinemets Ü (2023) Variation in leaf photosynthetic capacity within plant canopies: optimization, structural and physiological constraints and inefficiencies. Photosynth Res 158:131–149

Niinemets Ü, Lukjanova A (2003a) Needle longevity, shoot growth and branching frequency in relation to site fertility and within-canopy light conditions in *Pinus sylvestris*. Ann For Sci 60:195–208

Niinemets Ü, Lukjanova A (2003b) Total foliar area and average leaf age may be more strongly associated with branching frequency than with leaf longevity in temperate conifers. New Phytol 158:75–89

Niinemets Ü, Tenhunen JD (1997) A model separating leaf structural and physiological effects on carbon gain along light gradients for the shade-tolerant species *Acer saccharum*. Plant Cell Environ 20:845–866

Niinemets Ü, Valladares F (2006) Tolerance to shade, drought and waterlogging in the temperate dendroflora of the northern hemisphere: tradeoffs, phylogenetic signal and implications for niche differentiation. Ecol Monogr 76:521–547

Niinemets Ü, Zhang Z (2026) Online data accompanying Niinemets Ü, Zhang Z (2026) Pine photosynthesis: needle longevity, structural and physiological controls, needle and tree age dependent changes and plasticity. In: Johnson DA, Domec JC, Niinemets Ü (eds.) Physiology and ecology of Pinus, Tree Physiology. Springer, Berlin, in press [Data set]. Estonian University of Life Sciences. https://doi.org/10.15159/EDS.DT.26.02

Niinemets Ü, Ellsworth DS, Lukjanova A, Tobias M (2001) Site fertility and the morphological and photosynthetic acclimation of *Pinus sylvestris* needles to light. Tree Physiol 21:1231–1244

Niinemets Ü, Cescatti A, Lukjanova A, Tobias M, Truus L (2002a) Modification of light-acclimation of *Pinus sylvestris* shoot architecture by site fertility. Agric For Meteorol 111:121–140

Niinemets Ü, Ellsworth DS, Lukjanova A, Tobias M (2002b) Dependence of needle architecture and chemical composition on canopy light availability in three North American *Pinus* species with contrasting needle length. Tree Physiol 22:747–761

Niinemets Ü, Kull O, Tenhunen JD (2004a) Within canopy variation in the rate of development of photosynthetic capacity is proportional to integrated quantum flux density in temperate deciduous trees. Plant Cell Environ 27:293–313

Niinemets Ü, Sonninen E, Tobias M (2004b) Canopy gradients in leaf intercellular CO_2 mole fractions revisited: interactions between leaf irradiance and water stress need consideration. Plant Cell Environ 27:569–583

Niinemets Ü, Cescatti A, Rodeghiero M, Tosens T (2005) Leaf internal diffusion conductance limits photosynthesis more strongly in older leaves of Mediterranean evergreen broad-leaved species. Plant Cell Environ 28:1552–1566

Niinemets Ü, Tobias M, Cescatti A, Sparrow AD (2006) Size-dependent variation in shoot light-harvesting efficiency in shade-intolerant conifers. Int J Plant Sci 167:19–32

Niinemets Ü, Lukjanova A, Turnbull MH, Sparrow AD (2007) Plasticity in mesophyll volume fraction governs the light-acclimation in needle photosynthesis in two pines. Tree Physiol 27:1137–1151

Niinemets Ü, Díaz-Espejo A, Flexas J, Galmés J, Warren CR (2009) Role of mesophyll diffusion conductance in constraining potential photosynthetic productivity in the field. J Exp Bot 60:2249–2270

Niinemets Ü, Flexas J, Peñuelas J (2011) Evergreens favored by higher responsiveness to increased CO_2. Trends Ecol Evol 26:136–142

Niinemets Ü, García-Plazaola JI, Tosens T (2012) Photosynthesis during leaf development and ageing. In: Flexas J, Loreto F, Medrano H (eds) Terrestrial photosynthesis in a changing environment. A molecular, physiological and ecological approach. Cambridge University Press, Cambridge, pp 353–372

Niinemets Ü, Keenan TF, Hallik L (2015) Tansley review. A worldwide analysis of within-canopy variations in leaf structural, chemical and physiological traits across plant functional types. New Phytol 205:973–993

Oleksyn J, Tjoelker MG, Lorenc-Plucinska G, Konwinska A, Zhytkowiak R, Karolewski P, Reich PB (1997) Needle CO_2 exchange, structure and defence traits in relation to needle age in *Pinus heldreichii* Christ – a relict of Tertiary flora. Trees Struct Funct 12:82–89

Onoda Y, Wright IJ, Evans JR, Hikosaka K, Kitajima K, Niinemets Ü et al (2017) Physiological and structural tradeoffs underlying the leaf economics spectrum. New Phytol 214:1447–1463

Op de Beeck M, Gielen B, Jonckheere I, Samson R, Janssens IA, Ceulemans R (2010) Needle age-related and seasonal photosynthetic capacity variation is negligible for modelling yearly gas exchange of a sparse temperate Scots pine forest. Biogeosciences 7:199–215

Pan Y, Cieraad E, Armstrong J, Armstrong W, Clarkson BR, Colmer TD et al (2020) Global patterns of the leaf economics spectrum in wetlands. Nat Commun 11:4519

Pardos M, Climent J, Almeida H, Calama R (2014) The role of developmental stage in frost tolerance of *Pinus pinea* L. seedlings and saplings. Ann For Sci 71:551–562

Pausas JG, Keeley JE (2017) Epicormic resprouting in fire-prone ecosystems. Trends Plant Sci 22:1008–1015

Peguero-Pina JJ, Sancho-Knapik D, Flexas J, Galmés J, Niinemets Ü, Gil-Pelegrín E (2016) Light acclimation of photosynthesis in two closely related firs (*Abies pinsapo* Boiss. and *Abies alba* Mill.): the role of leaf anatomy and mesophyll conductance to CO_2. Tree Physiol 36:300–310

Peters J, Morales D, Soledad Jiménez M (2003) Gas exchange characteristics of *Pinus canariensis* needles in a forest stand on Tenerife, Canary Islands. Trees Struct Funct 17:492–500

Peters J, González-Rodríguez ÁM, Jiménez MS, Morales D, Wieser G (2008) Influence of canopy position, needle age and season on the foliar gas exchange of *Pinus canariensis*. Eur J For Res 127:293–299

Pons TL, Flexas J, von Caemmerer S, Evans JR, Genty B, Ribas-Carbo M, Brugnoli E (2009) Estimating mesophyll conductance to CO_2: methodology, potential errors, and recommendations. J Exp Bot 60:2217–2234

Preston RJ (1976) North American trees: exclusive of Mexico and tropical United States – a handbook designed for field use, with plates and distribution maps, 3rd edn. Iowa State University Press, Ames

Pumijumnong N, Palakit K (2020) Effects of climate variability on the annual and intra-annual ring formation of *Pinus merkusii* growing in Central Thailand. Environ Nat Resour J 18:234–248

R Development Core Team (2024) R: a language and environment for statistical computing. R 4.3.3. The R Foundation for Statistical Computing, Vienna

Reich PB, Uhl C, Walters MB, Ellsworth DS (1991) Leaf lifespan as a determinant of leaf structure and function among 23 Amazonian tree species. Oecologia 86:16–24

Reich PB, Walters MB, Ellsworth DS (1992) Leaf life-span in relation to leaf, plant, and stand characteristics among diverse ecosystems. Ecol Monogr 62:365–392

Reich PB, Kloeppel BD, Ellsworth DS, Walters MB (1995a) Different photosynthesis-nitrogen relations in deciduous hardwood and evergreen coniferous tree species. Oecologia 104:24–30

Reich PB, Koike T, Gower ST, Schoettle AW (1995b) Causes and consequences of variation in conifer leaf life-span. In: Smith WK, Hinckley TM (eds) Ecophysiology of coniferous forests, Physiological ecology. A series of monographs, texts, and treatises. Academic Press, Inc, San Diego, New York, Boston, London, Sydney, Tokyo, Toronto, pp 225–254

Reich PB, Oleksyn J, Modrzynski J, Tjoelker MG (1996) Evidence that longer needle retention of spruce and pine populations at high elevations and high latitudes is largely a phenotypic response. Tree Physiol 16:643–647

Reich PB, Walters MB, Ellsworth DS, Vose JM, Volin JC, Gresham C, Bowman WD (1998) Relationships of leaf dark respiration to leaf nitrogen, specific leaf area and leaf life-span: a test across biomes and functional groups. Oecologia 114:471–482

Reich PB, Ellsworth DS, Walters MB, Vose JM, Gresham C, Volin JC, Bowman WD (1999) Generality of leaf trait relationships: a test across six biomes. Ecology 80:1955–1969

Reich PB, Rich RL, Lu X, Wang YP, Oleksyn J (2014) Biogeographic variation in evergreen conifer needle longevity and impacts on boreal forest carbon cycle projections. Proc Natl Acad Sci USA 111:13703–13708

Ren L, Huang Y, Pan Y, Xiang X, Huo J, Meng D et al (2022) Differential investment strategies in leaf economic traits across climate regions worldwide. Front Plant Sci 13:798035

Richardson DM, Rundel PW (1998) Ecology and biogeography of *Pinus*: an introduction. In: Richardson DM (ed) Ecology and biogeography of *Pinus*. Cambridge University Press, Cambridge (UK), pp 3–46

Ross KM, Loik ME (2021) Photosynthetic sensitivity to historic meteorological variability for conifers in the eastern Sierra Nevada. Int J Biometeorol 65:851–863

Roth-Nebelsick A, Hacke UG, Voigt D, Schreiber SG, Krause M (2023) Foliar water uptake in Pinus species depends on needle age and stomatal wax structures. Ann Bot 131:287–300

Rundel PW, Patterson M, Boonpragob K, Watthana S (2001) Photosynthetic capacity in Thai conifers. Nat Hist Bull Siam Soc 49:295–303

Sáenz-Romero C, Beaulieu J, Rehfeldt GE (2011) Altitudinal genetic variation among *Pinus patula* populations from Oaxaca, México, in growth chambers simulating global warming temperatures. Agrociencia 45:399–411

Sáenz-Romero C, Rehfeldt GE, Soto-Correa JC, Aguilar-Aguilar S, Zamarripa-Morales V, López-Upton J (2012) Altitudinal genetic variation among *Pinus pseudostrobus* populations from Michoacán, México: two location shadehouse test results. Rev Fitotec Mex 35:111–121

Sandford AP, Jarvis PG (1986) Stomatal response to humidity in selected conifers. Tree Physiol 2:89–103

Sargent CP (1897) The Silva of North America. A description of the trees which grow naturally in North America exclusive Mexico. Volume XI. Coniferae (*Pinus*). The Riverside Press, Cambridge

Sargent CP (1922) Manual of the trees of North America (exclusive of Mexico), 2nd edn. The Riverside Press Cambridge, Boston and New York

Savitch LV, Leonardos ED, Krol M, Jansson S, Grodzinski B, Huner NPA, Öquist G (2002) Two different strategies for light utilization in photosynthesis in relation to growth and cold acclimation. Plant Cell Environ 25:761–771

Schlesinger WH, DeLucia EH, Billings WD (1989) Nutrient-use efficiency of woody plants on contrasting soils in the western Great Basin, Nevada. Ecology 70:105–113

Schmiege SC, Buckley BM, Stevenson D, Cuong TQ, Nam LC, Griffin KL (2021a) Contrasting physiological traits of shade tolerance in *Pinus* and Podocarpaceae native to a tropical Vietnamese forest: insight from an aberrant flat-leaved pine. Tree Physiol 41:223–239

Schmiege SC, Buckley BM, Stevenson DW, Heskel MA, Cuong TQ, Nam LC, Griffin KL (2021b) Respiratory temperature responses of tropical conifers differ with leaf morphology. Funct Ecol 35:1408–1423

Schoettle AW (1989) Potential effect of premature needle loss on the foliar biomass and nutrient retention of lodgepole pine. In: Olson RK, Lefohn AS (eds) Transactions of air pollution on western forests. Air & Waste Management Association, Anheim, pp 443–454

Schoettle AW (1990) The interaction between leaf longevity and shoot growth and foliar biomass per shoot in *Pinus contorta* at two elevations. Tree Physiol 7:209–214

Schoettle AW (1994) Influence of tree size on shoot structure and physiology of *Pinus contorta* and *Pinus aristata*. Tree Physiol 14:1055–1068

Schoettle AW, Fahey TJ (1994) Foliage and fine root longevity of pines. In: Gholz HL, Linder S, McMurtrie RE (eds) Environmental constraints on the structure and productivity of pine forest ecosystems: a comparative analysis. Ecological Bulletins, 43. Munksgaard International Booksellers and Publishers, Copenhagen, pp 136–153

Schoettle AW, Rochelle SG (2000) Morphological variation of *Pinus flexilis* (*Pinaceae*), a bird-dispersed pine, across a range of elevations. Am J Bot 87:1797–1806

Shelton MG, Switzer GL (1984) Variation in the surface area relationships of loblolly pine fascicles. For Sci 30:355–363

Sheriff DW, Mattay JP (1995) Simultaneous effects of foliar nitrogen, temperature, and humidity on gas exchange in *Pinus radiata*. Aust J Plant Physiol 22:615–626

Singh SP, Adhikari BS, Zobel DB (1994) Biomass, productivity, leaf longevity, and forest structure in the Central Himalaya. Ecol Monogr 64:401–421

Small E (1972) Photosynthetic rates in relation to nitrogen recycling as an adaptation to nutrient deficiency in peat bog plants. Can J Bot 50:2227–2233

Smith WK, Schoettle AW, Cui M (1991) Importance of the method of leaf area measurement to the interpretation of gas exchange of complex shoots. Tree Physiol 8:121–127

Smith L, Primack RB, Zipf L, Pardo S, Gallinat AS, Panchen ZA (2019) Leaf longevity in temperate evergreen species is related to phylogeny and leaf size. Oecologia 191:483–491

Song Y, Sterck F, Zhou X, Liu Q, Kruijt B, Poorter L (2022) Drought resilience of conifer species is driven by leaf lifespan but not by hydraulic traits. New Phytol 235:978–992

Sporbert M, Welk E, Seidler G, Jandt U, Aćić S, Biurrun I et al (2021) Different sets of traits explain abundance and distribution patterns of European plants at different spatial scales. J Veg Sci 32:e13016

Stangl ZR, Tarvainen L, Wallin G, Marshall JD (2022) Limits to photosynthesis: seasonal shifts in supply and demand for CO_2 in Scots pine. New Phytol 233:1108–1120

Stenberg P, Linder S, Smolander H (1995) Variation in the ratio of shoot silhouette area to needle area in fertilized and unfertilized Norway spruce trees. Tree Physiol 15:705–712

Stenberg P, Smolander H, Sprugel DG, Smolander S (1998) Shoot structure, light interception, and distribution of nitrogen in an *Abies amabilis* canopy. Tree Physiol 18:759–767

Stenberg P, Kangas T, Smolander H, Linder S (1999) Shoot structure, canopy openness, and light interception in Norway spruce. Plant Cell Environ 22:1133–1142

Stenberg P, Palmroth S, Bond BJ, Sprugel DG, Smolander H (2001) Shoot structure and photosynthetic efficiency along the light gradient in a Scots pine canopy. Tree Physiol 21:805–814

Steppe K, Niinemets Ü, Teskey RO (2011) Tree size- and age-related changes in leaf physiology and their influence on carbon gain. In: Meinzer FC, Dawson T, Lachenbruch B (eds) Size- and age-related changes in tree structure and function. Tree Physiology, 4. Springer, Berlin, pp 235–253

Strand M, Öquist G (1988) Effects of frost hardening, dehardening and freezing stress on *in vivo* chlorophyll fluorescence of seedlings of Scots pine (*Pinus sylvestris* L.). Plant Cell Environ 11:231–238

Sudworth GB (1908) Forest trees of the Pacific slope. U. S. Department of Agriculture, Forest Service, Government Printing Office, Washington, DC

Sun Y, Zhu J, Sun OJ, Yan Q (2016) Photosynthetic and growth responses of *Pinus koraiensis* seedlings to canopy openness: implications for the restoration of mixed-broadleaved Korean pine forests. Environ Exp Bot 129:118–126

Sung S-JS, Zarnoch SJ, Haywood JD, Leduc D, Sword-Sayer MA (2013) Developmental dynamics of longleaf pine seedling flushes and needles. In: Guldin JM (ed) Proceedings of the 15th biennial southern silvicultural research conference. e-Gen. Tech. Rep. SRS-GTR-175. Department of Agriculture, Forest Service, Southern Research Station, Asheville, NC, USA, pp 149–155

Sutinen ML, Repo T, Sutinen S, Lasarov H, Alvila L, Pakkanen TT (2000) Physiological changes in *Pinus sylvestris* needles during early spring under sub-arctic conditions. For Ecol Manag 135:217–228

Szaniawski RK, Wierzbicki B (1978) Net photosynthetic rate of some coniferous species at diffuse high irradiance. Photosynthetica 12:412–417

Tarvainen L, Lutz M, Rantfors M, Nasholm T, Wallin G (2016) Increased needle nitrogen contents did not improve shoot photosynthetic performance of mature nitrogen-poor Scots pine trees. Front Plant Sci 7:1051

Tarvainen L, Lutz M, Räntfors M, Näsholm T, Wallin G (2018) Temperature responses of photosynthetic capacity parameters were not affected by foliar nitrogen content in mature Pinus sylvestris. Physiol Plant 162:370–378

Teskey RO, Will RE (1999) Acclimation of loblolly pine (*Pinus taeda*) seedlings to high temperatures. Tree Physiol 19:519–525

Teskey RO, Whitehead D, Linder S (1994) Photosynthesis and carbon gain by pines. In: Gholz HL, Linder S, McMurtrie RE (eds) Environmental constraints on the structure and productivity of pine forest ecosystems: a comparative analysis. Ecological Bulletins, 43. Munksgaard International Booksellers and Publishers, Copenhagen, pp 35–49

Tjoelker MG, Oleksyn J, Lorenc-Plucinska G, Reich PB (2008) Acclimation of respiratory temperature responses in northern and southern populations of *Pinus banksiana*. New Phytol 181:218–229

Tomás M, Flexas J, Copolovici L, Galmés J, Hallik L, Medrano H et al (2013) Importance of leaf anatomy in determining mesophyll diffusion conductance to CO_2 across species: quantitative limitations and scaling up by models. J Exp Bot 64:2269–2281

Tosens T, Nishida K, Gago J, Coopman R, Cabrera HM, Carriquí M et al (2016) The photosynthetic capacity in 35 ferns and fern allies: mesophyll CO_2 diffusion as a key trait. New Phytol 209:1576–1590

Trueba S, Theroux-Rancourt G, Earles JM, Buckley TN, Love D, Johnson DM, Brodersen C (2022) The three-dimensional construction of leaves is coordinated with water use efficiency in conifers. New Phytol 233:851–861

Vanderklein D, Martínez-Vilalta J, Lee S, Mencuccini M (2007) Plant size, not age, regulates growth and gas exchange in grafted Scots pine trees. Tree Physiol 27:71–79

Veromann L-L, Tosens T, Laanisto L, Niinemets Ü (2017) Extremely thick cell walls and low mesophyll conductance: welcome to the world of ancient living! J Exp Bot 68:1639–1653

Villar Salvador P (2000) Estrategias ecológicas y funcionales del xilema en plantas leñosas mediterráneas. PhD thesis. Universitat de València

Wang X-R, Szmidt AE, Nguyên HN (2000) The phylogenetic position of the endemic flat-needle pine *Pinus krempfii* (Pinaceae) from Vietnam, based on PCR-RFLP analysis of chloroplast DNA. Plant Syst Evol 220:21–36

Wang N, Palmroth S, Maier CA, Domec JC, Oren R (2019) Anatomical changes with needle length are correlated with leaf structural and physiological traits across five *Pinus* species. Plant Cell Environ 42:1690–1704

Warren CR (2006) Why does photosynthesis decrease with needle age in *Pinus pinaster*? Trees Struct Funct 20:157–164

Wasowicz P, Óskarsdóttir G, Þórhallsdóttir ÞE (2025) Lodgepole pine (*Pinus contorta* Douglas ex Loudon) invasion in subarctic Iceland: evidence from a long-term study. NeoBiota 97:47–66

Wei Y-F, Fang J, Zhao X-Y, Li S-G (2011) Eco-physiological traits of different-age needles of *Pinus sylvestris* var. *mongolica* plantation in Horqin sandy land of China. Chin J Plant Ecol 35:1271–1280

Weidman RH (1939) Evidences of racial influence in a 25-year test of ponderosa pine. J Agric Res 59:855–887

Whitney GG (1982) A demographic analysis of the leaves of open and shade grown *Pinus strobus* L. and *Tsuga canadensis* (L.) Carr. New Phytol 90:447–453

Wieser G, Oberhuber W, Walder L, Spieler D, Gruber A (2010) Photosynthetic temperature adaptation of *Pinus cembra* within the timberline ecotone of the Central Austrian Alps. Ann For Sci 67:201

Williams CB, Reese Naesborg R, Dawson TE (2017) Coping with gravity: the foliar water relations of giant sequoia. Tree Physiol 37:1312–1326

Wilson KB, Baldocchi DD, Hanson PJ (2000) Quantifying stomatal and non-stomatal limitations to carbon assimilation resulting from leaf aging and drought in mature deciduous tree species. Tree Physiol 20:787–797

Wilson KB, Baldocchi DD, Hanson PJ (2001) Leaf age affects the seasonal pattern of photosynthetic capacity and net ecosystem exchange of carbon in a deciduous forest. Plant Cell Environ 24:571–583

Wright IJ, Groom PK, Lamont BB, Poot P, Prior LD, Reich PB et al (2004a) Leaf trait relationships in Australian plant species. Funct Plant Biol 31:551–558

Wright IJ, Reich PB, Westoby M, Ackerly DD, Baruch Z, Bongers F et al (2004b) The world-wide leaf economics spectrum. Nature 428:821–827

Wu H, Yin D, Rodríguez-Calcerrada J, Zhang J, Gil L, Zhang P, Shen H (2021) Cone-bearing effects on photosynthetic traits do not change with needle age in *Pinus koraiensis* trees. New For 53:607–626

Wyka TP, Żytkowiak R, Oleksyn J (2016) Seasonal dynamics of nitrogen level and gas exchange in different cohorts of Scots pine needles: a conflict between nitrogen mobilization and photosynthesis? Eur J For Res 135:483–493

Xiao Y (2003) Variation in needle longevity of *Pinus tabulaeformis* forests at different geographic scales. Tree Physiol 23:463–471

Yan CF, Han SJ, Zhou YM, Wang CG, Dai GH, Xiao WF, Li MH (2012) Needle-age related variability in nitrogen, mobile carbohydrates, and ?1[3C] within P*inus koraiensis* tree crowns. PLoS One 7:e35076

Yáñez MA, Fox TR, Seiler JR, Guerra F, Baettig RM, Zamudio F, Gyenge JE (2017) Within-crown acclimation of leaf-level physiological and morphological parameters in young loblolly pine stands. Trees Struct Funct 31:1849–1857

Yoder BJ, Ryan MG, Waring RH, Schoettle AW, Kaufmann MR (1994) Evidence of reduced photosynthetic rates in old trees. For Sci 40:513–527

Yuan SY, Zhang P, Shen HL (2023) Response of needle photosynthetic and anatomical characteristics of naturally regenerated *Pinus koraiensis* seedlings to different canopy densities. Ying Yong Sheng Tai Xue Bao 34:2314–2320

Zagirova SV (2001) Structure and CO_2 exchange in the needles of *Pinus sylvestris* and *Abies sibirica*. Russ J Plant Physiol 48:23–28

Zhang S-B (2014) Biomass partitioning affects the growth of *Pinus* species from different elevations. Plant Divers Resour 36:47–55

Zhang J-W, Wang L, D'Rozario A, Liang X-Q, Huang J, Zhou Z-K (2021) *Pinus leptokrempfii*, an oligocene relative of the flat-needled pine *Pinus krempfii* (Pinaceae) from China: implications for paleogeographic origin. Int J Plant Sci 182:389–400

Zhao MJ, Jin GZ, Liu ZL (2024a) Seasonal variation in the economic strategy of typical ferns in mixed broadleaved Korean pine forest. Ying Yong Sheng Tai Xue Bao 35:3315–3322

Zhao X, Liu Z, Jin G (2024b) Effects of nitrogen addition and leaf age on needle traits and the relationships among traits in *Pinus koraiensis*. Environ Exp Bot 223:105795

Zhelawski W, Szaniawski R, Dybczynski W, Piechurowski A (1973) Photosynthetic capacity of conifers in diffuse light of high illuminance. Photosynthetica 7:351–357

Zhu J, Wang K, Sun Y, Yan Q (2014) Response of *Pinus koraiensis* seedling growth to different light conditions based on the assessment of photosynthesis in current and one-year-old needles. J For Res 25:53–62

Zotz G, Wilhelm K, Becker A (2011) Heteroblasty – a review. Bot Rev 77:109–151

Doing Just Pine, Thank You Very Much: The Hydraulics of *Pinus*

Steven P. Augustine and Katherine A. McCulloh

Abstract Globally, many *Pinus* species occupy arid habitats yet rely on a vascular network to supply their leaves with water that is relatively vulnerable to loss of function. The main cause of the high vulnerability of the xylem seems to be relictual pores through the torus of the interconduit pit membranes, which in most species is impermeable to air. Because of their vulnerable xylem, pine species tend to avoid highly negative branch water potentials by closing their stomata at milder leaf water potentials and maintaining a lower leaf area-to-sapwood area ratio than most other conifer species. To mitigate the effects of the low hydraulic supply to their leaves, pines have the greatest leaf hydraulic conductance (K_{leaf}) across conifers, which allows for a greater ability to effectively replace water at the site of evaporation and maintain mild leaf water potentials. Pines' efficient K_{leaf} also permits them to maintain a high photosynthetic rate and grow rapidly. High K_{leaf} may also have permitted the evolution of very long needles in a few *Pinus* species that survive in habitats with very high fire return intervals by forming a grass stage as seedlings, in which the needles are thought to insulate the apical meristem during fires. The costs and benefits of the unusual physiology that pines exhibit requires further study to effectively predict the ecological consequences for this important group.

S. P. Augustine (✉)
Department of Botany, University of Wisconsin-Madison, Madison, WI, USA

Department of Biology, University of Central Arkansas, Conway, AR, USA
e-mail: saugustine@uca.edu

K. A. McCulloh
Department of Botany, University of Wisconsin-Madison, Madison, WI, USA

D. M. Johnson et al. (eds.), *Physiology and Ecology of Pines Worldwide*, Tree Physiology 9, https://doi.org/10.1007/978-3-032-23001-0_7

1 Pine Hydraulics and Water Relations: Distillation of Decades of Research

Pines, like all terrestrial plants, must transport water from their roots to their leaves to replace the water lost through transpiration, or risk desiccation and death. Much of the research exploring the hydraulics and water relations of pines (*Pinus*) has been in the context of their growth (Domec and Gartner 2003; Creese et al. 2011; Rosas et al. 2021) and stomatal regulation (West et al. 2007; McDowell et al. 2008; Woodruff et al. 2015; Garcia-Forner et al. 2016). Ecologically, pines tend to be early successional, with relatively high photosynthetic rates compared to other conifers (Brodribb and Feild 2008; Schmiege et al. 2021) and low shade tolerance (though with a few exceptions; Brodribb and Feild 2008). Physiologically, pines tend to be isohydric, showing tight stomatal regulation during periods of reduced soil moisture (Linton et al. 1998; Garcia-Forner et al. 2016; McDowell et al. 2016), which allows them to maintain relatively high (i.e., less negative) water potentials even as soil water potentials decrease (West et al. 2007; Breshears et al. 2009; Garcia-Forner et al. 2016). Given that pines rarely experience low water potentials, their hydraulic network is fairly vulnerable, with stems and leaves that lose half of their water transport capacity at relatively high water potentials (referred to as a P_{50}) when compared to other conifers (Fig. 1; Martínez-Vilalta et al. 2004; Brodribb et al. 2014).

Globally, most plants that have vulnerable stems and leaves tend to be restricted to more mesic environments (Choat et al. 2012; Larter et al. 2017; Sanchez-Martinez et al. 2020; Brodribb et al. 2014). This trend would suggest that the vulnerable hydraulic network of pines would restrict them to these environments and would exclude them from regions with severe drought stress. However, the ranges of *Pinus* species include some of the most arid of conifers, being rivaled only by several genera within Cupressaceae (e.g., *Juniperus*, *Callitris*, *Tetraclinus*, *Cupressus*) and *Cedrus* within Pinaceae (Fig. 1; genus level data not shown). This apparent disconnect between the hydraulic systems of pines and the environments they inhabit highlights their propensity to avoid drought, rather than tolerate it. This, in combination with their early successional nature, constrains their hydraulic systems and water relations in interesting ways and will be the discussion of the sections that follow. Much of this discussion will be based around data from the literature, which were obtained through a systematic meta-analysis and climate modeling (see Table 1, McCulloh et al. (2023), for additional details).

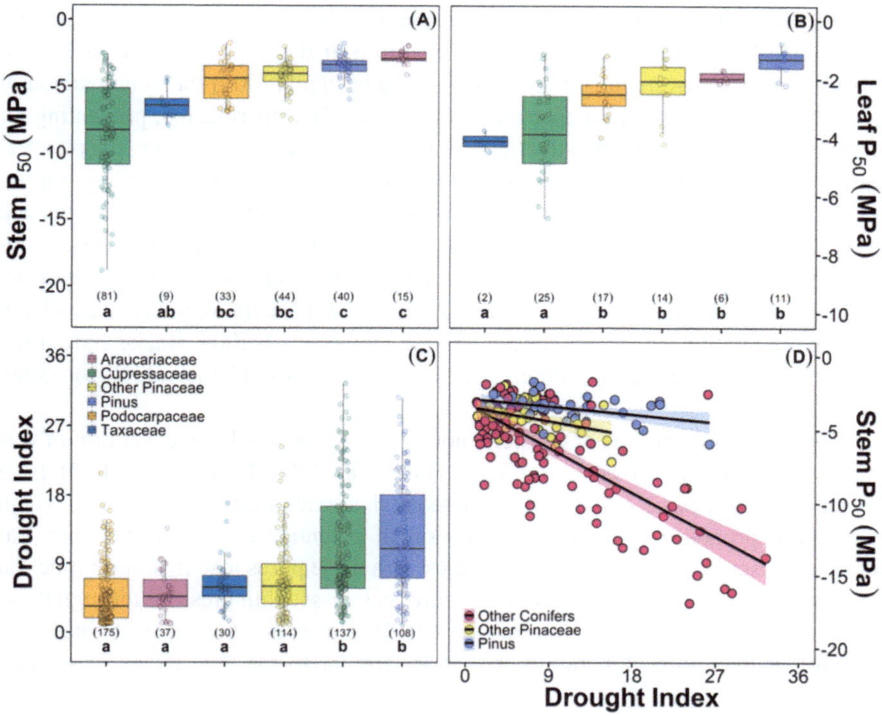

Fig. 1 Comparison between stem hydraulic vulnerability (**a**), leaf hydraulic vulnerability (**b**), and drought index (**c**) between groups, as well as how stem hydraulic vulnerability varies across drought index in pines, Pinaceae, and all other conifers (**d**). Each point represents the mean value of a species. Letters below bars in panels (**a**), (**b**), and (**c**) represent significant differences based on pairwise comparisons. Numbers of species with data in each group are indicated in parentheses. Groups are ranked from left to right in ascending order based on the mean of each group. All relationships in panel (**d**) were significant (*Pinus*: $p = 0.007$, $R^2 = 0.16$; other Pinaceae: $p = 0.007$, $R^2 = 0.15$; other Pinaceae: $p < 0.0001$, $R^2 = 0.53$). The drought index is a function of the minimum monthly soil water content fraction and monthly vapor pressure deficit (see McCulloh et al. (2023) for additional details)

2 Drought Avoidance in Pines: Causes, Consequences, and Constraints

2.1 The Role of Pit Level Variations in Hydraulic Vulnerability

The drought avoidance strategy of pines can be traced back to the microanatomy of their vascular system. In conifers, water moves through unicellular tracheids, which are connected by interconduit pits which allow water to flow between them. Unlike the homogenous pit membranes that connect the vessels of most angiosperms, the pit membranes of most conifers are heterogenous, composed of porous microfibrils in the distal regions (margo) surrounding a thickened central torus (see Jacobsen

(2021) for a detailed analysis). When a tracheid is air-filled (i.e., embolized), the pit membrane becomes aspirated to the pit aperture of the adjacent, water-filled tracheid. The torus covers the pit aperture (see Hacke et al. (2004) for a comparison of the size of the torus and pit aperture) and is typically impermeable, preventing air from entering the water-filled tracheid. However, when the torus covers the pit aperture, it forms an imperfect seal. This seal is thought to be the site of capillary failure of the air-water meniscus, allowing air bubbles through gaps between the torus and the pit wall at the edge of the torus (Cochard et al. 2009). As such, xylem that exhibits a greater ratio of the diameter of the torus to the diameter of the pit aperture (referred to as the torus overlap) tends to have greater drought tolerance, a trend that is seen both within and across species (Fig. 2; Domec et al. 2008; Hacke and Jansen 2009; Delzon et al. 2010; Pittermann et al. 2010; Bouche et al. 2014; Song et al. 2022).

However, some groups of conifers, notably Pinaceae (with the exception of certain genera: *Cedrus*, *Pseudotsuga*; see Domec et al. 2006; Jansen et al. 2012), have plasmodesmatal pores (12–144 nm diameter) that extend through their tori, which have been proposed to be another location of potential air-seeding between tracheids (Jansen et al. 2012). Indeed, species with these pores lose hydraulic function at water potentials consistent with the theoretical air-seeding pressure through these pores (−2.02 to −4.16 MPa; Jansen et al. 2012). This is notable because pines have a range of hydraulic vulnerabilities that are similar to the theoretical air-seeding

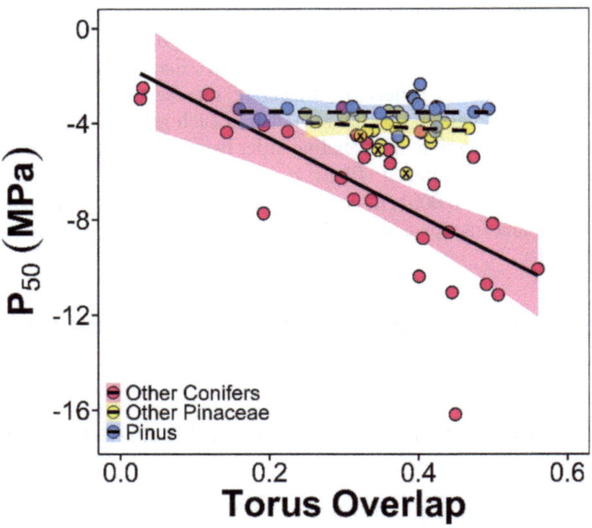

Fig. 2 Relationship between torus overlap and stem hydraulic vulnerability (P_{50}) across conifers. Each point represents the mean value of a species. Both *Pinus* and other Pinaceae species show no relationship between torus overlap and stem hydraulic vulnerability ($p > 0.52$; dashed lines), but other conifers do show this relationship ($p < 0.0001$, $R^2 = 0.47$; solid line). The three points with an "x" in the middle represent Pinaceae species within genera that are believed to have no plasmodesmatal pores in their tori (*Cedrus*, *Pseudotsuga*)

pressure (P_{50} ranging from −1.74 to −5.93 MPa), with 80% of species falling between the range (Fig. 2). Along with this, *Cedrus*, the most drought resistant genus in Pinaceae and one of the few genera within the family that lacks plasmodesmatal pores in its tori (Jansen et al. 2012), has much more resistant xylem (P_{50} ranging from −5.14 to −7.45 MPa) than pines while occupying similarly arid regions. While air-seeding through the plasmodesmatal pores has yet to be directly measured, these trends highlight that the pores within tori may limit the functional range of xylem pressures in pines more than torus overlap does. With the capacity for pines to prevent the spread of embolism apparently constrained by the presence of plasmodesmatal pores, it is interesting to speculate whether pines display such tight stomatal regulation due to constraints to building more resistant xylem.

Furthermore, it is interesting that torus overlap varies so much both within *Pinus* and Pinaceae given that there appears to be no interspecific increase in hydraulic safety associated with greater overlap within the family (Fig. 2). Notably, an increase in torus overlap does result in more negative P_{50} values in *Pseudotsuga menziesii*, which appears to have no plasmodesmatal pores (Domec et al. 2008), and there was a trade-off between overlap and both growth and longevity in *Pinus ponderosa*, which presumably has plasmodesmatal pores (Roskilly et al. 2019). With the torus being a major source of resistance to water flow (Domec et al. 2008; Schulte 2012; Schulte et al. 2015), increasing the torus overlap (which is mostly obtained by reductions in the aperture; Domec et al. 2008) reduces the hydraulic conductivity of the xylem. Further research is needed to understand the role of the torus and its plasmodesmatal pores in the hydraulic function of conifers. Although there is far less variation in P_{50} in *Pinus* and Pinaceae than in other conifers (Fig. 2), the lack of an interspecific relationship between P_{50} and torus overlap indicates that there must be other mechanisms that limit air-seeding. Moreover, the functional significance (if any) of tori with plasmodesmatal pores needs to be elucidated, as they are too small to cause any appreciative increase in pit hydraulic conductance.

2.2 Allometric Consequences of a Vulnerable Xylem

Having relatively vulnerable xylem, it is necessary for pines living in arid habitats to avoid extremely negative water potentials, and leaf water potentials measured in the field tend to be mild (West et al. 2007; Breshears et al. 2009; Garcia-Forner et al. 2016). To prevent more severe pressure gradients from forming, pines both reduce the amount of water being lost to evaporation and have an efficient water supply to leaves. Pines have stomata that close at high water potentials (Linton et al. 1998; Garcia-Forner et al. 2016; McDowell et al. 2016), preventing further water loss and inhibiting water potentials from dropping further. Hydraulically, on a sapwood-specific basis, pine hydraulic conductivity (k_s) is no more or less efficient than any other group of conifers (Fig. 3), which implies that the xylem network of pines is not more efficient than other groups of conifers. However, an additional mechanism

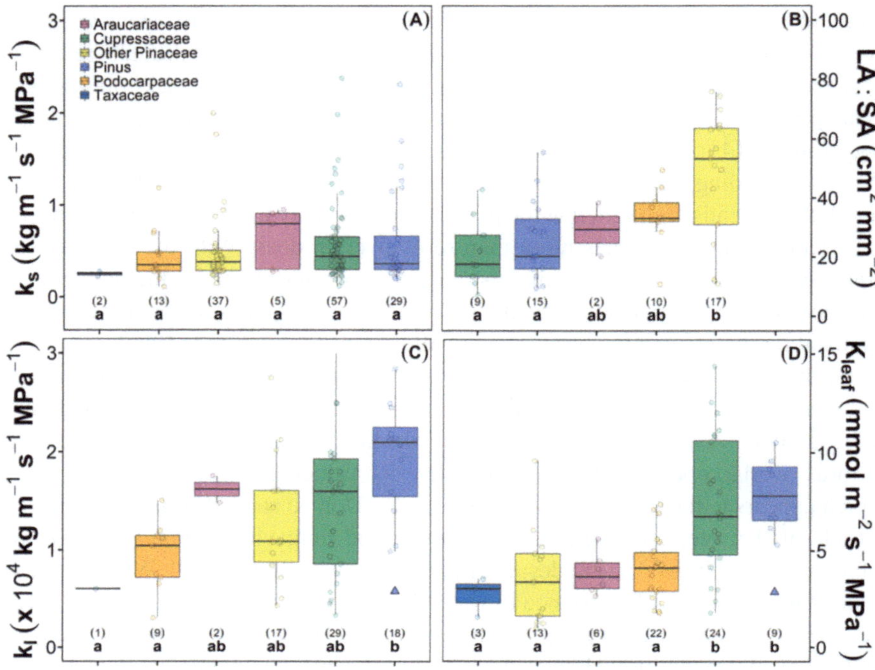

Fig. 3 Comparison between sapwood-specific hydraulic conductivity (**a**; k_s), leaf area-to-sapwood area ratio (**b**; LA/SA), leaf-specific hydraulic conductivity (**c**; k_l), and leaf hydraulic conductance (**d**; K_{leaf}) between groups. Letters below bars represent significant differences based on pairwise comparisons. Numbers of species with data in each group are indicated in parentheses. Groups are ranked from left to right in ascending order based on the mean of each group. We were unable to find LA/SA data from any Taxaceae species. The triangle in panels (**c**) and (**d**) represent the shade-tolerant *Pinus krempfii*

to ensure the supply of water to the site of evaporation is to reduce the transpiring area (i.e., leaf area) for a given conducting area (i.e., sapwood area; known as the leaf/sapwood area ratio or the inverse of Huber value; LA/SA). Low LA/SA increases the water flow for a given pressure gradient per unit transpiring area, which limits the development of large water potential gradients. While data on leaf area-specific conductivity (i.e., stem water supplied per unit leaf area; k_l) and LA/SA are limited in many conifers, the available data indicate pines have lower LA/SA and higher k_l compared to other species within Pinaceae (Fig. 3).

2.3 *Leaf Hydraulic Conductance and Potential Bottlenecks*

Hydraulically, adapting to arid regions typically requires both resistant xylem (Choat et al. 2012; Fig. 1d) and a relatively low LA/SA (Rosas et al. 2019; Sanchez-Martinez et al. 2020). However, pines appear to be unique among conifers in that

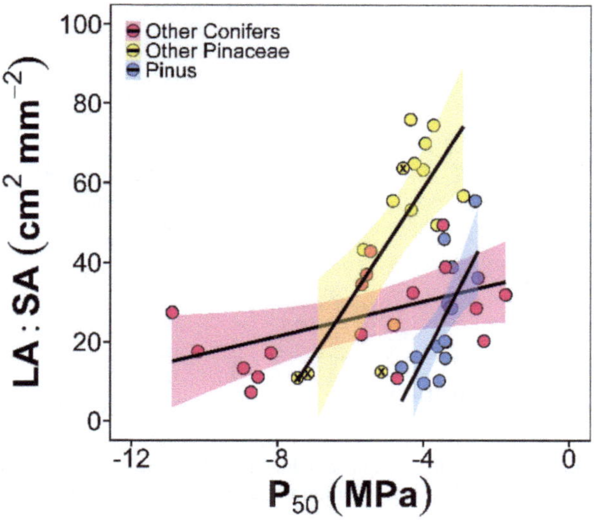

Fig. 4 Relationship between stem hydraulic vulnerability (P_{50}) and leaf area-to-sapwood area ratio (LA/SA) across conifers. All three groups showed significant positive relationships between P_{50} and LA/SA (*Pinus*: $p = 0.003$, $R^2 = 0.47$; other Pinaceae: $p = 0.0009$, $R^2 = 0.55$; other conifers: $p = 0.04$, $R^2 = 0.21$). The four points with an "x" in the middle represent Pinaceae species within genera that are believed to have no plasmodesmatal pores in their tori (*Cedrus*, *Pseudotsuga*). *Pseudotsuga* is the "x" with the greatest LA/SA, while the other three are *Cedrus*

they occupy arid regions with vulnerable xylem and a low LA/SA, suggesting that this trait is sufficient to overcome the limitations associated with having a vulnerable xylem. In contrast, conifers with similarly vulnerable xylem, particularly species in other Pinaceae genera (e.g., *Abies*, *Picea*, *Tsuga*), have much greater LA/SA for a given P_{50} when compared to pines and are limited to more mesic regions (Figs. 1 and 4). Consistent with this, the lower limit of LA/SA of other Pinaceae species (with the exception of the typically more arid *Cedrus* species: 11.5–12.4 $cm^2\ mm^{-2}$) is relatively high (e.g., *Larix occidentalis*: 24.4 $cm^2\ mm^{-2}$), with most pines having a lower LA/SA. The hydraulic and ecological limitations causing variations in the relationship between P_{50} and LA/SA deserve further study, and some potential mechanisms are discussed below.

Leaves are one of the greatest resistances to water flow along the soil-plant-atmosphere continuum (Sack et al. 2003; Domec et al. 2009a), specifically once the water exits the vascular system (Buckley et al. 2015). This extra-xylary movement of water constrains leaf hydraulic conductance (K_{leaf}) greatly, and differences in it are likely what cause K_{leaf} to vary so much within plants (Brodribb et al. 2007; Fig. 2). In conifer species whose leaves are plumbed by a single vein, the resistance to water flow is mostly a factor of the distance between the leaf vein and a stoma (Zwieniecki et al. 2004; Brodribb et al. 2007). Pines have narrow needles that position their vein much closer to their stomata, making the pathlength for extra-xylary water transport much shorter than many other conifers, especially those with broad

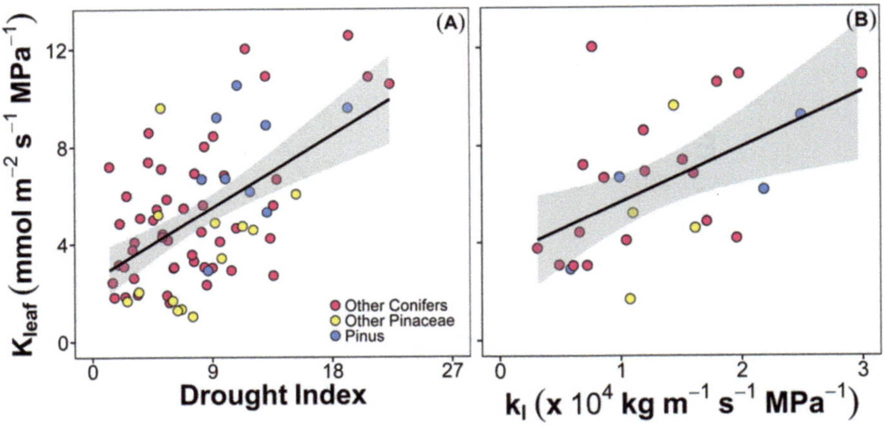

Fig. 5 Relationship between leaf hydraulic conductance (K_{leaf}) and (**a**) drought index ($p < 0.0001$, $R^2 = 0.29$) and (**b**) leaf-specific stem hydraulic conductivity (k_l; $p = 0.035$, $R^2 = 0.12$) across conifers

leaves (Brodribb et al. 2007; Brodribb and Feild 2008). The shorter pathlength reduces the overall hydraulic resistance to water flow, providing them with relatively efficient water transport in their needles. Thus, pines have the greatest K_{leaf} across conifers (Fig. 3) and a greater ability to effectively replace water at the site of evaporation and maintain mild leaf water potentials.

The relative inefficiency of other conifer needles may prevent those species from occupying arid regions. Conifers generally have higher K_{leaf} in more arid environments (Fig. 5), a trend that has been seen in several plant groups (Yao et al. 2020). For a given hydraulic capacitance, small increases in transpiration generate large water potential gradients in leaves with low K_{leaf} values (Brodribb et al. 2010), which could cause species with vulnerable xylem to lose hydraulic function. Given that most conifers have a similar hydraulic conductivity on a sapwood-specific basis, there is likely a trade-off between the K_{leaf} and the total leaf area. Meaning a given sapwood area can either supply a large leaf area with low K_{leaf} or a smaller leaf area with high K_{leaf} (Mencuccini et al. 2019). This likely underpins the relationship that is seen between K_{leaf} and k_l in conifers (Fig. 5) and restricts certain groups (e.g., *Abies*, *Picea*, *Tsuga*) with relatively vulnerable xylem to having many, inefficient needles and occupying more mesic environments.

Surprisingly, within Pinaceae, needle width in pines generally overlaps other genera (Fig. 6). Furthermore, pines have much longer needles for a given width (Fig. 6), which is only possible with efficient xylem that prevents large drops in water potential across the length of the needle (Zwieniecki et al. 2006; Wang et al. 2019). Compared with *Pinus*, other genera within the Pinaceae have shorter needles for a given width and low K_{leaf}, suggesting their needles may be constrained hydraulically, but it is still unclear as to the cause. One noticeable difference across genera in Pinaceae is the relative allocation to transfusion tracheids, which have been found to play a large role in leaf hydraulic conductance (Johnson et al. 2009;

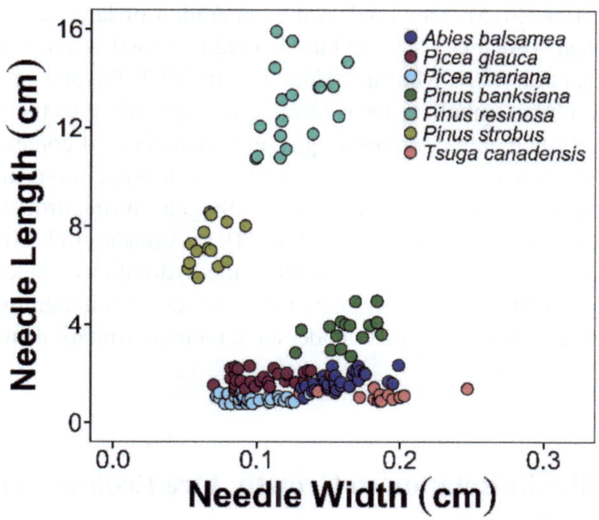

Fig. 6 Relationship between needle width and length of seven coexisting temperate conifers growing at the same site in Northern Wisconsin. (Data from Augustine & McCulloh, unpublished)

Charra-Vaskou et al. 2012; Zhang et al. 2014; Bouche et al. 2016). Pines allocate more to transfusion tracheids than many other Pinaceae genera (Trueba et al. 2021), and allocation to transfusion tracheids increases with increasing needle length in pines (Wang et al. 2019). Transfusion tracheids may allow pines to both occupy more arid environments and have longer needles. Further research is needed on the role of transfusion tracheids on hydraulic function, the trade-offs associated with their presence, and any potential evolutionary constraints to their development.

2.4 *Vulnerable Above, Vulnerable Below: Aspects of Pine Root Hydraulics*

Like their other organs, the roots of pines are much more vulnerable to embolism than other cooccurring conifers, but unfortunately there is a lack of root data which prevented a systematic review (but see Linton et al. 1998; Oliveras et al. 2003; Stout and Sala 2003; Plaut et al. 2012). Across species and within pines, roots also tend to be more vulnerable than other organs (e.g., trunk, branch; Stout and Sala 2003; Domec and Pruyn 2008; Domec et al. 2009b, 2010; Aguadé et al. 2015; Johnson et al. 2016). Being so vulnerable, the roots of pines will often lose much of their hydraulic conductance during the growing season (Domec et al. 2004; West et al. 2007), becoming the greatest resistance to water flow within the plant (Domec et al. 2009a, b; Poyatos et al. 2018). The embolism within and/or death of the roots can hydraulically isolate pines from the soil, causing them to have plant water potentials that are less negative than both the soil and other cooccurring conifers (Plaut et al.

2012; Pangle et al. 2015). This hinders the generation of large xylem tensions in more distal organs (West et al. 2007) but also reduces total water flow through the plant (Williams et al. 2001; Martínez-Vilalta et al. 2007; Poyatos et al. 2018). So, while this can inhibit embolism from forming and spreading to their more costly, but still relatively vulnerable, organs (e.g., trunk, branches), it comes at the cost of inhibiting gas exchange and causes pines to become dependent on carbon reserves, which can leave them vulnerable to pests and pathogens during drought (McDowell et al. 2008; Gaylord et al. 2013; Oliva et al. 2014; Aguadé et al. 2015; Anderegg et al. 2015). From their roots to their needles, the hydraulics of pines enable their characteristic drought avoidance strategy, but more research integrating above- and belowground hydraulics is needed to understand their coordination and relative role in the ecology of pines.

3 Hydraulics in Relation to Growth, Fire Ecology, and Shade Tolerance in *Pinus*

3.1 Hydraulics Facilitate the Rapid Growth of Pines and Potentially Their Dependence On Fire

Pines have long been lauded (or cursed) for their ability to grow rapidly compared to other conifers. This rapid growth has allowed them to become the most economically important tree group (Cavender-Bares et al. 2022). Their growth rate also has immense ecological implications, from causing them to be one of the most invasive groups globally (Richardson et al. 1994; Grotkopp et al. 2002) to facilitating their relationship with fire (Keeley 2012). Rapid growth requires high photosynthesis (Kruger and Volin 2006; Cernusak et al. 2008), which is dependent on the capacity of a plant to effectively transport water to the mesophyll and stomata (Hubbard et al. 2001; Nardini and Salleo 2003; Sack and Holbrook 2006). The high k_l and K_{leaf} values of pines (Fig. 3) permit them to maintain stomatal conductance and achieve higher photosynthesis relative to most other conifers (Brodribb and Feild 2008; Schmiege et al. 2021).

While less research has been conducted tying the hydraulics (and photosynthesis) of pines to their broader evolution, it is interesting to speculate how their unique hydraulics factor into their relationship with fire. The persistence of many pine species is dependent on surface or canopy fires excluding competitors and/or enabling reproduction (Schwilk and Ackerly 2001; He et al. 2012; Pausas 2015; Varner et al. 2022). While the evolution of fire-adapted traits is not limited to pines, with serotiny (Swarup et al. 2021), thick bark, and branch shedding (Stevens et al. 2020), being present in many other species (notably within Cupressaceae), pines are unique within conifers in the ecosystems that they inhabit. Few conifers outside of *Pinus* occupy savannas that burn at such short return intervals, which can be as short as 2 years (Stambaugh et al. 2011). Occupying such systems requires a suite of traits that is not seen in any other group, and many of these traits may only be possible because of the unusual combination of pine hydraulics traits.

The most unusual trait that many pines (e.g., *P. palustris*, *P. engelmannii*, *P. devoniana*) in savannas possess is the grass stage during the first few years of development (He et al. 2012; Pausas 2015), which protects the apical meristem from the fires (Jin et al. 2019). Presumably, densely packed, long needles characteristic of the grass stage (Pile et al. 2017) insulate the meristem from the fire's high temperatures. When species with a grass stage are mature, they have longer needles than all other groups of pines (Fig. 7), which may be an adaptation for coexisting with frequent fires. Long needle litter increases the intensity and flame length of fires (Stevens et al. 2020; Varner et al. 2022), both of which increase the speed with which a surface fire moves through an area. This high intensity but short duration fire allows the grass stage to protect the developing meristem while these species are in the sapling stage (Jin et al. 2019; Wilson et al. 2022). As stated earlier, having long needles is dependent on being able to supply water efficiently enough to prevent drops in water potential across its length (Zwieniecki et al. 2006). The long needles needed to both survive and propagate fire effectively may only be possible through the unique hydraulic systems that pines possess, allowing them to exist in relatively novel niches compared to most conifers.

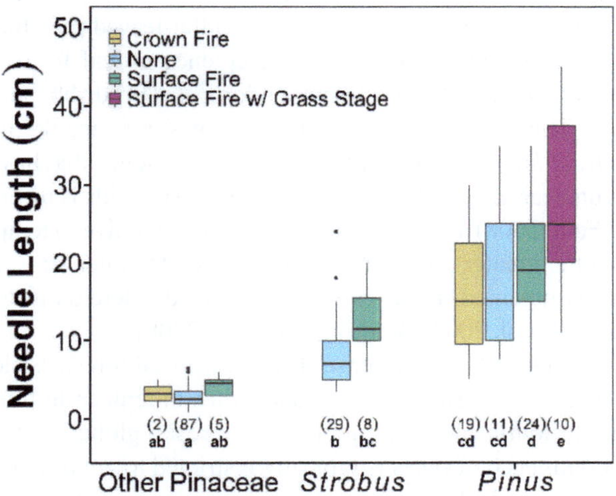

Fig. 7 Comparison of needle length within Pinaceae across different fire regimes. Pines are further divided by their subgenus (*Strobus* and *Pinus*). Letters below bars represent significant differences based on pairwise comparisons. Numbers of species in each group are indicated in parentheses. (Data of fire regimes comes from He et al. (2012) and needle length information is from Earle (2007))

3.2 Stuck in an Open Canopy: Shade Intolerance Within *Pinus*

While pines are abundant in most environments in the northern hemisphere, their presence, along with most other conifers, is notably absent from tropical rainforests (with a few exceptions, see below). In temperate forests, pines' abundance is often ephemeral as conditions become too shady once canopies close. Both their absence from tropical rainforests and early successional status are presumably related to their relative shade intolerance and have been studied in some detail, the former observation having a clear hydraulic mechanism while the latter being more ambiguous.

Most conifers native to tropical rainforests are within Podocarpaceae and Araucariaceae and are unique in having relatively broad leaves (see Brodribb (2011) for images of species within Podocarpaceae). Broader leaves in conifers are more efficient at harvesting light (Brodribb and Hill 1997), and economic theory predicts that broader leaves should be favored in shadier microsites (Givnish and Vermeij 1976). As discussed in the previous section, most conifers have a single vein that constrains needle width due to the inefficiency of extra-xylary water movement, restricting them from light limited environments. However, there are two hydraulic mechanisms that allow tropical conifers to construct relatively broad leaves. The first is having multiple veins, which is seen in Araucariaceae and some genera within Podocarpaceae. The second is to have a network of water-conducting sclereids, functioning effectively as veins, which is seen in many genera in Podocarpaceae. Both mechanisms allow leaves to increase in width without a consequent increase in the extra-xylary pathway (Brodribb et al. 2007; Brodribb and Feild 2008; Brodribb 2011). Notably, pines have one exception to the habitat generalities observed in most of the genus. *Pinus krempfii* is a flat-leaved pine found in tropical forests in Vietnam. It shows remarkable convergence in leaf form, having water-conducting sclereids like shade-tolerant podocarps (Brodribb and Feild 2008; Brodribb 2011). This exception highlights that pines are capable of being shade-tolerant, but why *P. krempfii* is the only member of the genus that has this hydraulic architecture is unclear. Likely the origin of hydraulic sclereids is recent within pines (Brodribb and Feild 2008) and is for the most part not adaptive to temperate and arid climates due to the greater vulnerability of sclereids to water stress and generally lower K_{leaf} that is obtained in leaves using elongated sclereids to transport water (Brodribb and Holbrook 2003; Brodribb and Feild 2008).

While their exclusion from the dense shade of tropical forests is clear, it is not as obvious why pines are restricted from shadier sites in temperate forests. Conifers, including pines, dominate many temperate forests globally (Givnish 2002). However, the dominance of pines is typically restricted to areas that maintain open canopies due to poor soils, low soil moisture, and/or prominent fire regimes (Keeley 2012). In areas where the canopies are more closed and fire is restricted, pines are typically replaced by more shade-tolerant conifers (e.g., *Tsuga*, *Abies*: Brown and Curtis 1952). Functionally, tolerating shade requires a suite of traits (outlined in Givnish (1988)) which is the antithesis of pines. Hydraulically, the traits that pines possess are at odds with maximizing light capture (pines have low LA/SA) and reducing metabolic costs (pines have high K_{leaf}, which is correlated to high photosynthesis and respiration; Reich 2014), both prerequisites to tolerating shade. Along with this, the construction of longer needles, which require higher K_{leaf} (e.g., greater allocation to transfusion tracheids; Wang et al. 2019), comes at the cost of reduced allocation to mesophyll tissue (Wang et al. 2019; Trueba et al. 2021), which would presumably reduce their ability to harvest light.

4 Takeaways, Next Steps in Pine Hydraulics

Over the past several decades, there has been considerable research on the hydraulics of pines. These studies have established that pines can grow quickly and have vulnerable xylem, which necessitates that pines exhibit stringent stomatal control to avoid dangerously low water potentials. However, how other traits are coordinated with the vulnerable hydraulic system to allow for such rapid growth and drought avoidance is poorly understood. The ecology and evolution of pines have been well studied, which makes them an ideal group to explore the form and function of the hydraulic network. For example, the hydraulic implications as to why the two pine subgenera, *Pinus* (the yellow, hard pines or diploxylon) and *Strobus* (the white, soft pines or haploxylon; Shaw 1924), are so ecologically and functionally (Fig. 7) different are an exciting area to explore. Expanding our knowledge of pine hydraulics can provide insight into the past evolution of this remarkable genus, as well as how it might fair in a changing climate.

References

Aguadé D, Poyatos R, Gómez M, Oliva J, Martínez-Vilalta J (2015) The role of defoliation and root rot pathogen infection in driving the mode of drought-related physiological decline in Scots pine (Pinus sylvestris L.). Tree Physiol 35:229–242

Anderegg WRL, Hicke JA, Fisher RA et al (2015) Tree mortality from drought, insects, and their interactions in a changing climate. New Phytol 208:674–683

Bouche PS, Larter M, Domec JC, Burlett R, Gasson P, Jansen S, Delzon S (2014) A broad survey of hydraulic and mechanical safety in the xylem of conifers. J Exp Bot 65:4419–4431

Bouche PS, Delzon S, Chost B et al (2016) Are needles of Pinus pinaster more vulnerable to xylem embolism than branches? New insights from X-ray computed tomography. Plant Cell Environ 39:860–870

Breshears DD, Myers OB, Meyer CW et al (2009) Tree die-off in response to global change-type drought: mortality insights from a decade of plant water potential measurements. Front Ecol Environ 7:185–189

Brodribb TJ (2011) A functional analysis of podocarp ecology. Smithson Contrib Bot 95:165–173

Brodribb TJ, Feild TS (2008) Evolutionary significance of a flat-leaved Pinus in Vietnamese rainforest. New Phytol 178:201–209

Brodribb T, Hill RS (1997) Light response characteristics of a morphologically diverse group of southern hemisphere conifers as measured by chlorophyll fluorescence. Oecologia 110:10–17

Brodribb TJ, Feild TS, Jordan GJ (2007) Leaf maximum photosynthetic rate and venation are linked by hydraulics. Plant Physiol 144:1890–1898

Brodribb TJ, Feild TS, Sack L (2010) Viewing leaf structure and evolution from a hydraulic perspective. Funct Plant Biol 37:488–498

Brodribb TJ, McAdam SAM, Jordan GJ, Martins SCV (2014) Conifer species adapt to low-rainfall climates by following one of two divergent pathways. Proc Natl Acad Sci 111:14489–14493

Brown RT, Curtis JT (1952) The upland conifer-hardwood forests of northern Wisconsin. Ecol Monogr 22:217–234

Buckley TN, John GP, Scoffoni C, Sack L (2015) How does leaf anatomy influence water transport outside the xylem? Plant Physiol 168:1616–1635

Cavender-Bares JM, Nelson E, Meireles JE (2022) The hidden value of trees: quantifying the ecosystem services of tree lineages and their major threats across the contiguous US. PLoS Sustain Transform 1:e0000010

Cernusak LA, Winter K, Aranda J, Turner BL (2008) Conifers, angiosperm trees, and lianas: growth, whole-plant water and nitrogen use efficiency, and stable isotope composition (d13C and d18O) of seedlings grown in a tropical environment. Plant Physiol 148:642–659

Charra-Vaskou K, Badel E, Burlett R, Cochard H, Delzon S, Mayr S (2012) Hydraulic efficiency and safety of vascular and non-vascular components in Pinus pinaster leaves. Tree Physiol 32:1161–1170

Choat B, Jansen S, Brodribb TJ et al (2012) Global convergence in the vulnerability of forests to drought. Nature 491:752–755

Cochard H, Hölttä T, Herbette S, Delzon S, Mencuccini M (2009) New insights into the mechanisms of water-stress-induced cavitation in conifers. Plant Physiol 151:949–954

Creese C, Benscoter AM, Maherali H (2011) Xylem function and climate adaptation in Pinus. Am J Bot 98:1437–1445

Delzon S, Douthe C, Sala A, Cochard H (2010) Mechanism of water-stress induced cavitation in conifers: bordered pit structure and function support the hypothesis of seal capillary-seeding. Plant Cell Environ 33:2101–2111

Domec JC, Gartner BL (2003) Relationship between growth rates and xylem hydraulic characteristics in young, mature and old-growth ponderosa pine trees. Plant Cell Environ 26:471–483

Domec JC, Pruyn ML (2008) Bole girdling affects metabolic properties and root, trunk and branch hydraulics of young ponderosa pine trees. Tree Physiol 28:1493–1504

Domec JC, Warren JM, Meinzer FC, Brooks JR, Coulombe R (2004) Native root xylem embolism and stomatal closure in stands of Douglas-fir and ponderosa pine: mitigation by hydraulic redistribution. Oecologia 141:7–16

Domec JC, Lachenbruch B, Meinzer FC (2006) Bordered pit structure and function determine spatial patterns of air-seeding thresholds in xylem of Douglas-fir (Pseudotsuga menziesii; Pinaceae) trees. Am J Bot 93:1588–1600

Domec JC, Lachenbruch B, Meinzer FC, Woodruff DR, Warren JM, McCulloh KA (2008) Maximum height in a conifer is associated with conflicting requirements for xylem design. Proc Natl Acad Sci 105:12069–12074

Domec JC, Noormets A, King JS, Sun G, McNulty SG, Gavazzi MJ, Boggs JL, Treasure EA (2009a) Decoupling the influence of leaf and root hydraulic conductances on stomatal conductance and its sensitivity to vapour pressure deficit as soil dries in a drained loblolly pine plantation. Plant Cell Environ 32:980–991

Domec JC, Warren JM, Meizner FC, Lachenbruch B (2009b) Safety factors for xylem failure by implosion and air-seeding within roots, trunks and branches of young and old conifer trees. IAWA J 30:101–120

Domec JC, Schäfer K, Oren R, Kim HS, McCarthy HR (2010) Variable conductivity and embolism in roots and branches of four contrasting tree species and their impacts on whole-plant hydraulic performance under future atmospheric CO2 concentration. Tree Physiol 30:1001–1015

Earle CJ (2007) The gymnosperm database. http://www.conifers.org/index.php

Garcia-Forner N, Adams HD, Sevanto S et al (2016) Responses of two semiarid conifer tree species to reduced precipitation and warming reveal new perspectives for stomatal regulation. Plant Cell Environ 39:38–49

Gaylord ML, Kolb TE, Pockman WT, Plaut JA, Yepez EA, Macalady AK, Pangle RE, McDowell NG (2013) Drought predisposes piñon–juniper woodlands to insect attacks and mortality. New Phytol 198:567–578

Givnish TJ (1988) Adaptation to sun and shade: a whole-plant perspective. Aust J Plant Physiol 15:63–92

Givnish TJ (2002) Adaptive significance of evergreen vs. deciduous leaves: solving the triple paradox. Silva Fenn 36:703–743

Givnish TJ, Vermeij CJ (1976) Sizes and shapes of liane leaves. Am Nat 110:743–778

Grotkopp E, Rejmánek M, Rost T (2002) Toward a causal explanation of plant invasiveness: seedling growth and life-history strategies of 29 pine (Pinus) species. Am Nat 159:396–419

Hacke UG, Jansen S (2009) Embolism resistance of three boreal conifer species varies with pit structure. New Phytol 182:675–686

Hacke UG, Sperry JS, Pittermann J (2004) Analysis of circular borded pit function II. Gymnosperm tracheids with torus-margo pit membranes. Am J Bot 91:386–400

He T, Pausas JG, Belcher CM, Schwilk DW, Lamont BB (2012) Fire-adapted traits of Pinus arose in the fiery Cretaceous. New Phytol 194:751–759

Hubbard RM, Ryan MG, Stiller RV, Sperry JS (2001) Stomatal conductance and photosynthesis vary linearly with plant hydraulic conductance in ponderosa pine. Plant Cell Environ 24:113–121

Jacobsen AL (2021) Diversity in conduit and pit structure among extant gymnosperm taxa. Am J Bot 108:559–570

Jansen S, Lamy JB, Burlett R, Cochard H, Gasson P, Delzon S (2012) Plasmodesmatal pores in the torus of bordered pit membranes affect cavitation resistance of conifer xylem. Plant Cell Environ 35:1109–1120

Jin S, Moule B, Yu D, Wang GG (2019) Fire survival of longleaf pine (Pinus palustris) grass stage seedlings: the role of seedling size, root collar position, and resprouting. Forests 10:1070

Johnson DM, Meinzer FC, Woodruff DR, McCulloh KA (2009) Leaf xylem embolism, detected acoustically and by cryo-SEM, corresponds to decreases in leaf hydraulic conductance in four evergreen species. Plant Cell Environ 32:828–836

Johnson DM, Wortemann R, McCulloh KA et al (2016) A test of the hydraulic vulnerability segmentation hypothesis in angiosperm and conifer tree species. Tree Physiol 36:983–993

Keeley JE (2012) Ecology and evolution of pine life histories. Ann For Sci 69:445–453

Kruger EL, Volin JC (2006) Reexamining the empirical relation between plant growth and leaf photosynthesis. Funct Plant Biol 33:421–429

Larter M, Pfautsch S, Domec JC, Trueba S, Nagalingum N, Delzon S (2017) Aridity drove the evolution of extreme embolism resistance and the radiation of conifer genus Callitris. New Phytol 215:97–112

Linton MJ, Sperry JS, Williams DG (1998) Limits to water transport in Juniperus osteosperma and Pinus edulis: implications for drought tolerance and regulation of transpiration. Funct Ecol 12:906–911

Martínez-Vilalta J, Sala A, Piñol J (2004) The hydraulic architecture of Pinaceae – a review. Plant Ecol 171:3–13

Martínez-Vilalta J, Korakaki E, Vanderklein D, Mencuccini M (2007) Below-ground hydraulic conductance is a function of environmental conditions and tree size in Scots pine. Funct Ecol 6:1072–1083

McCulloh KA, Augustine SP, Goke A, Jordan R, Krieg CP, O'Keefe K, Smith DD (2023) At least it is a dry cold: the global distribution of freeze–thaw and drought stress and the traits that may impart poly-tolerance in conifers. Tree Physiol 43:1–15

McDowell N, Pockman WT, Allen CD et al (2008) Mechanisms of plant survival and mortality during drought: why do some plants survive while others succumb to drought? New Phytol 178:719–739
McDowell NG, Williams AP, Xu C et al (2016) Multi-scale predictions of massive conifer mortality due to chronic temperature rise. Nat Clim Chang 6:295–300
Mencuccini M, Rosas T, Rowlands L et al (2019) Leaf economics and plant hydraulics drive leaf: wood area ratios. New Phytol 224:1544–1556
Nardini A, Salleo S (2003) Effects of the experimental blockage of the major veins on hydraulics and gas exchange of Prunus laurocerasus L. leaves. J Exp Bot 54:1213–1219
Oliva J, Stenlid J, Martínez-Vilalta J (2014) The effect of fungal pathogens on the water and carbon economy of trees: implications for drought-induced mortality. New Phytol 203:1028–1035
Oliveras I, Martínez-Vilalta J, Jimenez-Ortiz T, Lledó MJ, Escarré A, Piñol J (2003) Hydraulic properties of Pinus halepensis, Pinus pinea and Tetraclinis articulata in a dune ecosystem of Eastern Spain. Plant Ecol 169:131–141
Pangle RE, Limousin JM, Plaut JA et al (2015) Prolonged experimental drought reduces plant hydraulic conductance and transpiration and increases mortality in a piñon–juniper woodland. Ecol Evol 5:1618–1638
Pausas JG (2015) Evolutionary fire ecology: lessons learned from pines. Trends Plant Sci 20:318–324
Pile LS, Wang GG, Knapp BO, Liu G, Yu D (2017) Comparing morphology and physiology of southeastern US Pinus seedlings: implications for adaptation to surface fire regimes. Ann For Sci 75:68
Pittermann J, Choat B, Jansen S, Stuart SA, Lynn L, Dawson TE (2010) The relationships between xylem safety and hydraulic efficiency in the Cupressaceae: the evolution of pit membrane form and function. Plant Physiol 153:1919–1931
Plaut JA, Yepez EA, Hill J, Pangle R, Sperry JS, Pockman WT, McDowell NG (2012) Hydraulic limits preceding mortality in a piñon–juniper woodland under experimental drought. Plant Cell Environ 35:1601–1617
Poyatos R, Aguadé D, Martínez-Vilalta J (2018) Below-ground hydraulic constraints during drought-induced decline in Scots pine. Ann For Sci 75:100
Reich PB (2014) The world-wide "fast–slow" plant economics spectrum: a traits manifesto. J Ecol 102:275–301
Richardson DM, Williams PA, Hobbs RJ (1994) Pine invasions in the Southern Hemisphere: determinants of spread and invadability. J Biogeogr 21:511–527
Rosas T, Mencuccini M, Barba J, Cochard H, Saura-Mas S, Martínez-Vilalta J (2019) Adjustments and coordination of hydraulic, leaf and stem traits along a water availability gradient. New Phytol 223:632–646
Rosas T, Mencuccini M, Batlles C, Regalado Í, Saura-Mas S, Sterck F, Martínez-Vilalta J (2021) Are leaf, stem and hydraulic traits good predictors of individual tree growth? Funct Ecol 35:2435–2447
Roskilly B, Keeling E, Hood S, Giuggiola A, Sala A (2019) Conflicting functional effects of xylem pit structure relate to the growth-longevity trade-off in a conifer species. Proc Natl Acad Sci 116:15282–15287
Sack L, Holbrook NM (2006) Leaf hydraulics. Annu Rev Plant Biol 57:361–381
Sack L, Cowan PD, Jaikumar N, Holbrook NM (2003) The "hydrology" of leaves: co-ordination of structure and function in temperate woody species. Plant Cell Environ 26:1343–1356
Sanchez-Martinez P, Martínez-Vilalta J, Dexter KG, Segovia RA, Mencuccini M (2020) Adaptation and coordinated evolution of plant hydraulic traits. Ecol Lett 23:1599–1610
Schmiege SC, Buckley BM, Stevenson D, Cuong TQ, Nam LC, Griffin KL (2021) Contrasting physiological traits of shade tolerance in Pinus and Podocarpaceae native to a tropical Vietnamese forest: insight from an aberrant flat-leaved pine. Tree Physiol 41:223–239

Schulte PJ (2012) Computational fluid dynamics models of conifer bordered pits show how pit structure affects flow. New Phytol 193:721–729

Schulte PJ, Hacke UG, Schoonmaker AL (2015) Pit membrane structure is highly variable and accounts for a major resistance to water flow through tracheid pits in stems and roots of two boreal conifer species. New Phytol 208:102–113

Schwilk DW, Ackerly DD (2001) Flammability and serotiny as strategies: correlated evolution in pines. Oikos 94:326–336

Shaw GR (1924) Notes on the genus Pinus. J Arnold Arbor 5:225–227

Song Y, Poorter L, Horsting A, Delzon S, Sterck F (2022) Pit and tracheid anatomy explain hydraulic safety but not hydraulic efficiency of 28 conifer species. J Exp Bot 73:1033–1048

Stambaugh MC, Guyette RP, Marschall JM (2011) Longleaf pine (Pinus palustris Mill.) fire scars reveal new details of a frequent fire regime. J Veg Sci 22:1094–1104

Stevens JT, Kling MM, Schwilk DW, Varner JM, Kane JM (2020) Biogeography of fire regimes in western U.S. conifer forests: a trait-based approach. Glob Ecol Biogeogr 29:944–955

Stout DL, Sala A (2003) Xylem vulnerability to cavitation in Pseudotsuga menziesii and Pinus ponderosa from contrasting habitats. Tree Physiol 23:43–50

Swarup K, Contreras D, Leslie AB (2021) Serotiny and the evolution of seed cone size in Cupressaceae conifers. Int J Plant Sci 182:9–18

Trueba S, Théroux-Rancourt G, Earles JM, Buckley TN, Love D, Johnson DM, Brodersen C (2021) The three-dimensional construction of leaves is coordinated with water use efficiency in conifers. New Phytol 223:851–861

Varner JM, Shearman TM, Kane JM, Banwell EM, Jules ES, Stambaugh MC (2022) Understanding flammability and bark thickness in the genus Pinus using a phylogenetic approach. Sci Rep 12:7384

Wang N, Palmroth S, Maier CA, Domec JC, Oren R (2019) Anatomical changes with needle length are correlated with leaf structural and physiological traits across five Pinus species. Plant Cell Environ 42:1690–1704

West AG, Hultine KR, Jackson TL, Ehleringer JR (2007) Differential summer water use by Pinus edulis and Juniperus osteosperma reflects contrasting hydraulic characteristics. Tree Physiol 27:1711–1720

Williams M, Bond BJ, Ryan MG (2001) Evaluating different soil and plant hydraulic constraints on tree function using a model and sap flow data from ponderosa pine. Plant Cell Environ 24:679–690

Wilson LA, Spencer RN, Aubrey DP, O'Brien JJ, Smith AMS, Thomas RW, Johnson DM (2022) Longleaf pine seedlings are extremely resilient to the combined effects of experimental fire and drought. Fire 5:128

Woodruff DR, Meinzer FC, Marias DE, Sevanto S, Jenkins MW, McDowell NG (2015) Linking nonstructural carbohydrate dynamics to gas exchange and leaf hydraulic behavior in Pinus edulis and Juniperus monosperma. New Phytol 206:411–421

Yao GP, Nie ZF, Turner NC, Li FM, Gao TP, Fang XW, Scoffoni C (2020) Combined high leaf hydraulic safety and efficiency provides drought tolerance in Caragana species adapted to low mean annual precipitation. New Phytol 229:230–244

Zhang YG, Rockwell FE, Wheeler JK, Holbrook NM (2014) Reversible deformation of transfusion tracheids in Taxus baccata is associated with a reversible decrease in leaf hydraulic conductance. Plant Physiol 165:1557–1565

Zwieniecki MA, Boyce CK, Holbrook NM (2004) Functional design space of single-veined leaves: role of tissue hydraulic properties in constraining leaf size and shape. Ann Bot 94:507–513

Zwieniecki MA, Stone HA, Leigh A, Boyce CK, Holbrook NM (2006) Hydraulic design of pine needles: one-dimensional optimization for single-vein leaves. Plant Cell Environ 29:803–809

Data References

Augustine SP, McCulloh KA (2024) Physiological trait coordination and variability across and within three Pinus species. New Phytologist 244:451–463. https://doi.org/10.1111/nph.19859

Baker KV, Tai X, Miller ML, Johnson DM (2019) Six co-occurring conifer species in northern Idaho exhibit a continuum of hydraulic strategies during an extreme drought year. AoB Plants 11:plz056

Barros FV, Bittencourt PL, Eller CB, Signori-Müller C, Meireles LD, Oliviera RS (2022) Phytogeographical origin determines tropical montane cloud forest hydraulic trait composition. Funct Ecol 36:607–621

Beikircher B, Mayr S (2008) The hydraulic architecture of Juniperus communis L. ssp. communis: shrubs and trees compared. Plant Cell Environ 31:1545–1556

Beikircher B, Florineth F, Mayr S (2010) Restoration of rocky slopes based on planted gabions and use of drought-preconditioned woody species. Ecol Eng 36:421–426

Berry ZC, Johnson DM, Reinhardt K (2015) Vegetation-zonation patterns across a temperate mountain cloud forest ecotone are not explained by variation in hydraulic functioning or water relations. Tree Physiol 35:925–935

Bouche PS, Larter M, Domec JC, Burlett R, Gasson P, Jansen S, Delzon S (2014) A broad survey of hydraulic and mechanical safety in the xylem of conifers. J Exp Bot 65:4419–4431

Bouche PS, Delzon S, Chost B et al (2016a) Are needles of Pinus pinaster more vulnerable to xylem embolism than branches? New insights from X-ray computed tomography. Plant Cell Environ 39:860–870

Bouche PS, Jansen S, Sabalera JC, Cochard H, Burlett R, Delzon S (2016b) Low intra-tree variability in resistance to embolism in four Pinaceae species. Ann For Sci 73:681–689

Brodribb TJ, Cochard H (2009) Hydraulic failure defines the recovery and point of death in water-stressed conifers. Plant Physiol 149:575–584

Brodribb TJ, Feild TS (2000) Stem hydraulic supply is linked to leaf photosynthetic capacity: evidence from New Caledonian and Tasmanian rainforests. Plant Cell Environ 23:1381–1388

Brodribb TJ, Holbrook NM (2003) Stomatal closure during leaf dehydration, correlation with other leaf physiological traits. Plant Physiol 4:2166–73. https://doi.org/10.1104/pp.103.023879

Brodribb TJ, Feild TS (2008) Evolutionary significance of a flat-leaved Pinus in Vietnamese rainforest. New Phytol 178:201–209

Brodribb T, Hill RS (1999) The importance of xylen constraints in the distribution of conifer species. New Phytol 143:365–372

Brodribb TJ, Hill RS (2000) Increases in water potential gradient reduce xylem conductivity in whole plants. Evidence from a low-pressure conductivity method. Plant Physiol 123:1021–1028

Brodribb TJ, Holbrook NM, Zwieniecki MA, Palma B (2005) Leaf hydraulic capacity in ferns, conifers and angiosperms: impacts on photosynthetic maxima. New Phytol 165:839–846

Brodribb TJ, Feild TS, Jordan GJ (2007) Leaf maximum photosynthetic rate and venation are linked by hydraulics. Plant Physiol 144:1890–1898

Brodribb TJ, Bowman DJMS, Nichols S, Delzon S, Burlett R (2010) Xylem function and growth rate interact to determine recovery rates after exposure to extreme water deficit. New Phytol 188:533–542

Brodribb TS, McAdam SAM, Jordan GJ, Martins SCV (2014) Conifer species adapt to low-rainfall climates by following one of two divergent pathways. Proc Natl Acad Sci USA 111:14489–14493

Burgess SOS, Pittermann J, Dawson TE (2006) Hydraulic efficiency and safety of branch xylem increases with height in Sequoia sempervirens (D. Don) crowns. Plant Cell Environ 29:229–239

Cary KL, Ranieri GM, Pittermann J (2020) Xylem form and function under extreme nutrient limitation: an example from California's pygmy forest. New Phytol 226:760–769

Charra-Vaskou K, Badel E, Burlett R, Cochard H, Delzon S, Mayr S (2012a) Hydraulic efficiency and safety of vascular and non-vascular components in Pinus pinaster leaves. Tree Physiol 32:1161–1170

Charra-Vaskou K, Charrier G, Wortemann R, Beikircher B, Cochard H, Ameglio T, Mayr S (2012b) Drought and frost resistance of trees: a comparison of four species at different sites and altitudes. Ann For Sci 69:325–333
Cherif S, Ezzine O, Khouja ML, Nasr Z (2019) Hydraulic traits performances of three pine species in Tunisia. J Agric Sci 11:20–29
Clute T, Martin J, Looker N, Hu J (2021) Hydraulic traits of co-existing conifers do not correlate with local hydroclimate condition: a case study in the northern Rocky Mountains, U.S.A. Oecologia 197:1049–1062
Cochard H (1992) Vulnerability of several conifers to air embolism. Tree Physiol 11:73–83
Cochard H (2006) Cavitation in trees. C R Phys 7:1018–1026
Cochard H, Damour G, Bodet C, Tharwat I, Poirier M, Améglio T (2005) Evaluation of a new centrifuge technique for rapid generation of xylem vulnerability curves. Physiol Plant 124:410–418
Condo TK, Reinhardt K (2019) Large variation in branch and branch-tip hydraulic functional traits in Douglas-fir (Pseudotsuga menziesii) approaching lower treeline. Tree Physiol 39:1461–1472
Dai Y, Wang L, Wan X (2020) Frost fatigue and its spring recovery of xylem conduits in ring-porous, diffuse-porous, and coniferous species in situ. Plant Physiol Biochem 146:177–186
Delzon S, Douthe C, Sala A, Cochard H (2010) Mechanism of water-stress induced cavitation in conifers: bordered pit structure and function support the hypothesis of seal capillary-seeding. Plant Cell Environ 33:2101–2111
Dietrich L, Delzon S, Hoch G, Kahmen A (2018) No role for xylem embolism or carbohydrate shortage in temperate trees during the severe 2015 drought. J Ecol 107:334–349
Domec JC, Gartner BL (2002) Age- and position-related changes in hydraulic versus mechanical dysfunction of xylem: inferring the design criteria for Douglas-fir wood structure. Tree Physiol 22:91–104
Domec JC, Palmroth S, Ward E, Maier CA, Thérézien M, Oren R (2009) Acclimation of leaf hydraulic conductance and stomatal conductance of Pinus taeda (loblolly pine) to long-term growth in elevated CO2 (free-air CO2 enrichment) and N-fertilization. Plant Cell Environ 32:1500–1512
Domec JC, Rivera LN, King JS, Peszlen I, Hain F, Smith B, Frampton J (2013) Hemlock woolly adelgid (Adelges tsugae) infestation affects water and carbon relations of eastern hemlock (Tsuga canadensis) and Carolina hemlock (Tsuga caroliniana). New Phytol 199:452–463
Duan CY, Li MY, Fang LD et al (2022) Greater hydraulic safety contributes to higher growth resilience to drought across seven pine species in a semi-arid environment. Tree Physiol 42:727–739
Ducrey M, Huc R, Ladjal M, Guehl JM (2008) Variability in growth, carbon isotope composition, leaf gas exchange and hydraulic traits in the eastern Mediterranean cedars Cedrus libani and C. brevifolia. Tree Physiol 28:689–701
Dulamsuren C, Abilova SB, Bektayeva M, Eldarov M, Schuldt B, Leuschner C, Hauck M (2018) Hydraulic architecture and vulnerability to drought-induced embolism in southern boreal tree species of Inner Asia. Tree Physiol 39:463–473
Earle CJ (2007) The gymnosperm database. http://www.conifers.org/index.php
Ewers BE, Oren R, Sperry JS (2000) Influence of nutrient versus water supply on hydraulic architecture and water balance in Pinus taeda. Plant Cell Environ 23:1055–1066
Fan D, Wang X, Zhang W, Zhang X, Zhang S, Xie Z (2020) Does Cathaya argyrophylla, an ancient and threatened Pinaceae species endemic to China, show eco-physiological outliers to its Pinaceae relatives? Conserv Physiol 8:coaa094
Feng F, Losso A, Tyree M, Zhang S, Mayr S (2021) Cavitation fatigue in conifers: a study on eight European species. Plant Physiol 186:1580–1590
Garcia-Forner N, Adams HD, Sevanto S et al (2016) Responses of two semiarid conifer tree species to reduced precipitation and warming reveal new perspectives for stomatal regulation. Plant Cell Environ 39:38–49
Gonzalez-Benecke CA, Martin TA, Peter GF (2010) Hydraulic architecture and tracheid allometry in mature Pinus palustris and Pinus elliottii trees. Tree Physiol 30:361–375

Gyenge JE, Fernández ME, Dalla Salda G, Schlichter T (2005) Leaf and whole-plant water relations of the Patagonian conifer Austrocedrus chilensis (D. Don) Pic. Ser. et Bizzarri: implications on its drought resistance capacity. Ann For Sci 62:297–302

Hacke UG, Jansen S (2009) Embolism resistance of three boreal conifer species varies with pit structure. New Phytol 182:675–686

Hacke UG, Sperry JS, Ewers BE, Ellsworth DS, Schäfer KVR, Oren R (2000) Influence of soil porosity on water use in Pinus taeda. Oecologia 124:495–505

Hacke UG, Sperry JS, Pittermann J (2004) Analysis of circular borded pit function II. Gymnosperm tracheids with torus-margo pit membranes. Am J Bot 91:386–400

Hammond WM, Yu K, Wilson LA, Will RE, Anderegg WRL, Adams HD (2019) Dead or dying? Quantifying the point of no return from hydraulic failure in drought-induced tree mortality. New Phytol 223:1834–1843

He T, Pausas JG, Belcher CM, Schwilk DW, Lamont BB (2012) Fire-adapted traits of Pinus arose in the fiery Cretaceous. New Phytol 194:751–759

Hubbard RM, Ryan MG, Stiller RV, Sperry JS (2001) Stomatal conductance and photosynthesis vary linearly with plant hydraulic conductance in ponderosa pine. Plant Cell Environ 24:113–121

Jansen S, Pletsers A, Sano Y (2008) The effect of preparation techniques on SEM-imaging of pit membranes. IAWA J 29:161–178

Jansen S, Lamy JB, Burlett R, Cochard H, Gasson P, Delzon S (2012) Plasmodesmatal pores in the torus of bordered pit membranes affect cavitation resistance of conifer xylem. Plant Cell Environ 35:1109–1120

Jin Y, Wang C, Zhou Z (2019) Conifers but not angiosperms exhibit vulnerability segmentation between leaves and branches in a temperate forest. Tree Physiol 39:454–462

Johnson DM, Meinzer FC, Woodruff DR, McCulloh KA (2009a) Leaf xylem embolism, detected acoustically and by cryo-SEM, corresponds to decreases in leaf hydraulic conductance in four evergreen species. Plant Cell Environ 32:828–836

Johnson DM, Woodruff DR, McCulloh KA, Meinzer FC (2009b) Leaf hydraulic conductance, measured in situ, declines and recovers daily: leaf hydraulics, water potential and stomatal conductance in four temperate and three tropical tree species. Tree Physiol 29:879–887

Johnson DM, McCulloh KA, Meinzer FC, Woodruff DR, Eissenstat DM, Phillips N (2011) Hydraulic patterns and safety margins, from stem to stomata, in three eastern US tree species. Tree Physiol 31:659–668

Johnson DM, McCulloh KA, Woodruff DR, Meinzer FC (2012) Evidence for xylem embolism as a primary factor in dehydration-induced declines in leaf hydraulic conductance. Plant Cell Environ 35:760–769

Johnson DM, Wortemann R, McCulloh KA et al (2016) A test of the hydraulic vulnerability segmentation hypothesis in angiosperm and conifer tree species. Tree Physiol 36:983–993

Johnson DM, Berry ZC, Baker KV, Smith DD, McCulloh KA, Domec JC (2018) Leaf hydraulic parameters are more plastic in species that experience a wider range of leaf water potentials. Funct Ecol 32:894–903

Kilgore JS, Jacobsen A, Telewski FW (2020) Hydraulics of Pinus (subsection Ponderosae) populations across an elevation gradient in the Santa Catalina Mountains of southern Arizona. Madrono 67:218–226

Ladjal M, Huc R, Ducrey M (2005) Drought effects on hydraulic conductivity and xylem vulnerability to embolism in diverse species and provenances of Mediterranean cedars. Tree Physiol 25:1109–1117

Larter M, Brodribb TJ, Pfautsch S, Burlett R, Cochard H, Delzon S (2015) Extreme aridity pushes trees to their physical limits. Plant Physiol 168:804–807

Larter M, Pfautsch S, DOmec JC, Trueba S, Nagalingum N, Delzon S (2017) Aridity drove the evolution of extreme embolism resistance and the radiation of conifer genus Callitris. New Phytol 215:97–112

Laughlin DC, Delzon S, Clearwater MJ, Bellingham PJ, McGlone MS, Richardson SJ (2020) Climatic limits of temperate rainforest tree species are explained by xylem embolism resistance among angiosperms but not among conifers. New Phytol 226:727–740
Li Y, Sperry JS, Taneda H, Bush SE, Hacke UG (2008) Evaluation of centrifugal methods for measuring xylem cavitation in conifers, diffuse- and ring-porous angiosperms. New Phytol 177:558–568
Li MY, Leng QN, Hao GY (2021) Contrasting patterns of radial growth rate between Larix principis-rupprechtii and Pinus sylvestris var. mongolica along an elevational gradient are mediated by differences in xylem hydraulics. For Ecol Manag 497:119524
Linton MJ, Sperry JS, Williams DG (1998) Limits to water transport in Juniperus osteosperma and Pinus edulis: implications for drought tolerance and regulation of transpiration. Funct Ecol 12:906–911
López R, López de Heredia U, Collada C, Cano FJ, Emerson BC, Cochard H, Gil L (2013) Vulnerability to cavitation, hydraulic efficiency, growth and survival in an insular pine (Pinus canariensis). Ann Bot 111:1167–1179
Lu P, Biron P, Granier A, Cochard. (1996) Water relations of adult Norway spruce (Picea abies (L.) Karst) under soil drought in the Vosges mountains: whole-tree hydraulic conductance, xylem embolism and water loss regulation. Ann For Sci 53:113–121
Lusk CH, Jiménez-Castillo M, Salazar-Ortega N (2007) Evidence that branches of evergreen angiosperm and coniferous trees differ in hydraulic conductance but not in Huber values. Can J Bot 85:141–147
Maherali H, DeLucia EH (2000) Xylem conductivity and vulnerability to cavitation of ponderosa pine growing in contrasting climates. Tree Physiol 20:859–867
Maherali H, Moura CF, Caldeira MC, Willson CJ, Jackson RB (2006) Functional coordination between leaf gas exchange and vulnerability to xylem cavitation in temperate forest trees. Plant Cell Environ 29:571–583
Mantova M, Menezes-Silva PE, Badel E, Cochard H, Torres-Ruiz JM (2021) The interplay of hydraulic failure and cell vitality explains tree capacity to recover from drought. Physiol Plant 172:247–257
Martínez-Vilalta J, Piñol J (2002) Drought-induced mortality and hydraulic architecture in pine populations of the NE Iberian Peninsula. For Ecol Manag 161:247–256
Martínez-Vilalta J, Cochard H, Mencuccini M et al (2009) Hydraulic adjustment of Scots pine across Europe. New Phytol 184:353–364
Mayr S, Sperry JS (2010) Freeze–thaw-induced embolism in Pinus contorta: centrifuge experiments validate the "thaw-expansion hypothesis" but conflict with ultrasonic emission data. New Phytol 185:1016–1024
Mayr S, Wolfschwenger M, Bauer H (2002) Winter-drought induced embolism in Norway spruce (Picea abies) at the Alpine timberline. Physiol Plant 115:74–80
Mayr S, Gruber A, Bauer H (2003a) Repeated freeze-thaw cycles induce embolism in drought stressed conifers (Norway spruce, stone pine). Planta 217:436–441
Mayr S, Gruber A, Schwienbacher F, Dämon B (2003b) Winter-embolism in a krummholz shrub (Pinus mugo) growing at the alpine timberline. Aust J For Sci 1:29–38
Mayr S, Schwienbacher F, Bauer H (2003c) Winter at the alpine timberline. Why does embolism occur in Norway spruce but not in Stone pine? Plant Physiol 131:780–792
Mayr S, Hacke U, Schmid P, Schwienbacher F, Gruber A (2006) Frost drought in conifers at the alpine timberline: xylem dysfunction and adaptations. Ecology 87:3175–3185
McCulloh KA, Johnson DM, Meinzer FC, Woodruff DR (2014) The dynamic pipeline: hydraulic capacitance and xylem hydraulic safety in four tall conifer species. Plant Cell Environ 37:1171–1183
McElrone AJ, Pockman WT, Martínez-Vilalta J, Jackson RB (2004) Variation in xylem structure and function in stems and roots of trees to 20 m depth. New Phytol 163:507–517
Morcillo L, Muñoz-Rengifo JC, Torres-Ruiz JM, Delzon S, Moutahir H, Vilagrosa A (2022) Post-drought conditions and hydraulic dysfunction determine tree resilience and mortality across

Mediterranean Aleppo pine (Pinus halepensis) populations after an extreme drought event. Tree Physiol 42:1364–1376
Ning QR, Gong XW, Li MY, Hao GY (2022) Differences in growth pattern and response to climate warming between Larix olgensis and Pinus koraiensis in Northeast China are related to their distinctions in xylem hydraulics. Agric For Meteorol 312:108724
Ogle K, Barber JJ, Willson C, Thompson B (2009) Hierarchical statistical modeling of xylem vulnerability to cavitation. New Phytol 182:541–554
Olano JM, González-Muñoz N, Arzac A, Rozas V, von Arx G, Delzon S, García-Cervigón AI (2017) Sex determines xylem anatomy in a dioecious conifer: hydraulic consequences in a drier world. Tree Physiol 37:1493–1502
Olguin FY, Moretti AP, Pinazo MA, Graciano C (2019) Morpho-physiological acclimation to canopy coverage of Araucaria angustifolia during the establishment in the Atlantic Forest, Argentina. Bosque 40:323–333
Oliveras I, Martínez-Vilalta J, Jimenez-Ortiz T, Lledó MJ, Escarré A, Piñol J (2003) Hydraulic properties of Pinus halepensis, Pinus pinea and Tetraclinis articulata in a dune ecosystem of Eastern Spain. Plant Ecol 169:131–141
Olsen ME, Soriano D, Rosell JA et al (2018) Plant height and hydraulic vulnerability to drought and cold. Proc Natl Acad Sci 115:7551–7556
Peguero-Pina JJ, Sancho-Knapik D, Cochard H, Barredo G, Villarroya D, Gil-Pelegrín E (2012) Hydraulic traits are associated with the distribution range of two closely related Mediterranean firs, Abies alba Mill. and Abies pinsapo Boiss. Tree Physiol 31:1067–1075
Piñol J, Sala A (2000) Ecological implications of xylem cavitation for several Pinaceae in the Pacific Northern USA. Funct Ecol 14:538–545
Pittermann J, Sperry JS (2006) Analysis of freeze-thaw embolism in conifers. The interaction between cavitation pressure and tracheid size. Plant Physiol 140:374–382
Pittermann J, Sperry JS, Hacke UG, Wheeler JK, Sikkema EH (2006) Inter-tracheid pitting and the hydraulic efficiency of conifer wood: the role of tracheid allometry and cavitation protection. Am J Bot 93:1265–1273
Pittermann J, Choat B, Jansen S, Stuart SA, Lynn L, Dawson TE (2010) The relationships between xylem safety and hydraulic efficiency in the Cupressaceae: the evolution of pit membrane form and function. Plant Physiol 153:1919–1931
Pittermann J, Stuart SA, Dawson TE, Moreau A (2012) Cenozoic climate change shaped the evolutionary ecophysiology of the Cupressaceae conifers. Proc Natl Acad Sci 109:9647–9652
Pockman WT, Sperry JS (2000) Vulnerability to xylem cavitation and the distribution of Sonoran Desert vegetation. Am J Bot 87:1287–1299
Prendin AL, Mayr S, Beikircher B, von Arx G, Petit G (2018) Xylem anatomical adjustments prioritize hydraulic efficiency over safety as Norway spruce trees grow taller. Tree Physiol 38:1088–1097
Renninger HJ, Meinzer FC, Gartner BL (2006) Hydraulic architecture and photosynthetic capacity as constraints on release from suppression in Douglas-fir and western hemlock. Tree Physiol 27:33–42
Rosas T, Mencuccini M, Batlles C, Regalado Í, Saura-Mas S, Sterck F, Martínez-Vilalta J (2021) Are leaf, stem and hydraulic traits good predictors of individual tree growth? Funct Ecol 35:2435–2447
Sáenz-Romero C, Lamy JB, Loya-Rebollar E, Plaza-Aquilar A, Burlett R, Lobit P, Delzon S (2013) Genetic variation of drought-induced cavitation resistance among Pinus hartwegii populations from an altitudinal gradient. Acta Physiol Plant 35:2905–2913
Sáenz-Romero C, Larter M, González-Muñoz N et al (2017) Mexican conifers differ in their capacity to face climate change. J Plant Hydraul 4:e-003
Sancho-Knapik D, Peguero-Pina JJ, Flexas J, Herbette S, Cochard H, Niinemets U, Gil-Pelegrín E (2014) Coping with low light under high atmospheric dryness: shade acclimation in a Mediterranean conifer (Abies pinsapo Boiss.). Tree Physiol 34:1321–1333

Schoonmaker AL, Hacke UG, Landhäusser SM, Lieffers VJ, Tyree MT (2010) Hydraulic acclimation to shading in boreal conifers of varying shade tolerance. Plant Cell Environ 33:382–393
Song Y, Poorter L, Horsting A, Delzon S, Sterck F (2022) Pit and tracheid anatomy explain hydraulic safety but not hydraulic efficiency of 28 conifer species. J Exp Bot 73:1033–1048
Sperry JS, Ikeda T (1997) Xylem cavitation in roots and stems of Douglas-fir and white fir. Tree Physiol 17:275–280
Sperry JS, Sullivan JEM (1992) Xylem embolism in response to freeze-thaw cycles and water stress in ring-porous, diffuse-porous, and conifer species. Plant Physiol 100:605–613
Sperry JS, Tyree MT (1990) Water-stress-induced embolism in three species of conifers. Plant Cell Environ 13:427–436
Stout DL, Sala A (2003) Xylem vulnerability to cavitation in Pseudotsuga menziesii and Pinus ponderosa from contrasting habitats. Tree Physiol 23:43–50
Tognetti R, Longobucco A, Raschi A, Jones MB (2001) Stem hydraulic properties and xylem vulnerability to embolism in three co-occurring Mediterranean shrubs at a natural CO2 spring. Aust J Plant Physiol 28:257–268
Tyree MT, Dixon MA (1986) Water stress induced cavitation and embolism in some woody plants. Physiol Plant 66:397–405
Tyree MT, Patiño S, Becker P (1998) Vulnerability to drought-induced embolism of Bornean heath and dipterocarp forest trees. Tree Physiol 18:583–588
Urrutia-Jalabert R, Peña MP, Coopman RE et al (2018) Elucidating the hydraulic vulnerability of the longest-lived Southern Hemisphere conifer to aridification. For Ecol Manag 430:472–484
Vander Willigen C, Sherwin HW, Pammenter NW (2000) Xylem hydraulic characteristics of subtropical trees from contrasting habitats grown under identical environmental conditions. New Phytol 145:51–59
Wang J, Wang M, Zhang X, Sun S, Zhang A, Chen N, Zhao C (2018) Enhanced cell dehydration tolerance and photosystem stability facilitate the occupation of cold alpine habitats by a homoploid hybrid species, Picea purpurea. AoB Plants 10:ply053
Willson CJ, Jackson RB (2006) Xylem cavitation caused by drought and freezing stress in four co-occurring Juniperus species. Physiol Plant 127:374–382
Willson CJ, Manos PS, Jackson RB (2008) Hydraulic traits are influenced by phylogenetic history in the drought-resistant, invasive genus Juniperus (Cupressaceae). Am J Bot 95:299–314
Woodruff DR, McCulloh KA, Warren JM, Meinzer FC, Lachenbruch B (2007) Impacts of tree height on leaf hydraulic architecture and stomatal control in Douglas-fir. Plant Cell Environ 30:559–569
Woodruff DR, Meinzer FC, Lachenbruch B (2008) Height-related trends in leaf xylem anatomy and shoot hydraulic characteristics in a tall conifer: safety versus efficiency in water transport. New Phytol 180:90–99
Woodruff DR, Meinzer FC, Lachenbruch B, Johnson DM (2009) Coordination of leaf structure and gas exchange along a height gradient in a tall conifer. Tree Physiol 29:261–272
Woodruff DR, Meinzer FC, Marias DE, Sevanto S, Jenkins MW, McDowell NG (2015) Linking nonstructural carbohydrate dynamics to gas exchange and leaf hydraulic behavior in Pinus edulis and Juniperus monosperma. New Phytol 206:411–421
Zhang YG, Rockwell FE, Wheeler JK, Holbrook NM (2014) Reversible deformation of transfusion tracheids in Taxus baccata is associated with a reversible decrease in leaf hydraulic conductance. Plant Physiol 165:1557–1565

Evapotranspiration of Pine Forests: A Global Synthesis

Ge Sun, Maricar Aguilos, Ning Liu, Johnny Boggs, Devendra Amatya, Lulu Zhang, and Lu Hao

Abstract Pine (*Pinus* spp.) forests are found in many areas with a wide range of bioclimates, providing a myriad of ecosystem services. Pine trees have been widely planted for various purposes such as commercial timber production, soil erosion control and water quality improvement of source waters, urban forestry, wildlife habitat, and bioenergy for carbon sequestration. Quantifying these benefits requires a clear understanding of the hydrological cycle in pine forests, especially the evapotranspiration processes. This synthesis study aims at synthesizing published literature documenting the water balance components including evapotranspiration (ET), the sum of canopy/litter interception, tree transpiration, soil/understory evaporation, water yield as surface runoff and subsurface drainage, and soil water storage of pine

Supplementary Information The online version contains supplementary material available at https://doi.org/10.1007/978-3-032-23001-0_8.

G. Sun (✉) · J. Boggs
Eastern Forest Environmental Threat Assessment Center, Southern Research Station, USDA Forest Service, Durham, NC, USA
e-mail: ge.sun@usda.gov

M. Aguilos
Department of Forestry and Environmental Resources, North Carolina State University, Raleigh, NC, USA

N. Liu
CSIRO Environment, Black Mountain, Canberra, ACT, Australia

D. Amatya
Center for Forest Watershed Research, USDA Forest Service, Cordesville, SC, USA

L. Zhang
United Nations University, Institute for Integrated Management of Material Fluxes and of Resources, Dresden, Germany

L. Hao
Key Laboratory of Meteorological Disaster, Ministry of Education (KLME)/Jiangsu Key Laboratory of Agricultural Meteorology, Nanjing University of Information Science and Technology, Nanjing, China

D. M. Johnson et al. (eds.), *Physiology and Ecology of Pines Worldwide*, Tree Physiology 9, https://doi.org/10.1007/978-3-032-23001-0_8

forests, natural or managed, at stand or watershed levels. We conducted the analyses using the Budyko's water and energy balance framework to show how the aridity index, defined as potential evapotranspiration/precipitation ratio (PET/P), influences water use strategies and controls on evaporative efficiency (ET/P) of pine forests under different climatic regimes and management. We examine influences of tree and stand age, xylem anatomy, forest management such as thinning, species conversion (e.g., from loblolly pine to longleaf pine or deciduous to pine), and artificial drainage on pine forest ET and hydrology. We explain the large variability of forest ET and its components (i.e., canopy interception) and across a large global climatic gradient. We identify research gaps for fully accounting pine forest evapotranspiration such as water loss by understories and stress the importance of canopy interception and soil evaporation. Our review suggests that total ET rates of evergreen pine species generally are higher than native broadleaf deciduous forests under most climatic conditions. We demonstrate the success and challenges to use pine forests to meet multiple ecosystem services (soil erosion control, water supply, and timber yield) in a changing environment.

1 Introduction

Forests dominated by the genus *Pinus* are found over a large climatic and a nutrient gradient from the Arctic with a short growing season to the tropics with a year-round growing season and from nutrient-poor sandy soils to nutrient-rich organic soils (Knight et al. 1994; Richardson et al. 2007). For example, pines are naturally found in North America (the USA, Canada), Europe, and Asia, where that annual precipitation can vary from 300 to 1500 mm (e.g., Southern USA and China) with seasonal temperature extremes ranging from −40 to 20 °C (Adams et al. 2019). Pines are widely planted worldwide beyond their natural range because of their favorable long tracheids and wood properties for structural timber (Tor-ngern et al. 2017) and high growth rate (Fahey and Jackson 1997; Calder 2007) and bioenergy (Pisarello et al. 2022; Ruzol et al. 2022), often as monoculture in the past 50 years (Nichols et al. 2006). Because pines can survive in harsh environments (i.e., poor nutrients, drought, frequent burns), pines are widely used as pioneering trees for afforestation in soil erosion control or urban greening in China (Bao et al. 2004; Chen et al. 2007, 2023; Ma et al. 2018), rehabilitating degraded grass lands in Nepal (Ives 2006; Ghimire et al. 2013, 2014), and reforestation in India (Naudiyal and Schmerbeck 2017) and the Caribbean with a tropical climate (Francis et al. 2022). Pine forests vary greatly in leaf phenology, leaf morphology, canopy structure, and water use patterns, and their water balances can be rather complex (see review in Tor-ngern et al. 2017).

Pine plantations have greatly expanded since the nineteenth century, thanks to the advances of silviculture. For example, there were almost no pine plantations in the Southern USA in the 1950s, but planted pine, mostly loblolly pine (*Pinus taeda L*), increased from 8.5 million ha in 1985 to 13 million ha in 1999 with an estimated increase to 20 million ha by 2040 (Weir and Greis 2002). The growth of many pine plantations in the Southern USA is limited by soil nutrient availability (nitrogen and phosphorus) which limits leaf area production and therefore growth efficiency and stand production (Fox et al. 2007). Therefore, intensive forest management such as fertilization and minor drainage has been used to increase productivity, causing concerns of stream water quality and wetland degradations. In China, pine trees, native (*P. tabuliformis*; *P. massoniana*) (Wei et al. 2005) or introduced species from the USA (*P. taeda* and *P. elliottii*), have been widely used for massive reforestation for timber production and soil erosion controls (Ma et al. 2014; Xu 2011). Throughout the Caribbean, exotic pine trees such as *Pinus caribaea* or *Pinus oocarpa* have been used to reforest many degraded landscapes for their fast growth and resilience to drought (Francis et al. 2022).

Due to their wide distribution and large areas, pine forests are important for local economies, and their ecological implications, both benefits and risks, have been increasingly scrutinized globally. For example, intensive pine forest management in Southeastern USA in the 1980s that involved fertilization and artificial drainage substantially increased productivity (Jokela et al. 2010) but also caused concerns about stream water supply (Shepard 1994). A global review suggests plantation forests use more water and have lower biodiversity than native forests (Hua et al. 2022). Some other studies argued that plantation forests could provide net benefits for conservation purposes, but it is context-specific and depends on multiple factors, including controlled management (Carnus et al. 2006; Vanclay 2009). Coniferous forests generally have higher evapotranspiration than hardwoods under similar precipitation conditions (Ford et al. 2011). A watershed-scale forest manipulation experiment in the humid southern Appalachians Mountains shows that converting negative deciduous forests to pine plantations decreases water yield by over 30% (Swank and Douglass 1974). Swank and Douglass (1974) predicted that converting piedmont hardwood forests of the Southeastern USA to loblolly pine could reduce annual streamflow by about 4 acre-inches (100 mm) for each acre converted. Converting Englemann spruce (*Picea engelmanni*) to aspen (*Populus tremuloides*) forests alters snow distribution and transpiration and may result in 185 mm net loss of moisture available for streamflow and a loss of 71 mm when fir forests (*Abies lasiocarpa*) cover the watershed in the Colorado River Basin (Hibbert 1979; LaMalfa 2007; Zou et al. 2010). Thus, a succession of aspen to conifers would cause a decline of water yields in the Western USA (Gifford et al. 1984). Conversion of mixed evergreen forest to pine plantation in the northwestern South Island initially resulted in an increase in low flow, peak flow, and total flow of 310–340 mm (61–68%), but planting the harvested areas with pines caused the hydrology (low flow, high flow, total flow) to return to re-harvesting levels within 8–10 years (Fahey and Jackson 1997). Another study suggests that converting tussock grasslands to pine plantations in southern New Zealand decreased flow by 230 mm or 27% by year 10 (Fahey and Jackson 1997). At a site in northern Uruguay, the water yield reduction over the 3 years after pine plantation of a grassland watershed was 15%, with 50% reduction in peak flow rates 2 years after planting and by 75% 4 years after planting (Chescheir et al. 2009). Ramirez et al. (2022) compared the water balance of pine plantations and thorny forests in a Mediterranean landscape in Chile. They found that the ET rates of pines are 379 mm in the dry year and 616 mm in the wet year, while shrubland ET is 291 and 371 mm, respectively. Comparatively, pines in the dry year lost 88 mm more than shrubland on average and 245 mm more in the wet year. Similar studies in Japan (Komatsu et al. 2008) and West Virginia in the Eastern USA reported that total water loss from coniferous forests was higher than from deciduous forests due to higher canopy interception rates and total ET of conifer trees (Rau et al. 2022). It is generally believed that coniferous evergreen trees use more water than deciduous trees in Germany. For example, the mean annual canopy transpiration rate for Scots pine (*Pinus sylvestris*) is 320 mm, higher than that of European beech (*Fagus sylvatica L.*) forest (304 mm). A direct comparison of the neighboring pine/beech stands showed that the annual transpiration of pine was 30–80 mm higher than beech. Annual ET (transpiration + canopy interception) averaged 582 mm in the pine stands compared to 496 mm in the beech stands. The annual deep seepage ranged from 15 to 88 mm under the pine (mean 66 mm) and from 100 to 273 mm (mean 134 mm) under the beech, which equals 2–13% and 15–31% of annual precipitation, respectively (Leuschner et al. 2022). In Nepal, the hydrologic effects of afforestation with Chir pine (*Pinus roxburghii*) remain somewhat controversial. Local observations by farmers from the Jhikhu Khola region suggest that certain springs dried up after the afforestation program was implemented (Ives 2006). Limited forest hydrological studies in Nepal (Ghimire et al. 2013, 2014) do suggest higher water use of planted pines than degraded grass, but effects of tree planting on springs could be moderated by better forest and soil management or use native broadleaf trees promoting soil infiltration. In the tropics, exotic pine (*Pinus caribaea* or *Pinus oocarpa*) plantations also used more water than native secondary forests due to the higher canopy interception storage and the presence of understory in pines (Francis et al. 2022).

In addition to changes in water yield and carbon sequestration due to replacing hardwoods with pines, there are also concerns that pine planting may enhance climate warming and aridification due to changes in albedo in the north latitude. Based on measurements in paired pine and beech forests and the review of literature data, Leuschner et al. (2022) suggest that Scots pine (*Pinus sylvestris* L.) plantations in northern Germany had 4–6% lower albedo and 9% higher net radiation as well as higher canopy temperature and evapotranspiration in pine forests than beech forests. Consequently, Leuschner et al. (2022) conclude that large-scale planting of Scots pine is not climate-smart and environment friendly. However, little is known about how the albedo and net radiation contrasts translate to the effects on local air temperature and climate warming. There are arguments that forest with high leaf area and ET (i.e., latent heat) may have lower sensible heat transported to the boundary layer of the atmosphere, thus actually reducing global warming (Sun et al. 2010; Stoy 2018).

2 Objectives

This chapter aims to provide an overview of water balances (i.e., evapotranspiration, water yield) of pine forests at the stand-to-watershed scale by surveying published relevant literature. Such information is important for better forest management and meeting the needs of multiple ecosystem services from carbon sequestration to water supply.

The synthesis attempts to provide a summary of canopy interception, soil/understory evaporation, transpiration rate, and water yield. Data are put in the context of pine stand age and background climate using the Budyko framework that describes mean energy and water balances with respect to aridity index at a broad scale. We provide a summary of the effects of tree species conversions, such as from deciduous forests to pine, or from pine to hardwoods, or between pine species, such as from loblolly pine to longleaf pine on water balances.

3 Methods

We use the following water balance equation to guide our data compilation and synthesis at the stand or small watershed scale:

$$dS / dt = P(t) - \mathrm{ET}(t) - Q(t) \tag{1}$$

Equation (1) indicates that the change in soil water (unsaturated zone) and groundwater (saturated zone) storage (S) in a forest at any time t is a result of the balances of precipitation (P), evapotranspiration (ET), and total runoff or water yield (Q) or outflow or drainage in systems that are wet and subject to artificial drainage such as managed pine plantations (Amatya et al. 1996; Amatya and Skaggs 2011). The term Q includes deep seepage (DS) to groundwater on a watershed scale. ET is the sum of evaporation (E) from soil, litter, and understory (Es), canopy intercepted evaporation and litter interception (Ei), and plant transpiration (Et) (Tie et al. 2018).

$$\mathrm{ET}(t) = \mathrm{Es}(t) + \mathrm{Ei}(t) + \mathrm{Et}(t) \tag{2}$$

ET(t) is an important ecohydrological variable that links energy, water, and carbon fluxes (Sun et al. 2011; Fisher et al. 2017). Many variables affect Es, Ei, and Et; thus, ET is rather difficult to quantify accurately at a short timescale (Sun et al. 2016; Fisher et al. 2017). Meteorological variables, such as net solar radiation, air temperature, humidity, wind speed, atmospheric vapor pressure deficit, precipitation, and soil water availability, are dynamic and vary across time and space. Forest biophysical properties such as canopy structure, leaf area index (LAI), rooting depth, tree age and height, aerodynamic conductance, and leaf stomata conductance are major controls on vegetation ET and vary greatly between species (Sun et al. 2011).

The magnitude of *dS*/*dt* can be large on daily and even on monthly scale (Dymond et al. 2014). At a longer temporal scale (multiple years), *dS*/*dt* is negligible, and *Q*(*t*) can be expressed as follows:

$$Q = P - \mathrm{Es} - \mathrm{Ei} - \mathrm{Et} \tag{3}$$

We put the hydrological analyses in the Budyko water and energy balance framework (Budyko 1974; Chen and Sivapalan 2020) to show how the aridity index (potential evapotranspiration/precipitation) influences pine water use strategies and efficiency (ET/*P*) of pine forests in different climatic regions, demonstrating the role of climate in partitioning *P* into ET and water yield. There are many forms to describe ET/*P* (evaporative index) vs. PET/*P* (DI, dryness index) relationship. For example, based on global water balance data from 250 watersheds, Zhang et al. (2001) derive the following relationship to show that long-term mean forest ET is higher than grass ET.

$$\frac{ET}{P} = \frac{1 + w\dfrac{PET}{P}}{1 + w\dfrac{PET}{P} + \dfrac{P}{PET}} \tag{4}$$

For this model, *w* is found to be about 2.0 for forests and 0.5 for grasslands.

Other forms of the Budyko curve are available, including the widely used Fu model (Fu 1981; Zhang et al. 2004):

$$\frac{ET}{P} = 1 + \frac{PET}{P} - \left[1 + \left(\frac{PET}{P}\right)^{w}\right]^{1/w} \tag{5}$$

Using data from 470 watersheds, Zhang et al. (2004) found the *w* parameter value to be 2.84 to best fit data for forested and 2.55 for grass-dominated watersheds, respectively. These parameter values are close to 3.0 found by Amatya et al. (2002) for a managed pine forest site in coastal North Carolina in the USA.

In this study, PET was assumed to be equivalent to FAO grass reference ET (ETo), representing the local evaporation demand. To be consistent across sites, a global dataset ETo and *P* was used to calculate the mean dryness index, PET/*P* (Zomer et al. 2022).

We assume that annual forest ET rates estimated by Eqs. (4) and (5) represent the mean theoretical values for all forests in stable average climatic conditions. The two models provide differing ways of general approximations of ET rates and uncertainty of ET rates under different climatic regimes. So, any deviations from the calculated values under the average conditions suggest local observed ET may be influenced by factors beyond mean climatic regimes. In many cases, pine forest ET or Et values are only reported for the growing season since ET is low and Et is negligible in the nongrowing season for cold regions. For this type of sites, we discuss forest total ET or transpiration in the context of mean annual climate (*P* and PET) or growing season only.

Patterns of water yield (*Q*) from forests are influenced by precipitation, PET, forest ET, and soil properties (Eq. 3) and deep seepage in some cases. Changes in forest compositions and structure due to forest management or other disturbances (i.e., climate change) affect water yield *Q* by directly or indirectly altering Es, Ei, and Et. For example, thinning forests may increase Es but decrease Ei and Et, resulting in an increase or insignificant change in ET or *Q* due to offsets in components. For example, wildland fires may kill understory and/or overstory vegetation and reduce soil infiltration capacity, thus decreasing Et and Ei and increasing *Q* at least in the short term. However, if the after wildland fire or tree mortality, the increase in Es can exceed the decrease in Et, *Q* may decrease (Goeking and Tarboton 2022). Drainage (ditching) that has been used in establishing pine plantations in the southeast may facilitate subsurface flow, lower groundwater levels, and increase outflow (Skaggs et al. 2016, 2020). In such a management scenario, forest ET may be reduced due to likely lower soil moisture (Amatya et al. 1996).

This study compiles published literature on pine forest hydrology with a focus on measured or modeled ET and its components at either the annual or growing seasons. Datasets cover a wide range of climatic zones and 17 major pine species. We recognized that pine forests are often managed for different purposes and ecosystem benefits from bioenergy to timber production to water yield augmentation. So, in this assessment study, we reviewed and compared results from studies on the ET and water yield differences under various management conditions such as species conversion and thinning.

4 Results and Discussion

As discussed, forest ET and components vary greatly due to local climate, water availability, species, tree size, age, and leaf area, and any single factor cannot explain the spatial variability reported in the literature. We first present a comparison among pine forest canopy interception rates for major pine species globally (Fig. 1). Then, we examine characteristics of pine transpiration with other tree species using a sap flow dataset (105 sites) available in the USA. We also report total pine forest ET rates measured by the eddy covariance methods and other methods (e.g., modeling and lysimeters). Finally, we compare growing season pine ET and annual ET in the context of forest ET estimated by the theoretical Budyko framework to demonstrate how pine ET may differ from the mean ET of other forest ecosystems.

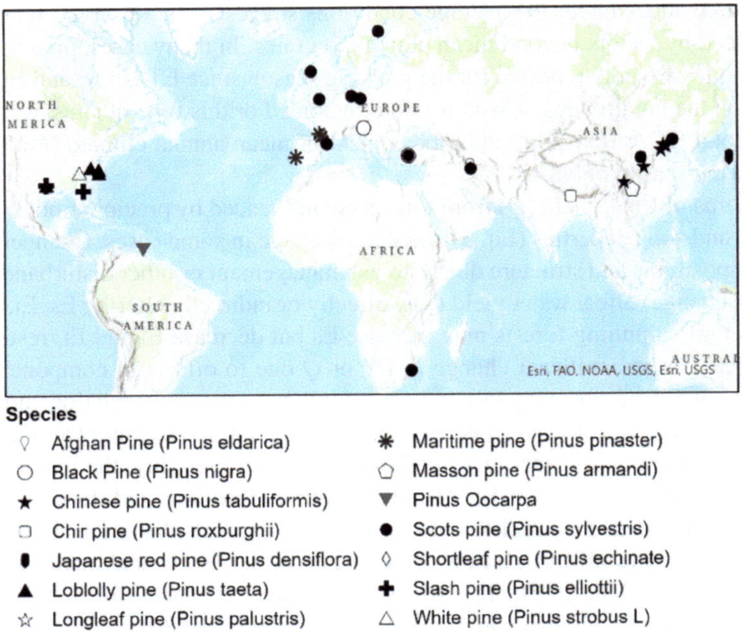

Fig. 1 Reported sites with studies on canopy interception for pine forests. (Data in Supplementary Table S1)

4.1 Interception by Pine Forests (Ei)

Forest interception (Ei), the sum of canopy (understory and overstory) and litter interception, represents intercepted precipitation, rainfall, or snow, evaporated from the surfaces of plant litter on forest floors, leaves, branches, and stems of forest during and after precipitation events. While transpiration dominates total ET (35–80% of ET), rainfall interception by plant canopy (10–50% of total precipitation) and litter of forest floor evaporation is (10–50% of throughfall) important as well. However, litter layer interception and soil surface evaporation have received relatively little attention in the literature (Domec et al. 2012; Gerrits et al. 2007; Sun et al. 2015). Because of the large surface of pine needles (Vose et al. 1994) and relatively high leaf area (up to all-sided LAI of 32 m^2/m^2 in *P. radiata* stand in New Zealand) (Grace 1987) and litterfall of pine stands, Ei can be a large component of the total ET and thus the water budgets for a watershed dominated by pine trees (Komatsu et al. 2008; Llorens and Domingo 2007; Sun et al. 2015). Ei can be even more important for water cycles in a dry climate with light rainfall intensity (Attarod et al. 2015). For example, an interception study on Afghan pine (*Pinus eldarica*) plantation shows that pine forest stand intercepted as high as 66% (36–88%) of rainfall on a storm basis and 44% on an annual accumulative basis in the arid region (mean annual precipitation = 1270 mm, PET = 2619 mm) in Iran (Attarod et al.

2015). Interception of understory along can be as long as 29% of precipitation in the pine plantations in the Caribbean (Francis et al. 2022).

Unique to forest hydrology, the large amounts of plant litter (including the $O_i + O_e$ horizons) are important for preventing soil erosion and water evaporative loss, storing water temporarily, and supplying soil nutrients and organic matter. Meanwhile, litter can intercept significant throughfall to reduce recharge to soils. For example, a study in South Africa (Bulcock and Jewitt 2012) documents that canopy interception by *E. grandis*, *A. mearnsii*, and *P. patula* was 14.9%, 27.7%, and 21.4% of gross precipitation, respectively, while litter interception was 8.5%, 6.6%, and 12.1% of gross precipitation, respectively, resulting in a decrease in recharge to soils. A litter removal experiment at the University of Tokyo Forests, Japan, shows that removing litters decreased litter layer interception in a watershed dominated by *Pinus densiflora* and *Quercus serrata*, increasing total flow (27 mm/year) and peak flow rates (1.4–1.5 times higher than control), particularly during relatively large flood runoff events. A study in the Southeastern USA on litterfall interception in a *P. elliottii* stand suggests that rainfall interception was 2–32% of rainfall and biomass materials related to phenological activity and meteorological disturbances affected the temporal variability of litter composition and litter interception processes. Litter interception is determined by both amounts of litter on the ground and its drying rate influenced by local climate. Hardwood forests in the humid southern Appalachians intercept about 2–6% of annual rainfall (Helvey and Patric 1965; Brantley et al. 2019). A summary of interception studies in North America indicates that conifer forest litters could lose 2–27% of gross precipitation (Helvey 1971).

Quantitative studies on forest canopy interception have a long history (Helvey 1971; Waring et al. 1981; Amatya et al. 1996; Sun et al. 2016; Qian et al. 2022), and mathematical models (e.g., Gash model) have been developed (Muzylo et al. 2009). Helvey (1971) provided the summary of early canopy and littler interception studies for conifer forests in the USA. A global review of canopy interception modeling studies in the tropics and extra-tropics is found in Wang et al. (2007). This study focuses on the literature on *Pinus* canopy interception to show its large variability due to differences in canopy characteristics that reflect species, age, stand structure, crown properties (roughness, density, leaf hydrophobicity), leaf size, and amount (leaf area index) and background climate as represented by precipitation and ET (Appendix Table 1).

In general, canopy interception rates increase with the stand age and leaf area (Brantley et al. 2019). For example, Ei for Chinese pine (*Pinus tabuliformis* Carr.) accounts for 20.4%, 24.8%, and 32.8% of gross rainfall in 40-, 50-, and 60-year-old, respectively (Dong et al. 2020), while 35-year-old Mongolian pine (*Pinus sylvestris* var. *mongolica*, MP) stand intercepts 29.5% of rainfall (Liu et al. 2016). Similarly, Brantley et al. (2019) found that the stand canopy interception as a percentage of precipitation (*P*) increased from 7.7% at age 2 to 26.4% at old growth (Age > 200) in a chronosequence study in the southern Appalachian Mountains in Southeastern USA. Conifers generally have higher canopy interception rates (Ei/*P*) than hardwoods that exhibit seasonal differences, but less than spruce-fir-hemlock forest type

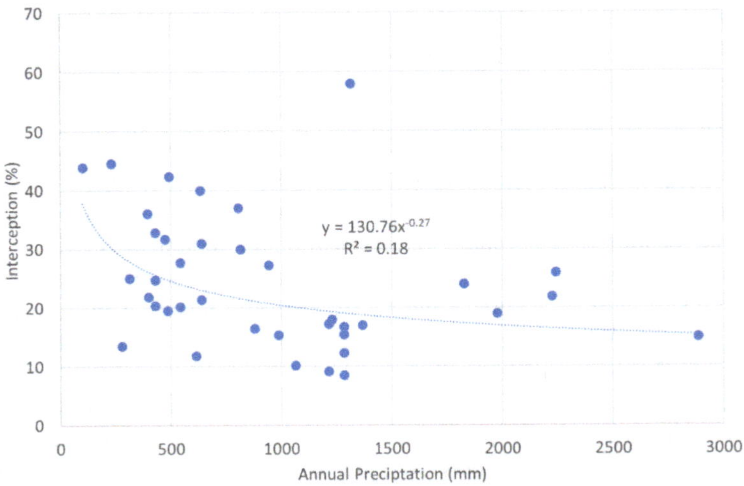

Fig. 2 There is a large variability in canopy interception rates among pine hydrology research sites. In general, annual accumulative pine forest interception rates decrease with an increase in precipitation. Each dot represents one global study site

(Waring et al. 1981). The interception rate decreases with an increase in storm size in the form of Log(Ei) = $a + b$ log(P) (Gavazzi et al. 2016) (a, b are parameters). Similarly, at the annual scale, cumulative interception rates are generally higher in arid regions than in humid regions (Fig. 2). This is especially true for arid regions with light and infrequent rains (Attarod et al. 2015). Reported annual mean interception rates can be as high as 72% in black pine (*Pinus nigra*) forests found in urban Slovenia (Zabret et al. 2018). This contrasts with a 28-site 1-year study on North American biomes that have interception rates as 10–25% (Allen et al. 2019).

Also, canopy interception rates vary seasonally due to tree leaf area dynamics and precipitation characteristics (Gavazzi et al. 2016). Reducing basal area and subsequent leaf area through thinning decreases canopy interception (McCarthy et al. 1991; Sun et al. 2015). Komatsu et al. (2008) summarize 12 studies on an interception from coniferous plantation forests in Japan and derive a positive correlation between stem density and interception ratio as Ei (%) = 0.00498 × (stem density in stems/ha) + 12.0.

Differences in canopy interception often explain the water yield change due to species conversion from deciduous forests to evergreen coniferous forests. For example, a species conversion study demonstrates that converting native deciduous forests with planted white pine (*Pinus strobus* L.) could reduce water yield by over 20% in the Appalachians Mountains. In this case, canopy interception by white pine can be as high as 585 mm/year or 26% of annual precipitation, representing 40–45% of total watershed ET (1291 mm) which is significantly higher than the ET from a deciduous forest (Ford et al. 2007). Another comparative study in Japan with succession from Japanese red pine (*Pinus densiflora*) to evergreen oak showed that the canopy interception reduced from 208 to 115 mm or from 17 to 9% of the annual gross rainfall due to reduced rainfall storage on tree surface (Iida et al. 2006).

4.2 *Tree Transpiration (Sap Flow)*

Water use by individual trees or transpiration (Et) is the dominant ET components of forest ecosystems. Previously reported global mean Et/(ET) from multiple independent sources, including satellite-based estimations, reanalysis, land surface models, and isotopic measurements, varies substantially from 24% to 90% (Wei et al. 2017). Combining global ET estimates and relationships between LAI and Et/ET for different vegetation types, Wei et al. (2017) show that Et accounts for about 57.2% (with standard deviation ± 6.8%) of global terrestrial ET. Another global study estimates Et/ET as 56–74% (median = 65%, mean = 64%), and most (65 ± 26%) of evaporation is from soils (Good et al. 2015). Et/ET varies among different types of vegetation, and generally, it increases with LAI (Hu et al. 2018). The average Et/ET in temperate coniferous forests is 55 ± 15%, while it is ca. 65 ± 18% for boreal forests (Schlesinger and Jasechko 2014). A global meta-analysis of arid ecosystems suggests Et/ET is about 50% (Sun et al. 2019) but variable depending on leaf area dynamics.

Transpiration from trees has been widely measured (Poyatos et al. 2016, 2021) globally through the SAPFLUXNET, thanks to the invention of sap flow methods that estimate either total sap flow through a tree's stem or sap flux density or the flow rate per unit sapwood area (Vandegehuchte and Steppe 2013; Granier 1987). Sap flow measurements provide insights into water use patterns at the species level and an understanding of environmental and ecological controls on water demand in extreme climates such as drought and heatwaves (Ahongshangbam et al. 2023). Sap flow measurements at the tree level are also helpful to answer bigger hydrological questions, such as managing forests for water supply under climate change (Caldwell et al. 2016) when the tree species level data are scaled up from hillslope to watershed and landscape scales (Boggs et al. 2015, 2021).

4.2.1 Transpiration Affected by Tree Age, Sapwood Area, and Height

Water use by trees increases with the increase in tree size as measured by tree diameter measured at the breast height (DBH) (Fig. 3). The amount of water transported through the sapwood area of a tree stem increases as the leaf area index (LAI) and sapwood area increase over time (Domec et al. 2012).

Tree water use increases with age initially from the seedling and young to mid-rotation and mature stage (Aguilos et al. 2021a), but transpiration rates may decrease as trees get older. Tree transpiration rates decrease as trees grow taller because it takes more energy to move water from roots to the top of the canopy in taller (older) trees with longer hydraulic pathways and lower whole-tree conductance trees.

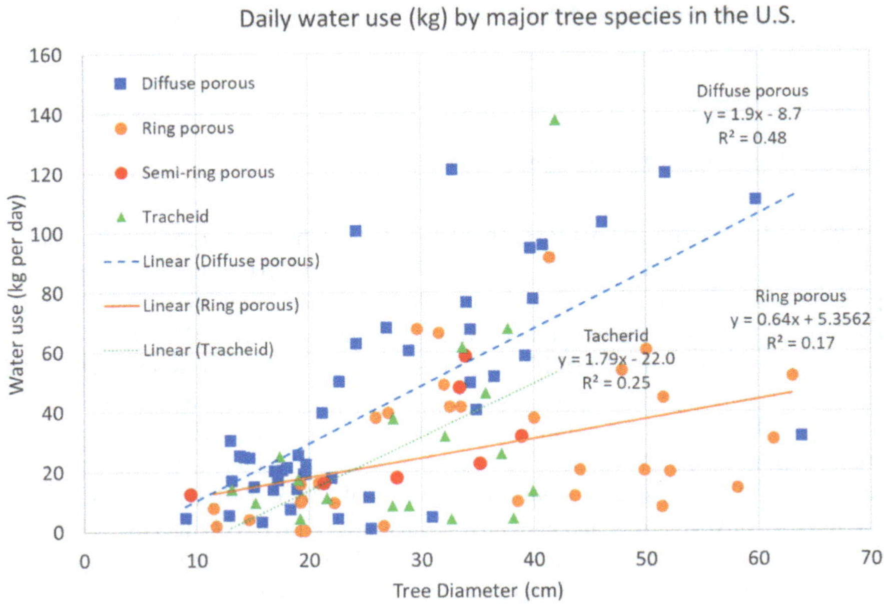

Fig. 3 Measured growing season daily mean sap flow rates (transpiration) grouped by xylem anatomy and tree diameter at breast height (DBH) for major species in the USA. *Diffuse-porous species* include red maple (*Acer rubrum*), sugar maple (*Acer saccharum*), birch (Betula spp.), hackberry (*Celtis laevigata*), dogwood (*Cornus florida*), American beech (*Fagus grandifolia*), European beech (*Fagus sylvatica*), sweetgum (*Liquidambar styraciflua*), tulip-poplar (*Liriodendron tulipifera*), black gum (*Nyssa sylvatica*), hop hornbeam (*Ostrya virginiana*), sycamore (*Platanus occidentalis*), American basswood (*Tilia americana*), catalpa (*Catalpa speciosa*). *Ring porous species* include green ash (*Fraxinus pennsylvanica*), oak (*Quercus* spp.). *Semi-porous species* include hickory (*Carya* spp.), aspen (*Populus* spp.), cherry (*Prunus serotina*). *Tracheid species* include fir (*Abies balsamea*), juniper (*Juniperus monosperma*), white spruce (*Picea glauca*), pine (*Pinus* spp.), northern white cedar (*Thuja occidentalis*), and Easter hemlock (*Tsuga canadensis*)

These patterns were found in ponderosa pine (*Pinus ponderosa*) forests (Ryan et al. 2000) and redwood (*Sequoia sempervirens*) (Ambrose et al. 2010) in Western USA. Measured tree sap flow rates (transpiration) for ~40-year-old 12-m-tall ponderosa pine trees were reported to be 45% higher than that for the ~290 years old 36-m-tall pines during a peak growing season (June–August) in an arid site (precipitation = 360 mm/year) in Oregon (Ryan et al. 2000). Similarly, Delzon and Loustau (2005) studied changes in water use of four even-aged maritime pine (*Pinus pinaster* Ait.) stands (10, 32, 54, and 91 years old) under similar environmental conditions in southwestern France. The highest transpiration rates were 3.8, 2.3, and 1.1 mm/day for 10-, 32-, and 54-year-old stands, respectively. Transpiration in the 91-year-old stand was similar to the 54-year-old stand, about 125 mm a year (or 16.4% of precipitation). The largest difference was observed between the 10- and 54-year-old stand. Annual total overstory stand transpiration declined with age from 508 mm/year (or 63% of precipitation) for the 10-year-old stand to 144 mm/year for the 54-year-old stand.

4.2.2 Transpiration and Xylem Anatomy

The xylem structure of tree stem and branches characterizes hydraulic conductivity and area of sapwood and drives transpiration rates (Ford et al. 2011). Tree species level transpiration studies in the Southeastern USA show that transpiration rates of diffuse-porous species like maples (*Acer* spp.) and tulip-poplars (*Liriodendron tulipifera*) are up to four times higher than similarly sized ring-porous species like oaks (*Quercus* spp.) (Ford et al. 2011) (Fig. 3). These differential xylem properties reflect tree water use strategies and how trees respond to drought (McQuillan et al. 2022).

Pine trees are characterized as having a tracheid xylem structure with relatively lower water use than the diffuse-porous group but are comparable to ring-porous or semi-ring-porous groups (Fig. 3). For example, under a similar climate in the humid Appalachian Mountains in western North Carolina, Eastern USA, a tulip-poplar tree (*Liriodendron tulipifera*) with a diameter of 40 cm transpires about 100 kg/day, while white pine (*Pinus strobus*) transpires only 13 kg with similar DBH. This contrasts with chestnut oak (*Quercus prinus*) (DBH = 52 cm) which belongs to the ring-porous group and transpires only 20 kg/day.

4.2.3 Transpiration Rates Influenced by Climate Regime

The large variability of mean daily water use at the tree level presented in Fig. 3 is unsurprising. Tree transpiration is known to be controlled by both soil water variability and atmospheric demand capacity that can be quantified by atmospheric vapor pressure deficit (VPD) (Shuttleworth and Wallace 1985; Novick et al. 2016) and a more complete climatic variable, potential evapotranspiration (PET), that integrates all meteorological factors often including net radiation and VPD (Whitehead and Kelliher 1991; Lu et al. 2005; Sun et al. 2011). We use regional PET/P (dryness index, DI) to represent the local climate regime and examine how background climate influences normalized water use by dividing daily mean water use per tree by its basal area (BA) (Fig. 4).

It appears that pine forest distributions in the USA have a large range in terms of climate with PET/P ranging from 0.75 (humid) to 1.25 (arid). Normalized water use rates generally increase with an increase in DI for the 0.75–1.1 dryness range, especially for deciduous trees characterized by a diffuse-porous xylem structure. However, water use rates for all forest types decrease when the climate gets drier (DI > 1.2). A sharp decrease in water use is found for diffuse-porous species, while water use is stabilized or moderately decreased. These patterns might suggest that trees growing in relatively dry regions (DI = 1.2–1.5) have lower transpiration and water stress. Transpiration rates are found to be the highest where DI is about 1.1 for most tree species including pines (Fig. 4). The extremely high value of water use of 180 g/day/cm^2 is for young quaking aspen (*Populus tremuloides*) (DBH = 9.4 cm).

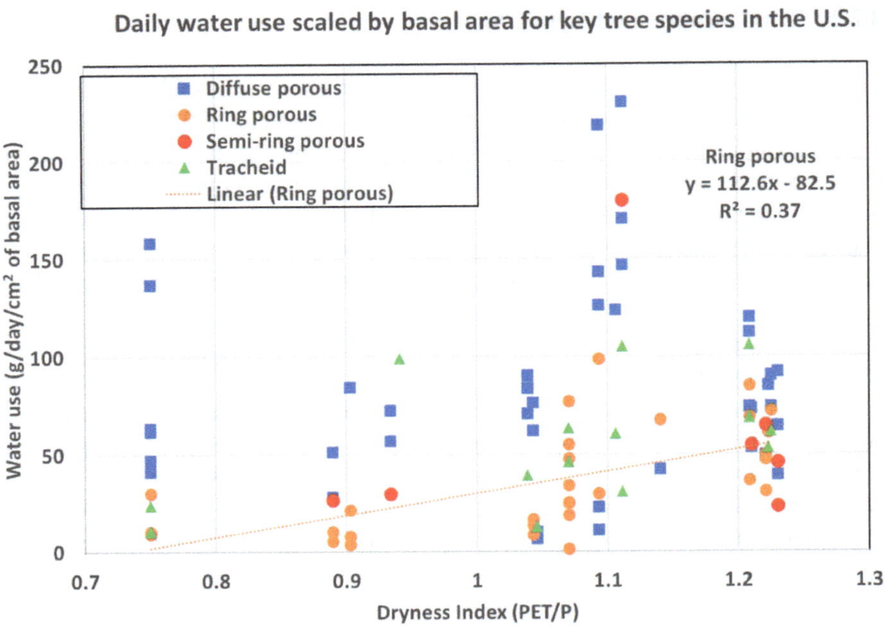

Fig. 4 Normalized water use (g/day/basal area cm^2) and climate dryness index (PET/P) for major tree species group by xylem structures in the USA

4.3 Total Evapotranspiration (ET)

As the sum of transpiration (Et), canopy and litter interception (Ei), and soil water evaporation (Es), total ET is affected by meteorological (e.g., net radiation, wind speed, VPD), and biological processes (e.g., tree hydraulics (root distribution and depth, LAI, and stomata conductance) and soils (soil water content or matric potential)). These processes may compensate, cancel, or enhance the ET rates contributed by each of the ET component. For example, an increase in LAI as trees mature generally increases stand transpiration and canopy interception, but it can also decrease soil and understory vegetation evaporation because of canopy shading resulting in reduced radiation and lower wind speed on forest floor. Thinning may reduce overstory transpiration and canopy interception, but it may increase transpiration from the understory or the remaining trees (Boggs et al. 2015), resulting in complex ET and water yield responses (del Campo et al. 2022). Studies suggest that forest ET is a conservative process (Roberts 1983)—tree transpiration rates tend to be stable in response to periodic climatic variability in precipitation (Oishi et al. 2010, 2018). Meteorological droughts and minor thinning do not necessarily result in a dramatic decline in forest total ET (Liu et al. 2018). Forest ET demand can also be met by variable soil water stores at deeper layers or rock fractures, the Earth's Critical Zone (Stocker et al. 2023).

Published pine forest ET data discussed below come from three primary sources, measurements by the eddy covariance method, gaged small watersheds, and simulated water balances, including annual or seasonal ET. In cold temperate or boreal regions, annual ET is dominated by the growing season, and ET measurements are often only available for the growing seasons. In such cases, we examine ET for the growing season in the context of the mean aridity index calculated at the annual scale to represent the local climatic regime and to be consistent with Budyko's analytic framework that characterizes mean conditions.

4.3.1 Climate and Pine Species Effects on Annual ET

The assembled datasets cover a large climatic gradient from the moist tropical Fiji (*P. caribaea*, $P = 2021$ mm, ETo = 1500 mm, aridity index ETo/P = 0.76) to extreme arid regions such as Iran ($P = 203$ mm, ETo = 2075 mm, ETo/P = 10.2) (Figs. 5 and 6). Some pine species are found in a large climatic range from arid to humid regimes. For example, Aleppo pine (*Pinus halepensis*) is found in Spain and Israel with a large annual precipitation range from less than 260 mm to more than 1500 mm. Pine trees can survive under a climate with low precipitation and cold temperature (ETo < 300 mm, *P. strobus* in Canada) or dry and hot environments (ETo > 2000 mm, *P. halepensis* and *P. eldarica*, in Israel and Iran, respectively).

Pine forest ET rates vary tremendously from less than 300 mm/year (*P. banksiana*) to more than 1800 mm/year (*P. caribaea*) (Fig. 7). Overall, annual ET increases with precipitation (Fig. 8). On average, pine forests lose about 66 ± 23% of annual precipitation through ET. The mean value is similar to those reported by Ford et al. (2011) (ET = 0.72P − 10.48) but is much higher than reported value (53.7%) for global forest lands (Oki and Kanae 2006). Due to climatic variability, pine forest

Fig. 5 A distribution of pine forest evapotranspiration research sites (total 158 sites). Numbers next to each color-coded symbol by pine species are aridity index values (ETo/P). Details of each study site are reported in Supplementary (Table S2)

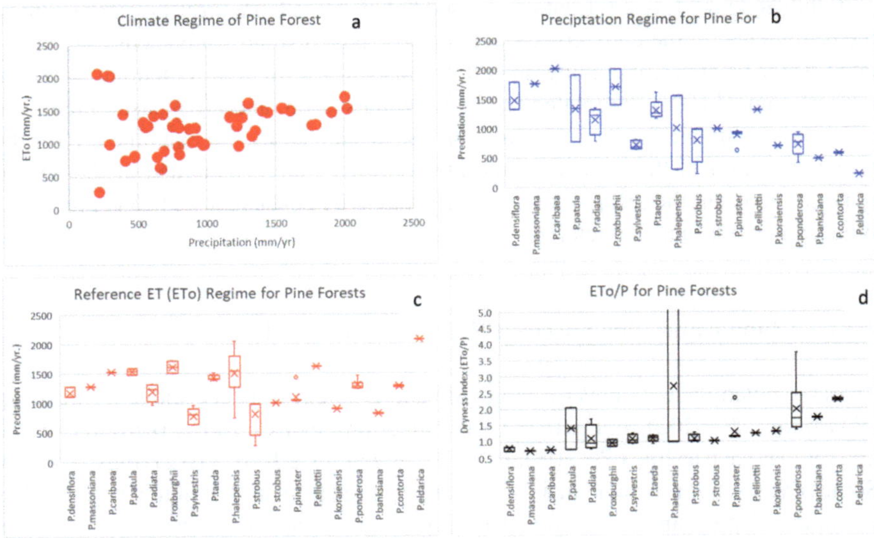

Fig. 6 Climate regimes of 18 pine forests. (**a**) Annual precipitation and reference ET (Eto), (**b**) precipitation by species, (**c**) ETo by pine species, and (**d**) dryness index (Eto/*P*) by pine species. ETo/*P* = 10.2 for *Pinus eldarica* not shown

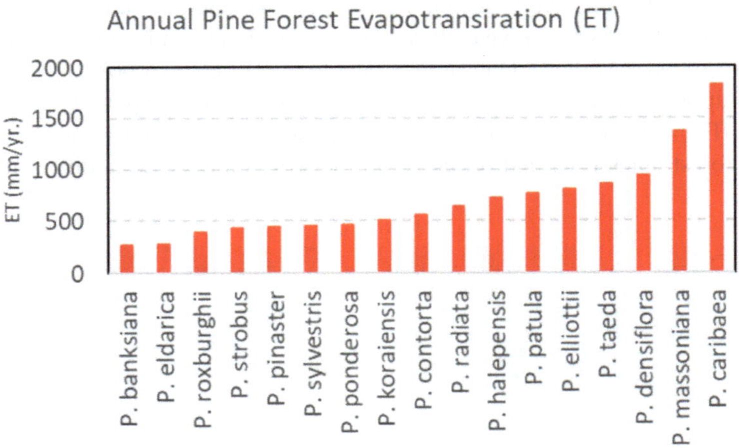

Fig. 7 A summary of reported values for annual evapotranspiration for 17 *Pinus* species

ET/*P* values vary tremendously from less than 40% in the cold areas (*P. strobus*) in Canada or wet regions (*P. radiata*) in Chile to more than 137% (*P. eldarica*) in an arid region. For the latter case, the *P. eldarica* stand likely used up all annual precipitation in the measurement year and accessed soil water storage.

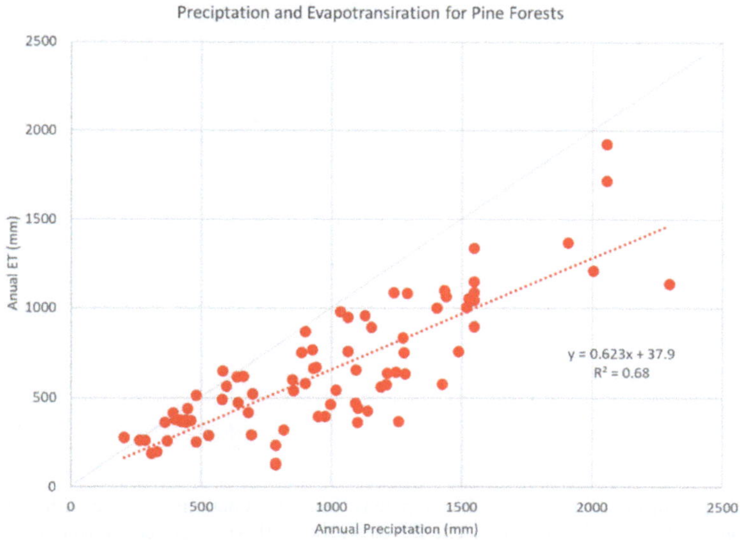

Fig. 8 Annual evapotranspiration increases with annual precipitation for 17 *Pinus* species

4.3.2 Stand Age Effects

In addition to the climatic regime (*P* and ETo) and species controls on ET rates, stand age also affects total forest ET. Forest ET rates increase with age as trees mature and leaf areas and sapwood areas increase. For example, long-term eddy flux measurements on loblolly pine (*P. taeda*) plantation forests on the coastal plain of North Carolina across an age chronosequence show that ET rates increase rapidly from 0 to 10 years and become stabilized around year 15 at the middle rotation stage (Fig. 9) (Aguilos et al. 2021b). However, age effects on ET can be masked by climatic variability. Changes in LAI and canopy structure over stand age may not necessarily cause a large change in annual ET in rather wet humid conditions such as forested wetlands. For example, annual ET estimates by the eddy covariance method in north central Florida averaged 959, 951, and 1110 mm along a chronosequence of slash pine (*P. elliottii*) for clear-cut, mid-rotation, full-rotation plantation stands, respectively (Gholz and Clark 2002). In such a case, total annual ET is more sensitive to environmental fluctuations than to stand age or management activities that affect both soil evaporation and transpiration. The role of vegetation in influencing shallow groundwater table and ET is most pronounced during the dry period (years) (Amatya et al. 1997; Sun et al. 2000; Skaggs et al. 2011; Riley et al. 2023).

Müller and Bolte (2009) quantified ET rates of Scots pine (*P. sylvestris*) trees using large lysimeters and found that the water consumption first increased with age and then decreased, reaching a peak around year 30 (Fig. 10). At age 30, Scots pine trees used up all water input from precipitation, resulting in no deep seepage from the rooting zone (Fig. 10).

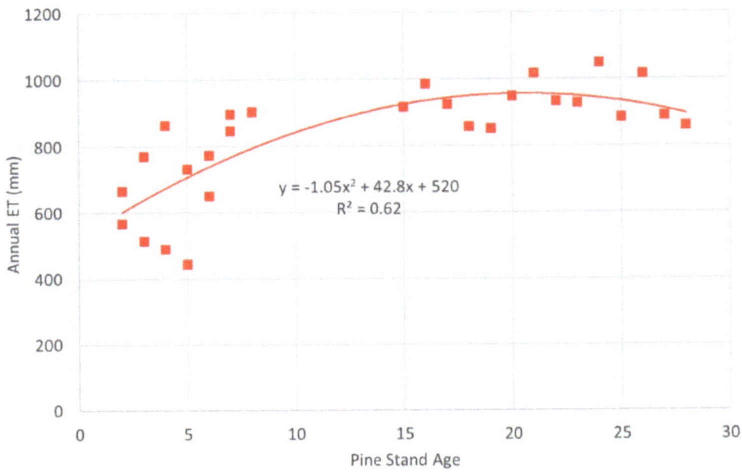

Fig. 9 Evapotranspiration rates of loblolly pine (*P. taeda*) plantations increase with stand, especially from seedlings to 15 years old. The ET data (2005–2018) are collected using an eddy flux method for stands with multiple ages at the lower coastal plain of North Carolina in the Southeastern USA (Sun et al. 2010; Domec et al. 2012; Aguilos et al. 2021a, 2021b)

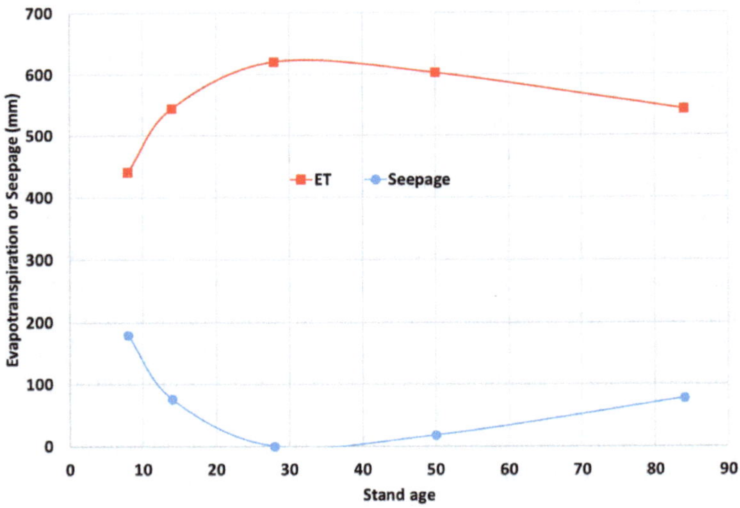

Fig. 10 Influences of Scots pine (*P. sylvestris*) growth stages on the water balances measured in lysimeters in Germany (annual precipitation = 620 mm) (Müller and Bolte 2009)

4.3.3 Effects of Management (Ditching and Prescribed Fires)

Globally, since the mid-eighteenth century, drainage has been used to increase timber yields on poorly drained peatlands and mineral soils in Northern Europe, Canada, and the Southern USA (Skaggs et al. 2016). In the Southeastern USA, large acreage of loblolly pine (*Pinus taeda* L.) and slash pine (*Pinus elliottii*) plantations have been established on pine flatwoods or formerly forested wetlands on the coastal plains. Silviculture practices often involve water management, such as free or controlled drainage (Amatya et al. 1996), to lower soil wetness and benefiting forest productivity. Studies document that tree growth is higher by 80%–1300% and as much as ten times greater wood production on drained compared to undrained loblolly pine plots (Skaggs et al. 2020). Amatya et al. (1996) report that controlled drainage using weirs reduced both drainage volumes (15–30%) and daily peak outflow rates and increased ET by 10–20%. However, a remote sensing study in North Carolina suggests that drained mature loblolly pine plantations have similar annual ET to adjacent natural forests without artificial drainage but have a much higher ET rate than young plantation forests (Yang et al. 2017).

Prescribed burning is also used in the Southern USA to control understory competition and improve nutrient and water availability and wildlife habitats (McLaughlin et al. 2013). These practices have important effects on water yield and groundwater recharge by reducing forest water use (ET). LAI is the dominant forest structural factor influencing both soil evaporation and interception rates; thus, it has been widely used as a major indicator for forest management to augment water yield (Acharya et al. 2022; Pisarello et al. 2022) and ET modeling (Sun et al. 2011; Malone et al. 2015).

4.3.4 Effects of Pine Species Conversions: The Case of Longleaf Pine Forest Restoration

Historically, longleaf pine (*Pinus palustris* Mill.) dominated much of the coastal plain and portions of the piedmont in the Southeastern USA (Peet et al. 2018). However, over the past century, management that favored fast-rotation wood production and fire suppression led to a drastic reduction in longleaf pine in favor of loblolly pine (*Pinus taeda* L.) plantations for timber production. Concerns about climate change and the rise of demand for multiple ecosystem services promoted large-scale efforts throughout the region to restore longleaf pine ecosystems that are believed to be more resilient to climate change (Franklin 2009; Brantley et al. 2018). One of the recognized ecological benefits of longleaf pine restoration is an increase in water yield. Compared to productive loblolly pine plantations, longleaf pine forests have lower ET because of their lower tree density, open canopies with C4 grass as an understory, lower total leaf areas, and leaf stomata conductance (Brantley et al. 2018; Qi et al. 2022). A remote sensing study across the southeastern region using a "paired stand" approach suggests that most longleaf pine stands have lower ET than the paired loblolly pine stands (Fig. 11). This contrast is explained by lower LAI and tree height but higher land surface temperature (LST) and albedo for longleaf pine stands than adjacent loblolly pine stands (Liu et al. 2025, Unpublished data). A synthesis study on eddy flux measurements of pine forests for the region also indicates that longleaf pine stands have lower ET and ET/P ratios (Liu et al. 2025; Aguilos et al. 2024).

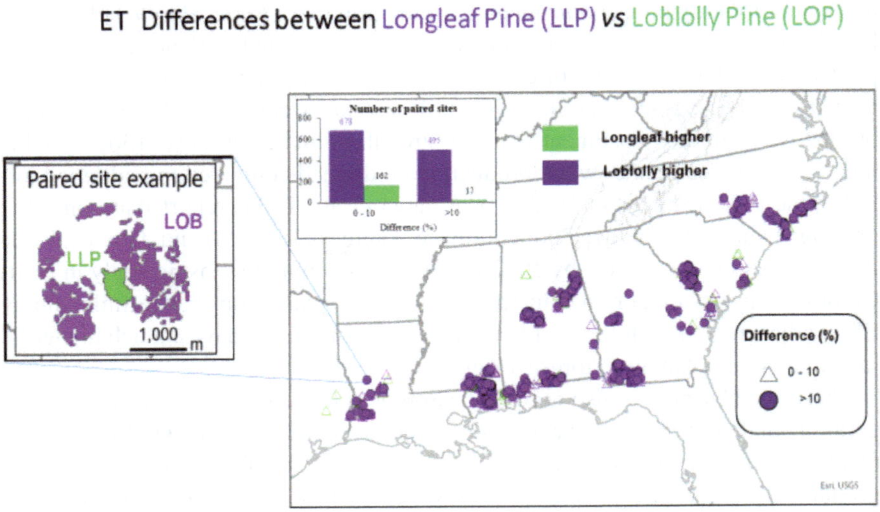

Fig. 11 A comparison of annual evapotranspiration (ET) between longleaf pine (LLP) stands and their adjacent loblolly pine (LOP) stands within a 2-km radius. Only sites with significant differences ($p < 0.05$ by paired-t test) are shown. ET is estimated using ECOSTRESS surface temperature data and an energy balance simulation model at a 30 m by 30 m resolution (https://lpdaac.usgs.gov/news/release-ecostress-usda-et-and-esi-products/) (Hook and Anderson 2019)

4.3.5 Global Pine Forest ET Under the Budyko's Framework

According to Budyko's water and energy balance theory, as quantitatively described in Eqs. (3) and (4), the evaporative index ET/P nonlinearly increases with the aridity index PET/*P*. Equations (3) and (4) have different forms, but they provide similar curves for PET/*P* vs. ET/*P* relationships of forest lands, especially under dryer conditions (PET/*P* > 1.0) or water-limited climatic regimes. Although the measured ET/*P* values follow the general trend as expected (Fig. 12), observed ET/*P* values at many sites deviate from ET/*P* values estimated by the theoretical two equations that represent the long-term means over a large spatial sale. Among the 84 pine forest sites examined, 35 sites (or 42%) exhibited higher annual ET/*P* values than predicted by the Budyko relationship, and 49 sites have lower ET/*P* values. These deviations from "the norm" offer insights into understanding how pine forest water uses respond to local climate (Creed et al. 2014) and other environmental factors such as soil and topography (Zhou et al. 2015).

We discuss these ET/*P* deviations in five major groups to illustrate the likely reasons causing the discrepancies.

Group 1 Growing season ET exceeds precipitation. These sites are dominated by maritime pine (*P. pinaster*) in France (Moreaux et al. 2011) with PET/*P* around 1.0 and annual precipitation of 900–1400 mm. These sites receive little precipitation (e.g., 20–35% of total annual) in the growing season when PET and ET are high. This type of climate also is common in the Pacific Northwest of the USA where

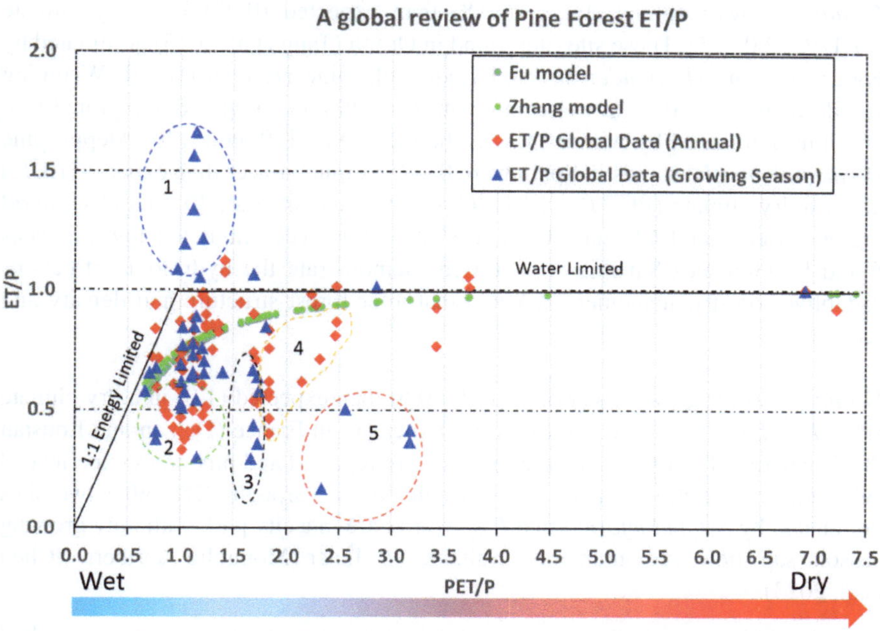

Fig. 12 A comparison between measured annual or growing season evaporative index (ET/*P*) values of pine forests and predicted by two theoretical models. ET/*P* rates are influenced by regional climate patterns and local environmental conditions and forest management (see Supplementary Table S2). The five circles represent large deviations from predicted ET/*P*

most of the precipitation falls in the winter (Segura et al. 2020). For these sites, nongrowing season precipitation and soil water storage are important for tree water use and growth in the summer months when the low flows present.

Group 2 Annual ET/P values for this group are much lower (<0.5) than expected (0.75), suggesting forest disturbances or soil water stress for tree growth. These are boreal forest sites with a short growing season. The Canadian sites (Tang et al. 2017) dominated by white pine (*P. strobus*) even-aged plantations including young, intermediate, and mature trees have warm summers and cold winters. The *P. sylvestri* stands in Finland (Ilvesniemi et al. 2010) had a boreal climate receiving precipitation dominated by winter snows.

Group 3 Lower ET in the growing season (May–September) than expected. These forests are found in Saskatchewan, Canada, that are dominated by jack pine (*P. banksiana*) under a cool climate (PET < 700 mm/year) with low precipitation (<500 mm/year) mostly in the growing season. The low growing season ET (117–313 mm) also reflects the impacts of wildfire and harvest disturbances (Mkhabela et al. 2009).

Group 4 Lower annual (ET/*P* < 0.8) than expected (0.93) in a dry climate (PET/*P* = 2.0–2.5). These sites are found in Orgon (Tang et al. 2015) dominated by ponderosa pine (*P. ponderosa*) and Lodgepole pine (*P. contorta*) in Wyoming (Biederman et al. 2014) (a wetter site with *P* = 840 mm compared to regional *P* of 560 mm dominated by winter snow). The modeled ET/*P* values for Aleppo pine (*Pinus halepensis*) stands in Spain were 0.4–0.61 much lower than expected (0.96) under a dry climate (PET/*P* = 1.84) (González-Sanchis et al. 2015) and semiarid regions (Bellot and Chirino 2013) (PET/*P* = 3.4). The simulated low ET rates found by González-Sanchis et al. (2015) demonstrate the hydrologic effects of "adaptive forest management" (AFM) that alter forest structure and density and reduced ET.

Group 5 Lower growing season (ET/*P* < 0.6) than expected (1.0) in a dry climate (PET/*P* = 2.3–2.5). These two of the four sites are in France (Delzon and Loustau 2005) where only sap flow measurements were reported and pine trees experienced substantial water stresses and loss of needles during drought. The other two sites dominated by *P. tabulaeformis* and *P. sylvestris* are in a city park with only growing season sap flow measurements available in Inner Mongolia northern (Chen et al. 2023).

It should be stressed that measured ET/P at the stand scale or even a watershed level is not likely to match the predicted by the two theoretical models due to the local conditions (i.e., climate, topography, and disturbances). Also, there is uncertainty in the empirical parameters derived from long-term water balance for many types of forests and watershed characteristics (Zhang et al. 2001, 2004).

5 Summary and Implications

Pine forests are found in all climatic zones on Earth and constantly impacted for global change (Richardson et al. 2007). Pine trees are increasingly planted for many purposes from urban greening and soil erosion control (e.g., in northern China and other dry regions) to commercial timber production (e.g., in Southern USA, southern China, tropics, and other humid regions). Pines as evergreen species may have lower albedo, higher leaf area, and longer growing seasons and thus often have higher canopy interception rates when compared to deciduous broadleaf forests.

Forest canopies (both overstory and understory) can intercept significant amount of precipitation influencing the precipitation partitioning and water yield from forests. Forest management practices, such as thinning and prescribed burning that reduce tree density and total stand leaf area index, can decrease total ET loss and increase water yield and groundwater recharge, thus having the potential to enhance forest resilience to drought. Stand age, height, structure, and plant species compositions affect canopy interception, litter interception, transpiration, and total ET from pine forests. As pine trees mature and get taller, ET may decrease. Growing season ET of pine forests can greatly exceed precipitation in the growing season; thus, nongrowing season precipitation is vital to meet water demand for ecological functions. The total annual ET of pine forests can be lower or higher than predicted from the Budyko type of general models, suggesting the complex effects of local climate, disturbance, and management on forest ET. Few studies have examined ET from understories in forests, hindering accurately modeling effects of forest management on total ET and water yield.

Our global review demonstrates a large variability of water use by pine trees and forests across a large climatic gradient, a combination of precipitation and potential ET. Pine forest water balances can be altered by changing tree density through thinning and prescribed burning, tree species conversion, artificial drainage, and climate change and variability. Tree planting with pines must consider their differential water demand to maximize ecosystem service goals (e.g., carbon, water, biodiversity, soil erosion control). This is especially important given the large uncertainty in projected future climate change, which is likely to exacerbate extreme events such as droughts that threaten forest sustainability in many parts of the globe.

Acknowledgments Lulu Zhang acknowledges the support of the German Academic Exchange Service (DAAD) PPP-Programme project "Assessing climate change impacts on forest carbon sequestration and water supply" (ID: 57510261). Lu Hao acknowledges the support of the National Natural Science Foundation of China (NSFC) and United Nations Environment Programme (UNEP) International Joint Research Project "Developing Nature-based Solutions for Nepal Following a Nexus Approach towards Sustaining Forestry, Water Resources and Livelihoods" (ID: 42061144004).

Appendix

Table 1 A comparison of canopy interception rates among key *Pinus* trees around the globe

Location	Species	Climate (annual P and ET) (mm)	Interception rate (I/P) (%)	References
USA 35°48′ N, 76°40′ W	Loblolly pine (*Pinus taeda*) (age, 13–23)	Humid (10 years, mean *P* = 1227)	14–23 (biweekly)	Gavazzi et al. (2016)
USA 34°50′ N, 76°39′ W	Loblolly pine (*Pinus taeda*) (age, 15)	Humid *P* = 2885 (storm events)	5–35 (un-thinned)	McCarthy et al. (1991)
USA 34°50′ N, 76°39′ W	Loblolly pine (*Pinus taeda*) (age, 15)	Humid *P* = 2885 (storm events)	5–25 (thinned)	McCarthy et al. (1991)
USA 36°20′ N, 79°28′ W	Loblolly pine (*Pinus taeda*) (age, 35–37)	*P* = 2224 (3 years)	21.9 (annual)	Tor-ngern et al. (2018)
USA	Longleaf pine (*Pinus palustris* Mill.)	Humid	20	Helvey (1971)

(continued)

Table 1 (continued)

Location	Species	Climate (annual *P* and ET) (mm)	Interception rate (I/P) (%)	References
USA 29°44′ N, 82°9′30″ W	Slash pine (mid-rotation age) *Pinus elliottii*	*P* = 1062, ET = 1057	10.2 (annual)	Gholz and Clark (2002)
USA 29°44′ N, 82°9′30″ W	Slash pine (rotation age) *Pinus elliottii*	*P* = 1288 ET = 1194	8.4 (annual)	Gholz and Clark (2002)
USA 35.07° N, 83.43° W	White pine (*Pinus strobus* L.) (age, 48–49 plantation)	*P* = 2240 ET = 1291	26 (annual)	Ford et al. (2007)
USA 31.498° N, 94.756° W	Longleaf pine (*Pinus palustris* Mill.) (age, 21)	*P* = 1282 (44 storms) 1 year	12.3	Roth II and Chang (1981)
USA 31.498° N, 94.756° W	Shortleaf pine (*Pinus echinata*) (age, 21)	*P* = 1282 44 storms 1 year	8.5	Roth II and Chang (1981)
USA 31.498° N, 94.756° W	Loblolly pine (*Pinus taeda*) (age, 19)	*P* = 1282 44 storms 1 year	15.5	Roth II and Chang (1981)
USA 31.498° N, 94.756° W	Slash pine (*Pinus elliottii*) (age, 21)	*P* = 1282 44 storms 1 year	16.8	Roth II and Chang (1981)
USA	Red pine (*Pinus resinosa* Ait.)	N/A	15.0 (annual)	Helvey (1971)
USA	Ponderosa pine (*Pinus ponderosa* Dougl. ex. Laws.)	N/A	14.0 (annual)	Helvey (1971)
Northern China (40°23′ N, 116°50′ E)	Chinese pine (*Pinus tabulaeformis*) (age, 33 plantation)	*P* = 1419 (3 years, 61 storms)	31.7 (storm)	Xiao et al. (2007)
Northern China (40°49′ N, 111°41′ E)	Chinese pine (*Pinus tabuliformis*)	May–September *P* = 240 ET = 393	23–48 (monthly scale)	Wang et al. (2022)
Northern China (41.7° N, 117.5° E)	Chinese pine (*Pinus tabuliformis*) (age, 28 plantation)	June–October *P* = 313 (1 year, 50 storms)	19.6–31.3	Dong et al. (1987)
Northern China (41.3° N, 118.52° E)	Chinese pine (*Pinus tabuliformis*)	429	20.4	Dong et al. (2020)
Northern China (41.3° N, 118.52° E)	(*Pinus tabuliformis*)	429	24.8	Dong et al. (2020)
Northern China (41.3° N, 118.52° E)	(*Pinus tabuliformis*)	429	32.8	Dong et al. (2020)

(continued)

Table 1 (continued)

Location	Species	Climate (annual *P* and ET) (mm)	Interception rate (I/P) (%)	References
Southwestern China (31°41′07″ N, 103°53′58″ E)	Chinese pine (*Pinus tabuliformis*) (age, 23 plantation)	*P* = 806 (136 storms)	37	Bao et al. (2004)
Southwestern China (29°38′ N, 106°41′ E)	Masson pine (*Pinus armandii*) (age, 50–70 plantation)	*P* = 1972 (2 years, 84 storms)	15.4	Wang et al. (2021)
Southwestern China 30°54′ N, 103°35′ E	Mixed *Pinus tabulaeformis* and *Pinus armandii*	*P* = 167 (24 storms)	51	Chang et al. (2006)
Northwestern China 35°52′ N, 110°45′ E	Chinese pine (*Pinus tabulaeformis*) (age, 25 plantation)	*P* = 486 (1 year, 5 storms)	19.6	Liu et al. (1991)
Northwestern China 38°20′11″ N, 109°42′54″ E	Mongolia Scots pine (*Pinus sylvestris* var. *mongolica*)	*P* = 395 (28 storms)	36.1 (modeled)	Deng (2020)
Northern China 120°43′ E, 42°55′ N	Mongolia Scots pine (*Pinus sylvestris* var. *mongolica*) (age, 27 plantation)	*P* = 279.4 (42 storms) 1 year	13.5	Sun et al. (2021)
Nepal 27°34′ N, 85°33′ E	Chir pine or longleaf Indian pine (*Pinus roxburghii*) (age, 25 plantation)	*P* = 878 1 year	16.5	Ghimire et al. (2012, 2013)
Spain 42.2° N, 1.8176° E	Scots pine (*Pinus sylvestris*)	*P* = 1825 30 months (152 storms)	24.0	Llorens et al. (1997)
Sweden 64°10′ N, 19°45′ E	Scots pine (*Pinus sylvestris*)	Growing season 3 years, mean *P* = 400	21.9	Tor-ngern et al. (2017)
Japan 36°07′ N, 140°06′ E	Japanese *red pine* (*Pinus densiflora*)	*P* = 1213 1 year (un-thinned)	17.2	Iida et al. (2005)
Japan 36°07′ N, 140°06′ E	Japanese *red pine* (*Pinus densiflora*)	*P* = 1245 (thinned)	9.2	Iida et al. (2005)
44°05′ N, 0°05′ W	Maritime pine (*Pinus pinaster*)	*P* = 613 1 year	11.9	Gash et al. (1995)
Portugal 38°50′ N, 8°51′ W	Maritime pine (*Pinus pinaster*) (age 60)	*P* = 1366.2 31 months	17.1	Valente et al. (1997)

(continued)

Table 1 (continued)

Location	Species	Climate (annual *P* and ET) (mm)	Interception rate (I/P) (%)	References
France 44°42′ N, 0°46′ W	Maritime pine (*Pinus pinaster*)	Total *P* = 1976 3 years	14–22	Loustau et al. (1992)
52°20′ N, 0°42′ W	Scots pine (*Pinus sylvestris*) (age 18)	Total *P* = 942.4 2 years	27.3	Gash (1979)
UK 57°41′ N, 3°29′ W	Scots pine (*Pinus sylvestris*)	*P* = 493 1 year	42.4	Gash et al. (1980)
Germany 52°26′34.55″ N, 13°26′1.97″ E	Scots pine (*Pinus sylvestris*)	*P* = 633, ET = 623 (8 years simulated)	40.0	Müller (2009)
Germany 52°58′ N, 10°19′ E	Scots pine (*Pinus sylvestris*)	*P* = 638 mm, ET = 588 (1991)	31.0	Leuschner et al. (2022)
Germany 52°58′ N, 10°19′ E	Scots pine (*Pinus sylvestris*)	*P* = 815 mm, ET = 738 (1992)	30.0	Leuschner et al. (2022)
Iran 36°25′ N, 50°55′ E	Afghan pine (*Pinus eldarica*)	*P* = 101	44.0 (46–80 by storms)	Attarod et al. (2015)
Iran 35°42′ N, 51°08′ E	Afghan pine (*Pinus eldarica*)	*P* = 232.5	44.6	Sadeghi et al. (2014)
Slovenia 46.04° N, 14.49° E	Black pine (*Pinus nigra*)	*P* = 1380	72	Zabret et al. (2018)
Turkey 39.155° N, 29.816° E	Black pine (*Pinus nigra*)	*P* = 542	27.7	Aydın et al. (2018)
Turkey 39.155° N, 29.816° E	Scots pine (*Pinus sylvestris*)	*P* = 542	20.2	Aydın et al. (2018)
South Africa 29°11′ S, 30°39′ E	Patula pine (*Pinus sylvestris*)	*P* = 692	21.4	Bulcock and Jewitt (2012)
Trinidad and Tobago 10°41′45.3″ N, 61°29′23.0″ W	*Pinus oocarpa*	*P* = 1314	58.0	Francis et al. (2022)

References

Acharya S, Kaplan DA, McLaughlin DL, Cohen MJ (2022) In-situ quantification and prediction of water yield from southern U.S. pine forests. Water Resour Res 58(5):e2021WR031020

Adams MB, Kelly C, Kabrick J, Schuler J (2019) Temperate forests and soils. In: Developments in soil science, vol 36. Elsevier, pp 83–108

Aguilos M, Sun G, Noormets A, Domec JC, McNulty S, Gavazzi M, Minick K, Mitra B, Prajapati P, Yang Y, King J (2021a) Effects of land-use change and drought on decadal evapotranspiration and water balance of natural and managed forested wetlands along the southeastern US lower coastal plain. Agric For Meteorol 303:108381

Aguilos M, Sun G, Noormets A, Domec JC, Mcnulty S, Gavazzi M, Prajapati P, Minick KJ, Mitra B, King J (2021b) Ecosystem productivity and evapotranspiration are tightly coupled in loblolly pine (Pinus taeda L.) plantations along the coastal plain of the southeastern US. Forests 12(8):1123

Aguilos M, Sun G, Liu N, Zhang Y, Starr G, Oishi AC, O'Halloran TL, Forsythe J, Wang J, Zhu M, Amatya D (2024) Energy availability and leaf area dominate control of ecosystem evapotranspiration in the southeastern US. Agric For Meteorol 349:109960

Ahongshangbam J, Kulmala L, Soininen J, Frühauf Y, Karvinen E, Salmon Y, Lintunen A, Karvonen A, Järvi L (2023) Sap flow and leaf gas exchange response to a drought and heatwave in urban green spaces in a Nordic city. Biogeosciences Discussions, pp 1–32

Allen ST, Fiorella R, Guo J, Good SP, Noone D, Anderegg W, Bowen GJ (2019) Canopy interception losses of rainfall in North American biomes. In: AGU Fall Meeting. Abstracts H52E-02. AGU

Amatya DM, Skaggs RW (2011) Long-term hydrology and water quality of a drained pine plantation in North Carolina. Trans ASABE 54(6):2087–2098

Amatya DM, Skaggs RW, Gregory JD (1996) Effects of controlled drainage on the hydrology of drained pine plantations in the North Carolina coastal plain. J Hydrol 181(1–4):211–232

Amatya DM, Skaggs RW, Gregory JD, Herrmann RB (1997) Hydrology of a drained forested pocosin watershed. J Am Water Resour Assoc 33(3):535–546

Amatya DM, Chescheir GM, Skaggs RW, Fernandez GP (2002) Hydrology of poorly drained coastal watersheds in eastern North Carolina. In 2002 ASAE Annual Meeting (p. 1). Am Soc Agric Biol Eng

Ambrose AR, Sillett SC, Koch GW, Van Pelt R, Antoine ME, Dawson TE (2010) Effects of height on treetop transpiration and stomatal conductance in coast redwood (*Sequoia sempervirens*). Tree Physiol 30(10):1260–1272

Attarod P, Sadeghi SMM, Pypker TG, Bagheri H, Bagheri M, Bayramzadeh V (2015) Needle-leaved trees impacts on rainfall interception and canopy storage capacity in an arid environment. New For 46:339–355

Aydın M, Güneş Şen S, Celik S (2018) Throughfall, stemflow, and interception characteristics of coniferous forest ecosystems in the western black sea region of Turkey (Daday example). Environ Monit Assess 190:1–10

Bao W, Bao W, He B, Ding D (2004) Redistribution effects of tree canopy of the artificial *Pinus tabulaeformis* forest on precipitation in the upper stream of Minjiang River. J Beijing For Univ 26(5):10–16

Bellot J, Chirino E (2013) Hydrobal: an eco-hydrological modelling approach for assessing water balances in different vegetation types in semi-arid areas. Ecol Model 266:30–41

Biederman JA, Harpold AA, Gochis DJ, Ewers BE, Reed DE, Papuga SA, Brooks PD (2014) Increased evaporation following widespread tree mortality limits streamflow response. Water Resour Res 50(7):5395–5409

Boggs J, Sun G, Domec JC, McNulty S, Treasure E (2015) Clearcutting upland forest alters transpiration of residual trees in the riparian buffer zone. Hydrol Process 29(24):4979–4992

Boggs JL, Sun G, Domec JC, McNulty SG (2021) Variability of tree transpiration across three zones in a southeastern US Piedmont watershed. Hydrol Process 35(10):e14389

Brantley ST, Vose JM, Wear DN, Band L (2018) Potential of longleaf pine restoration to mitigate water scarcity and sustain carbon sequestration: planning for an uncertain future. In: Kirkman LK, Jack SB (eds) Ecological restoration and management of longleaf pine forests. CRC Press/Taylor and Francis, pp 291–310

Brantley ST, Miniat CF, Bolstad PV (2019) Rainfall partitioning varies across a forest age chronosequence in the southern Appalachian Mountains. Ecohydrology 12(4):e2081

Budyko MI (1974) Climate and life. Academic Press, New York

Bulcock HH, Jewitt GPW (2012) Field data collection and analysis of canopy and litter interception in commercial forest plantations in the KwaZulu-Natal Midlands, South Africa. Hydrol Earth Syst Sci 16(10):3717–3728

Calder IR (2007) Forests and water—ensuring forest benefits outweigh water costs. For Ecol Manag 251(1–2):110–120

Caldwell PV, Miniat CF, Elliott KJ, Swank WT, Brantley ST, Laseter SH (2016) Declining water yield from forested mountain watersheds in response to climate change and forest mesophication. Glob Chang Biol 22(9):2997–3012

Carnus JM, Parrotta J, Brockerhoff E, Arbez M, Jactel H, Kremer A, Lamb D, O'Hara K, Walters B (2006) Planted forests and biodiversity. J For 104(2):65–77

Chang ZY, Bao WK, He BH, Yang YC, He QH (2006) Interception and distribution effects of mixed artificial *Pinus tabulaeformis* and *Pinus armandi* forests on precipitation in the upper reaches of Minjiang River. J Soil Water Conserv 20(6):37–40

Chen X, Sivapalan M (2020) Hydrological basis of the Budyko curve: data-guided exploration of the mediating role of soil moisture. Water Resour Res 56:e2020WR028221. https://doi.org/10.1029/2020WR028221

Chen L, Wei W, Fu B, Lü Y (2007) Soil and water conservation on the Loess Plateau in China: review and perspective. Prog Phys Geogr 31(4):389–403

Chen S, Chen Z, Feng Z, Kong Z, Xu H, Zhang Z (2023) Species difference of transpiration in three urban coniferous forests in a semiarid region of China. J Hydrol 617:129098

Chescheir GM, Skaggs RW, Amatya DM (2009) Quantifying the hydrologic impacts of afforestation in Uruguay: a paired watershed study. In: Presentation given at the XIII World Forestry Congress, Buenos Aires, Argentina. USDA Forest Service, pp 18–23

Creed IF, Spargo AT, Jones JA, Buttle JM, Adams MB, Beall FD, Booth EG, Campbell JL, Clow D, Elder K, Green MB (2014) Changing forest water yields in response to climate warming: results from long-term experimental watershed sites across North America. Glob Chang Biol 20(10):3191–3208

del Campo AD, Otsuki K, Serengil Y, Blanco JA, Yousefpour R, Wei X (2022) A global synthesis on the effects of thinning on hydrological processes: implications for forest management. For Ecol Manag 519:120324

Delzon S, Loustau D (2005) Age-related decline in stand water use: sap flow and transpiration in a pine forest chronosequence. Agric For Meteorol 129(3–4):105–119

Deng J (2020) Fitting the revised Gash analytical model of rainfall interception to Mongolian Scots pines in Mu Us Sandy Land, China. Trees For People 1:100007

Domec JC, Sun G, Noormets A, Gavazzi MJ, Treasure EA, Cohen E, Swenson JJ, McNulty SG, King JS (2012) A comparison of three methods to estimate evapotranspiration in two contrasting loblolly pine plantations: age-related changes in water use and drought sensitivity of evapotranspiration components. For Sci 58(5):497–512

Dong S, Guo S, Guo J (1987) Through-flow, stem flow and canopy interception in a *Pinus tabulaeformis* plantation in north China. J Beijing For Univ 9(1):58–68. https://doi.org/10.13332/j.1000-1522.1987.01.006

Dong L, Han H, Kang F, Cheng X, Zhao J, Song X (2020) Rainfall partitioning in Chinese pine (*Pinus tabuliformis* Carr.) stands at three different ages. Forests 11(2):243

Dymond SF, Kolka RK, Bolstad PV, Sebestyen SD (2014) Long-term soil moisture patterns in a northern Minnesota forest. Soil Sci Soc Am J 78(S1):S208–S216

Fahey B, Jackson R (1997) Hydrological impacts of converting native forests and grasslands to pine plantations, South Island, New Zealand. Agric For Meteorol 84(1–2):69–82

Fisher JB et al (2017) The future of evapotranspiration: global requirements for ecosystem functioning, carbon and climate feedbacks, agricultural management, and water resources. Water Resour Res 53:2618–2626

Ford CR, Hubbard RM, Kloeppel BD, Vose JM (2007) A comparison of sap flux-based evapotranspiration estimates with catchment-scale water balance. Agric For Meteorol 145(3–4):176–185

Ford CR, Hubbard RM, Vose JM (2011) Quantifying structural and physiological controls on variation in canopy transpiration among planted pine and hardwood species in the southern Appalachians. Ecohydrology 4(2):183–195

Fox TR, Lee Allen H, Albaugh TJ, Rubilar R, Carlson CA (2007) Tree nutrition and forest fertilization of pine plantations in the southern United States. South J Appl For 31(1):5–11

Francis JR, Wuddivira MN, Farrick KK (2022) Exotic tropical pine forest impacts on rainfall interception: canopy, understory, and litter. J Hydrol 609:127765

Franklin R (2009) Converting planted loblolly pine (or slash pine) to longleaf pine: an opportunity. Clemson Ext For Leafl 31:1–6

Fu BP (1981) On the calculation of the evaporation from land surface (in Chinese). Sci Atmos Sin 5:23–31

Gash JHC (1979) An analytical model of rainfall interception by forests. Q J R Meteorol Soc 105(443):43–55

Gash JHC, Wright IR, Lloyd CR (1980) Comparative estimates of interception loss from three coniferous forests in Great Britain. J Hydrol 48(1–2):89–105

Gash JH, Lloyd CR, Lachaud G (1995) Estimating sparse forest rainfall interception with an analytical model. J Hydrol 170(1–4):79–86

Gavazzi MJ, Sun G, McNulty SG, Treasure EA, Wightman MG (2016) Canopy rainfall interception measured over ten years in a coastal plain loblolly pine (*Pinus taeda L.*) plantation. Trans ASABE 59(2):601–610

Gerrits AMJ, Savenije HHG, Hoffmann L, Pfister L (2007) New technique to measure forest floor interception – an application in a beech forest in Luxembourg. Hydrol Earth Syst Sci 11(2):695–701

Ghimire CP, Bruijnzeel LA, Lubczynski MW, Bonell M (2012) Rainfall interception by natural and planted forests in the Middle Mountains of Central Nepal. J Hydrol 475:270–280. https://doi.org/10.1016/j.jhydrol.2012.09.051

Ghimire CP, Bonell M, Bruijnzeel LA, Coles NA, Lubczynski MW (2013) Reforesting severely degraded grassland in the Lesser Himalaya of Nepal: effects on soil hydraulic conductivity and overland flow production. J Geophys Res: Earth Surf 118(4):2528–2545

Ghimire CP, Bruijnzeel LA, Lubczynski MW, Bonell M (2014) Negative trade-off between changes in vegetation water use and infiltration recovery after reforesting degraded pasture land in the Nepalese Lesser Himalaya. Hydrol Earth Syst Sci 18(12):4933–4949

Gholz HL, Clark KL (2002) Energy exchange across a chronosequence of slash pine forests in Florida. Agric For Meteorol 112(2):87–102

Gifford GF, Humphries W, Jaynes RA (1984) A preliminary quantification of the impacts of aspen to conifer succession on water yield II: modeling results. J Am Water Resour Assoc 20(2):181–186

Goeking SA, Tarboton DG (2022) Variable streamflow response to forest disturbance in the western US: a large-sample hydrology approach. Water Resour Res 58(6):e2021WR031575

González-Sanchis M, Del Campo AD, Molina AJ (2015) Modeling adaptive forest management of a semi-arid Mediterranean Aleppo pine plantation. Ecol Model 308:34–44

Good SP, Noone D, Bowen G (2015) Hydrologic connectivity constrains partitioning of global terrestrial water fluxes. Science 349(6244):175–177

Grace JC (1987) Theoretical ratio between "one-sided" and total surface area for pine needles. N Z J For Sci 17(2/3):292–296

Granier A (1987) Evaluation of transpiration in a Douglas-fir stand by means of sap flow measurements. Tree Physiol 3(4):309–320

Helvey JD (1971) A summary of rainfall interception by certain conifers on North America. In: Proceedings of the 3rd international conference on hydrology professors, biological effects in the hydrologic cycle. Purdue University, West Lafayette, pp 103–113

Helvey JD, Patric JH (1965) Canopy and litter interception of rainfall by hardwoods of eastern United States. Water Resour Res 1:193–206

Hibbert AR (1979) Managing vegetation to increase flow in the Colorado River Basin, vol 66. Rocky Mountain Forest and Range Experiment Station, Forest Service, US Department of Agriculture, Fort Collins

Hook S, Anderson M (2019) ECOSTRESS evapotranspiration dis-ALEXI USDA daily L3 global 30 m V001 [Data set]. NASA EOSDIS Land Processes DAAC. https://doi.org/10.5067/ECOSTRESS/ECO3ETALEXIU.001

Hu H, Chen L, Liu H, Ali Khan MY, Tie Q, Zhang X, Tian F (2018) Comparison of the vegetation effect on ET partitioning based on eddy covariance method at five different sites of northern China. Remote Sens 10(11):1755

Hua F, Bruijnzeel LA, Meli P et al (2022) The biodiversity and ecosystem service contributions and trade-offs of forest restoration approaches. Science 376(6595):839–844

Iida SI, Tanaka T, Sugita M (2005) Change of interception process due to the succession from Japanese red pine to evergreen oak. J Hydrol 315(1–4):154–166

Iida SI, Tanaka T, Sugita M (2006) Change of evapotranspiration components due to the succession from Japanese red pine to evergreen oak. J Hydrol 326(1–4):166–180

Ilvesniemi H, Pumpanen J, Duursma R, Hari P, Keronen P, Kolari P, Kulmala M, Mammarella I, Nikinmaa E, Rannik Ü, Pohja T (2010) Water balance of a boreal Scots pine forest. Boreal Environ Res 15:375–396

Ives JD (2006) Water balances, floods and sediment transport in the Hindu Kush-Himalayas. Mt Res Dev 26(4):384–385

Jokela EJ, Martin TA, Vogel JG (2010) Twenty-five years of intensive forest management with southern pines: important lessons learned. J For 108(7):338–347

Knight DH, Vose JM, Baldwin VC, Ewel KC, Grodzinska K, Grodzinska K (1994) Contrasting patterns in pine forest ecosystems. Ecol Bull 43:9–19

Komatsu H, Kume T, Otsuki K (2008) The effect of converting a native broad-leaved forest to a coniferous plantation forest on annual water yield: a paired-catchment study in northern Japan. For Ecol Manag 255(3–4):880–886

LaMalfa EM (2007) Comparison of water dynamics in aspen and conifer: implications for ecology water yield augmentation. All Graduate Theses and Dissertations. 6603. https://digitalcommons.usu.edu/etd/6603

Leuschner C, Förster A, Diers M, Culmsee H (2022) Are northern German Scots pine plantations climate smart? The impact of large-scale conifer planting on climate, soil and the water cycle. For Ecol Manag 507:120013

Liu X, Wu Q, Zhao H (1991) A preliminary study on interceptive function of the canopy of *Pinus tabulaeformis* and *Populus davidiana* stands in the hilly areas on the Loess Plateau. Bull Soil Water Conserv 2:4–7, 42

Liu Y, Alamusa, Cao J (2016) Characteristics of rainfall partitioning by *Pinus sylvestris var. Mongolica* forest canopy in horqin sandy land, northern China. Chin J Ecol 35(8):2046–2055. https://doi.org/10.13292/j.1000-4890.201608.020

Liu X, Sun G, Mitra B, Noormets A, Gavazzi MJ, Domec JC, Hallema DW, Li J, Fang Y, King JS, McNulty SG (2018) Drought and thinning have limited impacts on evapotranspiration in a managed pine plantation on the southeastern United States coastal plain. Agric For Meteorol 262:14–23

Liu N, Sun G, Yang Y, Aguilos M, Starr G, O'Halloran T L, ... Trettin C (2025) Potential for augmenting water yield by restoring longleaf pine (Pinus palustris) forests in the southeastern United States. Water Resour Res 61(2):e2024WR037444

Llorens P, Domingo F (2007) Rainfall partitioning by vegetation under Mediterranean conditions. A review of studies in Europe. J Hydrol 335(1–2):37–54

Llorens P, Poch R, Latron J, Gallart F (1997) Rainfall interception by a Pinus sylvestris forest patch overgrown in a Mediterranean mountainous abandoned area I. Monitoring design and results down to the event scale. J Hydrol 199(3–4):331–345

Loustau D, Berbigier P, Granier A, Moussa FEH (1992) Interception loss, throughfall and stemflow in a maritime pine stand. I. Variability of throughfall and stemflow beneath the pine canopy. J Hydrol 138(3–4):449–467

Lu J, Sun G, McNulty SG, Amatya DM (2005) A comparison of six potential evapotranspiration methods for regional use in the southeastern United States 1. JAWRA J Am Water Resour Assoc 41(3):621–633

Ma Z, Hartmann H, Wang H, Li Q, Wang Y, Li S (2014) Carbon dynamics and stability between native Masson pine and exotic slash pine plantations in subtropical China. Eur J For Res 133:307–321

Ma J, Zha T, Jia X, Tian Y, Bourque CPA, Liu P, Bai Y, Wu Y, Ren C, Yu H, Zhang F (2018) Energy and water vapor exchange over a young plantation in northern China. Agric For Meteorol 263:334–345

Malone RW, Yagow G, Baffaut C, Gitau M, Qi Z, Amatya D, Parajuli P, Bonta J, Green T (2015) Parameterization guidelines and consideration for hydrologic models. Trans ASABE 58(6):1681–1703

McCarthy EJ, Skaggs RW, Farnum P (1991) Experimental determination of the hydrologic components of a drained watershed. Trans ASAE 34(5):2031–2039

McLaughlin DL, Kaplan DA, Cohen MJ (2013) Managing forests for increased regional water yield in the southeastern US coastal plain. JAWRA J Am Water Resour Assoc 49(4):953–965

McQuillan KA, Tulbure MG, Martin KL (2022) Forest water use is increasingly decoupled from water availability even during severe drought. Landsc Ecol 37(7):1801–1817

Mkhabela MS, Amiro BD, Barr AG, Black TA, Hawthorne I, Kidston J, McCaughey JH, Orchansky AL, Nesic Z, Sass A, Shashkov A (2009) Comparison of carbon dynamics and water use efficiency following fire and harvesting in Canadian boreal forests. Agric For Meteorol 149(5):783–794

Moreaux V, Lamaud É, Bosc A, Bonnefond JM, Medlyn BE, Loustau D (2011) Paired comparison of water, energy and carbon exchanges over two young maritime pine stands (Pinus pinaster Ait.): effects of thinning and weeding in the early stage of tree growth. Tree Physiol 31(9):903–921

Müller J (2009) Forestry and water budget of the lowlands in northeast Germany--consequences for the choice of tree species and for forest management. J Water Land Dev 13:133

Müller J, Bolte A (2009) The use of lysimeters in forest hydrology research in north-east Germany. Landbauforschung (vTI Agric For Res) 59(1):1–10

Muzylo A, Llorens P, Valente F, Keizer JJ, Domingo F, Gash JHC (2009) A review of rainfall interception modelling. J Hydrol 370(1–4):191–206

Naudiyal N, Schmerbeck J (2017) The changing Himalayan landscape: pine-oak forest dynamics and the supply of ecosystem services. J For Res 28(3):431–443

Nichols JD, Bristow M, Vanclay JK (2006) Mixed-species plantations: prospects and challenges. For Ecol Manag 233:383–390

Novick KA, Ficklin DL, Stoy PC, Williams CA, Bohrer G, Oishi AC, Papuga SA, Blanken PD, Noormets A, Sulman BN, Scott RL (2016) The increasing importance of atmospheric demand for ecosystem water and carbon fluxes. Nat Clim Chang 6(11):1023–1027

Oishi AC, Oren R, Novick KA, Palmroth S, Katul GG (2010) Interannual invariability of forest evapotranspiration and its consequence to water flow downstream. Ecosystems 13:421–436

Oishi AC, Miniat CF, Novick KA, Brantley ST, Vose JM, Walker JT (2018) Warmer temperatures reduce net carbon uptake, but do not affect water use, in a mature southern Appalachian forest. Agric For Meteorol 252:269–282

Oki T, Kanae S (2006) Global hydrological cycles and world water resources. Science 313(5790):1068–1072

Peet RK, Platt WJ, Costanza JK (2018) Fire-maintained pine savannas and woodlands of the southeastern United States coastal plain. In: Ecology and recovery of eastern old-growth forests. Island Press, Washington, DC, pp 39–62

Pisarello KL, Sun G, Evans JM, Fletcher RJ Jr (2022) Potential long term water yield impacts from pine plantation management strategies in the southeastern United States. For Ecol Manag 522:120454

Poyatos R, Granda V, Molowny-Horas R, Mencuccini M, Steppe K, Martínez-Vilalta J (2016) SAPFLUXNET: towards a global database of sap flow measurements. Tree Physiol 36(12):1449–1455

Poyatos R, Granda V, Flo V, Adams MA, Adorján B, Aguadé D, Aidar MP, Allen S, Alvarado-Barrientos MS, Anderson-Teixeira KJ, Aparecido LM (2021) Global transpiration data from sap flow measurements: the SAPFLUXNET database. Earth Syst Sci Data 13(6):2607–2649

Qi J, Brantley ST, Golladay SW (2022) Simulated longleaf pine (*Pinus palustris* Mill.) restoration increased streamflow—a case study in the Lower Flint River Basin. Ecohydrology 15(1):e2365

Qian Y, Shi C, Zhao T, Lu J, Bi B, Luo G (2022) Canopy interception of different rainfall patterns in the Rocky Mountain areas of northern China: an application of the revised Gash model. Forests 13(10):1666

Ramirez E, Galleguillos M, Seguel O, Faúndez C, Iribarne LR (2022) Dissecting the water balance of pine plantations and thorny forest in a Mediterranean landscape. In: Presentation, AGU Annual Fall Meeting. Chicago, IL. AGU

Rau BM, Adams MB, Kelly CN (2022) Conversion of an Appalachian hardwood forest to a Norway spruce plantation: effects on hydrologic processes and function. In: Latimer JS, Bosch DD, Faustini J, Lane CR, Trettin CC (eds) Enhancing landscapes for sustainable intensification and watershed resiliency—proceedings of the seventh interagency conference on research in the watersheds. November 16–19, 2020, Hosted virtually by the USDA-ARS and the University of Georgia, Tifton, GA. General technical report SRS–264. U.S. Department of Agriculture Forest Service, Southern Research Station. 217 p, Asheville. https://doi.org/10.2737/SRS-GTR-264

Richardson DM, Rundel PW, Jackson ST, Teskey RO, Aronson J, Bytnerowicz A, Wingfield MJ, Procheş Ş (2007) Human impacts in pine forests: past, present, and future. Annu Rev Ecol Evol Syst 38:275–297

Riley JW, Pangle LA, Aulenbach BT (2023) Evaluating the spatial and temporal variability of groundwater uptake by riparian vegetation in a humid southeastern US catchment. Ecohydrology 16:e2520

Roberts J (1983) Forest transpiration: a conservative hydrological process? J Hydrol 66:133–141

Roth FA II, Chang M (1981) Throughfall in planted stands of four southern pine species in east Texas. JAWRA J Am Water Resour Assoc 17(5):880–885

Ruzol R, Staudhammer CL, Younger S, Aubrey DP, Loescher HW, Jackson CR, Starr G (2022) Water use in a young *Pinus taeda* bioenergy plantation: effect of intensive management on stand evapotranspiration. Ecosphere 13(6):e4100

Ryan MG, Bond BJ, Law BE, Hubbard RM, Woodruff D, Cienciala E, Kucera J (2000) Transpiration and whole-tree conductance in ponderosa pine trees of different heights. Oecologia 124:553–560

Sadeghi SMM, Attarod P, Pypker TG, Dunkerley D (2014) Is canopy interception increased in semiarid tree plantations? Evidence from a field investigation in Tehran, Iran. Turk J Agric For 38(6):792–806

Schlesinger WH, Jasechko S (2014) Transpiration in the global water cycle. Agric For Meteorol 189:115–117

Segura C, Bladon KD, Hatten JA, Jones JA, Hale VC, Ice GG (2020) Long-term effects of forest harvesting on summer low flow deficits in the Coast Range of Oregon. J Hydrol 585:124749

Shepard JP (1994) Effects of forest management on surface water quality in wetland forests. Wetlands 14(1):18–26

Shuttleworth WJ, Wallace JS (1985) Evaporation from sparse crops-an energy combination theory. Q J R Meteorol Soc 111(469):839–855

Skaggs RW, Chescheir GM, Fernandez GP, Amatya DM, Diggs J (2011) Effects of land use on soil properties and hydrology of drained coastal plain watersheds. Trans ASABE 54(4):1357–1365

Skaggs RW, Tian S, Chescheir GM, Amatya DM, Youssef MA (2016) Forest drainage. In: Forest hydrology: processes, management and assessment. CABI, pp 124–140

Skaggs RW, Amatya DM, Chescheir GM (2020) Effects of drainage for silviculture on wetland hydrology. Wetlands 40:47–64

Stocker BD, Tumber-Dávila SJ, Konings AG, Anderson MC, Hain C, Jackson RB (2023) Global patterns of water storage in the rooting zones of vegetation. Nat Geosci 16:1–7

Stoy PC (2018) Deforestation intensifies hot days. Nat Clim Chang 8(5):366–368

Sun G, Riekerk H, Kornhak LV (2000) Ground-water-table rise after forest harvesting on cypress-pine flatwoods in Florida. Wetlands 20(1):101–112

Sun G, Noormets A, Gavazzi MJ, McNulty SG, Chen J, Domec JC, King JS, Amatya DM, Skaggs RW (2010) Energy and water balance of two contrasting loblolly pine plantations on the lower coastal plain of North Carolina, USA. For Ecol Manag 259(7):1299–1310

Sun G, Alstad K, Chen J, Chen S, Ford CR, Lin G, Liu C, Lu N, McNulty SG, Miao H, Noormets A (2011) A general predictive model for estimating monthly ecosystem evapotranspiration. Ecohydrology 4(2):245–255

Sun X, Onda Y, Kato H, Gomi T, Komatsu H (2015) Effect of strip thinning on rainfall interception in a Japanese cypress plantation. J Hydrol 525:607–618

Sun G, Amatya D, McNulty S (2016) Chapter 85: Forest hydrology. Part 7: Systems hydrology. In: Sing VV (ed) Handbook of applied hydrology. USDA Forest Service, pp 1–8

Sun X, Wilcox BP, Zou CB (2019) Evapotranspiration partitioning in dryland ecosystems: a global meta-analysis of in situ studies. J Hydrol 576:123–136. https://doi.org/10.1016/j.jhydrol.2019.06.022

Sun S, Liu XP, Wang CP, Zhang TH, He YH, Lu P (2021) Precipitation redistribution characteristics of *Pinus sylvestris var. mongolica* in semiarid sandy land. Arid Land Geogr 44(1):109–117

Swank WT, Douglass JE (1974) Streamflow greatly reduced by converting deciduous hardwood stands to pine. Science 185(4154):857–859

Tang Y, Jin S, Zhang X, Wang M, Busso CA (2015) Estimating the transpiration of *Pinus sylvestris* trees: from an individual to a stand scale. Biocell 39(1):14–23

Tang X, Li H, Ma M, Yao L, Peichl M, Arain A, Xu X, Goulden M (2017) How do disturbances and climate effects on carbon and water fluxes differ between multi-aged and even-aged coniferous forests? Sci Total Environ 599:1583–1597

Tie Q, Hu H, Tian F, Holbrook NM (2018) Comparing different methods for determining forest evapotranspiration and its components at multiple temporal scales. Sci Total Environ 633:12–29

Tor-ngern P, Oren R, Oishi AC, Uebelherr JM, Palmroth S, Tarvainen L, Ottosson-Löfvenius M, Linder S, Domec JC, Näsholm T (2017) Ecophysiological variation of transpiration of pine forests: synthesis of new and published results. Ecol Appl 27(1):118–133

Tor-ngern P, Oren R, Palmroth S, Novick K, Oishi A, Linder S, Ottosson-Löfvenius M, Näsholm T (2018) Water balance of pine forests: synthesis of new and published results. Agric For Meteorol 259:107–117

Valente F, David JS, Gash JHC (1997) Modelling interception loss for two sparse eucalypt and pine forests in central Portugal using reformulated Rutter and Gash analytical models. J Hydrol 190(1–2):141–162

Vanclay JK (2009) Managing water use from forest plantations. For Ecol Manag 257(2):385–389

Vandegehuchte MW, Steppe K (2013) Sap-flux density measurement methods: working principles and applicability. Funct Plant Biol 40:213–223

Vose JM, Dougherty PM, Long JN, Smith FW, Gholz HL, Curran PJ (1994) Factors influencing the amount and distribution of leaf area of pine stands. Ecol Bull 43:102–114

Wang D, Wang G, Anagnostou EN (2007) Evaluation of canopy interception schemes in land surface models. J Hydrol 347(3–4):308–318

Wang YH, Wang YH, Li ZH, Wang YJ (2021) The characteristics of rainfall interception and water storage of the natural secondary forests of Masson pine at Tieshanping of Chongqing. Acta Ecol Sin 41(16):6542–6551

Wang YN, Zhao Y, Zhang HD, Zhang N, Liu J (2022) Evapotranspiration characteristics of *Pinus tabuliformis* plantation in the arboretum of Inner Mongolia Academy of Forestry. J Inn Mong For Sci Technol 38–41

Waring RH, Rogers JJ, Swank WT (1981) Water relations and hydrologic cycles. In: Dynamic properties of forest ecosystems. Cambridge University Press, Cambridge, p 205

Wei X, Liu S, Zhou G, Wang C (2005) Hydrological processes in major types of Chinese forest. Hydrol Process 19(1):63–75

Wei Z, Yoshimura K, Wang L, Miralles DG, Jasechko S, Lee X (2017) Revisiting the contribution of transpiration to global terrestrial evapotranspiration. Geophys Res Lett 44(6):2792–2801

Weir DN, Greis JG (2002) The southern forest resource assessment–summary report. General technical report SRS-53. USDA Forest Service, Southern Research Station, Asheville

Whitehead D, Kelliher FM (1991) A canopy water balance model for a Pinus radiata stand before and after thinning. Agric For Meteorol 55(1–2):109–126

Xiao Y, Chen LH, Yu XX, Yan XB, Sun QY (2007) Influence on precipitation distribution of *Pinus tabuleaefomis* forest in Miyun reservoir. J Soil Water Conserv 21(3):154–157

Xu J (2011) China's new forests aren't as green as they seem. Nature 477:371

Yang Y, Anderson MC, Gao F, Hain CR, Semmens KA, Kustas WP, Noormets A, Wynne RH, Thomas VA, Sun G (2017) Daily Landsat-scale evapotranspiration estimation over a forested landscape in North Carolina, USA, using multi-satellite data fusion. Hydrol Earth Syst Sci 21(2):1017–1037

Zabret K, Rakovec J, Šraj M (2018) Influence of meteorological variables on rainfall partitioning for deciduous and coniferous tree species in urban area. J Hydrol 558:29–41

Zhang L, Dawes WR, Walker GR (2001) Response of mean annual evapotranspiration to vegetation changes at catchment scale. Water Resour Res 37(3):701–708

Zhang L, Hickel K, Dawes WR, Chiew FH, Western AW, Briggs PR (2004) A rational function approach for estimating mean annual evapotranspiration. Water Resour Res 40(2):W02502

Zhou G, Wei X, Chen X, Zhou P, Liu X, Xiao Y, Sun G, Scott DF, Zhou S, Han L, Su Y (2015) Global pattern for the effect of climate and land cover on water yield. Nat Commun 6(1):5918

Zomer RJ, Xu J, Trabuco A (2022) Version 3 of the global aridity index and potential evapotranspiration database. Sci Data 9:409. https://www.nature.com/articles/s41597-022-01493-1

Zou CB, Ffolliott PF, Wine M (2010) Streamflow responses to vegetation manipulations along a gradient of precipitation in the Colorado River Basin. For Ecol Manag 259(7):1268–1276

Hypothetical Responses of Pine Forests to Climate Change Based on Manipulation Experiments

Christopher A. Maier, David Medvigy, Ram Oren, Sari Palmroth, and Karina V. R. Schäfer

Abstract This chapter synthesizes the responses of pine forests, from leaf physiology to stand development and ecosystem processes, to climate change variables modified through experiments. We examine the effects of increased temperature and CO_2 enrichment, along with moisture manipulation (both increases and decreases), followed by research on interactions between these variables and other factors like nutrient availability. Many studies on the effects of these variables were performed on seedlings; however, responses of seedlings have shown to misrepresent the behavior and performance of larger individuals and stands. Thus, we referenced some reviews and original studies on seedlings for context but synthesized almost exclusively the results from studies performed on saplings or larger individuals, done mostly in accruing stands (Cermak et al. 2007).

The synthesis concludes with a modeling exercise that simulates the effects of these climate manipulations on pine forests focusing primarily on aboveground biomass accumulation and water yield available to downstream ecosystems and users.

Authors listed alphabetically.

C. A. Maier (✉)
Nicholas School of the Environment, Duke University, Durham, NC, USA
e-mail: chris.maier@duke.edu

D. Medvigy
Department of Biological Sciences, University of Notre Dame, Notre Dame, IN, USA

R. Oren
Nicholas School of the Environment & Pratt School of Engineering, Duke University, Durham, NC, USA

Department of Forest Sciences, University of Helsinki, Helsinki, Finland

S. Palmroth
Department of Forest Sciences, University of Helsinki, Helsinki, Finland

Nicholas School of the Environment, Duke University, Durham, NC, USA

K. V. R. Schäfer
Department of Earth and Environmental Sciences, Rutgers University, Newark, NJ, USA

D. M. Johnson et al. (eds.), *Physiology and Ecology of Pines Worldwide*, Tree Physiology 9, https://doi.org/10.1007/978-3-032-23001-0_9

We explain high-level model results (e.g., changes in aboveground biomass accumulation and water yield) in terms of changes in physiological processes, making *qualitative* comparisons with data representing two temperate *Pinus taeda* (loblolly pine) stands, one in warm-moist climate and the other in warm-dry climate, and one stand of boreal *P. sylvestris* (Scots pine). We selected these species because they are among the most planted worldwide and among the most studied in climate change manipulation experiments, thus providing sufficient information for both informing the modeling exercise and evaluating the predictions of responses to climate change scenarios. The goal was to generate hypotheses about what might have happened if experiments were continued longer in time.

1 Introduction

The genus *Pinus* consists of 106 species that are naturally concentrated in the northern hemisphere between about 70° and 13°N, with a few reaching to 2°S; most of the species are found in North America (66 species), followed by Asia (27) and Europe (12) (Nobis et al. 2012). The geographic distribution exposes the genus, and even single species (e.g., *Pinus sylvestris*), to a wide range of climatic and edaphic conditions. The strongest predictor for the occurrence of pines is latitude which is a proxy for the environmental conditions, mostly temperature and precipitation (Nobis et al. 2012). There is a very strong phylogenetic structured environmental component in the distribution of *Pinus*, and, as growing conditions change, the underlying phylogenetic traits will constrain morphological and functional acclimation (Nobis et al. 2012; Shao et al. 2019). In this chapter, we do not consider adaptation given the relative short duration of even the longest experiments. Some commercially important pine species are successfully cultivated, usually in plantations, well outside of their natural range. In this context, traits of *Pinus* investigated with respect to climate change variables may have a stronger phylogenetic driver (Nobis et al. 2012) than the experimental manipulation of these variables. This is reflected in varied physiological responses among species of the genus *Pinus*, complicating synthesis of response variables across species and growing conditions.

The distribution and growth of each pine species are constrained by water availability and temperature. Soil physical properties, soil texture (aeration, bulk density), relating to water holding capacity and the physiological and morphological traits of roots regulate water uptake. Soil texture has a large effect on the resistance to water uptake. Drying soil demands a steep increase in the amount of tension, or water potential gradient, that plants must withstand to extract soil water. Global climate change variables, i.e., increased atmospheric CO_2, changes in precipitation regimes, and increased temperature and vapor pressure deficits (VPD), may each and in combination affect the ability of plants to take up water in exchange for CO_2. Furthermore, future temperature and precipitation regimes may fall outside the historical range, creating novel environmental conditions (IPCC 2021) making it difficult to predict forest response. For example, there has been an increase in woody-plant mortality in many regions (Breshears et al. 2005; Allen et al. 2010; Matusick et al. 2013), and larger diameter trees appear to be more susceptible (Moore et al. 2016). The dieback is associated with increased frequency and intensity of soil and atmospheric drought accompanied by warmer temperatures (Allen et al. 2015). Tree mortality events have been recorded in forest ecosystems globally (Allen et al. 2010) particularly in xeric systems, but even in temperate systems, more frequent short-term droughts can reduce productivity (Klos et al. 2009). Mechanisms underlying pine response to climate change are tied to whole-plant water relations and carbon (C) balance. However, the mechanisms—hydraulic failure, C starvation, biotic agents, and the frequency and extent of mortality events—may differ with species, site, and region (McDowell et al. 2022).

This chapter synthesizes the responses of pine forests, from leaf physiology to stand development and ecosystem processes to climate change variables. We focus on the effects of temperature and CO_2, both expected to continue increasing in the foreseeable future, and atmospheric and soil moisture with less certain and more complex spatiotemporal forecasts. Most studies on the effects of these climate change-relevant variables on many species of pines were performed on seedlings. Yet, responses of seedlings have shown to misrepresent the behavior and performance of larger individuals and stands (McDowell et al. 2013).

Thus, we referenced some reviews and original studies on seedlings for context but synthesized almost exclusively the results from studies performed on saplings or larger individuals, done mostly in accruing stands.

We begin with synthesis of results from experiments on warming and CO_2 enrichment, followed by manipulations that increase or decrease soil moisture and atmospheric humidity. This is followed by results from experiments designed to quantify interaction effects among these climate change variables. The synthesis culminates with a modeling exercise simulating the effects of temperature, CO_2, and water manipulations on pine forests. The focus is on how these manipulations affect aboveground biomass accumulation over stand development, and the amount of water unused by the pine stands, available as yield to downstream ecosystems and users. We attempt to explain high-level model results (e.g., changes in aboveground biomass, AGB, accumulation) in terms of changes in physiological processes, making *qualitative* comparisons with data representing two temperate *Pinus taeda* (loblolly pine) stands, one in warm-moist climate and the other in warm-dry climate, and one stand of boreal *P. sylvestris* (Scots pine). We selected these species not only because they are among the most planted worldwide but also because among the pine species subjected to climate change-relevant manipulation experiments on saplings and larger individuals, these were the most intensively studied, thus providing sufficient information for both informing the modeling exercise and evaluating the predictions of responses to climate change scenarios. In effect, we used modeling to generate hypotheses about what might have happened if experiments were continued longer in time.

2 Over-view

[Photosynthesis (C source), respiration (sink), and growth (sink) are all temperature-dependent processes and likely affected by directional changes in growth temperature. We expect that, e.g., photosynthetic rate increases with temperature until the optimal temperature for maximum productivity (thermal optimum) is reached and that it declines thereafter. However, optimal temperature for maximum photosynthetic rate may shift with increasing growth temperature through the so-called thermal acclimation (Berry and Bjorkman 1980; Hikosaka et al. 2006). An increase in temperature optimum of photosynthesis in response to experimental warming has indeed been demonstrated across trees species from different biomes (Crous et al. 2022). Underlying thermal acclimations are potential adjustments in the biochemical efficiencies of photosynthetic processes (e.g., the maximum rate of carboxylation of Rubisco, V_{cmax}, and the maximum rate of electron transport, J_{max}) as well as mitochondrial respiration rate. The degree of acclimation can be assessed when the performance (here gas exchange rates) of trees grown in different temperatures is evaluated at a common temperature. Based on such comparisons, results from most warming experiments show no difference in V_{cmax}, J_{max}, or net photosynthetic rate, but a lower respiration rate (indicating acclimation) in trees that grew in warmer temperatures (Way and Oren 2010; Crous et al. 2022). When assessed at growth temperatures, net photosynthetic rate is often lower in trees grown under warmer conditions (Crous et al. 2022) but can also remain the same or increase (Dusenge et al. 2023). Moreover, based on their review of warming experiments on trees, (Chung et al. 2013) suggested that the effect of temperature on photosynthesis likely depends on the position of the leaf (within the crown) or individual (canopy or understory) in the forest. The temperature effects likely change with the time of year reflecting leaf phenology and aging, with shade or young leaves benefiting from warming more than sun-exposed or old leaves (Chung et al. 2013).

Global warming may also extend the season favorable for photosynthesis and growth, particularly in cooler climates. Where this is accompanied with earlier budburst and growth onset as well as later leaf senescence and growth cessation, photosynthetic C gain and biomass accumulation may increase (Chung et al. 2013). While most warming experiments confirm the lengthening of the physiologically active period, other factors such as day length, chilling requirement or conditions during previous growing season, often not manipulated in warming experiments, may also affect the realized increase in growing season length (Chung et al. 2013; Way and Montgomery 2015; Gu et al. 2022). In many temperate and boreal trees, photoperiod controls potential photosynthesis (Bauerle et al. 2012), growth cessation, and bud set, thus potentially limiting the extension of the growing season into warmer falls (Way and Montgomery 2015). At the same time, chilling requirements (low temperature sum) may not be fulfilled during warmer winters, such that trees may need even greater heat sums to force bud burst in the future.

A review of studies on warming effects (in isolation or in combination with other treatments) on individuals of conifers and broadleaved species, ranging from seedlings in growth chambers to open-grown trees (Way and Oren 2010), concluded that, where tree growth (measured as increment in height, diameter, or biomass) is temperature-limited (high latitude and altitude), growth may increase with some degree of warming whereas warm-adapted species may be negatively affected. Compared to trees grown in ambient temperature, those developing under and responding positively to experimentally elevated temperatures showed altered allocation of growth leading to higher shoot-to-root ratios (Way and Oren 2010). Moreover, growth enhancement was shown to be larger in deciduous species (angiosperms) compared to conifers, pointing at potential between-group differences in, e.g., partitioning of photosynthates to root exudates or secondary defense compounds (Way and Oren 2010). Similarly, a long-term open-field study with juveniles of 11 boreal and temperate species suggested that conifers showed no or negative response (in photosynthesis, survival, and growth) to higher temperature while growth of all the deciduous species in the study benefited from warming (Reich et al. 2022)].

[The effect of warming on physiological processes depends on other factors such as soil moisture, atmospheric moisture, and CO_2; some experiments quantified these interaction effects. Photosynthetic rate is influenced by the supply of CO_2 through stomata, or stomatal limitations, which can reduce the realized CO_2 uptake rate below the biochemical capacity of the leaf. Because water is lost through open stomata, stomatal conductance directly links leaf C gain to water supply and, thus, to the tree hydraulic structure providing water to leaves.

Although stomatal conductance may directly respond to temperature, there is little evidence from warming experiments, mostly done under well-watered conditions, that elevated temperature changes stomatal or other hydraulic traits, such as leaf and wood anatomy, leaf area-to-sapwood area ratio, or leaf-specific hydraulic conductance (Way and Oren 2010). However, gas exchange declined in all species when warming was combined with decreased soil water availability (Reich et al. 2022).

Temperature is increasing with atmospheric CO_2. Growth processes (sink activity) are directly driven by both the availability of C (source activity) and temperature. While changes in growth may not always follow those of photosynthetic rates (when growth is limited by other factors than source activity), the impact of elevated CO_2 and warming on growth likely depends on changes in photosynthesis, photorespiration, and respiration (Dusenge et al. 2019). For example, high CO_2 reduces photorespiration and photosynthetic light compensation point more at higher temperatures and is expected to increase the thermal optimum of net photosynthetic rate (Long 1991; Sage and Kubien 2007; Dusenge et al. 2023). At the same time, both elevated CO_2 and the high atmospheric vapor deficit accompanying elevated temperatures may decrease stomatal conductance and limit C uptake (Ainsworth and Long 2005; Franks and Farquhar 1999; Oren et al. 1999). Because thermal acclimation of respiration is commonly observed in warming experiments but the respiration response to elevated CO_2 is mixed, when the two factors are manipulated at the same time, effects of growth temperature are likely dominating respiration response (Dusenge et al. 2019). Based on an extensive review accompanied with a meta-analysis, (Dusenge et al. 2019) suggest that, to predict future ecosystem C cycling, understanding plant C flux responses to temperature (which vary among species) is more important than understanding the responses to CO_2 (which are more consistent across studies).

The limited number of experiments on the combined effects of elevated temperature (eT) and CO_2 (eCO_2) suggests that soil C cycling and biomass decomposition will likely increase (Pendall et al. 2004). In a whole-ecosystem study on a boreal ombrotrophic bog, warming increased plant available nutrients, while eCO_2 had no effect on the nutrient dynamics (Iversen et al. 2023). In most (but not all) warming experiments reviewed by (Chung et al. 2013), nitrogen (N) mineralization rate increased under warming, thus supporting enhanced plant growth and resulting in shifts in C partitioning from below- to aboveground; fewer roots (and their associated mycorrhizae) are needed to acquire N. A meta-analysis of results of warming experiments across vegetation types predicted that, considering increasing global temperature only, the increase in C influx would be slightly greater than the efflux, meaning terrestrial ecosystems would remain a sink in the future (Lu et al. 2013)].

3 Response of Pinus to Warming Experiments (eT)

3.1 Phenology

In cool/cold climates, increasing temperature will result in longer thermal growing season. Warmer spring temperatures advance budburst and the onset of shoot and diameter growth, while trees may also become more susceptible to late spring frosts. Warming may also increase the heat/temperature sum required for budbreak and growth onset (i.e., preventing or delaying the time for meeting chilling requirements), which may result in a delay in the start of the active period (realized growing season). Overall, the lengthening of the growing season will affect growth mostly through earlier and more active growth in spring; the gain in growth through extending the season into warmer falls will likely be small.

In boreal *P. sylvestris* trees growing in open-top chambers (in Mekrijärvi, Finland), elevated temperature (eT) caused needles to harden (marking onset of dormancy) later and deharden (release from dormancy) earlier than for trees growing under ambient temperatures (Repo et al. 1996). In other experiments, warming hastened both onset and cessation of height and diameter growth by a week or longer, increasing the temperature sums required for both events (Kilpeläinen et al. 2006). Where juvenile individuals of 11 boreal and temperate species (including one pine of each biome) were subjected to simultaneous air and soil warming (chamberless experiment with both soil and air warming; B4warmED in Minnesota, USA), eT advanced the date of budburst such that both eT and interannual changes in ambient temperature similarly moved budburst earlier in warmer springs for most species (Montgomery et al. 2020). However, the two pine species, *P. strobus* (temperate) and *P. banksiana* (boreal), delayed budburst under eT for a given mean spring temperature. In other words, juvenile pine trees under warming required more growing degree days (a higher temperature sum) to break bud, consistent with the results for *P. sylvestris*. In contrast to the findings from experiments in cool/cold climates, needle emergence of *P. edulis* in warm semiarid climate [open-top chamber; Los Alamos Survival–Mortality (SUMO) experiment in New Mexico, USA] did not respond to eT (Grossiord et al. 2017a). This suggests that species from colder environments may respond negatively to eT whereas those from warmer environments may show limited response to changes in temperature alone.

3.2 Leaf Structure and Function

Warming effects on leaf gas exchange rates varied across species/studies. None of the studies reviewed here reported on clear and consistent downregulation (or acclimation) of the biochemistry of photosynthesis under eT. The results on light-saturated net photosynthetic rate and corresponding stomatal conductance are more varied and indicate that these traits can either remain unchanged, increase, or decrease under eT. Most studies suggest that leaf dark respiration acclimates under eT; however, even with acclimation, leaf respiration was higher under eT compared to trees growing under control/ambient conditions.

Results from experiments on *P. sylvestris* (Mekrijärvi) suggest that eT had little effect on the biochemistry of photosynthesis when assessed at a reference temperature (carboxylation and RuBP regeneration capacity, $V_{\mathrm{cmax25\,°C}}$ and $J_{\mathrm{cmax25\,°C}}$, respectively), indicating no acclimation (Wang et al. 1995). Related photosynthetic traits, such as N concentration, Rubisco-N, maximal quantum yield of photosystem II (PSII), and chlorophyll concentration, were occasionally higher in warmed foliage compared to control conditions; however, the effect (including its sign) varied among years, needle cohorts, and even studies carried out in at that same site (Wang et al. 2003; Luomala et al. 2003).

Similarly, no downregulation of biochemical parameters but a decline in $V_{\mathrm{cmax25\,°C}}/J_{\mathrm{cmax25\,°C}}$ of warm-grown *P. strobus* and *P. banksiana* juveniles was detected (B4warmED, Stefanski et al. 2020; Bermudez et al. 2021).

Leaf gas exchange of *P. sylvestris* (Mekrijärvi) was affected by warming such that eT decreased leaf net photosynthetic rate (A_n) and light use efficiency, increased light compensation point, and dark respiration (R_{dark}, even after downregulation), while stomatal conductance (g_s) decreased somewhat along with its sensitivity to changes in vapor pressure deficit (Wang et al. 1995; Kellomäki and Wang 1996; Zha et al. 2002; Luomala et al. 2003). In two soil-warming experiments carried out with high-altitude, treeline species, *P. cembra* (soil covered with roofing; Tyrolean Alps, Austria) and *P. mungo* subsp. *uncinata* (heating cables under surface vegetation; Free Air CO_2 Enrichment, FACE, in Stillberg, Switzerland), no eT effect was detected on A_n and g_s (Streit et al. 2014a; Wieser et al. 2015). At B4warmed, A_n of three conifers, including *P. strobus*, was adversely influenced by warming, while that of the rest of the species, including *P. banksiana*, increased under eT (Reich et al. 2022). Respiration rate was shown to acclimate under eT similarly across all species (Reich et al. 2016). Warming had no effect on the daily mean *A* of temperate *P. taeda* shoots enclosed in branch bags (Teskey 1997). Likewise, A_n of temperate *P. edulis* was unaffected by eT in both saplings (SUMO) and mature trees (greenhouse, Biosphere 2, Arizona, USA) (Adams et al. 2013; Grossiord et al. 2017a; McDowell et al. 2019).

From studies exploring changes in leaf morphological traits and nonstructural C concentrations under eT, specific leaf area (SLA) of *P. sylvestris*, *P. cembra*, and *P. edulis* were not affected by warming (Luomala et al. 2003); (Gruber et al. 2018; Grossiord et al. 2017a). Moreover, in *P. sylvestris* foliage, starch and soluble sugar concentrations showed inconsistent changes under eT (Zha et al. 2002; Luomala et al. 2003), while the nonstructural C, sugar, or starch concentrations in *P. edulis* (Biosphere 2) were unaffected by eT (Adams et al. 2013). Warming had no effect on monoterpene concentrations and decreased their emissions from *P. sylvestris* (Räisänen et al. 2008a,b).

Overall, except for persistently higher dark respiration, no general pattern emerged in leaf gas exchange, morphology, or chemistry in response to warming.

3.3 *Tree Structure and Function*

Warming may increase tree water use. Most studies on pines reported no adjustments in hydraulic properties under eT, while the results on stomatal sensitivity to VPD vary. Overall, water use may increase, particularly in cool/cold and humid conditions, driven by higher leaf area, root biomass, and evaporative demand under eT.

Elevated temperature had little effect on xylem anatomy (e.g., lumen size) yet resulted in longer tracheids and a higher wood density, particularly of early wood of *P. sylvestris* trees (Merkijärvi; Kilpeläinen et al. 2005; Kilpeläinen et al. 2007). Compared to control conditions, annual sap flow per tree (E_T) was higher under eT driven by greater tree leaf area (maximum canopy conductance), lower stomatal sensitivity to VPD, and higher evaporative demand relative to the control treatments (Kellomäki and Wang 2000; Wang et al. 2005). In *P. cembra* (Tyrol), stomatal conductance was not affected by eT, and while leaf area was not quantified, the observed increase in tree water loss was attributed to increases in root production and root conductivity and a decrease in the viscosity of water under eT (Wieser et al. 2015). While most hardwood species exhibited increases in conduit diameters and specific hydraulic conductivity of the xylem under eT, no changes in hydraulic traits were found in the two pines, *P. strobus* and *P. banksiana* (B4warmED experiment; McCulloh et al. 2016). This is consistent with observations that hydraulic traits of *P. edulis* (SUMO; including leaf area-to-sapwood area ratio, leaf-specific hydraulic conductance, xylem wall thickness) and, thus, canopy conductance and tree sap flow did not respond to eT (Grossiord et al. 2017b; McDowell et al. 2019). Notably, eT reduced leaf specific conductance, canopy conductance, and sap flux in *Juniperus monosperma*, the other conifer included in the study (Grossiord et al. 2017b).

3.4 Stand-Scale and Ecosystem Processes

Warming may increase C uptake and growth in cooler locations. Annual shoot growth and radial growth in cool/cold humid conditions may increase or remain unaffected by warming, while no effect on or reduction of growth is likely in semi-arid, temperate conditions.

Warming likely stimulates C and nutrient cycling and may change community composition and abundance of ectomycorrhiza. Warming may, in some instances, lead to an imbalance between C inputs and outputs (enhancement of biomass growth and litterfall is smaller than that of decomposition), a decrease in soil organic C storage, and sustained stimulation of C and nutrient cycling. The effect of warming on ectomycorrhizal species richness and community composition is likely mixed and varies among biomes.

Modeled annual canopy net photosynthesis and aboveground respiration (wood and foliage) increased under eT (Wang 1996; Wang et al. 2002; Zha et al. 2005). Annual shoot growth and ring width of *P. sylvestris* trees were unaffected by eT (Peltola et al. 2002; Kilpeläinen et al. 2006). However, the cumulative diameter growth over several years was higher under eT resulting in bigger trees with larger number and diameter of branches and more leaf area per tree (Peltola et al. 2002; Wang et al. 2005; Volanen et al. 2006). At the alpine *P. cembra* ecosystem (Tyrol), warming did not increase radial growth of trees, but N additions and competition reductions (through understory removal) both resulted in enhanced growth, indicating stronger N compared to temperature limitation to tree growth at that site (Gruber et al. 2018). In contrast, warming enhanced coarse root growth of another alpine treeline species, *P. mungo* subsp. *uncinata* (Stillberg FACE), contributing to a higher total plant biomass in warmed plots (Dawes et al. 2015). In the same study, biomass accumulation of *Larix decidua* (and the understory species) was unaffected by warming, while total fine root biomass (all species included) was reduced under eT compared to control plots. In a nighttime warming experiment (reflective curtains) on Mediterranean *P. halepensis* in Spain, shoot elongation and biomass accumulation was unaffected (Prieto et al. 2009a), consistent with no observable effect of eT on needle and shoot growth of another temperature pine, *P. edulis* (SUMO) (Adams et al. 2015; Gehring et al. 2020).

While temperature sensitivity of forest soil CO_2 efflux (F_{soil}) declined under eT, the efflux remained consistently higher around *P. sylvestris* trees (Mekrijärvi; Niinistö et al. 2004). In contrast, no thermal acclimation of F_{soil} under eT was observed in *P. mungo* subsp. *uncinata*-*L. decidua* ecosystem (Stillberg FACE), resulting in sustained simulation of F_{soil} and microbial metabolic activity and causing a shift in microbial substrate use to older soil organic matter (Streit et al. 2014b). Changes in extracellular enzyme activities also indicated accelerated soil organic matter mineralization and associated release of N and phosphorous (P) under eT potentially leading to decrease in soil organic C storage and stimulation of nutrient cycling, plant growth, as well as nutrient loss (Souza et al. 2017; Dawes et al. 2017). In the same study, fungal composition changed under eT such that ectomycorrhizal fungi (ECM) taxa specialized for high N availability proliferated (Solly et al. 2017). At B4warmED, warming had no effect on species richness of the ECM fungal community, but together with species and canopy openness (closed vs. open), eT influenced the composition of the community (Mucha et al. 2018). In contrast, ECM species richness associated with *P. edulis* (SUMO) declined with eT without affecting soil microbial C. However, soil microbial N increased under both *P. edulis* and *J. monosperma*, as did the potential nitrification rates and soil nitrate concentration under *P. edulis* (Grossiord et al. 2018; Gehring et al. 2020).

Elevated temperature (with or without changes in rainfall) is likely to result in changes in community composition that will be influenced by the initial relative species abundances, and species- and population-specific responses to changes in climate, including their ability migrate outside their current location (Way and Montgomery 2015). Trees, particularly temperate and boreal species, may grow faster at temperatures warmer than in their current location. Local adaptation to cooler conditions may not reflect optimization of growth but survival and trees from cooler climates may be outcompeted by genotypes from warmer regions that show inherently faster growth rates (Rehfeldt et al. 1999; Way and Oren 2010). Thus, the magnitude and sign of the growth response to warming will likely differ depending on trees' location relative to the climate envelope of the species (Reich et al. 2022). That is, warming may negatively impact species at or near their current warmest ecotone boundary, while positive impacts (at least initially) could be seen for those that are currently in locations cooler than their historic, warmer range margins.

4 Effects of Elevated CO_2 (eCO_2)

In a global analysis of leaf-level responses to eCO_2, stomatal conductance was shown to be a homeostatic process, with photosynthesis increasing proportionally to the elevation in CO_2 (Cernusak et al. 2019), with consequent increase of water use efficiency (Cernusak et al. 2019; Keenan et al. 2013), corroborating experimental evidence from stepwise increase in eCO_2 (De Kauwe et al. 2013; Gardner et al. 2023). However, the increased forest C sink and enhanced water use efficiencies may be waning (Adams et al. 2020; Brienen et al. 2020; Büntgen et al. 2019; Hickler et al. 2015). Below we summarize results from eCO_2 experiments on pine species.

4.1 Leaf Structure and Function

The early experiments on saplings and tree responses to elevated CO_2 have primarily been conducted in open-top chambers (Kellomäki et al. 2000). This precludes an entire ecosystem scale assessment but enables quantifying leaf and tree-level responses. Part of the earliest experiments on *P. sylvestris* have shown varied responses of stomatal conductance, net photosynthesis, and chlorophyll content in different cohorts of needles. Whereas the first (and oldest) cohort of needles had higher stomatal conductance and transpiration under eCO_2, the third-year cohort showed a reduction in these variables (Luomala et al. 2003; Wang et al. 1995). In contrast, chamber studies in *P. taeda* consistently show that older needles had lower rates of net photosynthesis than current-year needles (Murthy et al. 1996; Maier et al. 2002) and eCO_2 increased branch flush length to a greater extent than needle length or fascicle number such that flush density (branch leaf area/flush length) decreased indicating a change in branch morphology (Murthy et al. 1997; Maier et al. 2002). Net photosynthesis was enhanced proportionally to the enhancement of CO_2 taking into account varying conductance responses of the different needle cohorts (Luomala et al. 2003). These results capture short-term responses to eCO_2 (3 years) observed on a few individuals (two to four saplings).

Growing under eCO_2, needle longevity may decrease, morphologically and anatomically different from needles growing under current concentrations; needles may become shorter and have smaller vascular tissue (Lin et al. 2001; Pritchard et al. 1998), and the changed morphology may be reflected in physiological changes as well. The majority of eCO_2 studies on *Pinus* spp. (*P. unicata*, *P. halapensis*, *P. sylvestris*, and *P. taeda*) did not detect an appreciable effect of elevated CO_2 on stomatal conductance (Dawes et al. 2013; Ellsworth et al. 2012; Handa et al. 2005; Pataki et al. 1998a), and because net assimilation scaled directly with the increase in eCO_2, these results suggest that Ci/Ca (leaf internal to atmospheric CO_2 concentration ratio) was unaffected as well (Gardner et al. 2023). However, in a study of *P. halepensis*, eCO_2 brought a slight reduction in stomatal conductance with a concomitant, yet less than proportional, decrease in net assimilation, suggesting lower Ci/Ca and resulting in increased water use efficiency (Birami et al. 2020; Gardner et al. 2023).

These gas exchange responses may differ among needle age classes (Luomala et al. 2003). For example, in contrast to current-year needles of *P. taeda*, the second-year cohort showed a slightly reduced stomatal conductance under eCO_2, but carboxylation capacity was unaffected (Ellsworth et al. 2012). In another study, downregulation of photosynthesis was observed in older but not in current needles of *P. koraiensis*, yet no such interaction between eCO_2 and needle age was observed in *P. sylvestriformis* (Zhou et al. 2011).

As more information is becoming available on the response of plants to climate change, our views continue to shift. For example, a review and meta-analysis by Medlyn et al. (2001) showed a marginal response of stomatal conductance of conifer species to eCO_2, evaluated across all enrichment methods (branch bags, open-top chambers, FACE) and study durations. However, a recent analysis altered these earlier findings by including the season of measurements (Lauriks et al. 2021). In this study, no discernable effects of eCO_2 were found in early spring, but a progressive decline in stimulation of net photosynthesis, particularly in older needles, was observed over the course of the growing season (Lauriks et al. 2021).

4.2 *Tree Structure and Function*

Elevated atmospheric CO_2 enhanced photosynthesis but had no effect on growth and aboveground biomass accumulation of *P. sylvestris* trees, yet more C was allocated belowground under nutrient poor conditions (Janssens et al. 2005). A similar lack of growth response was observed in high elevation *P. uncinata* trees over 4–9 years of eCO_2 (Swiss Alps) (Dawes et al. 2013; Dawes et al. 2011; Handa et al. 2006). Although in another study on *P. sylvestris*, tree height was also unaffected over 6 years of eCO_2, the radial growth was enhanced yet cellulose content decreases, suggesting future changes in the stem mechanical properties (Kilpeläinen et al. 2005). In contrast to the muted response observed in colder climates, a more pronounced growth response was observed in a *P. taeda* stand (Duke FACE, Southeastern USA (DeLucia et al. 1999)) that can even change the allometry of trees yielding higher biomass at a given diameter (Kim et al. 2020) with implication to both physiological studies and forestry (Huber value, site index, biomass equations). eCO_2 altered the hydraulic allometry index (sapwood area divided by the product of leaf area and height) and xylem hydraulic conductivity of *P. taeda* (Ward et al. 2013). The decreased hydraulic allometry index (Ward et al. 2013) and invariant sapwood-to-leaf area ratio (Huber value) of both saplings and trees growing in eCO_2 chambers and FACE, respectively (Pataki et al. 1998b; McCarthy et al. 2007), are consistent with a greater height at a given diameter or smaller taper of individuals growing under eCO_2 (Kim et al. 2020). In the same study, enhanced resin flow was found, suggesting more C allocation to defense especially where nutrients were limiting the growth response to eCO_2, suggesting a better protection against bark beetle if the insect population is unchanged and tree vigor is not reduced due to changes in other climate change variables (Novick et al. 2012).

There are three hypothetical growth trajectories for trees under elevated atmospheric CO_2: trees growing faster and dying earlier, trees growing faster exceeding heretofore seen size limits and dying at a similar age as before, and trees growing faster, thus reaching maturity and the same maximum size earlier and dying at a similar age (Büntgen et al. 2019). The lifespan of practically all species of *Pinus* exceeds a hundred years, and most live several hundred years, one (bristlecone pine, *Pinus aristata* Engelm.) reaching several thousand years of age (Brunstein and Yamaguchi 1992; Knowles and Grant 1983; Lanner and Connor 2001). None of the eCO_2 experiments lasted long enough to evaluate these hypothetical responses, and therefore models and historical data are used to garner insights. The longest running study on the effect of eCO_2 was on *P. taeda*, showing trees growing faster with variable growth responses mostly attributed to local nutrient availability (Drake et al.

2011; Finzi et al. 2006, 2007; Oren et al. 2001) and trees that are larger at a given diameter (Kim et al. 2020). Enhanced growth under elevated CO_2 has been shown to result in early maturity (LaDeau and Clark 2001) with enhanced pollen production (LaDeau and Clark 2006) and undiminished or even enhanced seed germination rates (Hussain et al. 2001; Way et al. 2010). However, a long running high elevation study on *P. uncinata* showed no growth enhancement (Dawes et al. 2013). Given the paucity of studies of relevant length and equivocal of the two longest, there seems to be insufficient evidence to support any of the three hypothetical responses, despite suggestions that the first is the most likely (Brienen et al. 2020; Büntgen et al. 2019; Voelker et al. 2006).

4.3 Stand-Scale and Ecosystem Responses

Data on mature trees at stand scale that allow assessment of the ecosystem level responses to eCO_2 are mostly derived from FACE (Free Air CO_2 Enrichment) experiments (Hendrey et al. 1999; Hendrey et al. 1993). The FACE experiments focusing on *Pinus* species are the Duke FACE on *P. taeda* (loblolly pine (Schlesinger et al. 2006)) and the Swiss Alps FACE (Dawes et al. 2013) on *P. uncinata* (Swiss mountain pine).

In the *P. taeda* study, ecosystem physiological responses were different from those that would have been predicted based on response at the leaf scale. For example, it was shown that, although direct effect of eCO_2 on mean canopy stomatal conductance was not observed when CO_2 was changed stepwise between 1.0 and 1.8 times aCO_2, leaf area index, L, increased under eCO_2 early in the study stabilizing above that under aCO_2 and resulting in proportionally lowered conductance, attributed to both changed hydraulic properties and increased shading (Tor-ngern et al. 2015). The reduction in mean canopy stomatal conductance was on average 13%, apparent only during half of the time during the study (Ward et al. 2013). Total canopy water use was unaltered under eCO_2 because the decrease in mean canopy stomatal conductance was compensated by the enhanced L (Tor-ngern et al. 2015); this was so in patches that were fertilized with nitrogen and reached somewhat higher L (Ward et al. 2018). As a consequence, water yield from similarly responding ecosystems will likely not increase under eCO_2.

The variation of L, concomitant with aboveground productivity, was not an effect of eCO_2 only (McCarthy et al. 2007); the variation reflected interactive response with soil fertility (McCarthy et al. 2006) and precipitation (Finzi et al. 2006; McCarthy et al. 2010; Norby et al. 2005; Oren et al. 2001). The increase in L and net primary production created more litter production and higher absolute turnover of litter (Finzi et al. 2001), supporting increases in both requirement for and uptake of nutrients (Finzi et al. 2007; Lichter et al. 2008). The stimulation of nutrient uptake is considered unsustainable (Johnson 2006), supported in part by increasing evidence of greater nitrogen mining from organic matter under eCO_2 with the help of the microbial community (i.e., increasing C:N (Drake et al. 2011)). Evidence for progressive nitrogen limitation at the *P. taeda* site (Finzi et al. 2006) have been suggested only by the shift in mycorrhizal communities that reduce nitrogen uptake capabilities (Steidinger et al. 2020), yet no indication of reduced net primary production (27%; McCarthy et al. 2006) response to eCO_2 had been detected over the 15–17 years of the experiment (Palmroth et al. 2024). Averaged over the study period, eCO_2 increased NPP by ~38% in pines and ~52% in broadleaved species, with nitrogen adding a further 11% to pine NPP. For pines, increased growth was driven by both higher production efficiency (PE = NPP/L) and increased L, which accounted for two-thirds of the response in later years. Additionally, pines grown under eCO_2 showed slightly higher biomass allocation to wood and reduced mortality (Palmroth et al. 2024).

At the alpine site, no appreciable growth response or alteration of canopy structure was observed, and in general, the ecosystem was more unresponsive (Dawes et al. 2013; Dawes et al. 2011). The large increase of L (~17%) observed in the overstory of the *P. taeda* forest under eCO_2 (McCarthy et al. 2007) resulted in light limitations in the understory, reducing understory stem density and total biomass (Kim et al. 2016). The impact of higher L was similar on both shade-tolerant and shade-intolerant species in the understory, retarding the development of advance regeneration and perhaps the succession to oak-hickory forest. Naturally, forest dynamics of, e.g., fire-dependent longleaf or ponderosa pine stands, representing different types of forest succession, may not respond to climate change processes in a similar manner to loblolly pine stands.

Moving to belowground, eCO_2 did not increase coarse root/shoot ratio (average 0.24 for roots >2 mm diameter) of *P. taeda* (Duke FACE); despite net primary production of coarse root biomass under eCO_2 averaging 49% higher than under aCO_2 (1997–2010), pine coarse root increased only 32% above the observed in aCO_2, largely accumulating in the top 30 cm of soil (Maier et al. 2022), and similar amount to the difference observed aboveground (Kim et al. 2020).

Faster belowground processes showed mixed results, whereby higher fine root production was found for *P. taeda* after long CO_2 exposure (Pritchard et al. 2008), but not for *P. uncinata* (Handa et al. 2008). Likewise, in the early stages of the experiment, no appreciable change in CO_2 efflux in CO_2 fumigated plots was found in *P. taeda* stands (Allen et al. 2000), yet an increase in soil respiration was found in *P. sylvestris* under elevated CO_2 in a closed chamber experiment (Niinistö et al. 2004). A synthesis concluded that litter decomposition rate declined with litter quality under eCO_2, altering biogeochemical cycling (Norby and Cotrufo 1998; Norby et al. 2001). In addition, although the fractional allocation of C belowground decreased within and among several FACE experiments as L and thus allocation aboveground increased (Palmroth et al. 2006), the absolute amount of C allocated belowground is higher under eCO_2, ultimately stabilizing in higher CO_2 efflux from the soil surface (Jackson et al. 2009; Suwa et al. 2004). Generally, a reallocation of C to the belowground is predicted when C is in oversupply relative to nutrient availability, hence stimulating belowground processes aimed at nutrient acquisition (Prescott et al. 2020). Indeed, adding nitrogen under eCO_2 attenuated C allocation belowground (Palmroth et al. 2006), reflecting decreased fine root and mycorrhizal biomass as well as specific respiration rate (Drake et al. 2008, Jackson et al. 2009) and thus reducing CO_2 efflux from the forest floor (Oishi et al. 2014). In contrast, where nutrient limitations develop, slowing down net primary production and potentially reducing *L*, gross primary production may decrease, and even if the proportional allocation of C belowground may increase, absolute belowground allocation and C sequestration may decline (Gill et al. 2006).

Soil microbial community may change under eCO_2 in *P. densiflora*; although the functional effects are not clear (Lee et al. 2021), the changed community might have contributed to observed changes in nutrient uptake dynamics (Bassirirad 2000). Such changes, including of mycorrhiza, were not observed in alpine *P. uncinate* stands under eCO_2 (Dawes et al. 2013). One critical consideration in belowground dynamics is the emerging concept of microbiomes. In general, new evidence is emerging that soil microbiota can have a positive impact for survival and resilience of trees under changing climatic conditions through the microbiota influence on plant adaptability to different environmental conditions (Afkhami 2023; Allsup et al. 2023; Krumins et al. 2015). Elevated CO_2 experiments have yet to investigate the potential effect of microbiome composition, which might affect the response of planted trees where a suitable microbiome is unavailable (Carter 1996; Matyas 1994). In a study on *P. taeda*, the soil microbial changes under eCO_2, and nitrogen addition showed a shift in community composition and an increase in heterotrophic respiration under nutrient limitation and a decrease in respiration when nutrients are in excess, highlighting the interactions and uncertainties of respiration responses to eCO_2 or N additions (Billings and Ziegler 2008).

As CO_2 continues to rise in the atmosphere, modeling results suggest that trees and stands will likely experience earlier death in the future due to a confluence of climate change factors (Brienen et al. 2020; Brienen et al. 2012) as has already been observed (Anderegg et al. 2022). Considering frequency, severity, and extent of fires, expected to increase with decreasing precipitation, physiological and ecological consequences of eCO_2 may have less of an effect on forests as C sink than the effects of eT and changes in precipitation.

5 Soil and Atmospheric Moisture

5.1 Overview

The physical and biological properties controlling leaf, whole tree, and stand-level water relations in pine species are generally well understood (Landsberg et al. 2017). Stomata are the active interface between leaves and the atmosphere and control the diffusion of CO_2 and water vapor into and out of the leaf and directly link plant carbon uptake to water supply. Tree water use or transpiration (E_t) is a function of the VPD, canopy stomatal conductance (G_s), and the hydraulic properties of the water conducting system (wood density, cell wall thickness, tracheid size) (Sterck et al. 2008). As Et increases, leaf water potential (Ψ_l) declines, and this drop in Ψ_l sets up the driving force for water movement under tension through a hydraulic continuum that extends from the soil to leaves often called the soil-plant-atmosphere continuum (SPAC) (Kramer and Boyer 1995):

$$E_t = (\psi_s - \psi_l)^* K_p \quad (1)$$

where Ψs is soil water potential and K_p is whole-plant hydraulic conductance. Hydraulic traits associated with K_p are coordinated with G_s to minimize water loss and to prevent hydraulic failure:

$$G_s = (K_p{}^* \Delta\psi) / \text{VPD}^* A_s / A_l \quad (2)$$

where $\Delta\Psi = \Psi_s - \Psi_l$ and A_s/A_l is the sapwood-to-leaf area ratio.

Stomata typically close at high VPD and the sensitivity of this response can be describe as follows:

$$G_s = G_{sref} + m \ln \text{VPD} \quad (3)$$

where G_{sref} is reference G_s at VPD = 1 kPa and m is the slope or sensitivity (Oren et al. 1999) Reference G_s and m are a function of K_p (Meinzer 2002) and Ψ_s (Novick et al. 2016) and differ between species (Whitehead and Jarvis 1981; Tardieu and Simonneau 1998; Mediavilla and Escudero 2003a). Plant growth and carbon gain are intrinsically coordinated to the efficiency of K_p. Within the SPAC, hydraulic resistance at the soil-root and leaf interfaces account for >70% of whole-plant hydraulic resistance (Hellkvist et al. 1980; West et al. 2007; Domec et al. 2009b) serving as primary bottlenecks that regulate leaf-level physiological responses (Mencuccini 2003; Rodriguez-Dominguez and Brodribb 2020). The relative importance between the two varies with soil moisture. In general, leaf hydraulic conductance (K_{leaf}) controls the sensitivity of G_s to VPD under high soil moisture, whereas root hydraulic conductance (K_{root}) is more important when soil moisture is low (Domec et al. 2009b,2010).

Relative to active role of stomata regulating E_t, K_p is static but generally increases with a reduction in canopy leaf area (i.e., increased A_s/A_l) or decreases with a loss of conducting area with xylem cavitation (Franks and Brodribb 2005). Tree A_s/A_l and resistance to embolism are traits important for long-term acclimation to changes in climate and resource availability (i.e., water and nutrients) (Whitehead et al. 1984; Whitehead 1998; McCulloh and Sperry 2005). Xylem cavitation and loss of conductivity begins to occur when water loss from Et exceeds water uptake from the soil. A fundamental limit to hydraulic failure is the stem Ψ that causes irreversible xylem cavitation (Tyree and Sperry 1988). Failure of the hydraulic system is a primary mechanism involved in tree mortality in many ecosystems (Choat et al. 2018), and *Pinus* species may be more vulnerable to cavitation than other species (Martínez-Vilalta et al. 2004). Cavitation likely occurs first in roots as the hydraulic resistance at the soil-/rhizosphere-root interface is a major limitation to K_p (West et al. 2007; Domec et al. 2009b).

In addition to tree growth and survival, the amount and frequency of rainfall can strongly affect above-/belowground C allocation, soil organic matter decomposition, and net ecosystem productivity (NEP). Pine forests can accumulate more soil organic carbon (organic layer and mineral soil) and have lower carbon turnover than deciduous forests (Diers et al. 2021; Jandl et al. 2021; Devi 2021). Soil CO_2 efflux (F_{soil}) is a major carbon flux in forests and accounts for greater than 65% of ecosystem respiration in pine forest (Law et al. 1999; Xu et al. 2001, Maier et al. 2004). Uncertainty regarding the effects of climate change on F_{soil} results from the many biotic (stand age, net primary productivity (NPP)) and abiotic factors (temperature and moisture) that influence the component processes root (R_a) and heterotrophic (R_h) respiration (Subke et al. 2006). Several studies have found that R_h may be more sensitive and more responsive to short-term changes in soil moisture than NPP and lead to somewhat counterintuitive result that a moderate experimental drought increased NEP while addition of water through irrigation decreased NEP (Maier et al. 2004; Ruehr et al. 2012; Noormets et al. 2021).

5.1.1 Experimental Approach

Manipulative experiments that examine effects of precipitation, alone or in combination with other limiting resources, have been conducted globally at various scales from individual seedlings, trees, and stands. Stand-level precipitation manipulation experiments aimed to reproduce drivers of ecosystem processes at spatial and temporal scales that occur under natural conditions. These experiments generally either supplement ambient precipitation through irrigation (IR) or reduce precipitation through rainfall removal at some percentage of throughfall, often referred to as throughfall removal (TR). In the latter case, some type of plastic sheeting positioned under the canopy collects and removes some proportion (30–100%) of throughfall during a rainfall event (Asbjornsen et al. 2018; Knapp et al. 2017). The power of these experiments is they allow the testing of changing soil water availability on ecosystem level processes (e.g., carbon, water, and nutrients fluxes, structure (e.g., leaf area index (LAI), biomass accumulation and allocation, and soil carbon storage) at the spatial scale of forest stands while keeping other covarying climate variables unaltered. Throughfall removal experiments can simulate extreme droughts and thus can identify species and processes that are sensitive to changes in precipitation patterns predicted from climate change and can define the climatic (water, temperature, and VPD) thresholds that induce tree mortality (McDowell et al. 2013). Stand-scale IR experiments in pine species are mostly aimed at understanding the limits on productivity or for disentangling the concurrent effects of water and nutrients on tree growth. The experiments are often done in combination with nutrient manipulations (e.g., Albaugh et al. 1998; Will et al. 2015). The scale and length of the experimental treatments varies widely across studies ranging from short-term (several months) on single trees to multiyear studies of forest stands. All studies looked at tree growth response, while some go further and examine ecosystem carbon and water fluxes reporting on one or more of the following attributes: NPP, gross primary productivity (GPP), net ecosystem productivity (NEP), canopy E_t and G_s and subcomponents thereof, understory vegetation, microbial biomass, activity, and diversity.

5.2 Leaf Structure and Function

Pines have evergreen needles and maintain E_t and net photosynthesis (A_n) throughout the year under favorable environmental conditions. Leaf gas exchange plays a central role in the ability of trees to respond to climate change. Plant control of g_s and Ψ_l can be described by two broad strategies: isohydry and anisohydry (Tardieu and Simonneau 1998). As VPD rises, trees typically close their stomata to control E_t (Oren et al. 1999). Isohydric species regulate g_s to maintain a stable midday Ψ_l and minimize the risk of hydraulic failure at the cost decreased A_n, whereas anisohydric species have less stomatal control allowing variable midday Ψ_l and higher E_t and A_n with increased risk of hydraulic failure (Martínez-Vilalta et al. 2014). The iso-/anisohydric classification is somewhat artificial as plants generally operate on a continuum between the two behaviors with considerable inter- and intraspecific variation depending on site conditions (xeric vs. mesic) and soil water availability (Domec and Johnson 2012; Klein 2014; Meinzer et al. 2016). Numerous studies that manipulate water availability and across a range of climates (xeric to mesic) generally show that *Pinus* species exhibit isohydric behavior (Whitehead et al. 1983; Gonzalez-Benecke et al. 2010); however, the degree of "isohydry" can vary among similar species (Héroult et al. 2013). For example, in a common garden experiment during a drought, *P. halepensis*, a Mediterranean species with anisohydric behavior, had higher g_s and A_n and growth than mesic species *P. nigra* and *P. sylvestris* with more conservative isohydric water use (Salazar-Tortosa et al. 2018).

Long-term TR and IR studies indicate that leaf physiological and morphological responses differ across species depending on their stomatal response to increasing soil and atmospheric aridity. Short-term TR experiments (weeks to months) generally found rapid, reversible shifts in leaf physiology, decreasing Ψ_l, leaf stomatal conductance (g_s), A_n, and increasing instantaneous leaf water use efficiency (iWUE), with all processes recovering to some extent following rewatering. Bengston (1980) conducted one of the earliest TR studies in *P. sylvestris* where leaf gas exchange was measured in a 20-year-old tree following 100% TR. After 4 months, TR reduced predawn Ψ_l (Ψ_{lp}) by 45% and lowered g_s and A_n. After rewatering, Ψ_l and g_s recovered to control levels within 1 day, but A_n took 5 days to recover, indicating that non-stomatal limitations of A_n are important during drought (Teskey et al. 1986). In a plot-level study, Tang et al. (2004) observed a 7, 14, and 15% reduction in Ψ_{lp}, g_s, and A_n, respectively, and a 30% reduction in whole-crown E_t following 100% TR for 7 months in an 18-year-old *P. taeda* stand.

Long-term TR and IR experiments (>3 years) have found varying plastic physiological and morphological responses related to species, habitat, and extent of the treatment. Reduced water availability can delay phenological development and leaf and branch morphological traits shifting to a drought tolerant strategy. For example, Limousin et al. (2013) compared leaf gas exchange and water relations in semiarid conifers *P. edulis* (isohydric) and *Juniper monosperma* (anisohydric) following 3 years of TR (−45%) and IR (+30%). Precipitation treatments changed the relationship between gas exchange and water stress and degree of acclimation in both species. Maximum g_s and A_n declined with Ψ_{lp}, and the slope of the relationships was steeper for pine than juniper. In pine, A_n reached zero at Ψ_{lp} of −2.55 MPa, whereas Ψ_{lp} at zero A_n in juniper ranged from −4.95 to −6.90 MPa, and both species showed acclimation to TR treatment by lowering Ψ_{lp} where $A_n = 0$. Furthermore, TR delayed pine needle emergence by 39 days and reduced shoot and needle growth, and both traits were correlated with seasonal Ψ_{lp}, a response not observed in *J. monosperma* (Adams et al. 2015).

Effects of TR are more varied in mesic pines generally affecting canopy leaf area, *L*, more than leaf-level physiology (Samuelson et al. 2014,2018; Maggard et al. 2016). In a 10-year-old *P. taeda* stand growing in a droughty sandy soil, IR to maintain 60% maximum available soil moisture increased fascicle number and flush length, which increased branch foliage biomass (46%) and *L* (42%) (Murthy et al. 1997). Irrigation treatment had no effect on g_s, but light-saturated A_n in 1-year-old and current-year foliage was higher in IR when available soil moisture in control plots fell below 50% (Murthy et al. 1996). Similarly, in a 7-year-old *P. taeda* stand following a −30% TR, Ψ_{lp}, g_s, and A_n decreased by 15, 20, and 13%, respectively, and g_s was correlated with Ψ_{lp}, but this response was only observed during a dry year when annual precipitation was 20% below normal (Samuelson et al. 2014,2018). A −40% TR in a young *P. palustris* plantation reduced whole-tree E_t, g_s, and A_n but had no effect on shoot and needle phenology, canopy development, or litter production (Mendonca et al. 2022).

5.3 *Tree Structure and Function*

Numerous studies have found a strong linear relationship between stem growth, NPP, E_t, and *L*, a proxy of absorbed radiation, in pine (Landsberg and Sands 2011; Tor-ngern et al. 2017). Climate, soil resource availability, moisture, and nutrients can strongly affect the slope of this relationship by altering stand structure (L, A_s/A_l), physiology, and above- and belowground C allocation.

5.3.1 Throughfall (TR) Experiments

Pines species in semiarid systems or populations along the edges of their range defined by low moisture availability may be particularly vulnerable to drought stress predicted from climate change (Breshears et al. 1997). For example, following an extended drought (2000–2003), a widespread regional die-off of pinon pine (*P. edulis*) occurred to a much greater extent than in juniper (*Juniperus monosperma*), codominant species in pinon-juniper woodlands that cover much of the Southwestern North American (Breshears et al. 2005). The mechanisms behind tree mortality and the difference between species were studied in stand-scale TR experiments in Arizona (Pangle et al. 2012; Plaut et al. 2012; Grossiord et al. 2017a). These species display contrasting stomatal behavior where juniper typically exhibits anisohydric stomatal behavior, has higher cavitation resistance, and was expected to maintain positive C gain and survival to a greater extent than isohydric pinon. After 3 years of - 45% TR, 67% of the pine died compared to zero mortality in the control and juniper in either treatment. Droughted pine experienced extended periods of high loss of xylem function where percent loss of conductivity (PLC) was >70% compared to an average 43% PLC in control and <40% PLC for Juniper in either treatment. Droughted pine had a 7-month period of near zero *E*t and had high incidence of infestation of bark beetles (*Ips confusus*) and bluestain fungus (*Ophiostoma* spp.).

Pinus palustris is one of the most drought-resistant pine species in the Southeastern USA (Samuelson et al. 2019). A 3-year throughfall study in a 13-year-old plantation found that a −40% TR had no impact on foliage or branch phenology or *L* (Mendonca et al. 2022). During an exceptionally dry growing season, E_t in droughted trees ceased for 2 months, but this did not result in increased mortality. However, aboveground NPP was reduced 34%. Tolerance to drought was achieved through stomatal control rather than in an adjustment of the hydraulic architecture as there was little treatment effect, A_l/A_s, *L*, and no leaf osmotic adjustment. Samuelson et al. (2019) concluded that reduced precipitation in the range predicted for the region under climate change will have little impact on phenology, canopy development, and litter production in established plantations. Moreno et al. (2021) measured tree water use in a Mediterranean mixed pine (*P. halepensis*)-oak (*Quercus ilex*) forest following 9 years of −30% TR. Both species experienced decreases in Ψ_l and E_t (i.e., sap flow) in the TR treatment; however, decreased growth was a function of reduced leaf area rather than xylem cavitation as estimated PLC <25% in both species. These studies indicate that in mesic, isohydric pines, stomatal limitations and extended periods of low K_p interact and are more important for reducing growth or inducing mortality than hydraulic failure (Sala et al. 2010; McDowell 2011).

The physiological mechanisms inducing mortality are much debated (Adams et al. 2017) but are generally group into three causes: (1) catastrophic hydraulic failure through partial or complete loss of xylem function from cavitation (embolized xylem) (Hoffmann et al. 2011) usually assessed as PLC; (2) carbon starvation (McDowell et al. 2008) where C demand for growth, defensive compounds, and osmotic regulation exceeds available supply; and (3) accelerated biotic attack in drought predisposed trees (Kolb et al. 2019). All these mechanisms probably play an important role and are highly interdependent spatially and temporally (Menezes-Silva et al. 2019). Hydraulic failure may be more important in anisohydric plants, whereas C starvation may play an important role in isohydric species such as pine (McDowell et al. 2011).

Adams et al. (2017) synthesize data from several experimental studies that examined drought induced tree mortality and found that hydraulic failure was the predominant cause of tree morality. In all the examined species, a 60% PLC was a threshold that caused significant mortality consistent with supply-demand hydraulic theory (Sperry and Love 2015; McDowell et al. 2016). They point out that reductions in NSC in combination with hydraulic failure played a role in mortality in some studies (48%), primarily in gymnosperms, indicating that hydraulic function and carbon metabolism are linked. Reinhardt et al. (2015) found in drought *P. flexilis* seedlings that mortality was directly linked to hydraulic conductivity, while NSC were observed to increase in needles and decreased in roots indicating impaired phloem transport. On the other hand, repeated nonlethal drought episodes, i.e., events that do not induce hydraulic failure, can deplete carbon reserves possibly leading to increased mortality (Galiano et al. 2011). The effects of repeated drought or compound climate stress events (high VPD) on tree mortality have been less studied. Repeated episodes of defoliation of canopy trees caused by compound climate events (drought, VPD, and temperature) can deplete tree carbon reserves eventually leading to reduced growth and mortality (Gazol and Camarero 2022). This suggests that the seasonality, frequency, and magnitude of droughts may be more important in influencing ecosystem processes and NPP than the absolute reduction in precipitation (Knapp et al. 2017; Felton et al. 2021).

5.3.2 Irrigation (IR) Experiments

Irrigation experiments generally show increased G_s, E_t, and tree growth and a decrease in the water use efficiency of production (WUEp = biomass increment/E_t); however, the response differs with species and treatment intensity and can interact with availability of other resources (e.g., nutrients, CO_2). A long-term experiment in the Pfyn forest in Switzerland studied the effects of 16 years of IR in 100-year-old *P. sylvestris* (Bose et al. 2022), a species that has experienced recent regional level drought related mortality (Rigling et al. 2013). The site has a Mediterranean climate with long summer droughts. Irrigation treatments were designed to double ambient rainfall. Aboveground responses occurred much sooner than belowground. Tree ring $\delta^{13}C$ decreased after 1 year of IR indicating a rapid physiological response in foliar gas exchange (Eilmann et al. 2010); however, changes in aboveground growth traits varied temporally. IR increased needle length (70%), shoot length (100%), ring width (120%), *L* (20%), and basal area (~30%), increases that were significant after 2, 4, 3, 9, and 9 years, respectively (Dobbertin et al. 2010; Bose et al. 2022). Compared to aboveground response, fine root biomass was only marginally increased with IR (Brunner et al. 2009), and effects were observed later, after 9 years, indicating a shift in carbon allocation decreasing root to shoot ratio (Herzog et al. 2014). The IR treatment also decreased tree mortality. The magnitude of the response to IR declined with time for some growth traits, but all remained positive throughout the experiment. When irrigation was stopped, physiological processes rapidly returned to control levels, whereas morphological traits, needle and shoot length, crown density, and stem growth took up to 4 years to fade (Zweifel et al. 2020).

5.3.3 TR, IR, and Nutrient Experiments

Many pine species are commercially important for wood supply and are often managed by improving water and nutrient availability. Canopy and root hydraulic traits are sensitive to water and nutrient supply; thus, silvicultural practices that affect resource supply (e.g., fertilization) may interact with climate change to affect water and carbon cycling in pine ecosystems. Stand-level responses to experimental manipulation of water and nutrients have been extensively studied in *P. sylvestris* (Persson 1980), *P. radiata* (Benson et al. 1992), *P. taeda* (Albaugh et al. 2004; Samuelson et al. 2004, Coyle et al. 2008), and *P. pinaster* (Trichet et al. 2008). The limitation of growth attributed to water and nutrient availability varied greatly among species, stand age, soil type, and climate.

In a long-term water and nutrient study in *P. sylvestris* in central Sweden, IR had a small effect on NPP (10%), but fertilization (FR) more than doubled NPP and stem biomass, and the combined treatment effects were additive (Axelsson and Axelsson 1986; Linder 1987). Similarly, several stand-scale factorial IR by FR studies in *P. taeda* in the Southern USA across a range of soil types show that IR had small and variable effects on tree growth, much less than observed with FR (Samuelson et al. 2001,2004; Coyle et al. 2008). For example, at the Southeast Tree Research and Education Site (SETRES), effects of IR and annual FR were studied in *P. taeda* growing on an excessively drained, nutrient poor sandy soil (Albaugh et al. 1998). After nine years (stand ages, 8–16), IR and FR treatments increased stem biomass 25 and 100%, stem production 23 and 119%, and stand growth efficiency (GE = stem mass increment/*L*) 15 and 32%, respectively, and the combined effects were additive (Albaugh et al. 1998; Albaugh et al. 2004). Several studies found that IR and/or FR had only small effects on above- and belowground biomass allocation and minor effects on leaf physiology, and although IR increased C allocation to fine roots (King et al. 1999; Coleman 2007; Samuelson et al. 2004; King et al. 2002), the increases were small relative to large increases in *L*. Similar results were reported for nutrient limited stands of *P. pinaster* where the response of *L*, growth, and biomass allocation was much larger in FR than IR (Trichet et al. 2008; Bakker et al. 2009).

The long-term IR and FR (>15 years) treatment effects on *P. radiata* productivity were studied at the Biology of Forest Growth Experiment in Australia (BFG, Benson et al. 1992). In contrast to the studies in *P. taeda* and *P. pinaster*, this study found that water was the primary limiting resource for aboveground growth, and there was a strong IR by FR interaction (Snowdon and Benson 1992) after 5 years of treatment. Fertilization alone did not affect NPP, whereas IR increased NPP by 60% and IR + FR increased NPP by 200–300% compared to control plots (Raison and Myers 1992). The treatment responses at the BFG site were sustained through 19 years of treatment although IR effects were attenuated presumably due to increased nutrient limitation as this response was not observed in IR + FR treatments (Waterworth et al. 2007). Waterworth et al. (2007) further showed that IR had a much larger effect on growth of large trees and increased stand diameter distribution. Climate may contribute to the differences in response to water and nutrients between these studies. The sites in Sweden and Southern USA receive ample rainfall (>1000 mm year), and growing season water deficits (potential minus actual evapotranspiration) are small (100–300 mm). In contrast, Australia receives less than 700 mm $year^{-1}$ and has a large growing season water deficit (>400 mm $year^{-1}$).

Higher temperatures and increased VPD associated with climate may be particularly stressful for intensively managed pine plantations. Large fertilizer-induced increases in leaf area relative to A_s and fine roots (Albaugh et al. 1998; Samuelson et al. 2004, 2008) will decrease A_s/A_l and root mass to leaf area ratio (R_m/A_l), which could compromise a tree's ability to tolerate decreased soil water availability during droughts (Ewers et al. 2000). In the SETRES study, Ewers et al. (2000) found the FR stands had roughly half the R_m/A_l and 50% lower G_s and K_{leaf} than control or IR stands. Root xylem in FR trees were 50% more cavitation resistant than non-fertilized trees; however, this did not increase the safety margin for hydraulic failure, indicating that large increase in leaf area may increase the sensitivity of fertilized trees to drought.

In an early TR by FR study in an 18-year-old *P. taeda* stand, the TR treatment (100% for 7 months) negated large increases in canopy photosynthesis (41%) and foliage biomass (26%) observed in FR plots (Tang et al. 2004) suggesting that even in regions with high rainfall, adequate precipitation is necessary to achieve increased productivity with fertilization (Lim et al. 2015). A range-wide TR by FR experiment aimed to define how young *P. taeda* plantations respond to predicted shifts in precipitation in Southeastern USA (PINEMAP, Will et al. 2015). Four sites were studied in Florida (FL), Georgia (GA), Virginia (VA), and Oklahoma (OK) that captured the regional variation in climate and productivity. Treatments were a factorial combination of a -30% TR versus ambient and FR from one-time application of optimum nutrients. The response to TR was site-specific and limited compared to the generally positive response to FR. At the GA, VA, and OK sites, TR and FR treatments reduced g_s, G_s, and E_t, but had no effect on A_n (Ward 2015; Bartkowiak et al. 2015; Maggard et al. 2016; Samuelson et al. 2018). Lower g_s in TR was associated with lower Ψ_{lp}, whereas in FR, decreased g_s was due to a reduced hydraulic allometry index ($A_h = A_s/A_l * 1/h$, where h is tree height) (Maggard et al. 2016; Samuelson et al. 2018). At the GA site, TR reduced G_s, E_t, and K_p and the sensitivity of G_{sref} to soil moisture but only in the FR treatment (Bartkowiak et al. 2015). Ward (2015) observed at the VA site that reduction in G_s and E_t in TR and FR treatments was of the same magnitude such that the combined treatment resulted in a 30% reduction in E_t compared to the control despite no treatment effect on stand L. Both studies occurred during a year with abundant rainfall; thus, TR effects may be muted;

however, fertilizer-induced changes in A_s/A_l, R_m/A_l (Ewers et al. 2000) or A_h (Samuelson et al. 2018) may make these stands more vulnerable during years of extreme drought. All sites responded to FR increasing average *L* (8–20%) and NPP (6–20%). Response of NPP to TR was small and site-specific ranging from no response at VA and FL to a −7 and −12% reduction at the OK and GA sites, respectively (Bracho et al. 2018). When both TR and FR effects were significant, they were additive such that NPP was similar between TR + FR treatment and control treatments (Maggard et al. 2016; Shephard et al. 2021). Results from the PineMap study suggest that predicted shifts in precipitation in the Southeastern USA (Konrad et al. 2013) will not significantly affect *P. taeda* growth; furthermore, fertilization management can sustain or improve NPP and WUEp under moderate drought (Bracho et al. 2018).

5.3.4 Vapor Pressure Deficit

Rising air temperature without concomitant rise in specific humidity will result in an exponential increase in atmospheric vapor pressure deficit (VPD) and increased evapotranspiration (Will et al. 2013). Few studies have looked directly at the effects of manipulated VPD on the physiology and production of pines as it is difficult to separate VPD effects from temperature and soil moisture under field conditions. In an elegant experiment, Whitehead et al. (1983) demonstrated isohydric behavior in seven *P. radiata* clones grown in a growth chamber at low (0.2 kPa), medium (1.0 kPa), and high (2.0 kPa) daytime VPD. While there was considerable temporal and clone variation, compared to the low VPD treatment, mean g_s was 42% and 55–71% lower for seedlings grown at 1.0 and 2.0 kPa, respectively. However, there was no VPD treatment effect on Ψ_l. Furthermore, when seedlings were transferred between treatments, rapid changes in g_s maintained similar Ψ_l among treatments. This experiment demonstrated that VPD had a strong control over leaf gas exchange on a daily timescale. In a growth chamber study, 3-year-old saplings of *Fagus sylvatica*, *Q. pubescens*, and *Q. ilex* representing a range of mesic to xeric species were exposed to a factorial combination of air temperature and VPD under unlimiting soil moisture conditions (Schönbeck et al. 2022). Increased VPD caused a significant decrease in stem hydraulic conductivity and stomatal conductance in mesic species (*Fagus sylvatica* and *Q. pubescens*), but not in xeric *Q. ilex*. In a small-scale study using open-top chambers, a 15% reduction in relative humidity over one growing reduced biomass production in *F. sylvatica* saplings by 30%, decreased leaf number, SLA, but had no effect on root/shoot ratio (Lendzion and Leuschner 2008). These studies generally show that changes in VPD can alter plant physiology and structure independent of soil moisture. Experimental effects of VPD at the stand scale are rare as it is difficult to manipulate VPD at larger scales (Breshears et al. 2013), however a stand-scale experiment manipulating VPD and TR has been initiated in the long-term irrigation experiment in *P. sylvestris* in the Pfyn forest (Bose et al. 2022; https://www.wsl.ch/en/projects/vpdrought/).

Novick et al. (2016) demonstrated that at short timescales (hourly to daily), effects of VPD and soil moisture (or low Ψ_s) on G_s are largely independent, and thus the primary driver of G_s could be separately studied. They showed that VPD was more important driver of G_s than soil moisture in mesic ecosystems and that declining soil moisture had small effects on the sensitivity of G_s to VPD (m in Eq. 3) in contrast to large changes in m with decreasing soil moisture in dry ecosystems. This is particularly important for isohydric pine species as earlier and more rapid stomatal closure with higher VPD could potentially offset benefits of elevated CO_2 or fertilization (Maggard et al. 2017) in managed plantations. For example, in a *P. palustris* stand, G_s was limited primarily by VPD at moderate to dry soil moisture and was severely limited only at soil moisture near permanent wilting point (Mendonca et al. 2022). In another analysis, McDowell and Allen (2015) used a hydraulic corollary to Darcy's (Eq. 2) to show that increased temperature with climate change will drive increased VPD decreasing G_s and canopy photosynthesis and increasing water stress. As a result, trees growing in semiarid environments that possess isohydric stomatal regulation, low K_p, and high leaf area are likely to experience increased mortality.

Changes in VPD at longer monthly or seasonal timescales could cause adjustments in hydraulic architecture. Delucia et al. (2000) using published values from the literature across several pine species showed that A_l/A_s declined with growing season maximum VPD and that within the same species, trees growing in semiarid environments had lower A_l/A_s than trees growing in mesic environments, suggesting that pine species may respond to increasing soil and atmospheric aridity by adjusting crown architecture to maintain crown water supply without risk to xylem cavitation. However, experimental evidence that high VPD causes changes in A_l/A_s is lacking. It is clear from these small-scale experiments and stand-level observations that increased air temperature and VPD can alter tree and stand structure and physiology independent of soil moisture. Large-scale long-term experiments that independently manipulate meteorological and atmospheric drought are needed to disentangle effect of soil moisture, air temperature, and VPD on forest function (Wright and Collins 2023).

5.4 *Stand-Scale and Ecosystem Processes*

Forest vegetation and soil respond differently to changes in soil water availability, and shifts in the frequency and magnitude of precipitation can potentially cause a fundamental shift in ecosystem C fluxes, C distribution in the soil, and storage. This is because soil moisture may have a larger and more immediate effect on soil microbial processes and heterotrophic respiration (R_h) than NPP. Most of the reviewed studies indicate that long-term IR generally increases fine root production and turnover, micro- and macrofauna biomass, F_{soil}, and R_h (Sohlenius and Wasilewska 1984; Maier et al. 2004; Li et al. 2005; Samuelson et al. 2009; Bose et al. 2022) and decomposition with a potential redistribution of soil organic carbon deeper in the soil profile (Guidi et al. 2022), while TR suppresses these processes (Bakker et al. 2009; Zhang et al. 2016). Furthermore, water and nutrients may differentially affect these processes. For example, in *P. taeda*, 7 years of IR increased fine root and microbial biomass, while FR increased microbial biomass but decreased fine root biomass (Samuelson et al. 2009).

Net ecosystem productivity (NEP) in a 12-year-old *P. taeda* stand (SETRES) subject to 4 years of IR and FR was tightly linked to increased NPP due primarily to a large increase in *L* from fertilization (Maier et al. 2004). However, there were significant and divergent effects of IR on GPP and F_{soil}, where compared to control treatments, IR increased GPP by 22% but increased S_r by 30% and ecosystem respiration ($R_e = F_{soil}$ plus aboveground autotrophic respiration) by 25% such that there was no significant IR effect on NEP (=GPP- R_e). To understand precipitation effects on NEP more fully, F_{soil} must be partitioned into its component processes R_h and root or autotrophic respiration (R_a) as these processes respond differently to resource supply (Hanson et al. 2000). In young *P. taeda* stands, R_h can account for >70% of F_{soil} (Bracho et al. 2018) but likely decreases with stand age as R_a from increased root biomass becomes a larger proportion of F_{soil} (McElligott et al. 2016). Noormets et al. (2021) showed that NPP and R_h in *P. taeda* (PineMap) had divergent responses to TR and FR where a −30% TR had no effect on NPP but decreased R_h by 13% leading to a 26% increase in NEP (NEP = NPP-R_h). Furthermore, there was a strong TR × FR interaction, where FR increased NPP but decreased R_h, such that NEP increased threefold under normal precipitation and by 48% under drought. Ruehr et al. (2012) observed in a seasonally dry young *P. ponderosa* stand that IR had a stronger positive effect on microbial respiration and soil decomposition than NPP resulting in a decrease in NEP. In contrast, studies in broadleaved evergreen forests observed that TR decreases F_{soil} primarily through decreased R_a more than R_h (Hinko-Najera et al. 2015; Huang et al. 2018). These conflicting results highlight the importance of quantifying asymmetric responses of root respiration and R_h to drought when predicting NEP response to climate change.

Changes in soil moisture regimes can alter forest structure and interspecific competition leading to long-term changes in community composition. Variation in response to IR and FR is a function of soil type, site quality, and stages of stand development.

Irrigation can increase understory biomass and alter understory composition (Axelsson and Axelsson 1986; Herzog et al. 2014). In the Pfyn forest study, 16 years of IR had no effect on *P. sylvestris* regeneration, the dominant species, while it increased understory hardwood regeneration (*Quercus pubescens*, *Betula pendula*, and *Sorbus aria*) and decreased moss and herb coverage (Herzog et al. 2014; Bose et al. 2022). In contrast, TR studies show reduced ground cover and understory vegetation (Ozolinčius et al. 2009).

6 Interaction Effects of Warming, Elevated Atmospheric CO_2, and Variation in Soil and Atmospheric Moisture

6.1 eT and e[CO_2]

P. sylvestris grown under eT and elevated CO_2 (eCO_2) in Mekrijarvi or in a greenhouse of the Climate Change Experiment (CLIMEX, Norway) showed no consistent (across years and foliage cohorts) reduction of leaf N or downregulation biochemical processes/parameters (Wang et al. 1995; Luomala et al. 2003; Beerling 1999). The only study on temperature responses of the biochemical processes of photosynthesis indicates slight shifts in the temperature optima of V_{cmax} and J_{cmax} in *P. sylvestris* (Wang et al. 1995). Earlier studies from Mekrijärvi reported a decline in stomatal conductance under eCO_2 (Kellomäki and Wang 1996; Kellomäki and Wang 2000). However, a later study from the same site (Luomala et al. 2003) and from two other experiments found no change in leaf g_s under eCO_2 or the combination of eT and eCO_2 in *P. sylvestris* (CLIMEX, Beerling 1999) and *P. mungo* subsp. *uncinata* (Sillberg FACE) (Streit et al. 2014a). While leaf R_{dark} in *P. sylvestris* decreased under eCO_2, it increased in trees grown under eCO_2+ eT, but less than under warming alone (Zha et al. 2002). Most studies report increases in leaf A_n under eCO_2 and eT (Beerling 1999; Teskey 1997; Streit et al. 2014a). At Mekrijärvi, *P. sylvestris* showed a greater CO_2 enhancement of A_n under eT, presumably due to inhibition of photorespiration under eCO_2 (Luomala et al. 2003). The treatment combination also increased total monoterpene emissions from *P. sylvestris* foliage (Räisänen et al. 2008b), whereas SLA and starch and soluble sugar concentrations showed no consistent changes compared to trees grown under control conditions (Zha et al. 2001; Luomala et al. 2003).

The CO_2 effect on E_T of *P. sylvestris* in Mekrijärvi varied from year to year, depending on the magnitude of the decline in canopy conductance relative to the enhancement in tree leaf area. The combination of eT and eCO_2 consistently increased E_T (similarly to warming alone), and there was no interaction effect between eCO_2 and warming, suggesting that temperature dominated the driving of tree-level transpiration rates (Kellomäki and Wang 2000; Wang et al. 2005). Modeled annual canopy net photosynthesis increased, and aboveground respiration declined under eCO_2, while both increased under eCO_2+ eT (Wang 1996; Wang et al. 2002; Zha et al. 2005), as did forest soil CO_2 efflux which increased additively under both eCO_2 and warming (Niinistö et al. 2004). Annual shoot growth of *P. sylvestris* was not affected by any of the treatments in Mekrijärvi, while ring width increased under eCO_2 only and by similar amount under eT and eCO_2 (Peltola et al. 2002; Kilpeläinen et al. 2006). At CLIMEX, while N mineralization was higher and understory biomass increased under eCO_2 and eT, height and radial growth of *P. sylvestris* were not enhanced by the treatments (van Breemen et al. 1998; Rasmussen et al. 2002). Similarly, eCO_2 did not enhance biomass accumulation rate of either *P. mungo* subsp. *uncinata* or

L. decidua at Stillberg FACE; however, the combined warming and eCO_2 resulted in higher plant biomass in *P. mungo* subsp. *uncinata* plots (Dawes et al. 2015).

6.2 eT and Variation in Soil and Atmospheric Moisture

At B4warmED, warming or rainfall reductions, or their combination, had no effect on V_{cmax} and J_{cmax} estimates for *P. strobus* and *P. banksiana*, but there was a decline in the ratio of the two parameters across treatments (Stefanski et al. 2020; Bermudez et al. 2021). Across all the species, A_n declined with soil moisture (under TR), under both temperature treatments, due to increasing stomatal limitation. However, species differed in their sensitivity to soil drying, with *P. strobus* and *P. banksiana* showing greater sensitivity of A_n to reduced rainfall. Moreover, compared to control conditions, *P. strobus*, showing lower A_n under eT, was among the three species (all conifers) with the greatest reductions in A_n and g_s under combined eT and TR (Reich et al. 2022). This contrasts with the findings from the SUMO experiment, where large reductions in A_n and g_s were found in *P. edulis* grown under TR (Grossiord et al. 2017a). Indeed, leaf-specific hydraulic conductance, canopy conductance, and E_T of *P. edulis* were not reduced further when TR was combined with eT (Grossiord et al. 2017b). Moreover, TR delayed needle emergence, decreased shoot elongation, and induced shorter needles with a lower SLA, with no greater effect when trees were simultaneously subjected to eT (Adams et al. 2015; Grossiord et al. 2017a).

Among species, differences in gas exchange at the boreal B4warmED were reflected in the growth and survival, such that *P. strobus* along with *Picea glauca* and *Abies balsamea* exhibited reduced growth and survival and became less abundant in the simulated warmer or drier climate (Reich et al. 2022). Overall, the results from the boreal B4warmED suggest that species currently abundant in southern boreal forests had the largest reductions in growth and survival under warming combined with rainfall reductions. Concerningly, temperate species responding positively to warming and experiencing little mortality are rare in these ecosystems and thus unlikely to rapidly increase in density and expand their distribution (Reich et al. 2022). At SUMO, however, despite the severe heat and drought treatments, there was little indication that trees approached thresholds for carbon starvation or hydraulic failure, and no tree of either *P. edulis* or *J. monosperma* died over the experimental period (McDowell et al. 2019). Based on water isotope ratio measurements and ecosystem water flux modeling, trees persisted under such severe hot drought by accessing semipermanent water source in the bedrock. Moreover, based on insect trapping, bark beetles that typically attack and kill drought-stressed trees were not found at the site (Grossiord et al. 2017c; McDowell et al. 2019). At the Biosphere 2 mortality experiment, *P. edulis* trees died earlier in the combined warming and drought treatments compared to those experiencing drought only, presumably because of changes in carbon metabolism indicated by an earlier onset of starch accumulation concurrent with increasing hydraulic stress, together leading to tree death (Adams et al. 2013). Finally, while shoot elongation of *P. halepensis* in Spain was unaffected by eT and drought treatment (Prieto et al. 2009a), their combination increased mortality and accumulation of dead biomass and, thus, fire risk in this shrubland. Moreover, drought decreases the competitive ability of *P. halepensis*, which is obligate seeder, in favor of resprouting shrubs (Prieto et al. 2009b).

6.3 *eCO_2 and Variation in Soil and Atmospheric Moisture and Nutrient Supply*

When drought occurred under eCO_2 experiments, stomatal conductance decreased but eCO_2 increased water use efficiency (Gardner et al. 2023; Tor-ngern et al. 2015). Likewise, under high soil moisture conditions, eCO_2 caused a reduction in stomatal conductance, but leaf hydraulic conductance was acclimated (Domec et al. 2009a). Under eCO_2 with nutrients or water provided, an enhancement in photosynthesis ensues with concomitant enhancement in growth (Crous et al. 2008). Increased availability of both CO_2 and N decreased leaf longevity in *Pinus* (Mediavilla and Escudero 2003b), consistent with the leaf economic spectrum (Khan et al. 2022; Reich et al. 1995; Wright et al. 2004).

Overall, water and nutrient availabilities may limit photosynthetic responses (Franklin 2007; Gojon et al. 2023), growth (Oren et al. 2001), and ecosystem responses to eCO_2 (Kim et al. 2016; McCarthy et al. 2007). The stimulation of photosynthesis by eCO_2 is not maintained if water and nutrients, particularly nitrogen, become limited (Penuelas et al. 2020). However, if eCO_2-induced enhancement of CO_2 uptake results in greater biomass production, tissue element concentration may decline, leading to a negative physiological feedback (e.g., dilution of N and Mg may reduce the amount and activity of the photosynthetic machinery (Gojon et al. 2023)) as has been demonstrated in *Pinus* as well (Crous et al. 2008; Kellomäki and Wang 1997; Pritchard et al. 1998; Rogers and Ellsworth 2002; Rogers and Humphries 2000; Tian et al. 2019). This may be more apparent as reallocation of elements from second-year needle of some species to meet the requirements of newer ones results in photosynthetic downregulation (Ellsworth et al. 2012; Luomala et al. 2003; Wang et al. 1995; Zhou et al. 2011) with potential effect on net primary production. Indeed, in a model simulation under limiting N availability, reduced N turnover from the needles to litter and soil prevented eCO_2-induced growth enhancement (Johnson 2006), consistent with the law of the minimum (Ebelhar et al. 2008).

The clearest interactive response of mean canopy stomatal conductance (G_S) of *P. taeda* to eCO_2 and moisture was observed at the Duke FACE experiment, when a period of controlled enrichment to different concentration coincided with a flash-drought at the site, a period in which both soil limitation to water uptake and atmospheric demand for moisture progressively increased (Fig. 1). The long-term indirect effect of eCO_2 on G_S was attributed to both changes in plant hydraulics and increased mutual shading with *L*. However, little additional effect of step changes of eCO_2

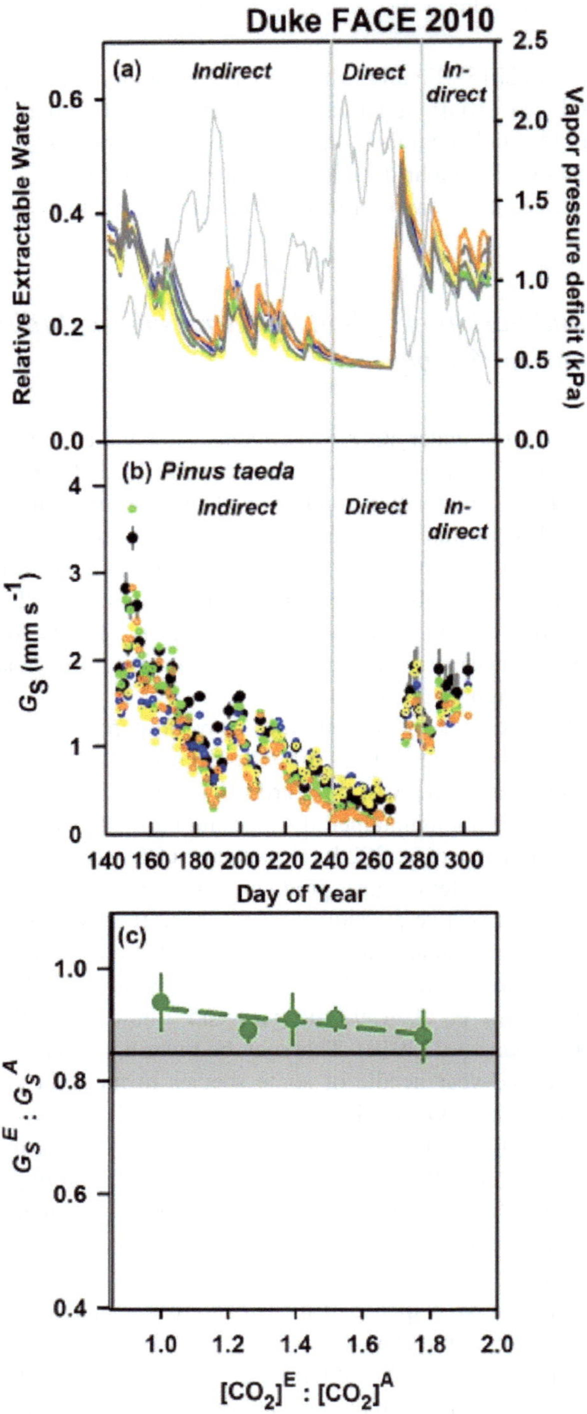

Fig. 1 The effect of declining soil moisture and increasing vapor pressure deficit (*D*) on mean canopy stomatal conductance (G_S) of *Pinus taeda* was greater than the effect of increasing eCO_2. (**a**) Relative extractable water (REW) and *D* over the final year of CO_2 enrichment broken into three periods; the first and last periods reflect the *Indirect* (acclimation) effect on G_S of eCO_2 ~1.5 X ambient concentrations

between ambient and 1.8 X ambient was observed, while the effect of flash-drought is readily observed in a drastic reduction of G_S. In the following section we produce hypothetical responses of several pine forests to changes in atmospheric forcings inspired by the simulations of climate change (IPCC AR6). Among them, a future precipitation scenario (IPCC 2021) calls for increased rainfall by 5% for our temperate pine stands and by 15% for our boreal stand, commensurate with 2–4 °K global warming above preindustrial temperatures. However, given the large uncertainty in predicting future precipitation, including the sign of the expected change and the observed greater sensitivity of G_S to increasing aridity compared to increasing atmospheric CO_2, we include scenarios where precipitation is decreased by similar fraction to the predicted increase.

7 Hypothetical Response of Pine Stands to Climate Change: Three Case Studies

7.1 Overview

Mechanistic model simulations can be used to complement experimental manipulations of pine forests. Some advantages of model simulations are that they can be logistically simpler, less costly, and much faster than field experiments. As a result, models can be used to investigate many potential scenarios, including ones not implemented in the field. Models also allow precise tuning of various forcings (e.g., temperature can be raised by exactly 2 °C throughout a simulation). Also, all the processes represented by the model can potentially be analyzed without any added costs. Finally, model simulations can be carried out over longer timescales than experiments. Of course, models are only simplifications and approximations, and so simulations necessarily vary in their fidelity to reality. Thus, it is typically important to consider model simulations in light of observations and experiments.

Fig. 1 (continued) caused by both the increased mutual shading with leaf area index (*L*, +17%) and reduced leaf-specific hydraulic conductivity, while a 3-week period was used to evaluate the *Direct* effect by changing the concentration in each plot to between ambient and ~1.8 X ambient for 5-day periods in which each eCO_2 plot was subjected to a different concentrations selected from one of three or one of four preset levels. (**b**) G_S declined to ~25% of the maximum with decreasing REW and increasing *D* in all plots before a recovery following drought-breaking rains in the middle of the *Direct* period (black circles and bars are mean and S.E. of $n = 4$ ambient plots; other colors represent one of the four eCO_2 plots). (**c**) The *Indirect* effect of long-term application of eCO_2 on G_S is shown as the ratio of G_S under elevated (G_S^E) relative to ambient (G_S^A) conditions depicted by a horizontal line with 1 S.E. gray band, amounted to ~15% reduction, reflecting 17% in *L*. The *Direct* effect, shown over a range of eCO_2 from 1.0 to 1.8 times ambient, was not significant ($P = 0.12$) and did not differ from the long-term *Indirect* effect. The similar relative increase in *L* and decrease in G_S under eCO_2 resulted in unchanging stand transpiration. The responses were the same regardless of nitrogen addition. A combination of three figures from Tor-ngern et al. (2015)

Here, we used the Ecosystem Demography 2 (ED2) model to carry out simulations of pine forests. Our objectives were to determine how changes in CO_2, temperature, and precipitation affect model results. The changes in the atmospheric forcings were inspired by the simulations of climate change reported in the IPCC AR6. Our analysis of model outputs focused on aboveground biomass (AGB) accumulation and water yield. In addition to these climate change simulations, we carried out comparisons between the model and experimental manipulations. In most cases, these comparisons were qualitative because we lacked the detailed information on initial conditions and environmental forcing needed to simulate the particular circumstances of each experiment.

7.2 *Modeling Approach*

7.2.1 Model Description

We used the ED2 model to carry out simulations (Medvigy et al. 2009; Longo et al. 2019a). ED2 falls within a class of models called "vegetation demographic models" (Fisher et al. 2018), which means that it explicitly simulates demographic variation. Thus, the model resolves competition between plant functional types, variation in tree growth rates within a stand, and variation of tree growth rates across stands of different densities. ED2 has been used to simulate temperate forests of the USA (Medvigy et al. 2009; Medvigy et al. 2010), boreal forests (Trugman et al. 2016), and tropical forests (Xu et al. 2016; Medvigy et al. 2019; Longo et al. 2019b; Schwartz et al. 2022). The model conceptualizes vegetation in terms of "cohorts." A cohort is a collection of plants of the same size and species located in the same resource environment. Cohorts compete for light, water, and nutrients, and the model computes rates of resource acquisition. After acquiring resources, cohorts both increase in size and reproduce. Size growth is characterized by increases in DBH, height, woody biomass, and maximum leaf biomass according to allometric relationships (Longo et al. 2019a).

As a demographic model, ED2 requires an initial condition that corresponds to the current vegetation state. The minimum initial condition consists of DBH and species assignment for all the trees in the stand. Additionally, meteorological data must be provided to drive the model over the time frame of the simulation. These data include temperature, precipitation, CO_2 concentration, specific (or relative) humidity, pressure, wind speed, downward shortwave radiation, and downward longwave radiation. These meteorological drivers are typically provided at the 30- or 60-min timescale. The minimum required soil information includes sand and clay content and %C.

7.2.2 Simulation Design

We selected three sites for our simulations. This site selection was inspired by manipulation experiments. First, Duke Forest was selected because of the FACE experiment that had been carried out there. As mentioned above, Duke Forest is dominated by loblolly pine in the overstory. An in situ weather station recorded the meteorological data needed to drive ED2. Detailed census data were also collected during the experiment and were available for our model initialization and validation. The second site that we selected was in Oklahoma, the site of a throughfall exclusion experiment mentioned above. This site is also dominated by loblolly pine, but the Oklahoma climate, at the dry end of the loblolly pine range, is much drier than the Duke Forest climate. Because in situ weather data contemporaneous with the experiment were not available, we drove our simulations with weather data from the ECMWF ERA-40 Reanalysis (Copernicus Climate Change Service 2022; Muñoz Sabater 2019). Forest composition was approximated using available information on average tree density and DBH. Finally, our third site, Mekrijärvi in Finland, was selected because it is dominated by Scots pine and was the site of a warming experiment. As we did for the Oklahoma simulations, weather data for Mekrijärvi was taken from reanalysis, and initial forest composition was approximated from the reported average tree density and diameter.

We designed simulations to identify the effects of climate change on these pine forests. Our focus is on qualitative rather than quantitative results for two reasons. First, even “current climate” simulations at Oklahoma and Mekrijärvi are not fully constrained because of the assumptions made with respect to the meteorological forcing and the initial condition.

Second, simulations of future climate vary depending on assumptions about future emissions, the particular climate model used, and stochastic variation within a particular climate model. Here, we developed a small number of “future climate” scenarios to probe the qualitative effects of climate change on these forests. Specifically, we sought to understand the impacts of changes in three drivers, CO_2 concentration, temperature, and precipitation. We evaluated the impacts when these three drivers were changed one-at-a-time and together.

For CO_2, we compared simulations forced with concentrations of about 365 ppm (“current CO_2”) to simulations forced with about 565 ppm (“future CO_2”). The concentration of 365 ppm corresponds to actual CO_2 concentrations in about the year 2000. The difference between the “future” and “current” scenarios is 200 ppm, corresponding to the CO_2 enrichment of the Duke FACE experiment. The concentration of 565 ppm also corresponds to a doubling of preindustrial CO_2 concentrations. Simulations from climate models disagree as to when (or if) CO_2 concentrations will reach 565 ppm, but it is plausible that this threshold will occur by 2100 under intermediate emissions scenarios (IPCC 2021).

We develop future temperature and precipitation scenarios using summary data presented in IPCC (2021). For future temperature, we assumed a climate sensitivity of 2.7, which would correspond to a global mean temperature increase of 2.7 °K above preindustrial temperatures accompanying a doubling in CO_2. Models disagree on the exact value of the climate sensitivity, but our value is well within the range of reported estimates (Knutti et al. 2017). By the year 2000, when we carried out our "current climate" simulations, global temperatures had already increased by about 1 °K over preindustrial temperatures, so we anticipate another 1.7 °K in global warming. This warming signal will not be evenly distributed over the globe. For Duke Forest and Oklahoma, we expect warming of about 2 °K, while for Mekrijärvi, we expect warming of about 4 °K, mainly because it is much closer to the pole where global warming is amplified (IPCC 2021). Thus, when we simulate Duke Forest and Oklahoma, we simply add 2 °K to the temperature time series from ca. 2000.

When we simulate Mekrijärvi, we add 4 °K. This simulated warming was also within the range of what was achieved at the Mekrijärvi warming experiment (2–6 °K) (Luomala et al. 2003).

Based on our past work with the ED2 model (e.g., Medvigy et al. 2010), we expected that our simulations would respond strongly to variation in relative humidity. However, how relative humidity will actually change in the future is still a matter of debate. Over the past few decades, relative humidity has remained approximately constant over the ocean (Trenberth 2011). But over land, relative humidity has declined somewhat, although specific humidity has increased (Trenberth 2011). We therefore considered two end-member cases, one in which current-temperature and future-temperature simulations had the same relative humidity and one in which they had the same specific humidity. Holding specific humidity constant resulted in a lower relative humidity in the future-temperature simulations than the current-temperature simulations between the higher temperatures in the future-temperature simulations.

We developed future precipitation scenarios using data from IPCC (2021) and from throughfall exclusion experiments (Will et al. 2015). The IPCC (2021) precipitation scenario corresponded to the average precipitation changes expected with 2–4 °K global warming above preindustrial temperatures. Specifically, we increased rainfall by 5% for Duke Forest and Oklahoma and by 15% for Mekrijärvi. Motivated by the throughfall exclusion experiments and the uncertainty in the IPCC precipitation projections, we also carried out precipitation reduction simulations in which we reduced precipitation by 15% from the control for all sites.

In summary, we qualitatively assessed the effects of climate change by using ten simulations at each site. One control simulation representing "current climate" based on meteorology from ca. 2000, one in which we increased CO_2 only, two in which we increased temperature only (corresponding to constant relative humidity and constant specific humidity), two in which we changed precipitation only, and four in which we simultaneously change CO_2, temperature (constant relative humidity and constant specific humidity) and precipitation. All simulations were run for 80 years, corresponding approximately to the maximum pine plantation rotation time and the timeframe of many IPCC projections.

Beyond developing "current" and "future" meteorological forcings, we also had to develop plant functional types (PFTs) for loblolly pine and for Scots pine. The parameterization for loblolly pine was taken from trait and allometric observations at Duke Forest (McCarthy et al. 2010; Kim et al. 2020). For the Scots pine PFT, leaf traits were taken from Xiao et al. (2006), the diameter-height relationship was taken from Sharma and Breidenbach (2015), biomass allometric equations were taken from Jagodziński et al. (2018), and wood traits were taken from Auty et al. (2014).

We also had to initialize our simulations. At Duke Forest, we used the detailed census information to develop two initializations. For model validation, we used the actual census information. But for our climate change simulations, we selected the plot that was most pine-dominated (96% of the basal area) in 1996, and we initialized the model with those pines.

The non-pine trees (4% of the actual basal area) were simply omitted from the simulation. Because detailed census information was not available for Oklahoma or Mekrijärvi, we only carried out climate change simulations. For Oklahoma, we initialized the model with a stand density of 1600 stems/ha with DBH of 3.5 cm. For Mekrijärvi, the initial stand density was 2500 stems/ha and DBH of 5.5 cm.

7.3 *Model Validation*

In the case of Duke Forest only, we were able to compare model simulations to observations because we had in situ meteorological data and detailed measurements of tree growth over time. In our validation simulations, the model simulates the increase in AGB over time in both the control plots and the elevated CO_2 plots (Fig. 2). The overall magnitude of the CO_2 effect is simulated reasonably well.

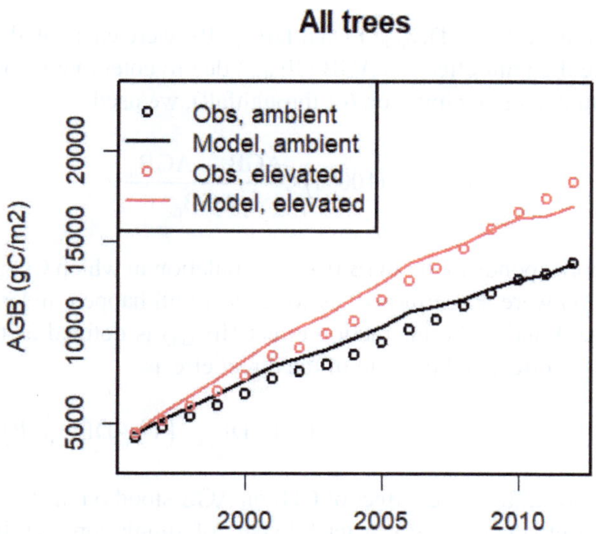

Fig. 2 Model validation at Duke Forest. For most of the time series, the model does a reasonable job of simulating AGB under both ambient and elevated CO_2 concentrations

However, it is also clear that the model overestimates biomass in the middle of the time series.

7.4 *Hypothetical Responses to Climate Change*

7.4.1 Aboveground Biomass

We examined how simulated aboveground biomass (AGB) varied in response to changes in CO_2, temperature, and precipitation. A control simulation, forced with ca. 2000 climate, was first carried out. The AGB resulting from this simulation is denoted *AGB*0. We then carried out simulations to determine three other kinds of effects: a direct effect, a joint effect, and an interaction effect. To calculate a direct effect, we carried out simulations in which just one climate variable was subject to climate change and the other climate variables were held at their ca. 2000 values. These simulations roughly correspond to manipulation experiments where only one climate variable was modified. We denote the AGB from the simulation where CO_2 was modified (but temperature and precipitation held at ca. 2000 values) as $AGB_1(CO_2)$. Similarly, the AGB from temperature-modification and precipitation-modification simulations were denoted $AGB_1(T)$ and $AGB_1(P)$, respectively. The direct effect of CO_2 increase on AGB (DEAGB(CO_2)) was calculated as a percentage difference:

$$DE_{AGB}(CO_2) = (100\%) * \frac{AGB_1(CO_2) - AGB_0}{AGB_0} \tag{4}$$

The other direct effects ($DE_{AGB}(T)$ and $DE_{AGB}(P)$) were calculated similarly.

To calculate the joint effect on AGB (JE_{AGB}) due to cotemporaneous changes in CO_2, temperature, and precipitation (or throughfall), we used

$$JE_{AGB} = (100\%) * \frac{AGB_3 - AGB_0}{AGB_0} \tag{5}$$

where AGB_3 corresponds to the AGB from a simulation in which CO_2, temperature, and precipitation were all perturbed, as we expect will happen under actual future climate change. Finally, the interaction effect (IE_{AGB}) is defined as the difference between the joint effect and the sum of the direct effects:

$$IE_{AGB} = JE_{AGB} - DE_{AGB}(CO_2) - DE_{AGB}(T) - DE_{AGB}(P) \tag{6}$$

For Duke Forest, the direct effect of CO_2 on AGB stood out as being larger than the other direct effects (Fig. 3a). After 80 years of simulation, the direct effect of CO_2 enhancement exceeded +50%. The direct effects of temperature increases and

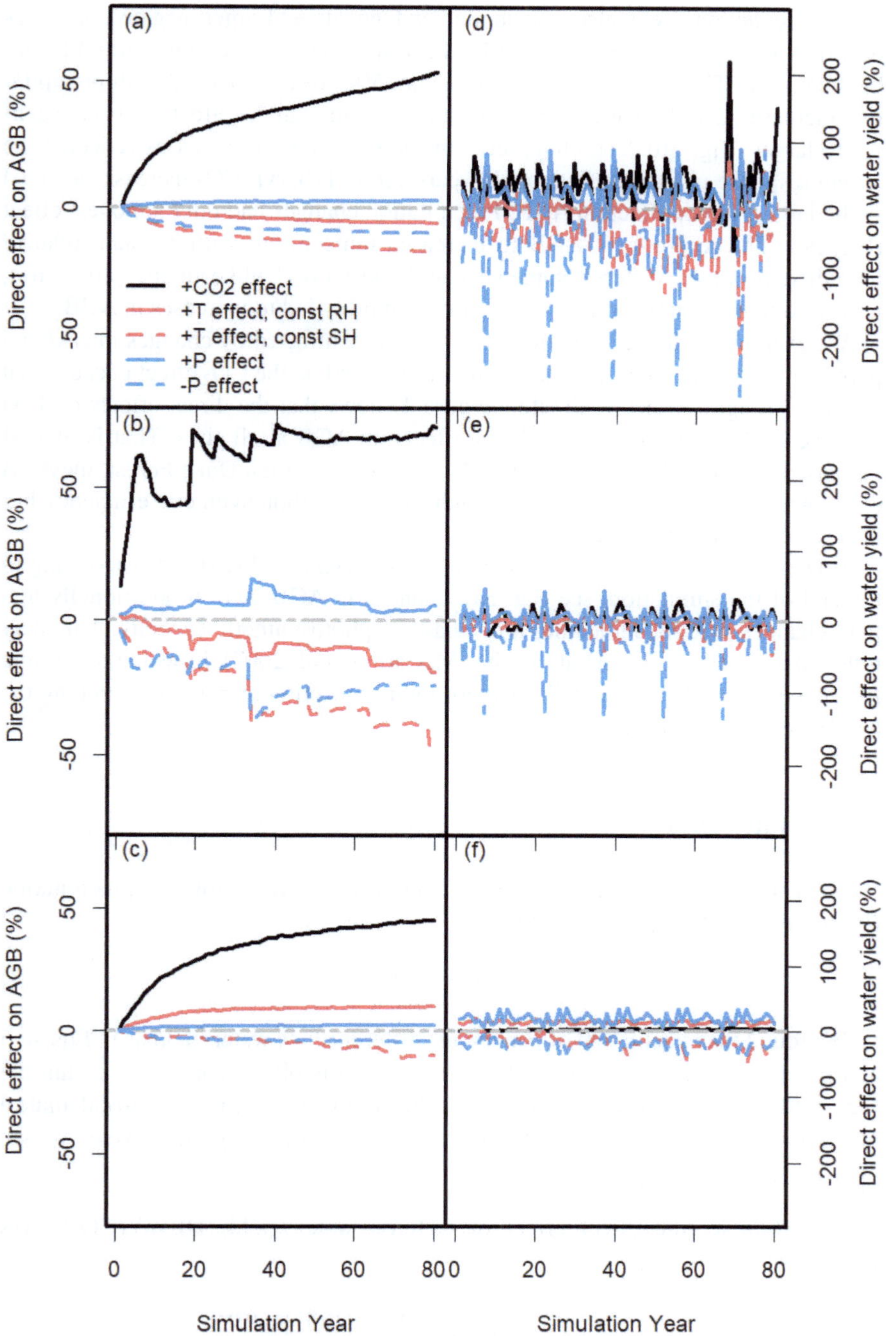

Fig. 3 Direct effects of CO_2, temperature (T), and precipitation (P) changes on aboveground biomass (AGB) (**a–c**) and water yield (**d–f**). The temperature effect was computed both assuming constant relative humidity (RH) and assuming constant specific humidity (SH). The "+P" effect corresponds to precipitation increases, and the "−P" effect corresponds to throughfall exclusion. Panels (**a**) and (**d**) show Duke Forest, panels (**b**) and (**e**) show Oklahoma, and panels (**c**) and (**f**) show Mekrijärvi. The dashed gray line is included for reference to indicate a zero difference

rainfall reduction were also substantial and negative. Larger negative changes occurred when specific humidity was held constant than when relative humidity was held constant. The smallest changes, less than 5%, were associated with precipitation increases. Similar patterns held in Oklahoma, though the effect sizes were generally larger (Fig. 3b). The choppiness in the Oklahoma time series reflected the strong response of this site to drought years. At Mekrijärvi, CO_2 increases also led to the largest direct effect on AGB. Precipitation changes had only a modest effect (Fig. 3c). The response of AGB to increased temperature with constant relative humidity, however, differed somewhat at Mekrijärvi. Unlike at the other sites, increased temperature and constant relative humidity led to increases in AGB.

We examined whether the effects on AGB were affected by changes in LAI and photosynthetic efficiency. For simplicity, we focused on the CO_2 effect because that was the strongest effect at all sites. Figure 4 shows that the direct effects on LAI were generally followed by the direct effects on AGB at all sites. That is, if LAI increased, so did AGB. At Oklahoma, Mekrijärvi, and often Duke Forest, the AGB effect was larger than the LAI effect, indicating that photosynthetic efficiency has also increased under elevated CO_2.

However, at Duke Forest for large changes in AGB and LAI (corresponding to the end of the simulation; see Fig. 3a), changes in AGB were proportionally less than changes in LAI, indicating a decrease in photosynthetic capacity. This may reflect the already high LAI at the site (McCarthy et al. 2007), in which increasing LAI beyond a certain value will increase light absorption marginally, reducing the average efficiency of the leaves.

7.4.2 Water Yield

We also considered changes in water yield (WY). WY was defined as precipitation (P) minus evapotranspiration (ET):

$$WY = P - ET \tag{7}$$

We found that calculation of percentage differences in water yield as in Eqs. 4, 5 and 6 was not helpful because the denominator was often near zero, causing the (percentage) direct effects to span many orders of magnitude. In order to calculate a nondimensional direct effect on water yield, we normalized by the 80-year average of WY0 (i.e., the ca. 2000 control) ($\overset{?}{WY_0}$).

For example, the direct effect of CO_2 on water yield (DEWY(CO_2)) was defined as:

$$DE_{WY}\left(CO_2\right) = \left(100\%\right) * \frac{WY_1\left(CO_2\right) - WY_0}{\overline{WY_0}} \tag{8}$$

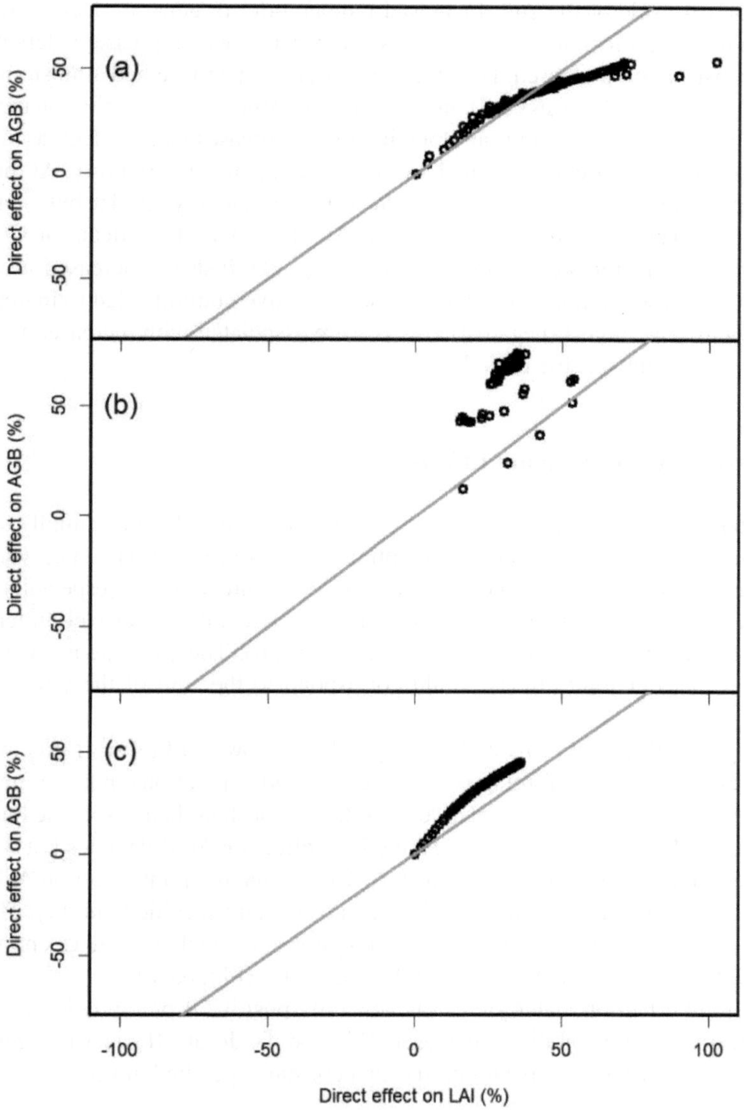

Fig. 4 Comparison of the direct effects of CO_2 on LAI and AGB at the three sites. Panel (**a**) shows Duke Forest, panel (**b**) shows Oklahoma, and panel (**c**) shows Mekrijärvi. Each panel has 80 points, corresponding to the 80 simulation years. The smaller effect sizes tend to correspond to the early years of the simulation and the larger effect sizes tend to correspond to the later years; see Fig. 2. The gray line is the 1:1 line and is included for reference

At Duke Forest, the CO_2 direct effect was a small increase in water yield on average, corresponding to reduced evapotranspiration with eCO_2 (Fig. 3d; Eq. 7). Considering the uncertainty in both model results and scaled observations at the site, this can be considered in agreement with estimates of no eCO_2 effect on the hydrological cycle at the site. Increased temperature, in general, decreased water yield when specific humidity was constant because vapor pressure deficits and evapotranspiration increased. However, increased temperature had almost no effect when relative humidity was constant. As expected from Eq. 7, precipitation increases caused increased water yield, and precipitation decreases caused decreased water yield. All of these effects exhibited substantial interannual variability. At the drier Oklahoma site, the effects on water yield were qualitatively similar but smaller in magnitude than at Duke Forest (Fig. 3e). At Mekrijärvi, the effects of CO_2 and throughfall exclusion were both very small (Fig. 3f). Instead, increased precipitation and increased temperature (with constant relative humidity) led to increases in water yield. The largest (negative) effects were associated with increased temperature with constant specific humidity.

7.4.3 Interaction and Joint Effects

The interaction effects are shown in Fig. 5. These are mostly substantially smaller than the direct effects. The major exception is the drier Oklahoma site, where the AGB interaction effect becomes comparable to the direct effects, especially in the drier years (Fig. 5b). Duke Forest also had 1 year when the water yield interaction effect was anomalously large in magnitude (Fig. 5d). Thus, for the most part, we expected that our joint effects would correspond to the sum of the direct effects (Eq. 3).

The joint effects on AGB and water yield are shown in Fig. 6. If precipitation increases, the dominant forcing is that of eCO_2, and simulations indicate increased AGB. AGB increases more if relative humidity is constant than if specific humidity is constant. If precipitation decreases, the joint effect on AGB declines but was still positive at Duke Forest and Mekrijärvi. At Oklahoma, the joint effect on AGB was often negative under throughfall exclusion and constant specific humidity. This site was also strongly impacted by periodic drought years, which imparted a choppiness on its time series (Fig. 6b) not seen at the other sites (Fig. 6a, c).

The joint effect on water yield was generally positive if precipitation increased and negative if precipitation decreased (Fig. 6d–f). Joint effects were generally larger under constant relative humidity than constant specific humidity.

7.5 *Synthesis*

Simulations replicating experimental conditions were reasonably successful in simulating the observed AGB accumulation at Duke Forest during the FACE experiment (Fig. 12). At the other two sites, we did not make strict comparisons between the simulations and the experiments because of the limited amount of data on stand initial conditions and in situ meteorology. However, a couple of general tendencies are noteworthy. In the Oklahoma throughfall exclusion

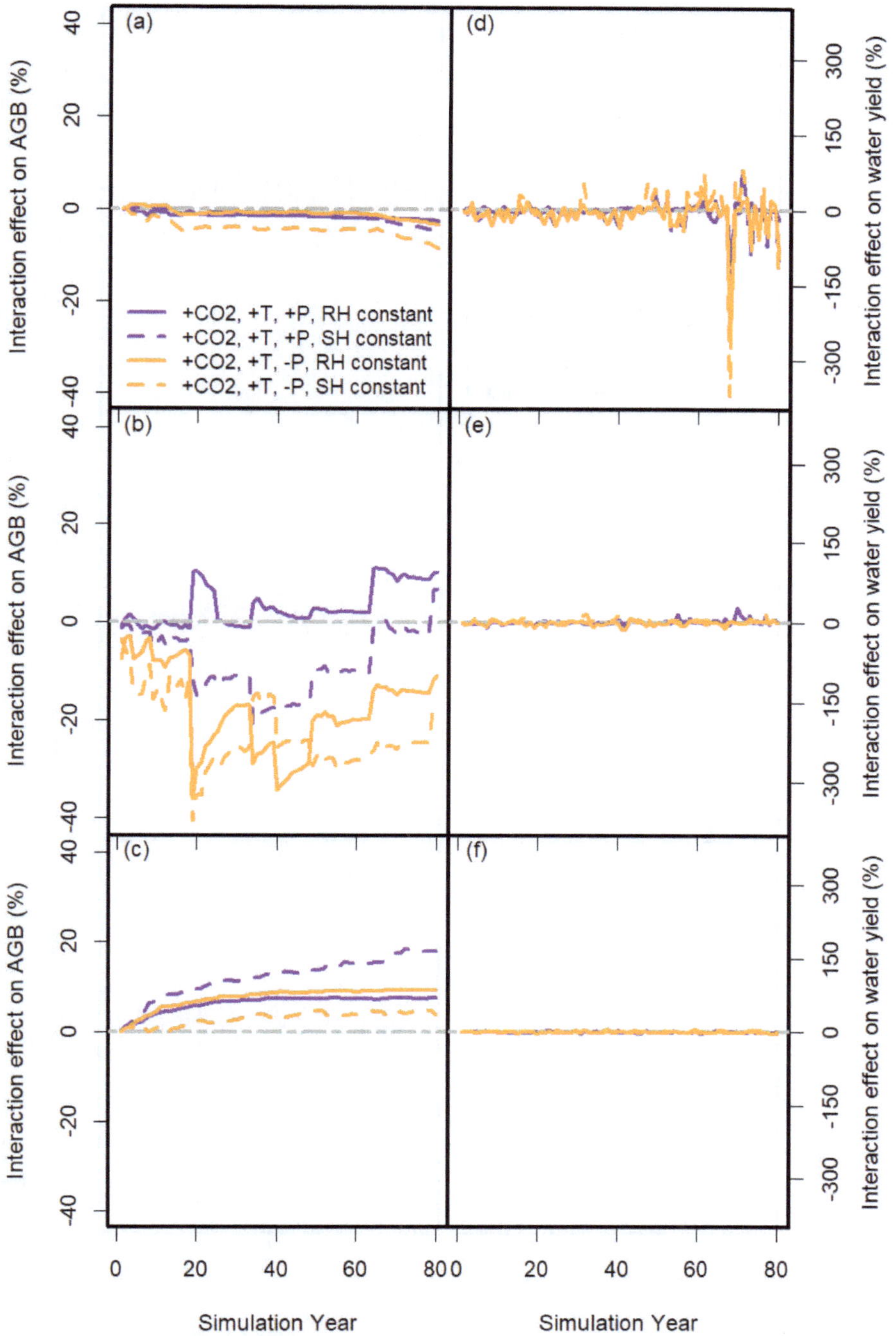

Fig. 5 Interaction effects of CO_2, temperature (T), and precipitation (P) changes on aboveground biomass (AGB) (**a–c**) and water yield (**d–f**). The temperature effect was computed both assuming constant relative humidity (RH) and assuming constant specific humidity (SH). The "+P" effect corresponds to precipitation increases, and the "−P" effect corresponds to throughfall exclusion. Panels (**a**) and (**d**) show Duke Forest, panels (**b**) and (**e**) show Oklahoma, and panels (**c**) and (**f**) show Mekrijärvi. The dashed gray line is included for reference to indicate a zero difference

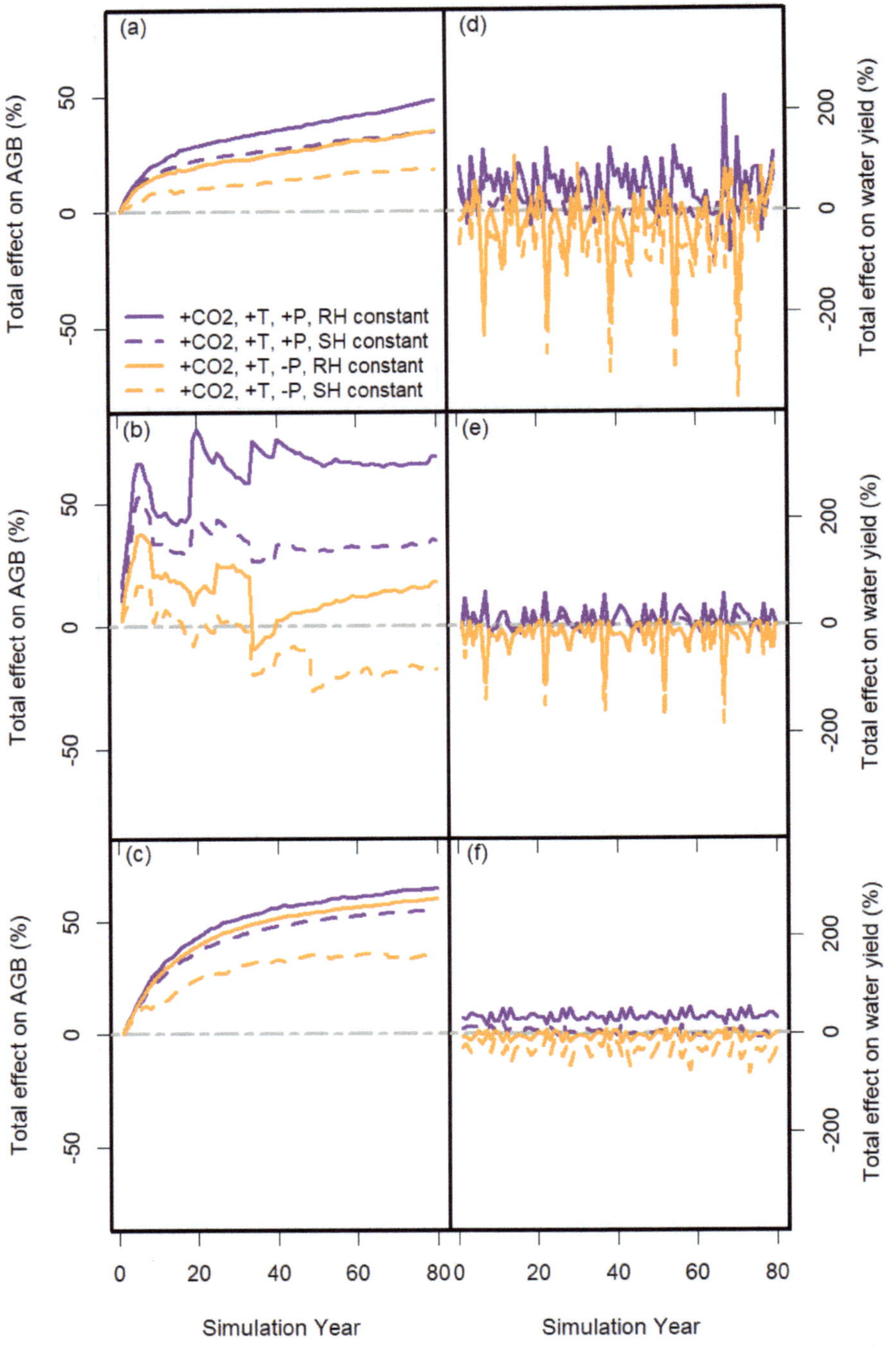

Fig. 6 Joint effects of CO_2, temperature (T), and precipitation (P) changes on aboveground biomass (AGB) (**a–c**) and water yield (**d–f**). The temperature effect was computed both assuming constant relative humidity (RH) and assuming constant specific humidity (SH). The "+P" effect corresponds to precipitation increases, and the "−P" effect corresponds to throughfall exclusion. Panels (**a**) and (**d**) show Duke Forest, panels (**b**) and (**e**) show Oklahoma, and panels (**c**) and (**f**) show Mekrijärvi. The dashed gray line is included for reference to indicate a zero difference

experiment, throughfall exclusion resulted in significantly decreased volume growth during the first year of the experiment. Similarly, simulations showed that throughfall exclusion resulted in reduced AGB accumulation (Fig. 3b). In the Mekrijärvi warming experiment, increased temperature led to decreased net photosynthesis; in our simulations, the sign of the effect of increased temperature on AGB depended on whether relative humidity or specific humidity was held constant (Fig. 3c). Since the actual experiment did not strictly hold either measure of humidity constant (though relative humidity likely decreased somewhat), we cannot evaluate the sign of the model's response. Again, strict model-data comparison would require the in situ meteorological data. The modeling results also underscore the need for greater understanding of how (or whether) humidity will change as temperatures increase. This analysis also applies to the case of water yield at Mekrijärvi: the experiment showed no significant effect of temperature increases on transpiration, while the sign of the simulated effect on water yield depended on whether relative humidity or specific humidity was held constant (Fig. 3f). Resolving these uncertainties should be a priority for future model-data comparisons (Medlyn et al. 2015).

For AGB on the 80-year time horizon, the model simulated a strong positive AGB response to increased CO_2 at all three sites. Because the other direct effects and the interaction effects were generally smaller, the joint effect of changes in CO_2, temperature, and precipitation tended to be dominated by the direct CO_2 effect (Fig. 6a–c). The direct CO_2 effect can be attributed to both changes in LAI and photosynthetic efficiency at all sites (Fig. 4).

Understanding the model assumptions that lead to either or both of these effects is a worthwhile area for future studies (Medlyn et al. 2015). Another commonality across sites was that AGB tended to be larger when temperature increased with constant relative humidity than with constant specific humidity. But there were also a few notable differences between the sites. The drier Oklahoma site responded in a more erratic way to climate change (Fig. 3b), and the colder Mekrijärvi site had a positive response to increased temperature with constant relative humidity (Fig. 3c). Water yield generally responded positively to CO_2 and precipitation, but negatively to temperature. The sign of the joint effect on water yield was controlled by whether precipitation increased or decreased. Thus, our modeling work suggests substantial similarities across sites in terms of water yield responses.

Acknowledgments We wish to thank Pantana Tor-ngern for her help in compiling Figure 1 from her published work (Tor-ngern et al. 2015). Financial support for RO and SP was provided by the ACCC Flagship, funded by the Academy of Finland (grant number 337549) through the University of Helsinki. Muñoz Sabater, J. (2019) was downloaded from the Copernicus Climate Change Service (C3S) Climate Data Store (2022). The results contain modified Copernicus Climate Change Service information 2020. Neither the European Commission nor ECMWF is responsible for any use that may be made of the Copernicus information or data it contains.

References

Adams HD, Germino MJ, Breshears DD, Barron-Gafford GA, Guardiola-Claramonte M, Zou CB, Huxman TE (2013) Nonstructural leaf carbohydrate dynamics of Pinus edulis during drought-induced tree mortality reveal role for carbon metabolism in mortality mechanism. New Phytol 197(4):1142–1151. https://doi.org/10.1111/nph.12102

Adams HD, Collins AD, Briggs SP, Vennetier M, Dickman LT, Sevanto SA, McDowell NG (2015) Experimental drought and heat can delay phenological development and reduce foliar and shoot growth in semiarid trees. Glob Chang Biol 21(11):4210–4220. https://doi.org/10.1111/gcb.13030

Adams HD, Zeppel MJB, Anderegg WRL, Hartmann H, Landhäusser SM, Tissue DT, McDowell NG (2017) A multi-species synthesis of physiological mechanisms in drought- induced tree mortality. Nat Ecol Evol 1(9):1285–1291. https://doi.org/10.1038/s41559-017-0248-x

Adams MA, Buckley TN, Turnbull TL (2020) Diminishing CO_2-driven gains in water-use efficiency of global forests. Nat Clim Chang 10:466–471

Afkhami ME (2023) Past microbial stress benefits tree resilience. Science 380:798–799

Ainsworth EA, Long SP (2005) What have we learned from 15 years of free-air CO_2 enrichment (FACE)? A meta-analytic review of the responses of photosynthesis, canopy properties and plant production to rising CO_2. New Phytol 165(2):351–372. https://doi.org/10.1111/j.1469-8137.2004.01224.x

Albaugh TJ, Allen HL, Dougherty PM, Kress LW, King JS (1998) Leaf area and aboveg- and belowground growth responses of loblolly pine to nutrient and water relations. For Sci 44:317–328. https://doi.org/10.1093/forestscience/44.2.317

Albaugh TJ, Lee Allen H, Dougherty PM, Johnsen KH (2004) Long term growth responses of loblolly pine to optimal nutrient and water resource availability. For Ecol Manag 192(1):3–19. https://doi.org/10.1016/j.foreco.2004.01.002

Allen AS, Andrews JA, Finzi AC, Matamala R, Richter DD, Schlesinger WH (2000) Effects of free-air CO_2 enrichment (FACE) on belowground processes in a Pinus taeda forest. Ecol Appl 10:437–448

Allen CD, Macalady AK, Chenchouni H, Bachelet D, McDowell N, Vennetier M, Cobb N (2010) A global overview of drought and heat-induced tree mortality reveals emerging climate change risks for forests. For Ecol Manag 259(4):660–684. https://doi.org/10.1016/j.foreco.2009.09.001

Allen CD, Breshears DD, McDowell NG (2015) On underestimation of global vulnerability to tree mortality and forest die-off from hotter drought in the Anthropocene. Ecosphere 6(8):art129. https://doi.org/10.1890/ES15-00203.1

Allsup CM, George I, Lankau RA (2023) Shifting microbial communities can enhance tree tolerance to changing climates. Science 380:835–840

Anderegg WRL, Chegwidden OS, Badgley G, Trugman AT, Cullenward D, Abatzoglou JT, Hicke JA, Freeman J, Hamman JJ (2022) Future climate risks from stress, insects and fire across US forests. Ecol Lett 25:1510–1520

Asbjornsen H, Campbell JL, Jennings KA, Vadeboncoeur MA, McIntire C, Templer PH, Rustad LE (2018) Guidelines and considerations for designing field experiments simulating precipitation extremes in forest ecosystems. Methods Ecol Evol 9(12):2310–2325. https://doi.org/10.1111/2041-210X.13094

Auty D, Achim A, Macdonald E, Cameron AD, Gardiner BA (2014) Models for predicting wood density variation in Scots pine. Forestry Int J For Res 87(3):449–458

Axelsson E, Axelsson B (1986) Changes in carbon allocation patterns in spruce and pine trees following irrigation and fertilization. Tree Physiol 2(1-2-3):189–204. https://doi.org/10.1093/treephys/2.1-2-3.189

Bakker MR, Jolicoeur E, Trichet P, Augusto L, Plassard C, Guinberteau J, Loustau D (2009) Adaptation of fine roots to annual fertilization and irrigation in a 13-year-old Pinus pinaster stand. Tree Physiol 29(2):229–238. https://doi.org/10.1093/treephys/tpn020

Bartkowiak SM, Samuelson LJ, McGuire MA, Teskey RO (2015) Fertilization increases sensitivity of canopy stomatal conductance and transpiration to throughfall reduction in an 8-year-old loblolly pine plantation. For Ecol Manag 354(Supplement C):87–96. https://doi.org/10.1016/j.foreco.2015.06.033

Bassirirad H (2000) Research review: kinetics of nutrient uptake by roots: responses to global change. New Phytol 147:155–169

Bauerle WL, Oren R, Way DA, Qian SS, Stoy PC, Thornton PE, Reynolds RF (2012) Photoperiodic regulation of the seasonal pattern of photosynthetic capacity and the implications for carbon cycling. Proc Natl Acad Sci 109(22):8612–8617. https://doi.org/10.1073/pnas.1119131109

Beerling DJ (1999) Long-term responses of boreal vegetation to global change: an experimental and modelling investigation. Glob Chang Biol 5(1):55–74. https://doi.org/10.1046/j.1365-2486.1998.00209.x

Bengston C (1980) Effects of water stress on Scotts pine. Ecol Bull 1:205–213

Benson ML, Landsberg JJ, Borough CJ (1992) The biology of forest growth experiment: an introduction. For Ecol Manag 52(1):1–16. https://doi.org/10.1016/0378-1127(92)90493-S

Bermudez R, Stefanski A, Montgomery RA, Reich PB (2021) Short- and long-term responses of photosynthetic capacity to temperature in four boreal tree species in a free-air warming and rainfall manipulation experiment. Tree Physiol 41(1):89–102. https://doi.org/10.1093/treephys/tpaa115

Berry J, Bjorkman O (1980) Photosynthetic response and adaptation to temperature in higher plants. Annu Rev Plant Physiol 31(1):491–543. https://doi.org/10.1146/annurev.pp.31.060180.002423

Billings SA, Ziegler SE (2008) Altered patterns of soil carbon substrate usage and heterotrophic respiration in a pine forest with elevated CO_2 and N fertilization. Glob Chang Biol 14:1025–1036

Birami B, Nägele T, Gattmann M, Preisler Y, Gast A, Arneth A, Ruehr NK (2020) Hot drought reduces the effects of elevated CO_2 on tree water-use efficiency and carbon metabolism. New Phytol 226:1607–1621

Bose AK, Rigling A, Gessler A, Hagedorn F, Brunner I, Feichtinger L, Schaub M (2022) Lessons learned from a long-term irrigation experiment in a dry Scots pine forest: Impacts on traits and functioning. Ecol Monogr 92(2):e1507. https://doi.org/10.1002/ecm.1507

Bracho R, Vogel JG, Will RE, Noormets A, Samuelson LJ, Jokela EJ, Martin TA (2018) Carbon accumulation in loblolly pine plantations is increased by fertilization across a soil moisture availability gradient. For Ecol Manag 424:39–52. https://doi.org/10.1016/j.foreco.2018.04.029

Breshears DD, Rich PM, Barnes FJ, Campbell K (1997) Overstory-imposed heterogeneity in solar radiation and soil moisture in a semiarid woodland. Ecol Appl 7(4):1201–1215. https://doi.org/10.1890/1051-0761(1997)007[1201:OIHISR]2.0.CO;2

Breshears DD, Cobb NS, Rich PM, Price KP, Allen CD, Balice RG, Meyer CW (2005) Regional vegetation die-off in response to global-change-type drought. Proc Natl Acad Sci 102(42):15144–15148. https://doi.org/10.1073/pnas.0505734102

Breshears D, Adams H, Eamus D, McDowell N, Law D, Will R, Zou C (2013) The critical amplifying role of increasing atmospheric moisture demand on tree mortality and associated regional die-off. Front Plant Sci 4(266). https://doi.org/10.3389/fpls.2013.00266

Brienen RJW, Gloor E, Zuidema PA (2012) Detecting evidence for CO_2 fertilization from tree ring studies: The potential role of sampling biases. Glob Biogeochem Cycles 26:GB1025

Brienen RJW, Caldwell L, Duchesne L, Voelker S, Barichivich J, Baliva M, Ceccantini G, Di Filippo A, Helama S, Locosselli GM, Lopez L, Piovesan G, Schöngart J, Villalba R, Gloor E (2020) Forest carbon sink neutralized by pervasive growth-lifespan trade-offs. Nat Commun 11:4241

Brunner I, Pannatier EG, Frey B, Rigling A, Landolt W, Zimmermann S, Dobbertin M (2009) Morphological and physiological responses of Scots pine fine roots to water supply in a dry climatic region in Switzerland. Tree Physiol 29(4):541–550. https://doi.org/10.1093/treephys/tpn046

Brunstein FC, Yamaguchi DK (1992) The oldest known Rocky Mountain Bristlecone Pine ristate aristata Engelm. Arct Alp Res 24:253–256

Büntgen U, Krusic PJ, Piermattei A, Coomes DA, Esper J, Myglan VS, Kirdyanov AV, Camarero JJ, Crivellaro A, Körner C (2019) Limited capacity of tree growth to mitigate the global greenhouse effect under predicted warming. Nat Commun 10:2171

Cermak J, Kučera J, Bauerle WL, Phillips N, Hinckley TM (2007) Tree water storage and its diurnal dynamics related to sap flow and changes in stem volume in old-growth Douglas-fir trees. Tree Physiol 27(2):181–198. https://doi.org/10.1093/treephys/27.2.181

Carter KK (1996) Provenance tests as Indicators of growth response to climate change in 10 north temperate tree species. Can J For Res 26:1089–1095

Cernusak LA, Haverd V, Brendel O, Le Thiec D, Guehl J-M, Cuntz M (2019) Robust response of terrestrial plants to rising CO_2. Trends Plant Sci 24:578–586

Choat B, Brodribb TJ, Brodersen CR, Duursma RA, López R, Medlyn BE (2018) Triggers of tree mortality under drought. Nature 558(7711):531–539. https://doi.org/10.1038/s41586-018-0240-x

Chung H, Muraoka H, Nakamura M, Han S, Muller O, Son Y (2013) Experimental warming studies on tree species and forest ecosystems: a literature review. J Plant Res 126(4):447–460. https://doi.org/10.1007/s10265-013-0565-3

Coleman M (2007) Spatial and temporal patterns of root distribution in developing stands of four woody crop species grown with drip irrigation and fertilization. Plant Soil 299(1):195–213. https://doi.org/10.1007/s11104-007-9375-5

Copernicus Climate Change Service (C3S) (2022) ERA5-Land hourly data from 1950 to present. Copernicus Climate Change Service (C3S) Climate Data Store (CDS). https://doi.org/10.24381/cds.e2161bac. Accessed 1 June 2023

Coyle DR, Coleman MD, Aubrey DP (2008) Above- and below-ground biomass accumulation, production, and distribution of sweetgum and loblolly pine grown with irrigation and fertilization. Can J For Res 38(6):1335–1348. https://doi.org/10.1139/X07-231

Crous KY, Walters MB, Ellsworth DS (2008) Elevated CO_2 concentration affects leaf photosynthesis–nitrogen relationships in Pinus taeda over nine years in FACE. Tree Physiol 28:607–614

Crous KY, Uddling J, De Kauwe MG (2022) Temperature responses of photosynthesis and respiration in evergreen trees from boreal to tropical latitudes. New Phytol 234(2):353–374. https://doi.org/10.1111/nph.17951

Dawes MA, Hättenschwiler S, Bebi P, Hagedorn F, Handa IT, Körner C, Rixen C (2011) Species-specific tree growth responses to 9 years of CO_2 enrichment at the alpine treeline. J Ecol 99:383–394

Dawes MA, Hagedorn F, Handa IT, Streit K, Ekblad A, Rixen C, Körner C, Hättenschwiler S (2013) An alpine treeline in a carbon dioxide-rich world: synthesis of a nine-year free- air carbon dioxide enrichment study. Oecologia 171:623–637

Dawes MA, Philipson CD, Fonti P, Bebi P, Hättenschwiler S, Hagedorn F, Rixen C (2015) Soil warming and CO_2 enrichment induce biomass shifts in alpine tree line vegetation. Glob Chang Biol 21(5):2005–2021. https://doi.org/10.1111/gcb.12819

Dawes MA, Schleppi P, Hättenschwiler S, Rixen C, Hagedorn F (2017) Soil warming opens the nitrogen cycle at the alpine treeline. Glob Chang Biol 23(1):421–434. https://doi.org/10.1111/gcb.13365

De Kauwe MG, Medlyn BE, Zaehle S, Walker AP, Dietze MC, Hickler T, Jain AK, Luo Y, Parton WJ, Prentice IC, Smith B, Thornton PE, Wang S, Wang YP, Warlind D, Weng E, Crous KY, Ellsworth DS, Hanson PJ, Seok Kim H, Warren JM, Oren R, Norby RJ (2013) Forest water use and water use efficiency at elevated CO_2: a model-data intercomparison at two contrasting temperate forest FACE sites. Glob Chang Biol 19:1759–1779

DeLucia EH, Hamilton JG, Naidu SL, Thomas RB, Andrews JA, Finzi A, Lavine M, Matamala R, Mohan JE, Hendrey GR, Schlesinger WH (1999) Net primary production of a forest ecosystem with experimental CO_2 enrichment. Science 284:1177–1179

Delucia EH, Maherali H, Carey EV (2000) Climate-driven changes in biomass allocation in pines. Glob Chang Biol 6(5):587–593. https://doi.org/10.1046/j.1365-2486.2000.00338.x

Devi AS (2021) Influence of trees and associated variables on soil organic carbon: a review. J Ecol Environ 45(1):5. https://doi.org/10.1186/s41610-021-00180-3

Diers M, Weigel R, Culmsee H, Leuschner C (2021) Soil carbon and nutrient stocks under Scots pine plantations in comparison to European beech forests: a paired-plot study across forests with different management history and precipitation regimes. For Ecosys 8(1):47. https://doi.org/10.1186/s40663-021-00330-y

Dobbertin M, Eilmann B, Bleuler P, Giuggiola A, Graf Pannatier E, Landolt W, Rigling A (2010) Effect of irrigation on needle morphology, shoot and stem growth in a drought-exposed Pinus sylvestris forest. Tree Physiol 30(3):346–360. https://doi.org/10.1093/treephys/tpp123

Domec J-C, Johnson DM (2012) Does homeostasis or disturbance of homeostasis in minimum leaf water potential explain the isohydric versus anisohydric behavior of Vitis vinifera L. cultivars? Tree Physiol 32:245–248

Domec JC, Palmroth S, Ward E, Maier CA, Therezien M, Oren R (2009a) Acclimation of leaf hydraulic conductance and stomatal conductance of *Pinus taeda* (loblolly pine) to long-term growth in elevated CO_2 (free-air CO_2 enrichment) and N-fertilization. Plant Cell Environ 32:1500–1512

Domec JC, Noormets A, King JS et al (2009b) Decoupling the influence of leaf and root hydraulic conductance on stomatal conductance and its sensitivity to vapor pressure deficit as soil dries in a drained loblolly pine plantation. Plant Cell Environ 32:980–991

Domec J-C, Schäfer K, Oren R, Kim HS, McCarthy HR (2010) Variable conductivity and embolism in roots and branches of four contrasting tree species and their impacts on whole-plant hydraulic performance under future atmospheric CO_2 concentration. Tree Physiol 30(8):1001–1015. https://doi.org/10.1093/treephys/tpq054

Drake JE, Stoy PC, Jackson RB, Delucia EH (2008) Fine-root respiration in a loblolly pine (*Pinus taeda* L.) forest exposed to elevated CO2 and N fertilization. Plant Cell Environ 31:1663–1672. https://doi.org/10.1111/j.1365-3040.2008.01869.x

Drake JE, Gallet-Budynek A, Hofmockel KS, Bernhardt ES, Billings SA, Jackson RB, Johnsen KS, Lichter J, McCarthy HR, McCormack ML, Moore DJP, Oren R, Palmroth S, Phillips RP, Pippen JS, Pritchard SG, Treseder KK, Schlesinger WH, DeLucia EH, Finzi AC (2011) Increases in the flux of carbon belowground stimulate nitrogen uptake and sustain the long-term enhancement of forest productivity under elevated CO_2. Ecol Lett 14:349–357

Dusenge ME, Duarte AG, Way DA (2019) Plant carbon metabolism and climate change: Elevated CO_2 and temperature impacts on photosynthesis, photorespiration and respiration. New Phytol 221(1):32–49. https://doi.org/10.1111/nph.15283

Dusenge ME, Warren JM, Reich PB, Ward EJ, Murphy BK, Stefanski A, Bermudez R, Cruz M, McLennan DA, King AW, Montgomery RA, Hanson PJ, Way DA (2023) Boreal conifers maintain carbon uptake with warming despite failure to track optimal temperatures. Nat Commun 14(1):Article 1. https://doi.org/10.1038/s41467-023-40248-3

Ebelhar SA, Chesworth W, Paris Q (2008) Law of the minimum. In: Chesworth W (ed) Encyclopedia of soil science. Springer, Dordrecht, pp 431–437

Eilmann B, Buchmann N, Siegwolf R, Saurer M, Cherubini P, Rigling A (2010) Fast response of Scots pine to improved water availability reflected in tree-ring width and $\delta^{13}C$. Plant Cell Environ 33(8):1351–1360. https://doi.org/10.1111/j.1365-3040.2010.02153.x

Ellsworth DS, Thomas R, Crous KY, Palmroth S, Ward E, Maier C, DeLucia E, Oren R (2012) Elevated CO_2 affects photosynthetic responses in canopy pine and subcanopy deciduous trees over 10 years: a synthesis from Duke FACE. Glob Chang Biol 18:223–242

Ewers BE, Oren R, Sperry JS (2000) Influence of nutrient versus water supply on hydraulic architecture and water balance in *Pinus taeda*. Plant Cell Environ 23(10):1055–1066. https://doi.org/10.1046/j.1365-3040.2000.00625.x

Felton AJ, Knapp AK, Smith MD (2021) Precipitation–productivity relationships and the duration of precipitation anomalies: an underappreciated dimension of climate change. Glob Chang Biol 27(6):1127–1140. https://doi.org/10.1111/gcb.15480

Finzi AC, Allen AS, DeLucia EH, Ellsworth DS, Schlesinger WH (2001) Forest litter production, chemistry, and decomposition following two years of free-air CO_2 enrichment. Ecology 82:470–484

Finzi AC, Moore DJ, DeLucia EH, Lichter J, Hofmockel KS, Jackson RB, Kim HS, Matamala R, McCarthy HR, Oren R, Pippen JS, Schlesinger WH (2006) Progressive nitrogen limitation of ecosystem processes under elevated CO_2 in a warm-temperate forest. Ecology 87:15–25

Finzi AC, Norby RJ, Calfapietra C, Gallet-Budynek A, Gielen B, Holmes WE, Hoosbeek MR, Iversen CM, Jackson RB, Kubiske ME, Ledford J, Liberloo M, Oren R, Polle A, Pritchard S, Zak DR, Schlesinger WH, Ceulemans R (2007) Increases in nitrogen uptake rather than nitrogen-use efficiency support higher rates of temperate forest productivity under elevated CO_2. Proc Natl Acad Sci 104:14014–14019

Fisher RA, Koven CD, Anderegg WR, Christoffersen BO, Dietze MC, Farrior CE, Holm JA, Hurtt GC, Knox RG, Lawrence PJ, Lichstein JW, Longo M, Matheny AM, Medvigy D, Muller-Landau HC, Powell TL, Serbin SP, Sato H, Shuman JK, Smith B, Trugman AT, Viskari T, Verbeeck H, Weng E, Xu C, Xu X, Zhang T, Moorcroft PR (2018) Vegetation demographics in Earth System Models: a review of progress and priorities. Glob Chang Biol 24(1):35–54

Franklin O (2007) Optimal nitrogen allocation controls tree responses to elevated CO_2. New Phytol 174:811–822

Franks P, Brodribb TJ (2005) 4 – Stomatal control and water transport in the xylem. In: Holbrook NM, Zwieniecki MA (eds) Vascular transport in plants. Academic, Burlington, pp 69–89

Franks PJ, Farquhar GD (1999) A relationship between humidity response, growth form and photosynthetic operating point in C3 plants. Plant Cell Environ 22(11):1337–1349. https://doi.org/10.1046/j.1365-3040.1999.00494.x

Galiano L, Martínez-Vilalta J, Lloret F (2011) Carbon reserves and canopy defoliation determine the recovery of Scots pine 4 yr after a drought episode. New Phytol 190(3):750–759. https://doi.org/10.1111/j.1469-8137.2010.03628.x

Gardner A, Jiang M, Ellsworth DS, MacKenzie AR, Pritchard J, Bader MK-F, Barton CVM, Bernacchi C, Calfapietra C, Crous KY, Dusenge ME, Gimeno TE, Hall M, Lamba S, Leuzinger S, Uddling J, Warren J, Wallin G, Medlyn BE (2023) Optimal stomatal theory predicts CO_2 responses of stomatal conductance in both gymnosperm and angiosperm trees. New Phytol 237:1229–1241

Gazol A, Camarero JJ (2022) Compound climate events increase tree drought mortality across European forests. Sci Total Environ 816:151604. https://doi.org/10.1016/j.scitotenv.2021.151604

Gehring C, Sevanto S, Patterson A, Ulrich DEM, Kuske CR (2020) Ectomycorrhizal and dark septate fungal associations of pinyon pine are differentially affected by experimental drought and warming. Front Plant Sci 11. https://doi.org/10.3389/fpls.2020.582574

Gill RA, Anderson LJ, Polley HW, Johnson HB, Jackson RB (2006) Potential nitrogen constraints on soil carbon sequestration under low and elevated atmospheric CO_2. Ecology 87:41–52

Gojon A, Cassan O, Bach L, Lejay L, Martin A (2023) The decline of plant mineral nutrition under rising CO_2: physiological and molecular aspects of a bad deal. Trends Plant Sci 28:185–198

Gonzalez-Benecke CA, Martin TA, Peter GF (2010) Hydraulic architecture and tracheid allometry in mature Pinus palustris and Pinus elliottii trees. Tree Physiol 30(3):361–375. https://doi.org/10.1093/treephys/tpp129

Grossiord C, Sevanto S, Adams HD, Collins AD, Dickman LT, McBranch N, Michaletz ST, Stockton EA, Vigil M, McDowell NG (2017a) Precipitation, not air temperature, drives functional responses of trees in semi-arid ecosystems. J Ecol 105(1):163–175. https://doi.org/10.1111/1365-2745.12662

Grossiord C, Sevanto S, Borrego I, Chan AM, Collins AD, Dickman LT, Hudson PJ, McBranch N, Michaletz ST, Pockman WT, Ryan M, Vilagrosa A, McDowell NG (2017b) Tree water dynamics in a drying and warming world. Plant Cell Environ 40(9):1861–1873. https://doi.org/10.1111/pce.12991

Grossiord C, Sevanto S, Dawson TE, Adams HD, Collins AD, Dickman LT, Newman BD, Stockton EA, McDowell NG (2017c) Warming combined with more extreme precipitation regimes modifies the water sources used by trees. New Phytol 213(2):584–596. https://doi.org/10.1111/nph.14192

Grossiord C, Gessler A, Reed SC, Borrego I, Collins AD, Dickman LT, Ryan M, Schönbeck L, Sevanto S, Vilagrosa A, McDowell NG (2018) Reductions in tree performance during hotter droughts are mitigated by shifts in nitrogen cycling. Plant Cell Environ 41(11):2627–2637. https://doi.org/10.1111/pce.13389

Gruber A, Oberhuber W, Wieser G (2018) Nitrogen addition and understory removal but not soil warming increased radial growth of Pinus cembra at Treeline in the Central Austrian Alps. Front Plant Sci 9. https://doi.org/10.3389/fpls.2018.00711

Gu H, Qiao Y, Xi Z, Rossi S, Smith NG, Liu J, Chen L (2022) Warming-induced increase in carbon uptake is linked to earlier spring phenology in temperate and boreal forests. Nat Commun 13(1):Article 1. https://doi.org/10.1038/s41467-022-31496-w

Guidi C, Frey B, Brunner I, Meusburger K, Vogel ME, Chen X, Hagedorn F (2022) Soil fauna drives vertical redistribution of soil organic carbon in a long-term irrigated dry pine forest. Glob Chang Biol 28(9):3145–3160. https://doi.org/10.1111/gcb.16122

Handa IT, Körner C, Hättenschwiler S (2005) A test of the treeline carbon limitation hypothesis by in situ CO_2 enrichment and defoliation. Ecology 86:1288–1300

Handa IT, Körner C, Hättenschwiler S (2006) Conifer stem growth at the altitudinal treeline in response to four years of CO_2 enrichment. Glob Chang Biol 12:2417–2430

Handa IT, Hagedorn F, Hättenschwiler S (2008) No stimulation in root production in response to 4 years of in situ CO_2 enrichment at the Swiss treeline. Funct Ecol 22:348–358

Hanson PJ, Edwards NT, Garten CT, Andrews JA (2000) Separating root and soil microbial contributions to soil respiration: a review of methods and observations. Biogeochemistry 48(1):115–146. https://doi.org/10.1023/A:1006244819642

Hellkvist J, Hillerdal-Hagströmer K, Mattson-Djos E (1980) Field studies of water relations and photosynthesis in scots pine using manual techniques. Ecol Bull 32:183–204

Hendrey GR, Lewin KF, Nagy J (1993) Free air carbon dioxide enrichment: development, progress, results. Vegetatio 104–105:17–31

Hendrey GR, Ellsworth DS, Lewin KF, Nagy J (1999) A free-air enrichment system for exposing tall forest vegetation to elevated atmospheric CO_2. Glob Chang Biol 5:293–309

Héroult A, Lin Y-S, Bourne A, Medlyn BE, Ellsworth DS (2013) Optimal stomatal conductance in relation to photosynthesis in climatically contrasting Eucalyptus species under drought. Plant Cell Environ 36(2):262–274. https://doi.org/10.1111/j.1365-3040.2012.02570.x

Herzog C, Steffen J, Graf Pannatier E, Hajdas I, Brunner I (2014) Nine years of irrigation cause vegetation and fine root shifts in a water-limited pine forest. PLoS One 9(5):e96321

Hickler T, Rammig A, Werner C (2015) Modelling CO_2 impacts on forest productivity. Curr For Rep 1:69–80

Hikosaka K, Ishikawa K, Borjigidai A, Muller O, Onoda Y (2006) Temperature acclimation of photosynthesis: mechanisms involved in the changes in temperature dependence of photosynthetic rate. J Exp Bot 57(2):291–302. https://doi.org/10.1093/jxb/erj049

Hinko-Najera N, Fest B, Livesley SJ, Arndt SK (2015) Reduced throughfall decreases autotrophic respiration, but not heterotrophic respiration in a dry temperate broadleaved evergreen forest. Agric For Meteorol 200:66–77. https://doi.org/10.1016/j.agrformet.2014.09.013

Hoffmann WA, Marchin RM, Abit P, Lau OL (2011) Hydraulic failure and tree dieback are associated with high wood density in a temperate forest under extreme drought. Glob Chang Biol 17(8):2731–2742. https://doi.org/10.1111/j.1365-2486.2011.02401.x

Huang S, Ye G, Lin J, Chen K, Xu X, Ruan H, Chen HYH (2018) Autotrophic and heterotrophic soil respiration responds asymmetrically to drought in a subtropical forest in the Southeast China. Soil Biol Biochem 123:242–249. https://doi.org/10.1016/j.soilbio.2018.04.029

Hussain M, Kubiske ME, Connor KF (2001) Germination of CO_2 enriched *Pinus taeda* L. seeds and subsequent seedling growth responses to CO_2 enrichment. Funct Ecol 15:344–350

IPCC (2021) Summary for policymakers. In: Masson-Delmotte V, Zhai P, Pirani A, Connors SL, Péan C, Berger S, Caud N, Chen Y, Goldfarb L, Gomis MI, Huang M, Leitzell K, Lonnoy E, Matthews JBR, Maycock TK, Waterfield T, Yelekçi O, Yu R, Zhou B (eds) Climate Change 2021: the physical science basis. Contribution of Working Group I to the Sixth Assessment Report of the Intergovernmental Panel on Climate Change. Cambridge University Press, Cambridge/New York, pp 3–32. https://doi.org/10.1017/9781009157896.001

Iversen CM, Latimer J, Brice DJ, Childs J, Vander Stel HM, Defrenne CE, Graham J, Griffiths NA, Malhotra A, Norby RJ, Oleheiser KC, Phillips JR, Salmon VG, Sebestyen SD, Yang X, Hanson PJ (2023) Whole-ecosystem warming increases plant-available nitrogen and phosphorus in an Ombrotrophic Bog. Ecosystems 26(1):86–113. https://doi.org/10.1007/s10021-022-00744-x

Jackson RB, Cook CW, Pippen JS, Palmer SM (2009) Increased belowground biomass and soil CO_2 fluxes after a decade of carbon dioxide enrichment in a warm-temperate forest. Ecology 90:3352–3366

Jagodziński AM, Dyderski MK, Gęsikiewicz K, Horodecki P, Cysewska A, Wierczyńska S, Maciejczyk K (2018) How do tree stand parameters affect young Scots pine biomass? – Allometric equations and biomass conversion and expansion factors. For Ecol Manag 409:74–83

Jandl R, Ledermann T, Kindermann G, Weiss P (2021) Soil organic carbon stocks in mixed-deciduous and coniferous forests in Austria. Front For Glob Change 4. https://doi.org/10.3389/ffgc.2021.688851

Janssens IA, Medlyn B, Gielen B, Laureysens I, Jach ME, Van Hove D, Ceulemans R (2005) Carbon budget of Pinus sylvestris saplings after four years of exposure to elevated atmospheric carbon dioxide concentration. Tree Physiol 25:325–337

Johnson DW (2006) Progressive N limitation in forests: review and implications for long-term responses to elevated CO_2. Ecology 87:64–75

Keenan TF, Hollinger DY, Bohrer G, Dragoni D, Munger JW, Schmid HP, Richardson AD (2013) Increase in forest water-use efficiency as atmospheric carbon dioxide concentrations rise. Nature 499:324

Kellomäki S, Wang K-Y (1996) Photosynthetic responses to needle water potentials in Scots pine after a four-year exposure to elevated CO_2 and temperature. Tree Physiol 16(9):765–772. https://doi.org/10.1093/treephys/16.9.765

Kellomäki S, Wang KY (1997) Effects of elevated O_3 and CO_2 concentrations on photosynthesis and stomatal conductance in Scots pine. Plant Cell Environ 20:995–1006

Kellomäki S, Wang K-Y (2000) Modelling and measuring transpiration from scots pine with increased temperature and carbon dioxide enrichment. Ann Bot 85(2):263–278. https://doi.org/10.1006/anbo.1999.1030

Kellomäki S, Wang K-Y, Lemettinen M (2000) Controlled environment chambers for investigating tree response to elevated CO_2 and temperature under boreal conditions. Photosynthetica 38:69–81

Khan A, Yan L, Mahadi Hasan M, Wang W, Xu K, Zou G, Liu X-D, Fang X-W (2022) Leaf traits and leaf nitrogen shift photosynthesis adaptive strategies among functional groups and diverse biomes. Ecol Indic 141:109098

Kilpeläinen A, Peltola H, Ryyppö A, Kellomäki S (2005) Scots pine responses to elevated temperature and carbon dioxide concentration: growth and wood properties. Tree Physiol 25(1):75–83. https://doi.org/10.1093/treephys/25.1.75

Kilpeläinen A, Peltola H, Rouvinen I, Kellomäki S (2006) Dynamics of daily height growth in Scots pine trees at elevated temperature and CO_2. Trees 20(1):16–27. https://doi.org/10.1007/s00468-005-0008-4

Kilpeläinen A, Gerendiain AZ, Luostarinen K, Peltola H, Kellomäki S (2007) Elevated temperature and CO_2 concentration effects on xylem anatomy of Scots pine. Tree Physiol 27(9):1329–1338. https://doi.org/10.1093/treephys/27.9.1329

Kim D, Oren R, Qian SS (2016) Response to CO_2 enrichment of understory vegetation in the shade of forests. Glob Chang Biol 22:944–956

Kim D, Medvigy D, Maier CA, Johnsen K, Palmroth S (2020) Biomass increases attributed to both faster tree growth and altered allometric relationships under long-term carbon dioxide enrichment at a temperate forest. Glob Chang Biol 26:2519–2533

King JS, Albaugh TJ, Allen HL, Kress LW (1999) Stand-level allometry in *Pinus taeda* as affected by irrigation and fertilization. Tree Physiol 19(12):769–778. https://doi.org/10.1093/treephys/19.12.769

King JS, Albaugh TJ, Allen HL, Buford M, Strain BR, Dougherty P (2002) Below- ground carbon input to soil is controlled by nutrient availability and fine root dynamics in loblolly pine. New Phytol 154:389–398

Klein T (2014) The variability of stomatal sensitivity to leaf water potential across tree species indicates a continuum between isohydric and anisohydric behaviours. Funct Ecol 28:1313–1320

Klos RJ, Wang GG, Bauerle WL, Rieck JR (2009) Drought impact on forest growth and mortality in the southeast USA: an analysis using Forest Health and Monitoring data. Ecol Appl 19(3):699–708. https://doi.org/10.1890/08-0330.1

Knapp AK, Avolio ML, Beier C, Carroll CJW, Collins SL, Dukes JS, Smith MD (2017) Pushing precipitation to the extremes in distributed experiments: recommendations for simulating wet and dry years. Glob Chang Biol 23(5):1774–1782. https://doi.org/10.1111/gcb.13504

Knowles P, Grant MC (1983) Age and size structure analyses of Engelmann spruce, Ponderosa pine, Lodgepole pine, and Limber pine in Colorado. Ecology 64:1–9

Knutti R, Rugenstein MA, Hegerl GC (2017) Beyond equilibrium climate sensitivity. Nat Geosci 10(10):727–736

Kolb T, Keefover-Ring K, Burr SJ, Hofstetter R, Gaylord M, Raffa KF (2019) Drought- mediated changes in tree physiological processes weaken tree defenses to bark beetle attack. J Chem Ecol 45(10):888–900. https://doi.org/10.1007/s10886-019-01105-0

Konrad, CE, Fuhrmann CM, Billiot A, Keim MC, Kunkel KE, Needham H, Shafer M, Stevens L (2013) Climate of the southeast USA: past, present, and future. In Climate of the southeast United States: Variability, change, impacts, and vulnerability (pp. 8–42). Washington, DC: Island Press/Center for Resource Economics

Kramer PJ, Boyer JS (1995) Water relations of plants and soils. Academic

Krumins JA, Goodey NM, Gallagher F (2015) Plant–soil interactions in metal contaminated soils. Soil Biol Biochem 80:224–231

LaDeau SL, Clark JS (2001) Rising CO_2 levels and the fecundity of forest trees. Science 292:95–98

LaDeau SL, Clark JS (2006) Pollen production by *Pinus taeda* growing in elevated atmospheric CO_2. Funct Ecol 20:541–547

Landsberg J, Sands P (eds) (2011) Physiological ecology of forest production. Academic, London

Landsberg J, Waring R, Ryan M (2017) Water relations in tree physiology: where to from here? Tree Physiol 37(1):18–32. https://doi.org/10.1093/treephys/tpw102

Lanner RM, Connor KF (2001) Does bristlecone pine senesce? Exp Gerontol 36:675–685

Lauriks F, Salomón RL, Steppe K (2021) Temporal variability in tree responses to elevated atmospheric CO_2. Plant Cell Environ 44:1292–1310

Law BE, Baldocchi DD, Anthoni PM (1999) Below-canopy and Soil CO_2 fluxes in a ponderosa pine forest. Agric For Meteorol 94:171–188

Lee H, Lee WY, Kang JW (2021) Effects of elevated atmospheric CO_2 on rhizosphere microbial community of *Pinus densiflora* (Korean red pine). For Sci Technol 17:16–22

Lendzion J, Leuschner C (2008) Growth of European beech (*Fagus sylvatica* L.) saplings is limited by elevated atmospheric vapour pressure deficits. For Ecol Manag 256:648–655

Li Q, Allen HL, Wollum AG (2005) Effects of irrigation and fertilization on soil microbial biomass and functional diversity. J Sustain For 20(4):17–35. https://doi.org/10.1300/J091v20n04_02

Lichter J, Billings SA, Ziegler SE, Gaindh D, Ryals R, Finzi AC, Jackson RB, Stemmler EA, Schlesinger WH (2008) Soil carbon sequestration in a pine forest after 9 years of atmospheric CO_2 enrichment. Glob Chang Biol 14:2910–2922

Lim H, Oren R, Palmroth S, Tor-ngern P, Mörling T, Näsholm T, Linder S (2015) Inter- annual variability of precipitation constrains the production response of boreal *Pinus sylvestris* to nitrogen fertilization. For Ecol Manag 348:31–45. https://doi.org/10.1016/j.foreco.2015.03.029

Limousin J-M, Bickford CP, Dickman LT, Pangle RE, Hudson PJ, Boutz AL, Mcdowell NG (2013) Regulation and acclimation of leaf gas exchange in a piñon– juniper woodland exposed to three different precipitation regimes. Plant Cell Environ 36(10):1812–1825. https://doi.org/10.1111/pce.12089

Lin J, Jach ME, Ceulemans R (2001) Stomatal density and needle anatomy of Scots pine (*Pinus sylvestris*) are affected by elevated CO_2. New Phytol 150:665–674

Linder S (1987) Responses to water and nutrients in coniferous ecosystems. In: Schulze E-D, Zwölfer H (eds) Potentials and limitations of ecosystem analysis. Springer, Berlin/Heidelberg, pp 180–202

Long SP (1991) Modification of the response of photosynthetic productivity to rising temperature by atmospheric CO_2 concentrations: Has its importance been underestimated? Plant Cell Environ 14(8):729–739. https://doi.org/10.1111/j.1365-3040.1991.tb01439.x

Longo M, Knox RG, Medvigy DM, Levine NM, Dietze MC, Kim Y, Swann AL, Zhang K, Rollinson CR, Bras RL, Wofsy SC, Moorcroft PR (2019a) The biophysics, ecology, and biogeochemistry of functionally diverse, vertically and horizontally heterogeneous ecosystems: The Ecosystem Demography model, version 2.2–Part 1: Model description. Geosci Model Dev 12(10):4309–4346

Longo M, Knox RG, Levine NM, Swann AL, Medvigy DM, Dietze MC, Kim Y, Zhang K, Bonal D, Burban B, Camargo PB, Hayek MN, Saleska SR, da Silva R, Rollinson, Bras RL, Wofsy SC, Moorcroft PR (2019b) The biophysics, ecology, and biogeochemistry of functionally diverse, vertically and horizontally heterogeneous ecosystems: The Ecosystem Demography model, version 2.2–Part 1: Model evaluation for tropical South America. Geosci Model Dev 12(10):4347–4374

Lu M, Zhou X, Yang Q, Li H, Luo Y, Fang C, Chen J, Yang X, Li B (2013) Responses of ecosystem carbon cycle to experimental warming: a meta-analysis. Ecology 94(3):726–738. https://doi.org/10.1890/12-0279.1

Luomala E-M, Laitinen K, Kellomäki S, Vapaavuori E (2003) Variable photosynthetic acclimation in consecutive cohorts of Scots pine needles during 3 years of growth at elevated CO_2 and elevated temperature. Plant Cell Environ 26(5):645–660. https://doi.org/10.1046/j.1365-3040.2003.01000.x

Maggard AO, Will RE, Wilson DS, Meek CR, Vogel JG (2016) Fertilization reduced stomatal conductance but not photosynthesis of *Pinus taeda* which compensated for lower water availability in regards to growth. For Ecol Manag 381:37–47. https://doi.org/10.1016/j.foreco.2016.08.046

Maggard AO, Will RE, Wilson DS, Meek CR, Vogel JG (2017) Fertilization can compensate for decreased water availability by increasing the efficiency of stem volume production per unit of leaf area for loblolly pine (*Pinus taeda*) stands. Can J For Res 47(4):445–457. https://doi.org/10.1139/cjfr-2016-0422

Maier CA et al (2002) Branch growth and gas exchange in 13 year-old loblolly pine (*Pinus taeda*) trees in response to elevated carbon dioxide concentration and fertilization. Tree Physiol 22:1093–1106. https://doi.org/10.1093/treephys/22.15-16.1093

Maier CA, Albaugh TJ, Lee Allen H, Dougherty PM (2004) Respiratory carbon use and carbon storage in mid-rotation loblolly pine (*Pinus taeda* L.) plantations: the effect of site resources on the stand carbon balance. Glob Chang Biol 10(8):1335–1350. https://doi.org/10.1111/j.1529-8817.2003.00809.x

Maier CA, Johnsen KH, Anderson PH, Palmroth S, Kim D, McCarthy HR, Oren R (2022) The response of coarse root biomass to long-term CO2 enrichment and nitrogen application in a maturing *Pinus taeda* stand with a large broadleaved component. Glob Chang Biol 28:1458–1476. https://doi.org/10.1111/gcb.15999

Martínez-Vilalta J, Sala A, Piñol J (2004) The hydraulic architecture of Pinaceae – a review. Plant Ecol 171(1):3–13. https://doi.org/10.1023/B:VEGE.0000029378.87169.b1

Martínez-Vilalta J, Poyatos R, Aguadé D, Retana J, Mencuccini M (2014) A new look at water transport regulation in plants. New Phytol 204(1):105–115. https://doi.org/10.1111/nph.12912

Matusick G, Ruthrof KX, Brouwers NC, Dell B, Hardy GSJ (2013) Sudden forest canopy collapse corresponding with extreme drought and heat in a mediterranean-type eucalypt forest in southwestern Australia. Eur J For Res 132(3):497–510. https://doi.org/10.1007/s10342-013-0690-5

Matyas C (1994) Modeling climate change effects with provenance test data. Tree Physiol 14:797–804

McCarthy HR, Oren R, Finzi AC, Johnsen KH (2006) Canopy leaf area constrains [CO_2]- induced enhancement of productivity and partitioning among aboveground carbon pools. Proc Natl Acad Sci USA 103:19356–19361

McCarthy HR, Oren R, Finzi AC, Ellsworth DS, Kim HS, Johnsen KH, Millar B (2007) Temporal dynamics and spatial variability in the enhancement of canopy leaf area under elevated atmospheric CO_2. Glob Chang Biol 13:2479–2497

McCarthy HR, Oren R, Johnsen KH, Gallet-Budynek A, Pritchard SG, Cook CW, Finzi AC (2010) Re-assessment of plant carbon dynamics at the Duke free-air CO_2 enrichment site: interactions of atmospheric [CO_2] with nitrogen and water availability over stand development. New Phytol 185(2):514–528. https://doi.org/10.1111/nph.2009.185.issue-2 https://doi.org/10.1111/j.1469-8137.2009.03078.x

McCulloh KA, Sperry JS (2005) Patterns in hydraulic architecture and their implications for transport efficiency. Tree Physiol 25(3):257–267. https://doi.org/10.1093/treephys/25.3.257

McCulloh KA, Petitmermet J, Stefanski A, Rice KE, Rich RL, Montgomery RA, Reich PB (2016) Is it getting hot in here? Adjustment of hydraulic parameters in six boreal and temperate tree species after 5 years of warming. Glob Chang Biol 22(12):4124–4133. https://doi.org/10.1111/gcb.13323

McDowell N, Pockman WT, Allen CD, Breshears DD, Cobb N, Kolb T, Yepez EA (2008) Mechanisms of plant survival and mortality during drought: why do some plants survive while others succumb to drought?. New Phytol 178(4):719–739. https://doi.org/10.1111/nph.2008.178.issue-4 https://doi.org/10.1111/j.1469-8137.2008.02436.x

McDowell NG (2011) Mechanisms linking drought, hydraulics, carbon metabolism, and vegetation mortality. Plant Physiol 155(3):1051–1059. https://doi.org/10.1104/pp.110.170704

McDowell NG, Beerling DJ, Breshears DD, Fisher RA, Raffa KF, Stitt M (2011) The interdependence of mechanisms underlying climate-driven vegetation mortality. Trends Ecol Evol 26(10):523–532. https://doi.org/10.1016/j.tree.2011.06.003

McDowell NG, Fisher RA, Xu C, Domec JC, Hölttä T, Mackay DS, Pockman WT (2013) Evaluating theories of drought-induced vegetation mortality using a multimodel–experiment framework. New Phytol 200(2):304–321. https://doi.org/10.1111/nph.12465

McDowell NG, Sapes G, Pivovaroff A et al (2022) Mechanisms of woody-plant mortality under rising drought, CO2 and vapour pressure deficit. Nat Rev Earth Environ 3:294–308. https://doi.org/10.1038/s43017-022-00272-1

McDowell NG, Williams AP, Xu C, Pockman WT, Dickman LT, Sevanto S, Koven C (2016) Multi-scale predictions of massive conifer mortality due to chronic temperature rise. Nat Clim Chang 6(3):295–300. https://doi.org/10.1038/nclimate2873

McDowell NG, Grossiord C, Adams HD, Pinzón-Navarro S, Mackay DS, Breshears DD, Allen CD, Borrego I, Dickman LT, Collins A, Gaylord M, McBranch N, Pockman WT, Vilagrosa A, Aukema B, Goodsman D, Xu C (2019) Mechanisms of a coniferous woodland persistence under drought and heat. Environ Res Lett 14(4):045014. https://doi.org/10.1088/1748-9326/ab0921

McElligott KM, Seiler JR, Strahm BD (2016) Partitioning soil respiration across four age classes of loblolly pine (*Pinus taeda* L.) on the Virginia Piedmont. For Ecol Manag 378:173–180. https://doi.org/10.1016/j.foreco.2016.07.026

Mediavilla S, Escudero A (2003a) Leaf life span differs from retention time of biomass and nutrients in the crowns of evergreen species. Funct Ecol 17:541–548

Mediavilla S, Escudero A (2003b) Stomatal responses to drought at a Mediterranean site: a comparative study of co-occurring woody species differing in leaf longevity. Tree Physiol 23(14):987–996. https://doi.org/10.1093/treephys/23.14.987

Medlyn BE, Barton CVM, Broadmeadow MSJ, Ceulemans R, De Angelis P, Forstreuter M, Freeman M, Jackson SB, Kellomaki S, Laitat E, Rey A, Roberntz P, Sigurdsson BD, Strassemeyer J, Wang K, Curtis PS, Jarvis PG (2001) Stomatal conductance of forest species after long-term exposure to elevated CO_2 concentration: a synthesis. New Phytol 149:247–264

Medlyn BE, Zaehle S, De Kauwe MG, Walker AP, Dietze MC, Hanson PJ, Hickler T, Jain AK, Luo Y, Parton W, Prentice IC (2015) Using ecosystem experiments to improve vegetation models. Nat Clim Chang 5(6):528–534

Medvigy D, Wofsy SC, Munger JW, Hollinger DY, Moorcroft PR (2009) Mechanistic scaling of ecosystem function and dynamics in space and time: Ecosystem Demography model version 2. J Geophys Res Biogeosci 114:G01002

Medvigy D, Wofsy SC, Munger JW, Moorcroft PR (2010) Responses of terrestrial ecosystems and carbon budgets to current and future environmental variability. Proc Natl Acad Sci 107(18):8275–8280

Medvigy D, Wang G, Zhu Q, Riley WJ, Trierweiler AM, Waring BG, Xu X, Powers JS (2019) Observed variation in soil properties can drive large variation in modelled forest functioning and composition during tropical forest secondary succession. New Phytol 223(4):1820–1833

Meinzer FC (2002) Co-ordination of vapour and liquid phase water transport properties in plants. Plant Cell Environ 25:265–274

Meinzer FC, Woodruff DR, Marias DE, Smith DD, McCulloh KA, Howard AR, Magedman AL, Penuelas J (2016) Mapping 'hydroscapes' along the iso- to anisohydric continuum of stomatal regulation of plant water status. Ecol Lett 19:1343–1352

Mencuccini M (2003) The ecological significance of long-distance water transport: short-term regulation, long-term acclimation and the hydraulic costs of stature across plant life forms. Plant Cell Environ 26(1):163–182. https://doi.org/10.1046/j.1365-3040.2003.00991.x

Mendonca CC, Samuelson LJ, Aspinwall MJ (2022) Experimental throughfall reduction has little effect on shoot and needle developmental patterns or leaf area dynamics in a young longleaf pine (*Pinus palustris* Mill.) plantation. For Ecol Manag 517:120246. https://doi.org/10.1016/j.foreco.2022.120246

Menezes-Silva PE, Loram-Lourenço L, Alves R, Sousa LF, Almeida S, Farnese FS (2019) Different ways to die in a changing world: Consequences of climate change for tree species performance and survival through an ecophysiological perspective. Ecol Evol 9(20):11979–11999. https://doi.org/10.1002/ece3.5663

Montgomery RA, Rice KE, Stefanski A, Rich RL, Reich PB (2020) Phenological responses of temperate and boreal trees to warming depend on ambient spring temperatures, leaf habit, and geographic range. Proc Natl Acad Sci 117(19):10397–10405. https://doi.org/10.1073/pnas.1917508117

Moreno M, Simioni G, Cailleret M, Ruffault J, Badel E, Carrière S, Martin-StPaul N (2021) Consistently lower sap velocity and growth over nine years of rainfall exclusion in a Mediterranean mixed pine-oak forest. Agric For Meteorol 308–309:108472. https://doi.org/10.1016/j.agrformet.2021.108472

Moore GW, Edgar CB, Vogel JG, Washington-Allen RA, March RG, Zehnder R (2016) Tree mortality from an exceptional drought spanning mesic to semiarid ecoregions. Ecol Appl 26(2):602–611. https://doi.org/10.1890/15-0330

Mucha J, Peay KG, Smith DP, Reich PB, Stefański A, Hobbie SE (2018) Effect of simulated climate warming on the ectomycorrhizal fungal community of boreal and temperate host species growing near their shared ecotonal range limits. Microb Ecol 75(2):348–363. https://doi.org/10.1007/s00248-017-1044-5

Muñoz Sabater J (2019) ERA5-Land hourly data from 1950 to present. Copernicus Climate Change Service (C3S) Climate Data Store (CDS). https://doi.org/10.24381/cds.e2161bac. Accessed 1 June 2023

Murthy R, Dougherty PM, Zarnoch SJ, Allen HL (1996) Effects of carbon dioxide, fertilization, and irrigation on photosynthetic capacity of loblolly pine trees. Tree Physiol 16:537–546

Murthy R, Zarnoch SJ, Dougherty PM (1997) Seasonal trends of light-saturated net photosynthesis and stomatal conductance of loblolly pine trees grown in contrasting environments of nutrition, water and carbon dioxide. Plant Cell Environ 20:558–568

Niinistö SM, Silvola J, Kellomäki S (2004) Soil CO_2 efflux in a boreal pine forest under atmospheric CO_2 enrichment and air warming. Glob Chang Biol 10(8):1363–1376. https://doi.org/10.1111/j.1365-2486.2004.00799.x

Nobis MP, Traiser C, Roth-Nebelsick A (2012) Latitudinal variation in morphological traits of the genus Pinus and its relation to environmental and phylogenetic signals. Plant Ecol Divers 5(1):1–11. https://doi.org/10.1080/17550874.2012.687501

Noormets A, Bracho R, Ward E, Seiler J, Strahm B, Lin W, Martin TA (2021) Heterotrophic respiration and the divergence of productivity and carbon sequestration. Geophys Res Lett 48(7):e2020GL092366. https://doi.org/10.1029/2020GL092366

Norby RJ, Cotrufo MF (1998) Global change: a question of litter quality. Nature 396:17–18

Norby RJ, Cotrufo MF, Ineson P, O'Neill EG, Canadell JG (2001) Elevated CO_2, litter chemistry, and decomposition: a synthesis. Oecologia 127:153–165

Norby RJ, DeLucia EH, Gielen B, Calfapietra C, Giardina CP, King JS, Ledford J, McCarthy HR, Moore DJP, Ceulemans R, De Angelis P, Finzi AC, Karnosky DF, Kubiske ME, Lukac M, Pregitzer KS, Scarascia-Mugnozza GE, Schlesinger WH, Oren R (2005) Forest response to elevated CO_2 is conserved across a broad range of productivity. Proc Natl Acad Sci USA 102:18052–18056

Novick KA, Katul GG, McCarthy HR, Oren R (2012) Increased resin flow in mature pine trees growing under elevated CO_2 and moderate soil fertility. Tree Physiol 32:752–763

Novick KA, Ficklin DL, Stoy PC, Williams CA, Bohrer G, Oishi AC, Phillips RP (2016) The increasing importance of atmospheric demand for ecosystem water and carbon fluxes. Nat Clim Chang 6(11):1023–1027. https://doi.org/10.1038/nclimate3114

Oishi AC, Palmroth S, Johnsen KH, McCarthy HR, Oren R (2014) Sustained effects of atmospheric [CO_2] and nitrogen availability on forest soil CO_2 efflux. Glob Chang Biol 20:1146–1160. https://doi.org/10.1111/gcb.12414

Oren R, Sperry JS, Katul GG, Pataki DE, Ewers BE, Phillips N, Schäfer KVR (1999) Survey and synthesis of intra- and interspecific variation in stomatal sensitivity to vapour pressure deficit. Plant Cell Environ 22(12):1515–1526. https://doi.org/10.1046/j.1365-3040.1999.00513.x

Oren R, Ellsworth DS, Johnsen KH, Phillips N, Ewers BE, Maier C, Schafer KVR, McCarthy H, Hendrey G, McNulty SG, Katul GG (2001) Soil fertility limits carbon sequestration by forest ecosystems in a CO_2-enriched atmosphere. Nature 411:469–472

Ozolinčius R, Stakėnas V, Varnagirytė-Kabašinskienė I, Buožytė R (2009) Artificial drought in scots pine stands: effects on soil, ground vegetation and tree condition. Ann Bot Fenn 46(4):299–307

Palmroth S, Oren R, McCarthy HR, Johnsen KH, Finzi AC, Butnor JR, Ryan MG, Schlesinger WH (2006) Aboveground sink strength in forests controls the allocation of carbon below ground and its [CO_2]-induced enhancement. Proc Natl Acad Sci 103:19362–19367

Palmroth S, Kim D, Maier CA, Medvigy D, Walker AP, Oren R (2024) Increased leaf area index and efficiency drive enhanced production under elevated atmospheric [CO_2] in a pine-dominated stand showing no progressive nitrogen limitation. Glob Chang Biol 30(2). https://doi.org/10.1111/gcb.v30.2 https://doi.org/10.1111/gcb.17190

Pangle RE, Hill JP, Plaut JA, Yepez EA, Elliot JR, Gehres N, Pockman WT (2012) Methodology and performance of a rainfall manipulation experiment in a piñon–juniper woodland. Ecosphere 3(4):art28. https://doi.org/10.1890/ES11-00369.1

Pataki DE, Oren R, Katul G, Sigmon J (1998a) Canopy conductance of *Pinus taeda*, *Liquidambar styraciflua* and *Quercus phellos* under varying atmospheric and soil water conditions. Tree Physiol 18:307–315

Pataki DE, Oren R, Tissue DT (1998b) Elevated carbon dioxide does not affect stomatal conductance of *Pinus taeda* L. Oecologia 117:47–52

Peltola H, Kilpeläinen A, Kellomäki S (2002) Diameter growth of Scots pine (*Pinus sylvestris*) trees grown at elevated temperature and carbon dioxide concentration under boreal conditions. Tree Physiol 22(14):963–972. https://doi.org/10.1093/treephys/22.14.963

Pendall E, Bridgham S, Hanson PJ, Hungate B, Kicklighter DW, Johnson DW, Law BE, Luo Y, Megonigal JP, Olsrud M, Ryan MG, Wan S (2004) Below-ground process responses to elevated CO_2 and temperature: a discussion of observations, measurement methods, and models. New Phytol 162(2):311–322. https://doi.org/10.1111/j.1469-8137.2004.01053.x

Penuelas J, Fernández-Martínez M, Vallicrosa H, Maspons J, Zuccarini P, Carnicer J, Sanders TGM, Krüger I, Obersteiner M, Janssens IA, Ciais P, Sardans J (2020) Increasing atmospheric CO_2 concentrations correlate with declining nutritional status of European forests. Commun Biol 3:125

Persson T (1980) Structure and function of northern coniferous forests – an ecosystem study. Ecological Bulletins (Stockholm) 32:604

Plaut JA, Yepez EA, Hill J, Pangle R, Sperry JS, Pockman WT, Mcdowell NG (2012) Hydraulic limits preceding mortality in a piñon–juniper woodland under experimental drought. Plant Cell Environ 35(9):1601–1617. https://doi.org/10.1111/j.1365-3040.2012.02512.x

Prescott CE, Grayston SJ, Helmisaari H-S, Kaštovská E, Körner C, Lambers H, Meier IC, Millard P, Ostonen I (2020) Surplus Carbon Drives Allocation and Plant-Soil Interactions. Trends Ecol Evol 35:1110–1118

Prieto P, Peñuelas J, Llusià J, Asensio D, Estiarte M (2009a) Effects of experimental warming and drought on biomass accumulation in a Mediterranean shrubland. Plant Ecol 205(2):179–191. https://doi.org/10.1007/s11258-009-9608-1

Prieto P, Peñuelas J, Lloret F, Llorens L, Estiarte M (2009b) Experimental drought and warming decrease diversity and slow down post-fire succession in a Mediterranean shrubland. Ecography 32(4):623–636. https://doi.org/10.1111/j.1600-0587.2009.05738.x

Pritchard SG, Mosjidis C, Peterson CM, Runion GB, Rogers HH (1998) Anatomical and morphological alterations in longleaf pine needles resulting from growth in elevated CO_2: interactions with soil resource availability. Int J Plant Sci 159:1002–1009

Pritchard SG, Strand AE, McCormack ML, Davis MA, Finzi AC, Jackson RB, Matamala R, Rogers HH, Oren RAM (2008) Fine root dynamics in a loblolly pine forest are influenced by free-air-CO_2-enrichment: a six-year-minirhizotron study. Glob Chang Biol 14:588–602

Räisänen T, Ryyppö A, Julkunen-Tiitto R, Kellomäki S (2008a) Effects of elevated CO_2 and temperature on secondary compounds in the needles of Scots pine (*Pinus sylvestris* L.). Trees 22(1):121–135. https://doi.org/10.1007/s00468-007-0175-6

Räisänen T, Ryyppö A, Kellomäki S (2008b) Effects of elevated CO_2 and temperature on monoterpene emission of Scots pine (*Pinus sylvestris* L.). Atmos Environ 42(18):4160–4171. https://doi.org/10.1016/j.atmosenv.2008.01.023

Raison RJ, Myers BJ (1992) The biology of forest growth experiment: linking water and nitrogen availability to the growth of *Pinus radiata*. For Ecol Manag 52(1):279–308. https://doi.org/10.1016/0378-1127(92)90506-5

Rasmussen L, Beier C, Bergstedt A (2002) Experimental manipulations of old pine forest ecosystems to predict the potential tree growth effects of increased CO_2 and temperature in a future climate. For Ecol Manag 158(1):179–188. https://doi.org/10.1016/S0378-1127(00)00677-0

Rehfeldt GE, Ying CC, Spittlehouse DL, Hamilton DA Jr (1999) Genetic responses to climate in Pinus Contorta: niche breadth, climate change, and reforestation. Ecol Monogr 69(3):375–407. https://doi.org/10.1890/0012-9615(1999)069[0375:GRTCIP]2.0.CO;2

Reich PB, Koike T, Gower ST, Schoettle AW (1995) 8 – Causes and consequences of variation in conifer leaf life-span. In: Smith WK, Hinckley TM (eds) Ecophysiology of coniferous forests. Academic, San Diego, pp 225–254

Reich PB, Sendall KM, Stefanski A, Wei X, Rich RL, Montgomery RA (2016) Boreal and temperate trees show strong acclimation of respiration to warming. Nature 531(7596):Article 7596. https://doi.org/10.1038/nature17142

Reich PB, Bermudez R, Montgomery RA, Rich RL, Rice KE, Hobbie SE, Stefanski A (2022) Even modest climate change may lead to major transitions in boreal forests. Nature 608(7923):Article 7923. https://doi.org/10.1038/s41586-022-05076-3

Reinhardt K, Germino MJ, Kueppers LM, Domec J-C, Mitton J (2015) Linking carbon and water relations to drought-induced mortality in *Pinus flexilis* seedlings. Tree Physiol 35(7):771–782. https://doi.org/10.1093/treephys/tpv045

Repo T, Hänninen H, Kellomäki S (1996) The effects of long-term elevation of air temperature and CO on the frost hardiness of Scots pine. Plant Cell Environ 19(2):209–216. https://doi.org/10.1111/j.1365-3040.1996.tb00242.x

Rigling A, Bigler C, Eilmann B, Feldmeyer-Christe E, Gimmi U, Ginzler C, Dobbertin M (2013) Driving factors of a vegetation shift from Scots pine to pubescent oak in dry Alpine forests. Glob Chang Biol 19(1):229–240. https://doi.org/10.1111/gcb.12038

Rodriguez-Dominguez CM, Brodribb TJ (2020) Declining root water transport drives stomatal closure in olive under moderate water stress. New Phytol 225(1):126–134. https://doi.org/10.1111/nph.16177

Rogers A, Ellsworth DS (2002) Photosynthetic acclimation of *Pinus taeda* (loblolly pine) to long-term growth in elevated pCO_2 (FACE). Plant Cell Environ 25:851–858

Rogers A, Humphries SW (2000) A mechanistic evaluation of photosynthetic acclimation at elevated CO_2. Glob Chang Biol 6:1005–1011

Ruehr NK, Martin JG, Law BE (2012) Effects of water availability on carbon and water exchange in a young ponderosa pine forest: above- and belowground responses. Agric For Meteorol 164:136–148. https://doi.org/10.1016/j.agrformet.2012.05.015

Sage RF, Kubien DS (2007) The temperature response of C3 and C4 photosynthesis. Plant Cell Environ 30(9):1086–1106. https://doi.org/10.1111/j.1365-3040.2007.01682.x

Sala A, Piper F, Hoch G (2010) Physiological mechanisms of drought-induced tree mortality are far from being resolved. New Phytol 186(2):274–281

Salazar-Tortosa D, Castro J, Rubio de Casas R, Viñegla B, Sánchez-Cañete EP, Villar-Salvador P (2018) Gas exchange at whole plant level shows that a less conservative water use is linked to a higher performance in three ecologically distinct pine species. Environ Res Lett 13(4):045004. https://doi.org/10.1088/1748-9326/aab18f

Samuelson L, Stokes T, Cooksey T, McLemore P III (2001) Production efficiency of loblolly pine and sweetgum in response to four years of intensive management. Tree Physiol 21(6):369–376. https://doi.org/10.1093/treephys/21.6.369

Samuelson LJ, Johnsen K, Stokes T (2004) Production, allocation, and stemwood growth efficiency of *Pinus taeda* L. stands in response to 6 years of intensive management. For Ecol Manag 192(1):59–70. https://doi.org/10.1016/j.foreco.2004.01.005

Samuelson L, Mathew R, Stokes T, Feng Y, Aubrey D, Coleman M (2009) Soil and microbial respiration in a loblolly pine plantation in response to seven years of irrigation and fertilization. For Ecol Manag 258(11):2431–2438. https://doi.org/10.1016/j.foreco.2009.08.020

Samuelson LJ, Pell CJ, Stokes TA (2014) Two-year throughfall and fertilization effects on leaf physiology and growth of loblolly pine in the Georgia Piedmont. For Ecol Manag 330:29–37

Samuelson LJ, Kane MB, Markewitz D, Teskey RO, Akers M, Stokes TA, Qi J (2018) Fertilization increased leaf water use efficiency and growth of *Pinus taeda* subjected to five years of throughfall reduction. Can J For Res 48(2):227–236. https://doi.org/10.1139/cjfr-2017-0357

Samuelson LJ, Stokes TA, Ramirez MR, Mendonca CC (2019) Drought tolerance of a *Pinus palustris* plantation. For Ecol Manag 451:117557. https://doi.org/10.1016/j.foreco.2019.117557

Schlesinger WH, Bernhardt ES, DeLucia EH, Ellsworth DS, Finzi AC, Hendrey GR, Hofmockel KS, Lichter J, Matamala R, Moore D, Oren R, Pippen JS, Thomas RB (2006) The Duke Forest FACE experiment: CO_2 enrichment of a Loblolly Pine Forest. In: Nösberger J, Long SP, Norby

RJ, Stitt M, Hendrey GR, Blum H (eds) Managed ecosystems and CO_2: case studies, processes, and perspectives. Springer, Berlin/Heidelberg, pp 197–212

Schönbeck LC, Schuler P, Lehmann MM, Mas E, Mekarni L, Pivovaroff AL, Turberg P, Grossiord C (2022) Increasing temperature and vapour pressure deficit lead to hydraulic damages in the absence of soil drought. Plant Cell Environ 45:3275–3289

Schwartz NB, Medvigy D, Tijerin J, Pérez-Aviles D, Rivera-Polanco D, Pereira D, Werden L, Du D, Arnold L, Powers JS (2022) Intra-annual variation in microclimatic conditions in relation to vegetation type and structure in two tropical dry forests undergoing secondary succession. For Ecol Manag 511:120132

Sharma RP, Breidenbach J (2015) Modeling height-diameter relationships for Norway spruce, Scots pine, and downy birch using Norwegian national forest inventory data. For Sci Technol 11(1):44–53

Shao J, Yuan T, Li Z, Li N, Liu H, Bai SH, Zhou X (2019) Plant evolutionary history mainly explains the variance in biomass responses to climate warming at a global scale. New Phytol 222(3):1338–1351. https://doi.org/10.1111/nph.2019.222.issue-3. https://doi.org/10.1111/nph.15695

Shephard NT, Joshi O, Meek CR, Will RE (2021) Long-term growth effects of simulated-drought, mid-rotation fertilization, and thinning on a loblolly pine plantation in southeastern Oklahoma, USA. Forest Ecology and Management, 494, p.119323

Snowdon P, Benson ML (1992) Effects of combinations of irrigation and fertilisation on the growth and above-ground biomass production of *Pinus radiata*. For Ecol Manag 52(1):87–116. https://doi.org/10.1016/0378-1127(92)90497-W

Sohlenius B, Wasilewska L (1984) Influence of irrigation and fertilization on the nematode community in a Swedish pine forest soil. J Appl Ecol 21(1):327–342. https://doi.org/10.2307/2403057

Solly EF, Lindahl BD, Dawes MA, Peter M, Souza RC, Rixen C, Hagedorn F (2017) Experimental soil warming shifts the fungal community composition at the alpine treeline. New Phytol 215(2):766–778. https://doi.org/10.1111/nph.14603

Souza RC, Solly EF, Dawes MA, Graf F, Hagedorn F, Egli S, Clement CR, Nagy L, Rixen C, Peter M (2017) Responses of soil extracellular enzyme activities to experimental warming and CO_2 enrichment at the alpine treeline. Plant Soil 416(1):527–537. https://doi.org/10.1007/s11104-017-3235-8

Sperry JS, Love DM (2015) What plant hydraulics can tell us about responses to climate- change droughts. New Phytol 207(1):14–27. https://doi.org/10.1111/nph.13354

Stefanski A, Bermudez R, Sendall KM, Montgomery RA, Reich PB (2020) Surprising lack of sensitivity of biochemical limitation of photosynthesis of nine tree species to open-air experimental warming and reduced rainfall in a southern boreal forest. Glob Chang Biol 26(2):746–759. https://doi.org/10.1111/gcb.14805

Steidinger BS, Bhatnagar JM, Vilgalys R, Taylor JW, Qin C, Zhu K, Bruns TD, Peay KG (2020) Ectomycorrhizal fungal diversity predicted to substantially decline due to climate changes in North American Pinaceae forests. J Biogeogr 47:772–782

Sterck FJ, Zweifel R, Sass-Klaassen U, Chowdhury Q (2008) Persisting soil drought reduces leaf specific conductivity in Scots pine (*Pinus sylvestris*) and pubescent oak (*Quercus pubescens*). Tree Physiol 28(4):529–536. https://doi.org/10.1093/treephys/28.4.529

Streit K, Siegwolf RTW, Hagedorn F, Schaub M, Buchmann N (2014a) Lack of photosynthetic or stomatal regulation after 9 years of elevated [CO_2] and 4 years of soil warming in two conifer species at the alpine treeline. Plant Cell Environ 37(2):315–326. https://doi.org/10.1111/pce.12197

Streit K, Hagedorn F, Hiltbrunner D, Portmann M, Saurer M, Buchmann N, Wild B, Richter A, Wipf S, Siegwolf RTW (2014b) Soil warming alters microbial substrate use in alpine soils. Glob Chang Biol 20(4):1327–1338. https://doi.org/10.1111/gcb.12396

Subke JA, Inglima I, Cotrufo MF (2006) Trends and methodological impacts in soil CO_2 partitioning: a metaanalytical review. Glob Chang Biol 12:921–943

Suwa M, Katul GG, Oren R, Andrews J, Pippen J, Mace A, Schlesinger WH (2004) Impact of elevated atmospheric CO_2 on forest floor respiration in a temperate pine forest. Glob Biogeochem Cycles 18:GB2013. https://doi.org/10.1029/2003GB002182

Tang Z, Sword MA, Chambers JL, Barnett JP (2004) Interactive effects of fertilization and throughfall exclusion on the physiological responses and whole-tree carbon uptake of mature loblolly pine. Can J Bot 82(6):850–861. https://doi.org/10.1139/b04-064

Tardieu F, Simonneau T (1998) Variability among species of stomatal control under fluctuating soil water status and evaporative demand: modelling isohydric and anisohydric behaviours. J Exp Bot 49:419–432

Teskey RO (1997) Combined effects of elevated CO_2 and air temperature on carbon assimilation of *Pinus taeda* trees. Plant Cell Environ 20(3):373–380. https://doi.org/10.1046/j.1365-3040.1997.d01-75.x

Teskey RO, Fites JA, Samuelson LJ, Bongarten BC (1986) Stomatal and nonstomatal limitations to net photosynthesis in *Pinus taeda* L. under different environmental conditions. Tree Physiol 2(1-2-3):131–142. https://doi.org/10.1093/treephys/2.1-2-3.131

Tian D, Reich PB, Chen HYH, Xiang Y, Luo Y, Shen Y, Meng C, Han W, Niu S (2019) Global changes alter plant multi-element stoichiometric coupling. New Phytol 221:807–817

Tor-ngern P, Oren R, Ward EJ, Palmroth S, McCarthy HR, Domec J-C (2015) Increases in atmospheric CO_2 have little influence on transpiration of a temperate forest canopy. New Phytol 205:518–525

Tor-ngern P, Oren R, Oishi AC, Uebelherr JM, Palmroth S, Tarvainen L, Näsholm T (2017) Ecophysiological variation of transpiration of pine forests: synthesis of new and published results. Ecol Appl 27(1):118–133. https://doi.org/10.1002/eap.1423

Trenberth KE (2011) Changes in precipitation with climate change. Clim Res 47(1-2):123–138

Trichet P, Loustau D, Lambrot C, Linder S (2008) Manipulating nutrient and water availability in a maritime pine plantation: effects on growth, production, and biomass allocation at canopy closure. Ann For Sci 65(8):814–814. https://doi.org/10.1051/forest:2008060

Trugman AT, Fenton NJ, Bergeron Y, Xu X, Welp LR, Medvigy D (2016) Climate, soil organic layer, and nitrogen jointly drive forest development after fire in the North American boreal zone. J Adv Model Earth Syst 8(3):1180–1209

Tyree MT, Sperry JS (1988) Do woody plants operate near the point of catastrophic xylem dysfunction caused by dynamic water stress? 1: Answers from a model. Plant Physiol 88(3):574–580. https://doi.org/10.1104/pp.88.3.574

van Breemen N, Jenkins A, Wright RF, Beerling DJ, Arp WJ, Berendse F, Beier C, Collins R, van Dam D, Rasmussen L, Verburg PSJ, Wills MA (1998) Impacts of elevated carbon dioxide and temperature on a boreal forest ECOSYSTEM (CLIMEX Project). Ecosystems 1(4):345–351. https://doi.org/10.1007/s100219900028

Voelker SL, Muzika R-M, Guyette RP, Stambaugh MC (2006) Historical CO_2 growth enhancement declines with age in Quercus and Pinus. Ecol Monogr 76:549–564

Volanen V, Peltola H, Rouvinen I, Kellomäki S (2006) Impacts of long-term elevation of atmospheric carbon dioxide concentration and temperature on the establishment, growth and mortality of boreal Scots pine branches. Scand J For Res 21(2):115–123. https://doi.org/10.1080/02827580600604795

Wang K-Y (1996) Canopy CO_2 exchange of Scots pine and its seasonal variation after four-year exposure to elevated CO_2 and temperature. Agric For Meteorol 82(1):1–27. https://doi.org/10.1016/0168-1923(96)02342-8

Wang K, Kellomäki S, Laitinen K (1995) Effects of needle age, long-term temperature and CO_2 treatments on the photosynthesis of Scots pine. Tree Physiol 15(4):211–218. https://doi.org/10.1093/treephys/15.4.211

Wang KY, Zha T, Kellomäki S (2002) Measuring and simulating crown respiration of scots pine with increased temperature and carbon dioxide enrichment. Ann Bot 90(3):325–335. https://doi.org/10.1093/aob/mcf191

Wang K-Y, Kellomäki S, Zha T (2003) Modifications in photosynthetic pigments and chlorophyll fluorescence in 20-year-old pine trees after a four-year exposure to carbon dioxide and temperature elevation. Photosynthetica 41(2):167–175. https://doi.org/10.1023/B:PHOT.0000011948.00870.db

Wang K-Y, Kellomäki S, Zha T, Peltola H (2005) Annual and seasonal variation of sap flow and conductance of pine trees grown in elevated carbon dioxide and temperature. J Exp Bot 56(409):155–165. https://doi.org/10.1093/jxb/eri013

Ward JS (2015) Improving competitive status of oak regeneration using stand management and prescribed fires. J Sustain For 34(1-2):105–124. https://doi.org/10.1080/10549811.2014.973606

Ward EJ, Oren R, Sigurdsson BD, Jarvis PG, Linder S (2008) Fertilization effects on mean stomatal conductance are mediated through changes in the hydraulic attributes of mature Norway spruce trees. Tree Physiol 28:579–596

Ward EJ, Oren R, Bell DM, Clark JS, McCarthy HR, Kim H-S, Domec J-C (2013) The effects of elevated CO_2 and nitrogen fertilization on stomatal conductance estimated from 11 years of scaled sap flux measurements at Duke FACE. Tree Physiol 33:135–151

Ward EJ, Oren R, Seok Kim H, Kim D, Tor-ngern P, Ewers BE, McCarthy HR, Oishi AC, Pataki DE, Palmroth S, Phillips NG, Schäfer KVR (2018) Evapotranspiration and water yield of a pine-broadleaf forest are not altered by long-term atmospheric [CO_2] enrichment under native or enhanced soil fertility. Glob Chang Biol 24:4841–4856

Waterworth R, Raison RJ, Brack C, Benson M, Khanna P, Paul K (2007) Effects of irrigation and N fertilization on growth and structure of *Pinus radiata* stands between 10 and 29 years of age. For Ecol Manag 239(1):169–181. https://doi.org/10.1016/j.foreco.2006.12.003

Way DA, Montgomery RA (2015) Photoperiod constraints on tree phenology, performance and migration in a warming world. Plant Cell Environ 38(9):1725–1736. https://doi.org/10.1111/pce.12431

Way DA, Oren R (2010) Differential responses to changes in growth temperature between trees from different functional groups and biomes: a review and synthesis of data. Tree Physiol 30(6):669–688. https://doi.org/10.1093/treephys/tpq015

Way DA, Ladeau SL, McCarthy HR, Clark JS, Oren RAM, Finzi AC, Jackson RB (2010) Greater seed production in elevated CO_2 is not accompanied by reduced seed quality in *Pinus taeda* L. Glob Chang Biol 16:1046–1056

West AG, Hultine KR, Jackson TL, Ehleringer JR (2007) Differential summer water use by *Pinus edulis* and *Juniperus osteosperma* reflects contrasting hydraulic characteristics. Tree Physiol 27(12):1711–1720. https://doi.org/10.1093/treephys/27.12.1711

Whitehead D (1998) Regulation of stomatal conductance and transpiration in forest canopies. Tree Physiol 18(8–9):633–644. https://doi.org/10.1093/treephys/18.8-9.633

Whitehead D, Jarvis PG (1981) Coniferous forests and plantations water deficits and plant growth, vol 6. Academic, pp 49–152. (Reprinted from: In File)

Whitehead D, Sheriff DW, Greer DH (1983) The relationship between stomatal conductance, transpiration rate and tracheid structure in *Pinus radiata* clones grown at different water vapour saturation deficits. Plant Cell Environ 6(9):703–710. https://doi.org/10.1111/1365-3040.ep11589331

Whitehead D, Edwards WRN, Jarvis PG (1984) Conducting sapwood area, foliage area, and permeability in mature trees of Piceasitchensis and Pinuscontorta. Can J For Res 14(6):940–947. https://doi.org/10.1139/x84-166

Wieser G, Grams TEE, Matyssek R, Oberhuber W, Gruber A (2015) Soil warming increased whole-tree water use of *Pinus cembra* at the treeline in the Central Tyrolean Alps. Tree Physiol 35(3):279–288. https://doi.org/10.1093/treephys/tpv009

Will RE, Wilson SM, Zou CB, Hennessey TC (2013) Increased vapor pressure deficit due to higher temperature leads to greater transpiration and faster mortality during drought for tree seedlings common to the forest–grassland ecotone. New Phytol 200(2):366–374. https://doi.org/10.1111/nph.12321

Will RE, Fox T, Akers M, Domec J-C, González-Benecke C, Jokela EJ, Martin TA (2015) A range-wide experiment to investigate nutrient and soil moisture interactions in loblolly pine plantations. Forests 6(6):2014–2028

Wright AJ, Collins SL (2023) Drought experiments need to incorporate atmospheric drying to better simulate climate change. BioScience

Wright IJ, Reich PB, Westoby M, Ackerly DD, Baruch Z, Bongers F, Cavender-Bares J, Chapin T, Cornelissen JHC, Diemer M, Flexas J, Garnier E, Groom PK, Gulias J, Hikosaka K, Lamont BB, Lee T, Lee W, Lusk C, Midgley JJ, Navas M-L, Niinemets U, Oleksyn J, Osada N, Poorter H, Poot P, Prior L, Pyankov VI, Roumet C, Thomas SC, Tjoelker MG, Veneklaas EJ, Villar R (2004) The worldwide leaf economics spectrum. Nature 428:821–827

Xiao CW, Janssens IA, Curiel Yuste J, Ceulemans R (2006) Variation of specific leaf area and upscaling to leaf area index in mature Scots pine. Trees 20:304–310

Xu M, DeBiase TA, Qi Y, Goldstein A, Liu Z (2001) Ecosystem respiration in a young ponderosa pine plantation in the Sierra Nevada Mountains, California. Tree Physiol 21:309–318

Xu X, Medvigy D, Powers JS, Becknell JM, Guan K (2016) Diversity in plant hydraulic traits explains seasonal and inter-annual variations of vegetation dynamics in seasonally dry tropical forests. New Phytol 212(1):80–95

Zha T, Ryyppö A, Wang K-Y, Kellomäki S (2001) Effects of elevated carbon dioxide concentration and temperature on needle growth, respiration and carbohydrate status in field-grown Scots pines during the needle expansion period. Tree Physiol 21(17):1279–1287. https://doi.org/10.1093/treephys/21.17.1279

Zha T, Wang K-Y, Ryyppö A, Kellomäki S (2002) Impact of needle age on the response of respiration in Scots pine to long-term elevation of carbon dioxide concentration and temperature. Tree Physiol 22(17):1241–1248. https://doi.org/10.1093/treephys/22.17.1241

Zha T-S, Kellomäki S, Wang K-Y, Ryyppö A (2005) Respiratory responses of Scots pine stems to 5 years of exposure to elevated CO_2 concentration and temperature. Tree Physiol 25(1):49–56. https://doi.org/10.1093/treephys/25.1.49

Zhang Y, Vogel JG, Meek C, Will R, Wilson D, West J (2016) Wood decomposition by microbes and macroinvertebrates, and soil CO_2 efflux vary in response to throughfall reduction and fertilization in a loblolly pine (*Pinus taeda* L.) plantation. For Ecol Manag 382:10–20. https://doi.org/10.1016/j.foreco.2016.09.049

Zhou Y-M, Wang C-G, Han S-J, Cheng X-B, Li M-H, Fan AN, Wang X-X (2011) Species-specific and needle age-related responses of photosynthesis in two Pinus species to long-term exposure to elevated CO_2 concentration. Trees 25:163–173

Zweifel R, Etzold S, Sterck F, Gessler A, Anfodillo T, Mencuccini M, Rigling A (2020) Determinants of legacy effects in pine trees – implications from an irrigation-stop experiment. New Phytol 227(4):1081–1096. https://doi.org/10.1111/nph.16582

"Pine Death in the Anthropocene"

William M. Hammond, Medelin E. Kant, Marylou Mantova, Gerard Sapes, María Elena Fernández, Javier Gyenge, Gabriel Gatica, Rosana López Rodríguez, Yakir Preisler, Cuauhtémoc Saenz-Romero, Andreas P. Wion, and Craig D. Allen

Abstract Earth's warming climate has caused pulses of forest mortality, associated with hotter-drought stress. Further warming is expected to exacerbate this phenomenon, with maximum temperatures approaching an intensity that is hot enough to kill. A recently assembled global database of climate-induced forest mortality events found that of the 157 species to have experienced rapid pulses of elevated mortality (e.g., die-off events) since 1970, 26 belong to the genus *Pinus*, in the family *Pinaceae*. Earth's oldest living organisms are pine trees, yet *Pinus* may be one of the most vulnerable tree genera on Earth in the Anthropocene, a proposed geologic epoch characterized by rapid climate warming from human emissions. Gymnosperms broadly, and *Pinus* especially, have had a high number of species impacted by hotter-drought triggered pulses of forest mortality in recent decades, with massive conifer mortality potential

W. M. Hammond (✉) · M. Mantova · G. Sapes
Agronomy Department, Institute of Food and Agricultural Sciences, University of Florida, Florida, USA
e-mail: williamhammond@ufl.edu; gsapes@ufl.edu

M. E. Kant
School of Natural Resources and Environment, Institute of Food and Agricultural Sciences, University of Florida, Florida, USA
e-mail: mkant@ufl.edu

M. E. Fernández · J. Gyenge · G. Gatica
Instituto de Innovación para la Producción Agropecuaria y el Desarrollo Sostenible, UEDD INTA CONICET. LIA Forestia (INTA-Argentina, INRAE-Francia)., Buenos Aires, Argentina
e-mail: fernandez.maria@inta.gob.ar; gyenge.javier@inta.gob.ar; mggatica@unsj-cuim.edu.ar

R. L. Rodríguez
Departamento de Sistemas y Recursos Naturales, ETSI Montes, Forestal y del Medio Natural, Universidad Politécnica de Madrid, Madrid, Spain
e-mail: rosana.lopez@upm.es

D. M. Johnson et al. (eds.), *Physiology and Ecology of Pines Worldwide*, Tree Physiology 9, https://doi.org/10.1007/978-3-032-23001-0_10

under further anthropogenic warming. However, pine tree mortality events may be more reported due to their typically temperate-to-boreal distribution, occurrence in highly studied areas, economic importance, and dominating relatively large geographic areas, in conditions considered marginal for many other genera of trees. Here, we synthesize global observations of pine mortality events, the reported climatic and biotic drivers, and physiological thresholds known for *Pinus* species. We present case studies in five regions documenting pine mortality mechanisms and patterns during an ongoing megadrought in the Southwestern USA, at the epicenter of *Pinus* diversity in Mexico, alongside multiple land-use legacies in the Mediterranean, and in planted pine forests of both Argentina and Israel. Of the 116 pine species, drought was the most frequently reported driver of mortality events, followed by hotter droughts. Of measured survival traits, drought tolerance traits are more widely known, while by contrast, hot thermal limits have been published for few *Pinus* species, highlighting a major unknown for the response of pines to further warming and hotter droughts. Evidence from manipulative experiments and climate models suggest that the coming decades may become hot enough to kill pine trees across parts of their native ranges, suggesting heat extremes represent an emerging threat to *Pinus* survival. Understanding the interaction between hydraulic and thermal limits will be critical for predicting the fate of pines in coming decades. We identify critical research questions for understanding pine death in the Anthropocene. While *Pinus* are likely to persist and remain ecologically, economically, and culturally important across many regions, their distribution and role in Earth's forests may undergo dramatic reorganization by the end of this century. Every action to further limit anthropogenic warming will likely collectively influence the fate of this iconic genus on our changing planet.

Y. Preisler
School of Engineering and Applied Science, Harvard University, Cambridge, MA, USA

ARO, Volcani institute, Rishon LeZion, Israel

C. Saenz-Romero
Instituto de Investigaciones sobre los Recursos Naturales, Universidad Michoacana de San Nicolás de Hidalgo, Morelia, Michoacán, Mexico
e-mail: cuauhtemoc.saenz@umich.mx

A. P. Wion
U.S. Geological Survey, New Mexico Landscapes Field Station, Santa Fe, NM, USA
e-mail: awion@usgs.gov

C. D. Allen
Department of Geography & Environmental Studies, University of New Mexico, Albuquerque, NM, USA
e-mail: craigdallen@unm.edu

1 Introduction

Earth's warming climate has caused events of forest mortality, associated with the recent emergence of greater hotter-drought stress during the Anthropocene. The Anthropocene, a proposed geologic epoch not yet accepted by stratigraphers, is associated with the rapid warming of Earth's atmosphere since ~1950 associated with anthropogenic emissions. Further anthropogenic warming is expected to exacerbate the phenomenon of tree mortality in coming decades (Hammond et al. 2022; Mcdowell and Allen 2015). A recently assembled global database of climate-induced forest mortality found that of the 157 species to have experienced rapid events of elevated mortality since 1970, e.g., observed mortality greatly exceeding background mortality rates (Hammond et al. 2022), 25 belong to the genus *Pinus*, in the family Pinaceae. Pines evolved 150 million years ago and have persisted through dramatic periods of change in the past (Richardson 2000). Earth's oldest living organisms today are pine trees (Salzer et al. 2019)—yet *Pinus* seems to be one of the tree genera most subject to amplified tree mortality globally in recent warmer decades. Gymnosperms broadly, and *Pinus* especially, may be at risk under further anthropogenic warming, with massive conifer die-offs expected this century both regionally (McDowell et al. 2016) and globally. This eventually could lead to ecosystem tipping points, after which many pine-dominated forests may differ greatly from what we know today in terms of species composition (McDowell et al. 2020; Sáenz-Romero 2024). Perhaps *Pinus* tree mortality has been disproportionately documented due to widespread pine distributions and dominance across large temperate-to-boreal geographic areas with low tree diversity, their economic importance, and thus their overall prominence in highly studied regions with strong forestry and ecological research institutions. Yet in any case, substantial drought-induced pine mortality has been reported for over one in five species of *Pinus* (26 of the total 116 species) (Leslie et al. 2018, Fig. 1) in the last five decades alone (Hammond et al. 2022). While there are many drivers, mechanisms, and inciting factors associated with pine mortality (McDowell et al. 2022a; Hartmann et al. 2022), this book chapter will focus especially on pine death associated with abiotic hotter-drought stress.

The spatial distribution of *Pinus* mortality events published in the global tree mortality database spans all forested continents and biomes where *Pinus* naturally occur, which is restricted to the northern hemisphere (Fig. 1). Mean annual temperatures at these sites range from −2.1 to 21.9 °C, mean annual precipitation ranges from 207 to 1528 mm, and elevation ranges from 16 to 3284 m asl (Fig. 2; data source: Fick and Hijmans 2017). Despite the broad climatic and elevational distribution of *Pinus* globally, it appears that eventually, hotter drought may be able to cause pulses of elevated mortality wherever *Pinus* species occur.

The genus *Pinus* is divided into two recognized subgenera: subgenus *Strobus* and subgenus *Pinus*. While members of the *Strobus* subgenus are typically thought of as more drought tolerant—with morphological and functional traits that may confer greater tolerance to water deficits (Salazar-Tortosa et al. 2020, but see also Fig. 4

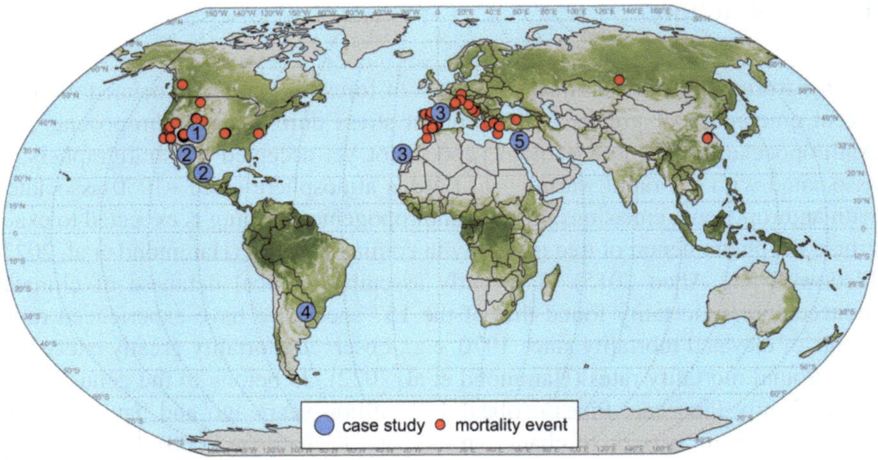

Fig. 1 Locations of Earth's pine tree mortality events. Red circles indicate a mortality event in the global tree mortality database (Hammond et al. 2022), while blue dots indicate mortality events which are detailed further and are numbered as case studies in their order of appearance later in this chapter. The five case studies are (1) *Pinus* vulnerability to hotter drought (observations from the Southwest USA); (2) mortality at the global maximum of *Pinus* species richness (observations from Mexico); (3) pine mortality across the Mediterranean; (4) planted pine mortality in Argentina; and (5) pine mortality in Earth's driest *Pinus* forest—Yatir Forest, Israel. All except Case Study 4 occur within the natural distributions of *Pinus* species, while that case study involves mortality of planted *Pinus* under hotter-drought stress

later in this chapter)—both subgenera have been afflicted by climate-induced pulses of forest mortality (Fig. 3). Compared to other tree genera, *Pinus* has both the highest number of unique species represented in the global tree mortality database (Fig. 3) and the highest proportion of species in any genus having documented mortality events. Regarding the relatedness of *Pinus* and their mortality events, no obvious patterns or clustering emerges when looking across a phylogeny of the genus (Fig. 3), with species that have experienced pulses of mortality broadly distributed. Of note, the *Pinaster* subsection of the subgenus *Pinus* did have mortality events observed in five of seven species (*P. heldreichii*, *P. pinaster*, *P. halepensis*, *P. brutia*, *P. pinea*, *P. canariensis*, and *P. roxburghii*).

In this chapter, we present case studies of globally distributed *Pinus* mortality linked to anthropogenic hotter-drought disturbances, often including climate-amplified biotic agents (Allen et al. 2015). While each of these case studies presents observations on a unique biome, species, and/or ecology, they are unified by unexpected pulses of *Pinus* die-off that occurred over relatively short periods of time. Following the case studies, we present a current understanding of the climatic drivers and the tree physiological mechanisms associated with such pulses of *Pinus* mortality and provide an outlook for *Pinus* on our warming planet. We conclude by proposing critical research questions regarding emergent patterns of *Pinus* mortality and survival in the Anthropocene.

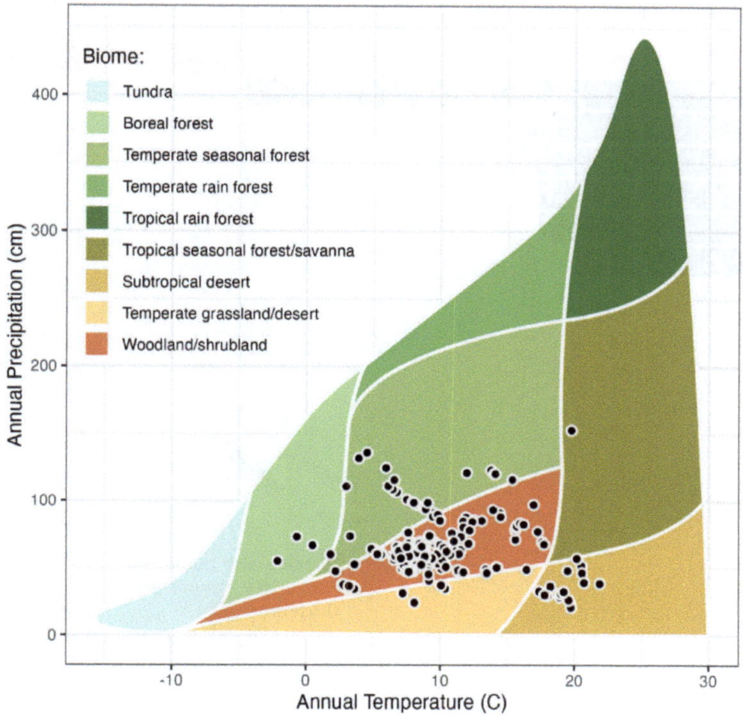

Fig. 2 Whittaker biome plot of pine mortality locations from the global tree mortality database (Hammond et al. 2022), showing the mean annual temperature (°C) and mean annual precipitation (cm). Each point represents a plot where climate-induced pine mortality occurred (Fig. 1). Climate data for mean annual precipitation and mean annual temperature are from TerraClimate (Abatzoglou et al. 2018) and represent the average over 1970–2000

2 Pine Mortality Mechanisms

2.1 Defining Death

The mechanisms of tree mortality are complex, interactive, and often difficult to distinguish for a particular mortality event. As a field, we have used different methods attempting to determine tree death. One strategy determines death by the tree's inability to produce or regenerate organs or tissues during the next vegetative season (Rice et al. 2004; Lloret et al. 2004; Brodribb and Cochard 2009; Anderegg et al. 2012; Barigah et al. 2013; O'Brien et al. 2014; Hammond et al. 2019; Mantova et al. 2021). Another strategy visually assesses tissue color beneath tree bark to determine death (O'Brien et al. 2014; Partelli-Feltrin et al. 2020; Gomez-Gallego et al. 2022). Perhaps the most widespread assessment of mortality in *Pinus* has been the changing of canopy color from green to brown (Adams et al. 2009; Breshears et al. 2021), which often lags many months behind the onset of dying (Blackman et al. 2019; Hammond et al. 2019). It can be difficult to determine when a tree begins the

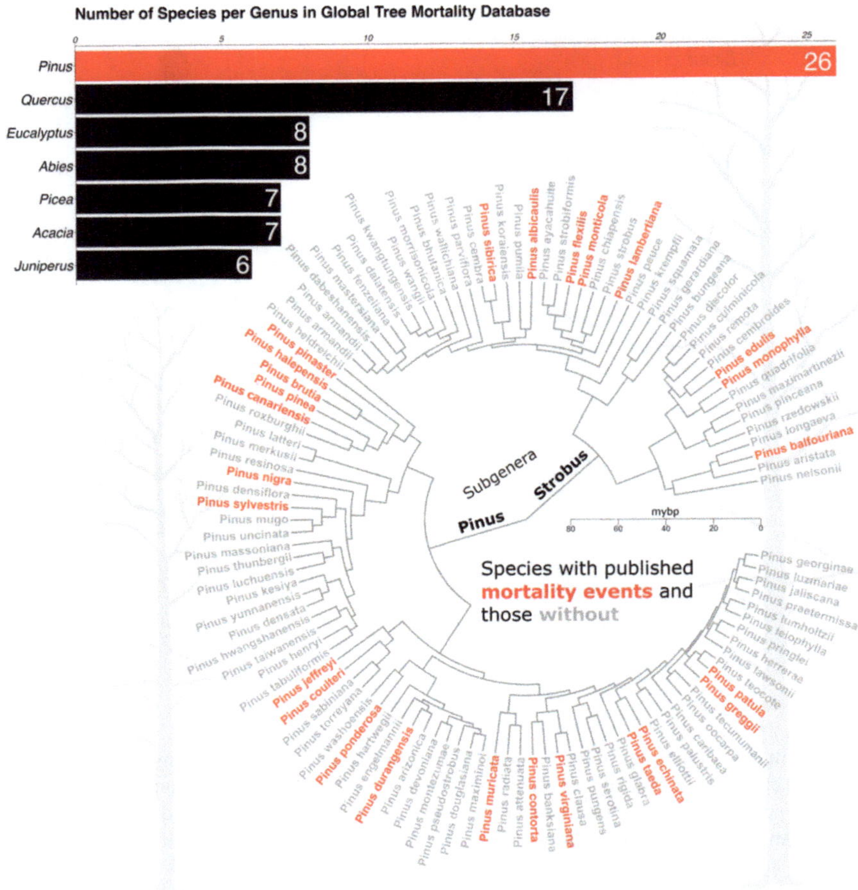

Fig. 3 Phylogenetic tree of *Pinus*, showing species with mortality events observed in the global database or case studies shown in red, while those without mortality events appear in gray. Above the phylogeny, a barplot shows the count of species per genus with recorded mortality events in the global tree mortality database; *Pinus* has more species having experienced mortality, despite being considerably less diverse than *Quercus* and *Eucalyptus*, the second and third genera in terms of total species with reported mortality events, respectively. Phylogeny was made with V.Phylomaker2 (Jin and Qian 2022)

irreversible process of dying that will eventually lead to its death (Hammond et al. 2019).

Studying tree mortality requires a clear definition of tree death. Tree death has been defined as "a thermodynamic equilibrium between the organism and its environment, in which plants no longer have energy gradients to drive metabolism or regenerate" (McDowell et al. 2011), "a complex system failure due to lack of water resources" (Anderegg et al. 2012), or "a tree's [in]ability to recover its physiological functions, such as exchanges of matter and energy with its environment, or to resprout" (Mantova et al. 2022). While definitions differ, for *Pinus*, the lack of recovery and complete arrest of metabolism following stress seems universal in determining tree death.

2.2 *Mechanisms of Drought-Induced Mortality*

Mechanisms driving death include hydraulic failure, observed to be ubiquitous in recent synthesis of global observations, including in three species of *Pinus*, while carbon starvation (reduced nonstructural carbohydrates) was often coincident, but never isolated (Adams et al. 2017). The water potential inducing 50% loss of xylem hydraulic conductance, P_{50}, had long been regarded as a critical threshold for tree mortality risk in conifers (Brodribb and Cochard 2009; Choat et al. 2012), but when directly tested, P_{50} was not observed as lethal for *Pinus* (Hammond et al. 2019), as trees always recovered from P_{50}—with mortality more likely than survival only after passing 80% loss of hydraulic conductivity. Some have argued hydraulic failure by xylem embolism is the symptom rather than the mechanism of tree death, highlighting the importance of plant water pools.

Pinus survive extreme drought stress by maintaining hydraulic supply to their vascular cambium (Hammond et al. 2019)—the secondary meristem critical for development of new xylem to support recovery. Other key traits modulating lethal levels of dehydration are the turgor loss point (TLP), beyond which stomatal closure is at its maximum, and the residual conductance (g_{res}) or rate of water loss after the turgor loss point is reached. Upon reaching TLP, a tree's water loss is reduced to g_{res}, and the time until reaching lethal thresholds of dehydration will be largely determined by the available water pool and g_{res} (Delzon and Cochard 2014; Blackman et al. 2019; Hammond and Adams 2019). Maintaining water pools and transport capacity are linked to the vitality of primary and secondary meristems necessary for recovery and survival.

2.3 *Why Do Pine Trees Die during Hotter-Drought Events?*

2.3.1 Tree-Level Mechanisms

Pine trees are modular organisms and can continue functioning even when many of their modules (leaves, branches, stems, roots, fine roots) are damaged or dead. Losing modules may at times increase tree survival likelihood due to reduced metabolic and water costs, like in ancient, strip-barked bristlecone pines (*P. longaeva*). While some plant tissues are built to maximize performance such as leaf mesophyll, others are built to prevent dysfunction (e.g., leaf cuticles and stem bark that prevents water loss). A lack of water flow into a tree will not necessarily kill it as long as cells remain hydrated (i.e., greatly reduced rates of water loss, the residual conductance). The concepts of drought and heat stress are tied to physiological and morphological characteristics and plastic capacity of each tree. Physiological drought stress generally occurs when water supply does not meet atmospheric demand (i.e., water deficit). This happens when either soil-water supply is too low, atmospheric demand is too high, or both. Physiological heat stress generally occurs when tissue temperature raises beyond $T_{critical}$, the temperature at which cell vitality starts to decline. Whether tissues reach or exceed this temperature depends on the surrounding air or soil temperatures, the boundary layer conductance, and the evaporative cooling capacity of plant tissues via transpiration, which inherently depends on plant water pools and flows (Gauthey et al. 2023).

To survive hotter droughts, trees must remain hydrated while at the same time avoiding potentially damaging temperatures. Dehydration rates and time to death become a function of tissue water content, residual conductance of tissues, cell tolerance to heat, low water potentials, and depleted water pools. Tree residual water loss may be further reduced when the surface area of the plant decreases, often as modular dieback of distal organs, including leaves and fine roots and even the mortality of entire canopy branches. This modular mortality may arise due to a lack of available carbon to meet demands of respiration, observed in manipulative experiments asymmetrically shading lower *Pinus* canopies (Schoonmaker et al. 2014). Additionally, distal organ vulnerability to drought-induced dysfunction (e.g., the hydraulic vulnerability segmentation hypothesis) is especially strong in conifers, including *Pinus* (Johnson et al. 2016), where vulnerability of leaves and distal roots has been observed higher than that of proximal organs like the trunk. While both mechanisms may reduce crown area via modular mortality of organs, dieback will also severely impact future carbon assimilation potential and the tree's ability to outcompete others during times of resource abundance. Compared to other conifers, species in the genus *Pinus* are relatively vulnerable to hydraulic failure, with higher P50 values than many other gymnosperm lineages (Martínez-Vilalta et al. 2004). Even some of the most hydraulically resistant *Pinus* species (e.g., those with lower P50 values) have endured climate-driven mortality events in recent decades—including *P. balourfiana*, *P. edulis*, *P. monophylla*, *P. monticola*, *P. brutia*, and *P. halepensis* (Fig. 4). While the values in Fig. 4 are for *Pinus* growing in their native ranges, it should be noted that often *Pinus* being grown outside their home range for timber can be significantly more vulnerable, with higher P50s (Bulfe and Fernández 2016).

Under heat extremes, pine trees must remain below critical thresholds of cell membrane destabilization (Teskey et al. 2015). The main physiological mechanism of heat tolerance in pine trees is heat dissipation through evapotranspiration. While evapotranspiration can limit warming of leaf tissues, it remains unclear the extent to which this occurs during heatwaves, and evidence that leaves are generally warmer than the air suggests that while cooling occurs, it cannot necessarily keep leaves below air temperature (Urban et al. 2017; Still et al. 2022; Muller et al. 2021; Gauthey et al. 2023), and thus it is more accurate to say that evapotranspiration merely limits warming. Extreme heat events can kill plant tissues even if water is abundant (Still et al. 2023), a phenomenon that may become more prevalent with further anthropogenic climate change. This risk may be increased for *Pinus* species,

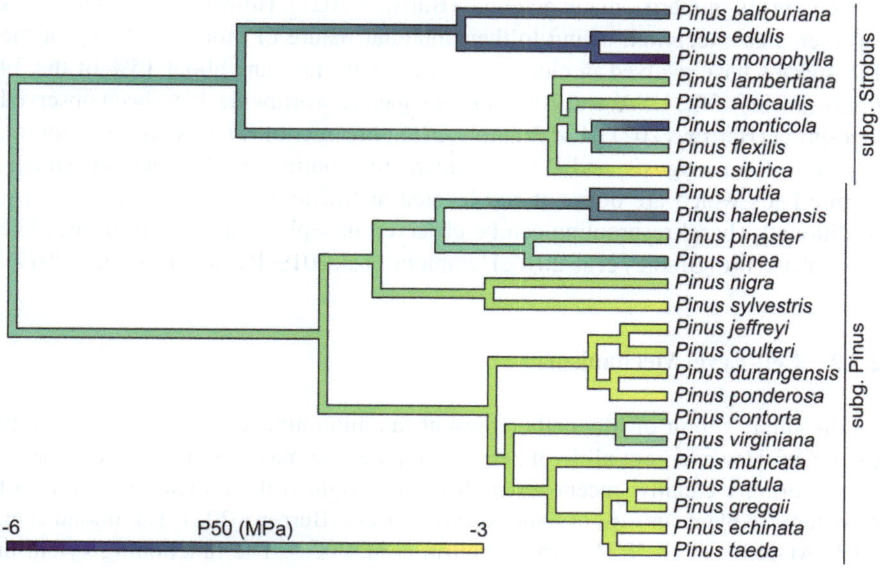

Fig. 4 Phylogeny mapped xylem vulnerability for mortality-afflicted *Pinus* species. Two subgenera (subgenus *Strobus* and subgenus *Pinus*) are labeled on the right of the tree. Branches are colored according to continuous trait evolution mapping of P50, the water potential inducing a 50% loss of conductivity in the xylem (trait values from xylemfunctionaltraits.org). With this method, the branch tips are the observed values, while estimates are generated for prior states. Trait evolution mapping and phylogeny visualization were conducted using the R package "phytools" (Revell 2012)

with leaf morphologies that make their surfaces more strongly coupled to air temperatures than other species (Ewers and Oren 2000). While this strong coupling is a benefit under a wide range of temperatures, at temperature extremes, this may lead to increased leaf death for conifers (Still et al. 2023), and especially pines—as the net benefit of coupling with air temperature becomes a net liability under increasing intensity and frequency of extreme heat events. Under compound heat and drought stress, trees face a dangerous dichotomy: retaining water to stay hydrated while risking lethal overheating or losing water to remain cool at the risk of lethal dehydration. It is unclear what factors, including plant physiological responses and physical and biophysical phenomena, determine how drought-stressed trees might respond with heat dissipation versus hydration (Mills et al. 2024). However, a pine tree that has partially lost the ability to transport water, or that has abscised transpiring tissues, will be less capable of reducing further tissue warming. Unfortunately, relatively little is known about how thermal limits (including $T_{critical}$) may vary across the genus *Pinus* or within a species across its natural range, ontogeny, or both.

When canopies suffer significant dieback during hotter-drought stress, tree survival may depend upon the ability to resprout from dormant meristems. Conifers are distinct in that resprouting is relatively rare when compared with woody angiosperms (Pausas and Keeley 2017). Their poor resprouting capabilities could be related to lack of meristematic potential (Burrows 2021). However, the genus *Pinus* represents an exception, owing to their modular nature of mortality. Many of the 116 species have evolved in close association with fire, and about 15% of the 39 *Pinus* species of the USA, and 20% of *Pinus* species worldwide, have been observed to resprout (Burrows 2021). For example, in mature trees of *P. leiophylla*, *P. seronita*, *P. patula*, *P. canariensis*, and *P. montezumae*, resprouting usually occurs from epicormic buds which are dormant and located at leaf axils throughout the canopy. Additionally, basal resprouting can be observed in saplings after disturbances like drought and fire (Bradley et al. 2016; Hammond et al. 2019; Pausas and Keeley 2017).

2.3.2 Cell-Level Mechanisms

As the smallest unit of life, cells represent the minimum scale at which mortality can occur. However, not all living cells of a pine tree may contribute to organism-level mortality equally; recent research has highlighted the critical importance of meristems in plant survival of hotter-drought stress (Burrows 2021; Hammond et al. 2019; Mantova et al. 2022; Partelli-Feltrin et al. 2023). The functioning xylem in *Pinus* surviving extreme hydraulic failure (e.g., > 80% loss of conductivity) has been observed directly adjacent to the vascular cambium, on which restoring hydraulic conductivity depends (Hammond et al. 2019).

When cells exceed critical thresholds in water content depletion, plasma membranes may rupture (Guadagno et al. 2017), an early sign of cell necrosis (van Doorn 2011; van Doorn and Woltering 2005). Plasma membrane rupture can be the result of biochemical reactions or physical constraints during drought (Mantova et al. 2022; McDowell et al. 2022a, b). Membrane rupture during cellular dehydration may be initiated by lipid peroxidation resulting from reactive oxygen species accumulation (Crowe and Crowe 1992; Cruz de Carvalho et al. 2017). Physical processes include cytorrhysis, when the cell wall deforms resulting in permanent and irreparable shrinkage of the cell (Oertli 1971), or cell cavitation, when a gas bubble forms inside the cell when a critical pressure is reached (Sakes et al. 2016). Both physical phenomena, in interaction with biochemical processes, could lead to the death of all plant cells, including meristematic ones. Presently, it remains unknown if drought-induced cellular mortality in trees, including *Pinus*, primarily results from oxidative stress, cytorrhysis, or cell cavitation.

Under other conditions, for example, when a pine tree faces lethal high temperatures during fire (Partelli-Feltrin et al. 2023) or a heatwave, meristem mortality is also likely to result from membrane damage. When facing extreme heat, oxidative stress could also be amplified, resulting in increased membrane damage (Teskey et al. 2015). However, the thermolability of the lipids constituting the plasma membrane may also directly lead to membrane dysfunction or leakage of cellular contents culminating in cellular death (Savchenko et al. 2002; Wahid et al. 2007). In addition, protein denaturation upon exceeding $T_{critical}$ could lead to cascading malfunctions that create a lethal energy imbalance (Nishad and Nandi 2021). As with drought-induced cellular mortality mechanisms, the relative contribution of oxidative stress, membrane fluidity, and protein denaturation during heat-induced cellular death is presently unknown.

2.4 *Biotic Agents, Other Climate Stressors, and Disturbances Driving Pine Death*

The role of biotic agents in tree mortality, especially for the genus *Pinus*, cannot be understated. Interactions between insects, pathogens, and plant stress have been noted across many forested systems, including those dominated by *Pinus*. Recent reviews and research on this topic, worth further reading, include McDowell et al. 2022a, b and Hartmann et al. 2022. Similarly, the emergence of changes in the timing and severity of frost events (Bojórquez et al. 2019) and sea level rise and salinization of coastal forests (McDowell et al. 2022a, b) and wildfires increasing in frequency, extent, and severity (Jones et al. 2022; Liu et al. 2010), along with storms (Yanoviak et al. 2020), other human disturbances are outside of this chapter's scope but are important drivers of pine death globally. In the mortality event case studies that follow, the first three case studies highlight important biotic agents that were linked to observed *Pinus* mortality events driven by hotter-drought stress, while Case Study 3 highlights interactions with fire.

3 Regional Case Studies of *Pinus* Mortality

Case Study 1 *Pinus* vulnerability to hotter drought—observations from the Southwest USA (Andreas P. Wion and Craig D. Allen)

In the Southwest USA, the mountains and mesas of Arizona (AZ) and New Mexico (NM) are expansively covered by 12.9 million hectares (ha) of conifer-dominated forests and woodlands (Goeking and Menlove 2017; Shaw et al. 2018), including *Juniperus*, *Cupressus*, *Pseudotsuga*, *Abies*, *Picea*, and several species of *Pinus*. Two pines are particularly widespread and prominent here: *Pinus edulis* (two-needle piñon or Colorado piñon) and *Pinus ponderosa* (ponderosa pine). Collectively, these two pine species dominate the ecosystem services provided by forests and woodlands in the Southwest USA but face accelerating tree mortality threats from drought, heat, and associated biotic agents (Fig. 5).

Colorado piñon is the most widespread and abundant conifer in both states, with forest inventories estimating about 2.34 billion piñon trees across seven million hectares (Goeking and Menlove 2017; Shaw et al. 2018). This small-statured, short-needled pine dominates broad expanses of low-elevation semiarid woodlands in the Southwest USA, often mixed with one of several codominant juniper species across a range of ecosystems (Romme et al. 2009). These southwestern low-elevation piñon-dominated woodlands generally grade upslope into adjoining savannas and

Fig. 5 Observed mortality of *Pinus* in New Mexico, USA. In panel (**a**), primarily *Pinus edulis* mortality is observed (with some *P. ponderosa*) in a photograph taken from the eastern Jemez Mountains in the winter of 2002–2003.; in panel (**b**), standing dead *P. edulis* at Bandelier National Monument, New Mexico, USA, taken July 2007, 4.5 years after being killed by hotter drought. Photo credits: Craig D. Allen

forests of large-statured *Pinus ponderosa*, a wide-ranging species across the western USA (Oliver and Ryker 1990). Long-needled ponderosa pine (*Pinus ponderosa*) is particularly well-adapted to and favored by the historical high-frequency surface-fire regimes that characterized pre-1900 mid-elevation dry conifer forests in the Southwest USA (Swetnam and Baisan 1996). Ponderosa pine stand compositions range from monotypic ponderosa savanna and open forests at dry lower and mid-elevations settings, transitioning upward into more mesic mixed-species conifer forests with a significant ponderosa component (Allen et al. 2002). As a result, ponderosa pine is the most extensively distributed and numerous large-sized forest tree in this region, with about 1.2 billion trees across two million hectares in Arizona and New Mexico (Goeking and Menlove 2017; Shaw et al. 2018).

Growth and direct drought-induced mortality of these two widespread Southwest US pine species—as well as associated regional bark beetle irruptions and forest fires—are all closely tied to low winter precipitation and warmer growing season temperatures (Williams et al. 2013). Climate in AZ and NM prominently includes a seasonally bimodal annual precipitation pattern, with widespread cyclonic winter storms bringing snow to upland conifer forests and woodlands, hot and dry springs—in contrast the summer monsoon season is characterized by localized convective thunderstorms, with a regional summer moisture gradient that weakens from southeast to northwest (Jacobs et al. 2008). Southwest US precipitation is highly variable on all timescales—winter precipitation is positively correlated with high-frequency oscillations (2–5 years) in the El Niño-Southern Oscillation (Swetnam and Betancourt 1998) and lower frequency multi-decadal oscillations in the Pacific Decadal Oscillation.

Episodes of drought and heat are globally recognized as natural drivers of tree physiological stress and tree mortality (Allen et al. 2010; Adams et al. 2017; Fensham et al. 2019). Severe decadal-scale droughts in the past have caused regional forest stress (Williams et al. 2013) and elevated tree mortality of both piñon and ponderosa pine populations in the Southwest USA—examples include the 1580s megadrought (Swetnam and Betancourt 1998) and the 1950s drought (Allen and Breshears 1998). Most recently, the emergence in the Southwest USA of strong warming since 1990 (e.g., Gutzler and DuBois 2022), combined with persistent drought since ca. 2000, has resulted in the most severe regional megadrought since at least 800 CE (Williams et al. 2022). This long-lasting megadrought (2000–2022) has amplified multiple tree-killing disturbance processes—interactions among physiological drought stress, bark beetle outbreaks, and high-severity wildfires—to historically unprecedented levels (Williams et al. 2013; Allen 2014), as exemplified by widespread pulses of recent forest die-off in essentially all Southwest US conifer species (Appendix A in Allen et al. 2015), including piñon and ponderosa pines (Fig. 6).

From 1999 to 2021, annual aerial forest health surveys mapped a cumulative regional total of 1,039,108 hectares of piñon mortality, along with 1,929,222 hectares of ponderosa pine mortality, that were attributed to combinations of bark beetles and drought stress (Fig. 6). Note that these pulses of regional pine mortality (Fig. 6) correspond with, and slightly lag behind (see Hammond et al. 2022), multi-year periods of hotter-drought stress in 2000–2004, 2011–2013, and 2019–2022 (Dannenberg et al. 2022; Williams et al. 2022).

Southwest US land-use history of intensive livestock grazing and fire suppression since ca. 1880 CE have contributed to widespread densification of regional ponderosa forests (Allen et al. 2002) and piñon woodlands (Gottfried et al. 1995) during multi-decadal wet periods in the twentieth century (Swetnam and Betancourt 1998; Allen 2014). This historically anomalous increase in the density, leaf area, and aboveground woody biomass of Southwest piñon and ponderosa ecosystems likely resulted in a "structural overshoot" (Jump et al. 2017) beyond levels that are sustainable during inevitable drought periods, further amplifying recent forest drought stress and vulnerability to tree mortality during the ongoing twenty-first-century regional megadrought (Allen 2014).

Similar patterns and trends of greater hotter-drought stress, bark-beetle outbreaks, and *Pinus* mortality have been documented across western North America forests over the past 25 years (Bentz et al. 2010), from British Columbia (Meddens et al. 2012) through California (Fettig et al. 2019) to Mexico (Gómez-Pineda et al. 2022). Larger pine trees often are disproportionately killed during hot droughts for multiple reasons, ranging from the physics of internal tree water transport (Mcdowell and Allen 2015) to host-tree size selection by particular bark beetle taxa (Stephenson et al. 2019).

Given current regional climate projections of further warming, reduced winter precipitation, increased climate extremes, and interannual variability (USGCRP et al. 2018; Gutzler and DuBois 2022), the resultant aridification of the Southwest USA is projected to have pronounced effects on future piñon and ponderosa

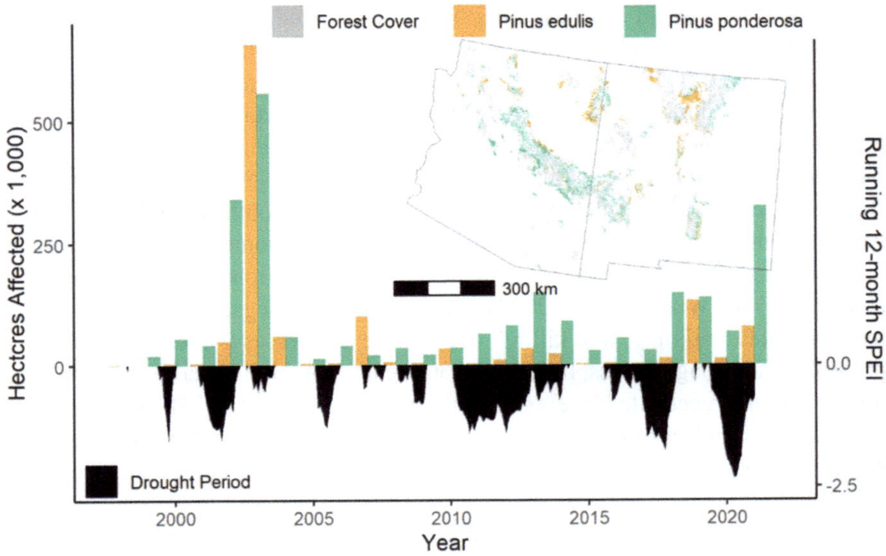

Fig. 6 Southwestern US mortality of *P. ponderosa* and *P. edulis* over time in relation to hotter droughts. Hot drought was quantified using a 12-month rolling monthly Standardized Precipitation Evapotranspiration Index (SPEI) averaged across Arizona and New Mexico for the period 1999–2021, acquired from https://spei.csic.es/spei_database/. Aerial Detection Survey (ADS) data from US Forest Service Region 3 (Arizona and New Mexico, inset map) of the annual mapped area affected by bark beetle-caused tree mortality of piñon and ponderosa pines in Arizona and New Mexico between 1999 and 2021 (USDA FS annual ADS reports; 'Detection Surveys'). Data only include observations of mortality (not canopy dieback), with the total area of each polygon and point observation of tree mortality summed by species for each year. Note that aerial field mapping covers most but not all forest and woodland areas, precise field routes vary between years, and mapped mortality areas typically include some live trees—so while these mapped mortality areas generally provide good estimates of overall spatial patterns and temporal trends, there is some unquantified precision in the actual magnitude of annual tree mortality. For example, particularly note that the initial emergence of massive *P. edulis* mortality in 2002 was greatly underassessed by ADS, especially in New Mexico (CD Allen, field observations), because the spatially extensive but commercially low-value piñon woodlands were not widely flown by ADS surveyors until 2003, justified by the dramatic piñon die-offs of 2002. These surveys rely on visual indicators of mortality (e.g., orange needles), which are known to lag months behind the actual onset of mortality for *Pinus*

ecosystem structures, compositions, and distributions. In particular, diverse empirical and process models project further increases in mortality of Southwest piñon and ponderosa pine in coming decades (Williams et al. 2013; McDowell et al. 2016; Wion et al. 2022). Importantly, impacts of host size selection by biotic agents (Stephenson et al. 2019) and differences in rooting or water table depth may play important parts in determining which trees survive and which die (Trugman et al. 2019).

Despite substantial recent increases in *Pinus* mortality in the Southwest USA, with limited growth recovery or regeneration during the current ongoing regional megadrought, large populations of several billion piñon and ponderosa pine trees currently persist in Arizona and New Mexico (Goeking and Menlove 2017; Shaw et al. 2018). While both *P. edulis* and *P. ponderosa* population dry-edge range margins may be undergoing low-elevation range shifts and ecotone retreats (e.g., Allen and Breshears 1998), so far these species have rarely been extirpated entirely from sites—even in Bandelier National Monument, many small *P. edulis* trees survived the extreme 2002–2003 mortality pulse that killed almost all adult piñon trees. To date, the large, ongoing increases in pulsed die-off events of Southwest US pines reflect necessary, unavoidable, and incremental ecosystem reorganization in response to ongoing climate warming (Allen 2014; McDowell et al. 2020)—arguably the tree mortality effects to date have been beneficial to ecosystem adaptation and resilience to increasingly unprecedented drought and heat extremes by fostering lower forest stand densities, reduced leaf areas, and genetic selection of more drought-tolerant tree populations. Still, given current projections of regional *Pinus* vulnerability to likely decadal warming scenarios, and the uncertainties in realistically modeling the nonlinearity of many climate change-related forest disturbance processes (Allen et al. 2015), there certainly are substantial reasons to be concerned about the potential for tipping-point increases and surprises in the rate and scale of increasingly severe piñon and ponderosa pine die-offs in the Southwest USA with further global warming (Hartmann et al. 2022).

Case Study 2 Mortality at the global maximum of *Pinus* species richness—observations from Mexico (Cuauhtemoc Saenz-Romero)

Mexico is a megabiodiverse country and the center of pine speciation, with 50 native *Pinus* species, ~43% of *Pinus* species globally (Farjon and Styles 1997; Sánchez-González 2008; Sáenz-Ceja et al. 2022). Mexican pines are primarily distributed on the principal mountain ranges, including the Sierra Madre Occidental, Sierra Madre Oriental, and Trans-Mexican Volcanic Belt, but also extend to other regions. They are adapted to a broad range of climates, from the lower altitudinal limit of the pine-oak forest where *P. oocarpa* grows in the tropical dry forest ecotone (~1200 m asl) to the timberline formed by *P. hartwegii* (~4000 m asl). A large variety of relatively short-statured, edible-nut bearing "piñon" pines occupy extremely dry climates, including the broadly distributed *P. cembroides* across vast semiarid regions. This also includes other rare, endemic piñon pines with incredibly narrow distributions, such as *P. nelsoni*, *P. caterinae*, *P. maximartinezii*, *P. culminicola*, *P. pinceana*, and *P. coulteri*, among others. On the wet extreme, fast-growing *P. patula*, globally planted for timber (Dvorak et al. 2000), are naturally distributed along the narrow upper limit of the Mexican cloud forest biome (Perry Jr 1991).

Mexico's immense diversity of native pines species is threatened by increasing climate stress, due to the combination of increased temperatures and intra-annual precipitation variation (larger drought season but also a concentration of precipitation resulting in wettest months becoming wetter; Murray-Tortarolo 2021). Specifically, the typically dry and warm season from March to May is becoming increasingly drier than usual, with warming exceeding all instrumental records. The dry and cold season from November to February is also becoming drier, with milder winter temperatures (Murray-Tortarolo 2021). The combination of a warmer and drier dry season (November to May) threatens pine tree growth and survival before the return of rain during the wet season from June to October (Cruzado-Vargas et al. 2021). Under severe water stress, this can lead to reduced tree growth, defoliation, and a progressive dying of branches and upper crowns (Sáenz-Romero et al. 2020).

Widespread pine tree mortality directly driven by hotter-drought stress is not very common yet in Mexico. Instead, the mortality mechanism appears to be a two-step process. First, trees are physiologically stressed during the exceedingly warm and dry March to May spring season. Second, tree defense mechanisms are weakened during physiological stress, resulting in high susceptibility to lethal biotic agent attacks (e.g., bark beetles, parasitic mistletoe, and other pests, as in Sáenz-Romero et al. 2020, 2023). One way to assess pine forest bark beetle outbreaks at the national scale is analyzing federally issued phytosanitary clear-cut permits for sanitation felling. The number of such authorizations, the extent of the affected area, and the timber volume authorized to be removed are all increasing steadily, indicating widespread pine forest mortality in Mexico and urgently indicating the need for different forest management practices (Gómez-Pineda et al. 2022; CONAFOR 2023).

The pine populations most vulnerable to hotter droughts are located at the xeric limit (sensu Mátyás 2010) of the natural range distribution of each species, either at the extreme lower altitudinal limit or at the southern latitudinal limit. In such locations, pine populations of a given species may already be at their adaptive limit. As climate becomes warmer and drier under further climatic change, tree phenotypic plasticity and ecophysiological mechanisms to avoid or tolerate drought stress may become exhausted (Mátyás 2010; Allen et al. 2015; Sáenz-Romero et al. 2020). An example is the decline of *P. pseudostrobus* stands, the most economically important and fast-growing timber species of the Trans-Mexican Volcanic Belt (central-southern Mexico). At the lower altitudinal limit *of P. pseudostrobus* (~2100–2300 m), some stands show clear signs of defoliation (Fig. 7a; López-Toledo et al. 2017) and are sometimes followed by outbreaks of bark beetles (Rubin-Aguirre et al. 2015). The most frequent and extensive pine beetle outbreaks on *P. pseudostrobus* stands are occurring at the species' xeric limit, where trees are likely weakened by hotter droughts. Recent studies have found that Palmer's Drought Severity Index (a measure of drought stress) indicated significantly drier conditions at the pine beetle outbreak sites preceding and during the year of pine mortality, as assessed through phytosanitary logging permits and long-term climate data (Gómez-Pineda et al. 2022).

At the upper altitudinal limit, even if the pine populations are expanding as minimum temperatures are less limiting (Carrillo-Arizmendi et al. 2022), they are expanding with a small number of individuals (Astudillo-Sánchez et al. 2017, 2019). This expansion is proceeding at a much slower pace than what is needed to keep up with the rate of climatic change. Also, the conicity of the mountains makes that the area available at higher elevations much smaller than at lower elevations. Thus, the total area gained and potentially available to pine populations at higher elevations is much lower than the area lost and/or at risk at lower elevations, as

Fig. 7 (**a**) *Pinus pseudostrobus* trees showing defoliation at the lower altitudinal limit (xeric limit) of its natural range distribution, after a warmer and drier than usual spring season. This site is at 2300 m of altitude, the lower altitudinal limit of the species on the Purépecha Plateau, Michoacán state. (**b**) Trees of *Pinus hartwegii* killed by *Dendroctonus adjunctus* after warmer and drier years (2011–2012 and 2016–2017). Site at 3670 m of altitude (near the timberline of ~4000 m), Nevado de Colima volcano (summit in background). Both (**a**) and (**b**) sites are in the Trans-Mexican Volcanic Belt, central-west Mexico

predicted by suitable climatic habitat modeling along Mexican elevational gradients (Gómez-Pineda et al. 2020).

Unfortunately, pine forest decline due to hotter-drought amplified bark beetle outbreaks is also occurring in very high altitude forests (~4000 m asl), in populations of *P. hartwegii*. The mortality of climate-stressed *P. hartwegii* due to bark beetle *Dendroctonus adjunctus* presents as larger outbreaks in natural protected areas (Fig. 7b). These protected areas may have undergone a period of increasing tree density, in contrast to managed and/or thinned forests, that could increase vulnerability to climate-induced mortality events (Allen and Breshears 1998). This problem may be worsened by a lack of phytosanitary logging, as a regionally accepted idea is that conservation in situ includes, to some extent, human inaction.

The challenge that represents the increasing number and size of bark beetle outbreaks on the Mexican pine forest is perhaps one of the most complicated silvicultural challenges that will be faced by Mexican forest managers in the coming years. Vapor pressure deficit (VPD), a measure of dryness of the air, is very high (annual mean of VPD $\geq$ 1 kPa) during hotter droughts. Even after stomatal closure, trees can lose high amounts of water (via residual conductance) leaving trees highly susceptible to bark beetle attacks, especially for those more aggressive bark beetle species, such as *Dendroctonus frontalis* (Sáenz-Romero et al. 2023). From a traditional ecological perspective, bark beetles usually performed the role of eliminating the older, weakened, and/or declining trees, which could be seen as a positive dynamic to make possible a natural renewal of the older stands. However, when the proportion of trees weakened in a forest becomes the majority, not due to age or high stand density, but to climatic change, this mortality should not be seen as a "normal" process and will likely require active human intervention. Will forest managers try to aggressively contain the bark beetle outbreaks by phytosanitary clear-cuts? And if so, until reaching what point of forest density? It is likely that the speed at which bark beetle outbreaks and subsequent silvicultural responses are taken will drive the speed at which "historical" Mexican pine forests may change to other types of forests, able to survive under future climates.

If climate-induced mortality of Mexican pine trees continues progressing, it will represent not only a serious loss for the Mexican forest biomes and local economies but also for the world. Mexico is Earth's center of *Pinus* biodiversity, and as so, the country presents a potential source of endemic species and exceptionally diverse forest genetic resources (Sáenz-Ceja et al. 2022).

Case Study 3 Pine mortality across the Mediterranean. (Rosana López Rodríguez)

Mediterranean pines have evolved in environments with high climate variability, intense and recurrent droughts, frequent heat waves, and, in some cases, intense fires (Hernández-Serrano et al. 2013). As a result of these strong selective pressures, populations are usually adapted to local conditions but have lower genetic diversity than widely distributed European pines (Grivet et al. 2013). High differentiation among populations has also been favored by their scattered distribution ranges, a result of intense land-use changes, long-term forest management, and low competitive capacity in comparison with broadleaved species mainly from the genus *Quercus*, which are more vigorous resprouters and are also more shade tolerant (Ruiz-Benito et al. 2013).

Mediterranean pine forests cover 25% of forested areas in the Mediterranean basin and have been subject to strong anthropogenic pressure for millennia (Blondel 2006). This long history of forest exploitation for timber and firewood, along with deforestation for agriculture and grazing, historically triggered the disappearance of many pine populations (Morales-Molino et al. 2017). In the last century, drastic societal changes reversed this trend, as pines were extensively used for afforestation of previously degraded lands, likely extending their lower altitudinal and latitudinal limits beyond climate optima (Ruiz-Benito et al. 2013). Increased drought stress in these warmer, arid edges of the distribution range of the species and the use of nonlocal seed sources have sometimes resulted in poor adaptation, reduced tree growth and recruitment, and increased mortality (Gil et al. 2009). Numerous examples can be found throughout the Mediterranean basin. In Israel, *Pinus halepensis* was planted in regions outside of its native range with annual precipitation lower than 450 mm (Klein et al. 2019; see Case Study 5 in this chapter). In the Canary Islands, *P. canariensis* (Fig. 8a–c) was used for reforestation of xeric volcanic slopes, which experienced higher rainfall in the past but now receive less than 300 mm of precipitation annually (López et al. 2021). Other species were extensively planted for restoration and to supply the growing demand for timber, resin, and pine nuts as occurred with one million ha of silviculturally planted *P. pinaster* in the Iberian Peninsula (Rubio-Cuadrado et al. 2021) and *P. pinea*, which has more than doubled its distributed area in the last hundred years (Mutke et al. 2019). From the 1950s, the rural exodus to cities also led to extensive abandonment of farmland, which was soon colonized by frugal and pioneer species including pine trees (Vadell

Fig. 8 *Pinus* mortality throughout the Mediterranean basin. (**a**) Mortality of *P. halepensis in* E Spain in 2024 after the ten driest months on record. (**b**) Resprouting *P. canariensis* after the eruption of Tajogaite volcano in La Palma (Canary Islands, Spain) in 2023. (**c**) Dead *P. canariensis* after drought and fire on Santa Cruz de Tenerife. (**d**) Dead *P. uncinata* in the Spanish Pyrenees. Photo credit: Rosana López Rodríguez

et al. 2016). In Spain alone, the area occupied by pine forests doubled from 1940 to 1993 (Rubio-Cuadrado et al. 2021). Simultaneously, management activities decreased and many natural areas were strictly protected, following a nonintervention policy similar to that mentioned in Case Study 2 of this chapter. This resulted in stands with higher density and fuel load than natural stands, leading to long-term consequences for vegetation dynamics and stand resilience (Ruiz-Benito et al. 2013). Denser even-aged forests, whether because of unmanaged plantings or by forest expansion and encroachment, often show lower regeneration, are less diverse than natural pine forests, and are prone to perturbations of higher intensity, particularly in less fertile and more xeric areas (Lloret et al. 2002). On the other hand, most pines are obligate seeders with high light requirements, hampering recruitment in dense forests, where increasing pine canopy closure has facilitated the establishment of later-successional species in areas with deeper soils and more precipitation (Vadell et al. 2016), as occurred with *Quercus calliprinos* or *Q. ilex* in the understory of *P. halepensis* forests, *Q. pyrenaica*, or *Q. suber* in the understory of *P. pinaster* or *Q. pubescens* in *P. sylvestris* forests (Sheffer 2012).

Competition often exacerbates the negative effects of hotter-drought stress and the recently observed aridification trend in the entire Mediterranean basin associated with climate change (Cook et al. 2016), as it has been previously discussed in Case Study 1. Moreover, warming in the region is likely to increase in the future as climate models predict an increment of warming up to 6 °C and a decrease in annual precipitation up to 20% by the end of this century (Stocker et al. 2013). Prolonged and recurrent summer droughts and warmer springs (as in Case Study 2) may cause defoliation of Mediterranean pines (Sánchez-Salguero et al. 2016), which often exhibit a slower recovery than coexisting angiosperms (DeSoto et al. 2020). As a result, the productivity and vigor of many pine stands have rapidly decreased during the last four decades (Carnicer et al. 2011; Gea-Izquierdo et al. 2019; López et al. 2021), challenging the resilience and recovery capacity of these forests following major disturbances (Julio Camarero et al. 2018; Vieira et al. 2022). Mortality events have peaked after years with severe and long droughts in spring and summer and periods of heat extremes. At least 14 extreme heat episodes have occurred since the mid 1990s, most recently in 2022, which was both the warmest and driest year in the western Mediterranean countries since instrumental record (Herrero et al. 2021). A myriad of examples is found not only in temperate pine species with drier distribution ranges in the Mediterranean basin such as *P. sylvestris*, the most widely distributed pine species in the world (Camarero et al. 2015), or *P. nigra* (Sánchez-Salguero et al. 2012), but also in more typically drought-resistant Mediterranean species (Caudullo and Barredo 2019; Rubio-Cuadrado et al. 2021). For instance, cumulative drought stress after 2 years of hotter drought during 2010–2011 in the Iberian Peninsula and the Canary Islands extensively affected forests of *P. halepensis*, *P. pinaster*, and *P. canariensis* in the subsequent years (Camarero et al. 2015; López et al. 2021). Widespread episodes of canopy dieback and mortality of *P. pinaster* occurred in central Spain in stands with sandy, shallow soils with low water holding capacity and where aquifer overuse for agricultural irrigation has deepened the water table (Gea-Izquierdo et al. 2019). Together, *P. halepensis* and *P. brutia* were involved in over half of the tree mortality events in Israel since 1948 (Klein et al. 2019). Planted monospecific pine forests were highly vulnerable, as shown later in this chapter in Case Study 5. Conversely, in areas with more anthropogenic pressure in the southern and part of the eastern Mediterranean basin, and in stands of *P. pinea*, there are fewer reported mortality events and less canopy dieback, probably due to lower tree densities (Rubio-Cuadrado et al. 2021).

Widespread forest mortality associated with climate-induced stress has been exacerbated by interaction with other processes and will continue to be amplified by ongoing and further warming. Drier environments and higher fuel loads have also increased fire recurrence and total burned area in Mediterranean pine forests (Pausas and Fernández-Muñoz 2012). Since the early 1970s, annual fire frequency has doubled and the burned area has increased almost tenfold, even negatively affecting thick-bark pines adapted to fire like *P. pinaster* or species with fire-dependent serotinous cones such as *P. halepensis*. These increased fire intensities have replaced entire *Pinus* spp. stands with shrublands or resprouting oaks (Karavani et al. 2018). These changes in species composition are consistent with paleo records not only in the Mediterranean basin (Rubiales et al. 2010) but also in the west coast of the USA during past interglacial climate warming and increased fire frequency (Adam and

West 1983). Additionally, rising temperatures have impacted insect population dynamics and the spread of fungi (Desprez-Loustau et al. 2006), placing additional pressures on pine survival. Abiotic stressors can weaken tree defenses to biotic agents. By reducing carbon allocation to the production of secondary metabolites used in defense (López et al. 2021), climate stress can increase vulnerability to biotic attack from insects and/or pathogens (Desprez-Loustau et al. 2006). A higher incidence of outbreaks of bark beetles, such as *Matsucoccus* or *Curcullionidae*, mistletoe infections, and root rot fungi spreading has been observed in many Mediterranean pine forests (Gea-Izquierdo et al. 2019; Klein et al. 2019).

Both drought and fire have historically shaped forests across the Mediterranean basin, but intense warming-induced changes occurring in the last four decades coupled with strong forest management legacies could reduce the range of Mediterranean pine forests due to increased pine mortality, reduced tree growth, and recruitment limitation (Carnicer et al. 2011; Karavani et al. 2018).

Case Study 4 Planted pine mortality in Argentina (María Elena Fernández, Javier Gyenge, and Gabriel Gatica)

In Argentina, pines are the main exotic species introduced by the forestry sector, representing 60% of forested areas (SAGyP. 2022). Humid subtropical climate, in the Mesopotamian ecoregion of the northeast of Argentina, supports fast growth of intensively managed pines (*P. taeda* and *P. elliottii*) and hybrids (*P. elliottii x P. caribea*), growing up to 25–32 m^3 ha^{-1} $year^{-1}$ (SAGyP. 2022). At higher latitudes, temperate-climate adapted pines (*P. ponderosa*, *P. radiata*, and *P. contorta*) are planted in the Patagonia and Pampean regions. The most economically important forest industry region in Argentina is located in Mesopotamia, with pine wood contributing the majority of raw material consumed by the roundwood industry.

Typically, stand management, including initial planting densities and thinning, has been credited with relatively low mortality in managed stands of pines compared to natural stands (Klockow et al. 2020). However, a high magnitude event of pine mortality recently occurred in planted forest stands across the Mesopotamian region (Fig. 9), raising high concern in the forestry sector. The distribution of affected areas estimated by remote sensing analysis (based on the Normalized Difference Vegetation Index, NDVI) can be visualized on an interactive web-based mapping platform (SAGyP. 2022). Pine mortality has also been observed in Patagonian silvicultural plantings in recent years, but of a lower magnitude. During 2022, at least 10,500 ha of pine crops were severely affected by tree mortality (i.e., more than 90% mortality), and almost 25,000 ha were partially affected in Corrientes province within the Mesopotamian region (Reporte Nacional 2022). This report concluded the most affected species was *P. taeda*, the most planted species in that ecoregion and Argentina overall. The affected stands were mature (Fig. 9), with an age of 10 years or more, close to the rotation period of the area. These abrupt and regionally distributed tree mortality observations were associated with an anomalous heat wave during the 2021–2022 warm season (> 30 days with maximum daily temperatures between 40 and 45 °C, i.e., 10–15 °C above the historical average of

Fig. 9 Images of the pine mortality event occurred in Corrientes province, NE Argentina. The left panel shows a Sentinel-2 scene from March 23, 2022, in true-color band combination (red = B4, green = B3, and blue = B2). The three images are different views of the same area. Aerial and ground photographs of the mortality event were kindly provided by Mr. Fernando Dalla Tea, Masisa Forestry Company-Argentina

around 30 °C), which coincided with the last of three consecutive years of pronounced water stress (Gatica et al. 2024). Of note, biotic agents were excluded as the primary factor triggering tree rapid browning and death by field surveys of specialists carried out immediately after the first mortality observations. There were no visible signs of diseases in leaves, stems, and root collar of the trees, and although galleries of borer-insects and bark beetles (Scolytidae) were observed in some trees, the magnitude of their incidence cannot explain the observed mortality. In other conifers, it is worth noting that such rapid foliar browning has been observed only under extreme heat (Still et al. 2023), in contrast to monthslong discoloration of pine needles during drought stress alone (Hammond et al. 2019).

To disentangle environmental factors that may have contributed to the observed mortality event, a first study investigated one of the areas affected, a pine planting of ~20,000 ha covering a large variation of soil conditions typical of the region. At the local scale, the canopy mortality observed in *P. taeda* was spatially and temporally aggregated (Fig. 10). In these locations, the canopy mortality was quantified from existing mortality detection maps created from Sentinel-2 image satellite data, which was validated using field data (Gatica et al. 2024). This map estimates canopy mortality (>80% of individuals with canopy die-off) with an accuracy of 97% at a spatial resolution of 10 m (Fig. 10; Gatica et al. 2024). Of the areas studied, the spatiotemporal pattern of the proportion of canopy mortality (% pixels with more than 80% tree death) at the forest stand level was strongly associated with site quality, as operationally defined by the forestry company based on potential annual

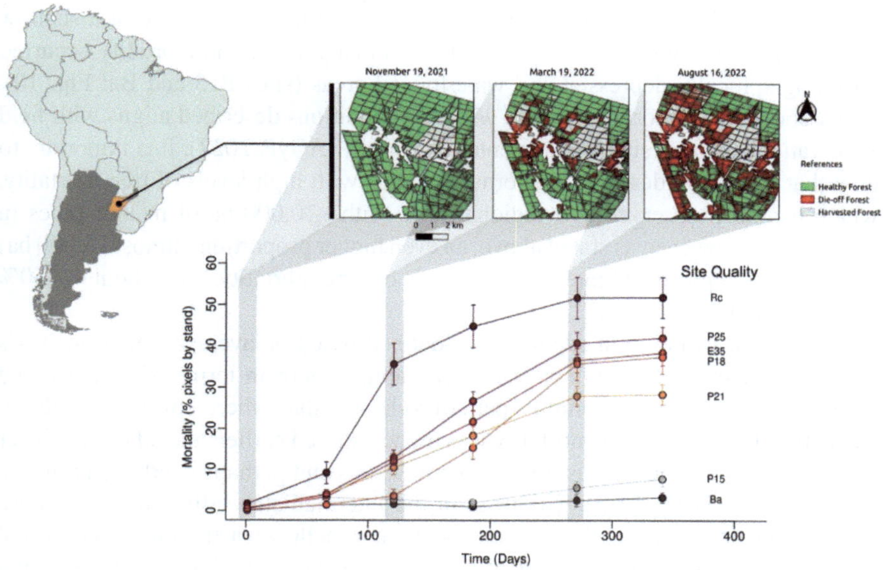

Fig. 10 Spatiotemporal pattern of *Pinus taeda* mortality in a study case of Corrientes province, Argentina, occurred in the summer 2021–2022. Map showing the location of Corrientes province (orange area) in Argentina and South America. Upper panel: Maps with increasing mortality proportion in different dates within a general area of about 20,000 ha of silviculturally planted pines. The pixels with >80% tree mortality (die-off) have been estimated with a combination of spectral indices explained in Gatica et al. (2024). Lower panel: Temporal pattern of the proportion of area with pixels presenting massive tree mortality (>80% of dead trees) within the silvicultural forests located in different site qualities (see text for general descriptions of the sites). Date 0 is November 19, 2021, when no mortality had been visually detected yet

productivity of different forestry species (Fig. 10). These site qualities were classified by the presence of rocky outcrops (Rc), pine growth potential in m^3 ha^{-1} $year^{-1}$ (P15, P21, and P25), high *Eucalyptus* spp. growth potential at a rate of 35 m^3 ha^{-1} $year^{-1}$ (E35), or waterlogged lowlands (Ba). Higher mortality proportion, earlier appearance, and more rapidly advancing canopy mortality were observed in the Rc sites (Fig. 10). Located in the highest areas of the landscape, Rc sites have soils characterized by low permeability, high bulk density, and an effective soil depth of less than 50 cm due to a lithic (rocky) horizon. Intermediate levels of canopy mortality were observed in sites with soils characterized by moderate to low permeability, slow runoff and drainage, and the presence of a textural clay horizon between 50 and 80 cm depth (site qualities P18, P21, P25, and E35; Fig. 10). These intermediate-mortality levels were related to a significantly more delayed and slower rate of mortality than in Rc sites (Fig. 10).

Within intermediate-mortality locations, P21 sites had lower accumulated mortality. Field observations suggest deeper root development (below 1 m depth) compared to trees in sites E35 (until 30 cm depth)—which experienced relatively higher mortality. Both site types have deep soils; however, in the E35 sites, the soils presented signs of anoxia in the upper layers indicating the presence of a shallow water table under typical climatic conditions. Finally, low to no mortality occurred on topographically depressed and waterlogged areas (sites P15 and Ba; Fig. 10). The association of mortality events with soil conditions described aligns with field observations in other sites of Corrientes province (SAGyP. 2022). It is important to note that this study described patterns of stands with high levels of tree mortality, which has been observed at a regional level with ~10,000 ha of mature pines in recent years. However, as stated above, an even higher proportion (almost 25,000 ha) of planted area presented intermediate levels of pine mortality, with about 40–50% tree death at the stand level.

The spatial variation in site quality—defined based on average tree growth—is likely to reflect changes in soil-water availability, both in terms of quantity and duration. The observed mortality pattern indicates that where soil-water holding capacity is limited, canopy mortality was higher, started earlier, and advanced faster likely due to the rapid exhaustion of water reserves and perhaps overheating of tree canopies (Still et al. 2023). The trees in the intermediately affected sites—those stands growing in areas with deep soils but with a shallow water table under normal climatic conditions (site quality E35)—may not have been able to maintain high water use due to their poorly developed root system. This could indicate that the mechanism of structural overshoot (Jump et al. 2017) played a role in the collapse of the most shallow-rooted trees, as canopy growth from relatively wet and mild years preceding the hotter-drought years developed an imbalance between canopy area and water supply. Of note, this agrees with global observations of the "hotter-drought fingerprint" for forest mortality similarly have shown relatively cool, wet period ~4 years preceding massive mortality events when climate anomalies reversed to a hotter-drought extreme (Hammond et al. 2022).

Under further global warming, hotter-drought events are expected to increase in intensity and frequency worldwide in the next 50 years (IPCC et al. 2021), suggesting that similar situations to recent events in NE Argentina may become more common even in planted pine forests. Moreover, societal fiber and wood demands are driving silvicultural forest expansions in regions of relatively marginal soils (i.e., with lower water holding capacity) due to land-use competition. Particularly in developing countries of Africa, Latin America, and parts of Asia, expansion of forestry crops competes with a still expanding agricultural sector (Smith et al. 2010) and/or faces restrictions to replace native forests imposed by local policies and/or international markets and regulations (e.g., EU Regulation 2023/1115). Events such as those described in this section suggest that even when average productivity could be high and economically attractive in the short term, climatic risk may be also high and deserves special consideration. Even when rare, extreme hotter-drought events could severely impact silvicultural forests by affecting tree survival through increased atmospheric demand, physiological damage caused by heat stress, and reduced soil-water content (McDowell et al. 2019; Hammond et al. 2022). Notably, both this case study and the one that follows consider silviculturally planted pines outside the genus' native range, suggest that extreme hotter-drought events may lead pines to death with no need of interacting biotic (insect or disease) agents. This holds true under very different climatic contexts: in the arid conditions of Israel and in the humid climate of NE Argentina. As a whole, the observed patterns in planted pine forests indicate that those sites where soil-water availability is expected to be higher and longer-lasting, and where trees can develop a deep and/or dense root system, could act as "hydraulic refugia," assisting to tolerate atmospheric demands and minimizing risks of physiological damage (McDowell et al. 2019; Kannenberg et al. 2021). From the forestry perspective, these hydraulic refugia could reduce the vulnerability to hotter-drought-induced tree and canopy mortality (McDowell et al. 2019) and represent areas with lower risk for production.

Case Study 5 Pine mortality in Earth's driest *Pinus* forest—Yatir Forest, Israel (Yakir Preisler)

Pines, primarily the Aleppo pine (*Pinus halepensis*), are the dominant forest species in Israel. These pines were introduced through extensive planting starting in the early twentieth century as part of a large-scale afforestation process. While planting played a major role in establishing forests, there were also preexisting local populations of pine trees documented in certain areas before afforestation began (Liphschitz and Biger 2001). The planted pine forests now cover 34% of the total forest area in Israel (39,890 hectares) and have experienced several drought-related mortality events over the last 15 years, which have been exacerbated by climate change (Dorman et al. 2015; Klein et al. 2019; Preisler et al. 2019; Hammond et al. 2022).

The Yatir Forest, located at the edge of the Negev desert, was planted in the early 1960s as part of Israel's afforestation, desertification prevention, and land conservation efforts. It stands as Israel's largest planted pine forest, covering 3800 hectares, and is the driest pine forest in the world. The ~60 years. old predominant Aleppo pine forest has a mean stand density of 300 trees per hectare, with an average tree height of 12 meters and a diameter at breast height (DBH) of 18 centimeters. Yatir Forest endures a prolonged dry season characterized by an average precipitation of 280 mm ± 90 mm MAP, which typically falls between November and April. Drought-related pine mortality events in the Yatir Forest primarily arise from diminished precipitation and changes in rainfall regime both leading to depleted soil-water content (SWC), limiting water availability needed to survive the prolonged dry season (Raz-Yaseef et al. 2010; Klein et al. 2012; Preisler et al. 2019). The survival or mortality of trees in the Yatir Forest depends on microsite variability and the intra- and interannual rainfall distribution. Currently, little to no natural regeneration (Osem et al. 2013; Pozner et al. 2022) has led both researchers and forest managers to develop improved seedling survival techniques such as runoff harvesting. The most conspicuous large-scale mortality event occurred between 2009 and 2011 during a prolonged dry season, spanning 350 consecutive rainless days. This event killed trees in patches (~50,000 trees in total) interspersed among the living forest trees (Preisler et al. 2019). Patches where trees died were characterized by deeper soils, lower soil stoniness (Fig. 11), and lower rocky cover compared to the areas where trees survived. These microsite variations contributed to a distinct response to the severe drought, with higher soil stoniness and rocky cover resulting in marginally improved water availability for the plants during the extended dry season; a conceptual diagram of this can be seen in Preisler et al. 2019. Even minor alterations in water availability, triggered by interannual fluctuations in the precipitation regime or protracted dry periods, serve as key factors in diminishing growth rates and, eventually, initiating mortality events (Raz-Yaseef et al. 2010; Haile 2019; Preisler et al. 2019).

The fate of the pine trees in this semiarid forest relies on the quantity of transpirable water, which represents the water accessible for tree transpiration as inferred from the soil-water retention curve (Klein et al. 2012). Toward the end of the dry season, this measure of transpirable water becomes pivotal in determining whether a particular patch of forest may suffer an elevated mortality event. Soil-water availability, influenced by both microsite characteristics and the precipitation regime, even if severely limited and verging on the threshold of transpirable SWC, decisively influences the chances of tree survival.

However, the mortality mechanism at the tree scale and which specific factors contributed to climate-induced tree mortality at Yatir remain unclear. Is it solely the limited water availability or could the high evaporative demand (high VPD) have resulted in excessive hydraulic failure? Two additional studies conducted at Yatir Forest may reveal mechanisms driving tree mortality at the tree and (needle) leaf scales. First, a study involving seven individual trees, each instrumented with stem dendrometers and sap flow sensors, revealed visual signs of mortality in spring (April) following an above-average wet season (357 mm). However, upon closer examination of the data, it was discovered that these trees had actually begun dying

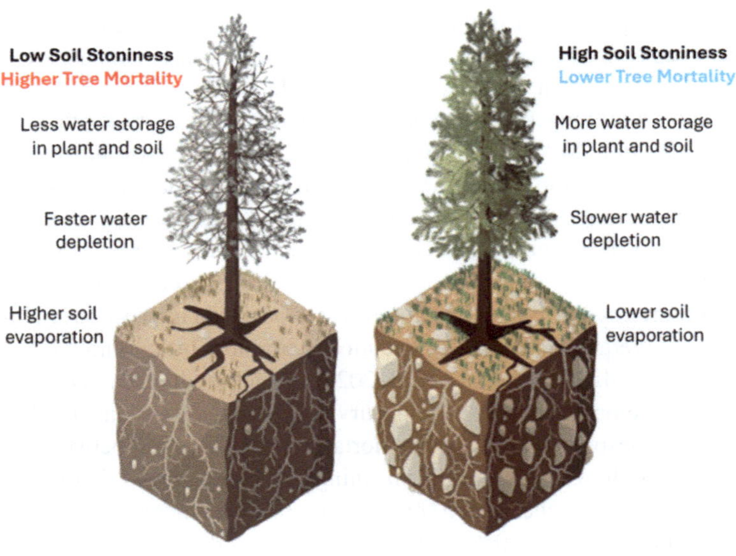

Fig. 11 Conceptual diagram showing differences in areas with low soil stoniness (left) and high soil stoniness (right). Higher tree mortality of *P. halepensis* was associated with low soil stoniness microclimates, which had less water storage in trees and soil, faster rates of water depletion, and higher soil evaporation. Figure modified from Preisler et al. 2019

several months prior to the onset of visible symptoms, as noted by other studies of *Pinus* mortality (Blackman et al. 2019; Hammond et al. 2019). From September to October, 6 months prior to visible signs of mortality, trees exhibited no growth, no radial water flow, and a cessation of phloem and cambial activity (Hammond et al. 2021; Preisler et al. 2021). Subsequently, during the rainy season in January and February, despite the high soil-water content, these trees were unable to conduct water axially through the xylem, as evidenced by the absence of sap flow. Once trees are unable to transport water radially, perhaps due to phloem, cambial, or other radial symplastic cellular damage, xylem function collapses, resulting in rapid water potential declines and accumulation of embolism.

The second study focused on the leaf scale and investigated the impact of vapor pressure deficit (VPD), or atmospheric demand for moisture, on hydraulic failure of leaves under contrasting soil moisture conditions. High VPD levels contributed to increased embolism rates, even when soil-water content (SWC) was relatively high, as observed during spring heatwaves. These in situ measurements of embolism formation, along with the direct relationship with transient high VPD events, revealed that vascular failures can occur even when soil water is available (Wagner et al. 2022; Schönbeck et al. 2022). However, even under extremely high VPD (>5 kPa), embolism in the needles did not exceed 12% loss of conductivity, suggesting that while VPD promotes embolism formation, it is not a direct mechanism of climate-induced tree mortality. Since *P. halepensis* leaf lifespan is typically 3 years, retained VPD-induced vascular failure may initiate a legacy of dysfunction (Hammond 2020). This could result in reduced carbon assimilation, growth, and production of defensive compounds necessary to promote tree survival during subsequent stress events (Hartmann et al. 2018). Warming-driven high VPD levels also accelerate depletion of soil water by amplifying evapotranspiration and soil evaporation, further disrupting the tree water uptake and internal water storage recovery capacity (Preisler et al. 2022). Interestingly, a key feature that helps the Yatir Forest tolerate heat stress is the convective cooling effect, as described by Rotenberg and Yakir (2010) and Kröniger et al. (2018). This effect is largely due to the pine needles' ability to dissipate heat through non-evaporative cooling, as shown by Muller et al. (2021). Pine needles have a high surface area and are well-suited for convective heat loss, which reduces leaf temperatures and minimizes heat stress, even under low water availability.

In conclusion, the primary challenge faced by *P. halepensis* in Yatir Forest is the depletion of soil-water content, which progressively pushes trees closer to a state where they can no longer adequately hydrate living tissues. High-VPD heatwaves act as an additional stressor, compounding drought stress factors and driving water potential below capacitance recharge and xylem air-seeding thresholds, initiating a decline that can ultimately result in mortality. Furthermore, the association between high VPD and increased embolism risk may also explain impaired radial water flow observed in large mature trees that failed to recover when rains returned (Preisler et al. 2019). Although planted, Yatir Forest provides a window into the future of many of Earth's naturally occurring *Pinus* species growing at their xeric limits—where further intensification of drought, accompanied by chronic and acute warming, may lead to similar mechanisms of forest failure. Despite the forest's low regeneration rate and the extreme conditions in which it was established, its ability to survive and grow efficiently provides valuable insights for future afforestation efforts in drylands.

3.1 Pinus Survival in the Anthropocene

Given observed globally distributed *Pinus* mortality events detailed in our five case studies, and documented in Fig. 1, the outlook for *Pinus* on our warming planet is uncertain. While some species may become less at risk of mortality from warming (especially at boreal latitudes), intensifying extremes of drought and heat, often interacting with biotic agents (as noted in Case Studies 1, 2, and 3), present clear challenges to survival for many *Pinus* species in the Anthropocene. Across most of the natural distribution of *Pinus*, and in many regions where the species are commercially planted outside their natural ranges, climate models project further increasing chronic warming, along with increasing frequency, intensity, and duration of acute heat and drought extremes (often, concurrently). Given observed climate drivers and mechanisms of pine mortality, we suggest that future research on pine survival focus on the following ten questions regarding pine mortality driven by hotter-drought stress:

1. How plastic are the traits involved in climate stress resistance along tree ontogeny and the mature tree lifespan?
2. Given that *Pinus* are some of the largest and longest-lived trees on the planet, how will compound hotter-drought extremes and overall warming impact growth and survival of early developmental stages of *Pinus*? (Will early stage vulnerability become a bottleneck for populations?)
3. Can the phylogeny of pines help us generalize and improve global monitoring, risk prediction, and management of entire pine lineages under future climates sensu Griffith et al. 2020, given that some traits of *Pinus* may vary with phylogeny more than environment?
4. What are the thermal limits of *Pinus* species, of what magnitude is thermal trait plasticity, and is wide interspecific variation observed, as with hydraulic limits?
5. Given observed mortality across the genus, what are the prospects for reestablishment of *Pinus*-dominated forests under warming and/or drought conditions of the Anthropocene?
6. Which are the trade-offs between productivity, defense, and survival in *Pinus* species, including those with commercial value?
7. To what extent can climate-resilient growth and survival traits be independently selected, and what are the prospects for commercially planted pine species?
8. How should pine forests and commercially planted pines be managed to increase their resistance and resilience to extreme climatic events?

9. What is the plastic potential of a *Pinus* species or provenance to rapidly changing climate within its lifespan, to conduct proactive adaptive management, including assisted migration (e.g., seedlings surviving present-day frost risk and later surviving warming and drying during decades of maturation)?
10. Facing the possibility that some *Pinus* species could go extinct due to further climatic change, and lacking resources to save *all* species at *any* cost, what criteria should guide extreme measures of ex situ conservation?

4 Conclusion

Earth's first *Pinus* trees emerged in the Cretaceous forests growing alongside dinosaurs; since then, *Pinus* species have evolved, diversified, and survived mass extinctions, including relatively rapid changes in Earth's climate. While pine death in the Anthropocene is concerning, including in the case studies listed in this chapter, it is virtually certain that *Pinus* will continue to be an ecologically, economically, and culturally important genus across much of Earth in its natural and cultivated contexts—however, the ways in which *Pinus*-dominated ecosystems may reorganize, collapse, adapt, or otherwise be different in future climates remain highly uncertain. While drivers and mechanisms of pine mortality indicate that this problem is likely to intensify in the coming decades, active research and advances are improving our understanding and ability to both predict and plan for the future of one of Earth's most iconic tree genera. In all of our case studies—and indeed, across most of the observed global tree mortality events—warming has played a critical role. Every human action to limit further warming will together likely decide the fate of *Pinus* on our planet.

Acknowledgments Thanks to Oscar Trejo-Ramírez (Mexican Secretaría de Medio Ambiente y Recursos Naturales) and Erika Gómez-Pineda (CIGA-UNAM Morelia, Mexico), for providing data and valuable comments about bark beetle outbreaks. Argentinean researchers thank Mr. Fernando Dalla Tea for his valuable comments about the studied silviculturally planted pines and for providing site quality data. W.M.H., M.E.K., M.M., and G.S. were supported in part by the intramural research program of the US Department of Agriculture, National Institute of Food and Agriculture, AFRI #2023-67013-39535.

References

Abatzoglou JT, Dobrowski SZ, Parks SA, Hegewisch KC (2018) TerraClimate, a high-resolution global dataset of monthly climate and climatic water balance from 1958–2015. Sci Data 5(1):1–12

Adam DP, West GJ (1983) Temperature and precipitation estimates through the last glacial cycle from Clear Lake, California, pollen data. Science 219:168–170

Adams HD, Guardiola-Claramonte M, Barron-Gafford GA, Villegas JC, Breshears DD, Zou CB, Troch PA, Huxman TE (2009) Temperature sensitivity of drought-induced tree mortality

portends increased regional die-off under global-change-type drought. Proc Natl Acad Sci 106:7063–7066
Adams HD, Zeppel MJB, Anderegg WRL, Hartmann H, Landhäusser SM, Tissue DT, Huxman TE, Hudson PJ, Franz TE, Allen CD et al (2017) A multi-species synthesis of physiological mechanisms in drought-induced tree mortality. Nat Ecol Evol 1:1285–1291
Allen CD (2014) Forest ecosystem re-organization underway in the southwestern United States: a preview of widespread forest changes in the Anthropocene. In: Proceedings. RMRS-P-71. US Department of Agriculture, Forest Service. Rocky Mountain Research Station, Fort Collins, pp 103–123
Allen CD, Breshears DD (1998) Drought-induced shift of a forest–woodland ecotone: rapid landscape response to climate variation. Proc Natl Acad Sci 95:14839–14842
Allen CD, Savage M, Falk DA, Suckling KF, Swetnam TW, Schulke T, Stacey PB, Morgan P, Hoffman M, Klingel JT (2002) Ecological restoration of southwestern ponderosa pine ecosystems: a broad perspective. Ecol Appl 12:1418–1433
Allen CD, Macalady AK, Chenchouni H, Bachelet D, McDowell N, Vennetier M, Kitzberger T, Rigling A, Breshears DD, Hogg ET (2010) A global overview of drought and heat-induced tree mortality reveals emerging climate change risks for forests. For Ecol Manag 259:660–684
Allen CD, Breshears DD, McDowell NG (2015) On underestimation of global vulnerability to tree mortality and forest die-off from hotter drought in the Anthropocene. Ecosphere 6:art129
Anderegg WR, Berry JA, Smith DD, Sperry JS, Anderegg LD, Field CB (2012) The roles of hydraulic and carbon stress in a widespread climate-induced forest die-off. Proc Natl Acad Sci 109:233–237
Astudillo-Sánchez CC, Villanueva-Díaz J, Endara-Agramont AR, Nava-Bernal GE, Gómez-Albores MÁ (2017) The influence of climate on *Pinus hartwegii* Lindl. Recruitment at the alpine tree line ecotone in Monte Tlaloc, Mexico. Agrociencia 51(1):105–118
Astudillo-Sánchez CC, Fowler MS, Villanueva-Díaz J, Endara-Agramont AR (2019) Recruitment and facilitation in *Pinus hartwegii*, a Mexican alpine treeline ecotone, with potential responses to climate warming. Trees 33:1087–1100
Barigah TS, Charrier O, Douris M, Bonhomme M, Herbette S, Améglio T, Fichot R, Brignolas F, Cochard H (2013) Water stress-induced xylem hydraulic failure is a causal factor of tree mortality in beech and poplar. Ann Bot 112:1431–1437
Bentz BJ, Régnière J, Fettig CJ, Hansen EM, Hayes JL, Hicke JA, Kelsey RG, Negrón JF, Seybold SJ (2010) Climate change and bark beetles of the western United States and Canada: direct and indirect effects. Bioscience 60:602–613
Blackman CJ, Creek D, Maier C, Aspinwall MJ, Drake JE, Pfautsch S, O'Grady A, Delzon S, Medlyn BE, Tissue DT et al (2019) Drought response strategies and hydraulic traits contribute to mechanistic understanding of plant dry-down to hydraulic failure. Tree Physiol 39:910–924
Blondel J (2006) The 'design'of Mediterranean landscapes: a millennial story of humans and ecological systems during the historic period. Hum Ecol 34:713–729
Bojórquez A, Álvarez-Yépiz JC, Búrquez A, Martínez-Yrízar A (2019) Understanding and predicting frost-induced tropical tree mortality patterns. Glob Chang Biol 25(11):3817–3828
Bradley JC, Will RE, Stewart JF, Nelson CD, Guldin JM (2016) Post-fire resprouting of shortleaf pine is facilitated by a morphological trait but fire eliminates shortleaf × loblolly pine hybrid seedlings. For Ecol Manag 379:146–152
Breshears DD, Fontaine JB, Ruthrof KX, Field JP, Feng X, Burger JR, Law DJ, Kala J, Hardy GESJ (2021) Underappreciated plant vulnerabilities to heat waves. New Phytol 231:32–39
Brodribb TJ, Cochard H (2009) Hydraulic failure defines the recovery and point of death in water-stressed conifers. Plant Physiol 149:575–584
Bulfe NML, Fernández ME (2016) Morpho-physiological response to drought of progenies of Pinus taeda L. contrasting in mean growth rate. New For 47:431–451
Burrows GE (2021) Gymnosperm resprouting—a review. Plants 10:2551
Camarero JJ, Gazol A, Sangüesa-Barreda G, Oliva J, Vicente-Serrano SM (2015) To die or not to die: early warnings of tree dieback in response to a severe drought. J Ecol 103:44–57

Carnicer J, Coll M, Ninyerola M, Pons X, Sanchez G, Penuelas J (2011) Widespread crown condition decline, food web disruption, and amplified tree mortality with increased climate change-type drought. Proc Natl Acad Sci 108:1474–1478

Carrillo-Arizmendi L, Pérez-Suárez M, Vargas-Hernández JJ, Rozenberg P, Martínez-Campos AR (2022) Warming effects on tree-ring variables in *P. hartwegii* Lindl. At the extremes of its natural elevational distribution in Central Mexico. Agric For Meteorol 324:109109

Caudullo G, Barredo JI (2019) A georeferenced dataset of drought and heat-induced tree mortality in Europe. One Ecosystem 4:e37753

Choat B, Jansen S, Brodribb TJ, Cochard H, Delzon S, Bhaskar R, Bucci SJ, Feild TS, Gleason SM, Hacke UG et al (2012) Global convergence in the vulnerability of forests to drought. Nature 491:752–755

CONAFOR (Comisión Nacional Forestal). 2023. Sistema Integral de Vigilancia y Control Fitosanitario Forestal.. *Estadísticas Nacionales de Afectación*

Cook BI, Anchukaitis KJ, Touchan R, Meko DM, Cook ER (2016) Spatiotemporal drought variability in the Mediterranean over the last 900 years. J Geophys Res Atmos 121:2060–2074

Crowe LM, Crowe JH (1992) Anhydrobiosis: a strategy for survival. Adv Space Res 12:239–247

Cruz de Carvalho R, Catalá M, Branquinho C, Marques da Silva J, Barreno E (2017) Dehydration rate determines the degree of membrane damage and desiccation tolerance in bryophytes. Physiol Plant 159(3):277–289

Cruzado-Vargas A, Blanco-García A, Lindig-Cisneros R, Gómez-Romero M, Lopez-Toledo L, De La Barrera E, Sáenz-Romero C (2021) Reciprocal common garden altitudinal transplants reveal potential negative impacts of climate change on Abies religiosa populations in the monarch butterfly biosphere reserve overwintering sites. Forests 12:69

Dannenberg MP, Yan D, Barnes ML, Smith WK, Johnston MR, Scott RL, Biederman JA, Knowles JF, Wang X, Duman T (2022) Exceptional heat and atmospheric dryness amplified losses of primary production during the 2020 US southwest hot drought. Glob Chang Biol 28:4794–4806

Delzon S, Cochard H (2014) Recent advances in tree hydraulics highlight the ecological significance of the hydraulic safety margin. New Phytol 203:355–358

DeSoto L, Cailleret M, Sterck F, Jansen S, Kramer K, Robert EM, Aakala T, Amoroso MM, Bigler C, Camarero JJ (2020) Low growth resilience to drought is related to future mortality risk in trees. Nat Commun 11:545

Desprez-Loustau M-L, Marçais B, Nageleisen L-M, Piou D, Vannini A (2006) Interactive effects of drought and pathogens in forest trees. Ann For Sci 63:597–612

Dorman M, Svoray T, Perevolotsky A, Moshe Y, Sarris D (2015) What determines tree mortality in dry environments? A multi-perspective approach. Ecol Appl 25:1054–1071

Dvorak WS et al. (2000) Conservation and testing of tropical and subtropical forest tree species by the CAMCORE Cooperative. Conservation and testing of tropical and subtropical forest tree species by the CAMCORE cooperative

Ewers BE, Oren R (2000) Analyses of assumptions and errors in the calculation of stomatal conductance from sap flux measurements. Tree Physiol 20:579–589

Farjon A, Styles BT (1997) Flora Neotropica. Pinus (Pinaceae). Flora Neotropica. Pinus (Pinaceae)

Fensham RJ, Laffineur B, Allen CD (2019) To what extent is drought-induced tree mortality a natural phenomenon? Glob Ecol Biogeogr 28:365–373

Fettig CJ, Mortenson LA, Bulaon BM, Foulk PB (2019) Tree mortality following drought in the central and southern Sierra Nevada, California, U.S. For Ecol Manag 432:164–178

Fick SE, Hijmans RJ (2017) WorldClim 2: new 1-km spatial resolution climate surfaces for global land areas. Int J Climatol 37:4302–4315

Gatica G, Gyenge J, Bulfe NML, Pahr N, Dalla Tea F, Fernández ME (2024) Unprecedented mortality induced by extreme hot-drought in mature planted Pinus spp stands is locally modulated by soil characteristics. New Forests 55(5):1333–1347

Gauthey A, Bachofen C, Deluigi J, Didion-Gency M, D'Odorico P, Gisler J et al (2023) Absence of canopy temperature variation despite stomatal adjustment in Pinus sylvestris under multidecadal soil moisture manipulation. New Phytol 240(1):127–137

Gea-Izquierdo G, Férriz M, García-Garrido S, Aguín O, Elvira-Recuenco M, Hernandez-Escribano L, Martin-Benito D, Raposo R (2019) Synergistic abiotic and biotic stressors explain widespread decline of Pinus pinaster in a mixed forest. Sci Total Environ 685:963–975

Gil L, López R, García-Mateos Á, González-Doncel I (2009) Seed provenance and fire-related reproductive traits of Pinus pinaster in Central Spain. Int J Wildland Fire 18:1003–1009

Goeking SA, Menlove J (2017) New Mexico's Forest resources, 2008–2014. United States Department of Agriculture, Forest Service, Rocky Mountain

Gomez-Gallego M, Galiano L, Martínez-Vilalta J, Stenlid J, Capador-Barreto HD, Elfstrand M, Camarero JJ, Oliva J (2022) Interaction of drought-and pathogen-induced mortality in Norway spruce and scots pine. Plant Cell Environ 45:2292–2305

Gómez-Pineda E, Sáenz-Romero C, Ortega-Rodríguez JM, Blanco-García A, Madrigal-Sánchez X, Lindig-Cisneros R, Lopez-Toledo L, Pedraza-Santos ME, Rehfeldt GE (2020) Suitable climatic habitat changes for Mexican conifers along altitudinal gradients under climatic change scenarios. Ecol Appl 30(2):e02041. https://doi.org/10.1002/eap.2041

Gómez-Pineda E, Hammond WM, Trejo-Ramirez O, Gil-Fernández M, Allen CD, Blanco-García A, Sáenz-Romero C (2022) Drought years promote bark beetle outbreaks in Mexican forests of Abies religiosa and Pinus pseudostrobus. For Ecol Manag 505:119944

Gottfried GJ, Swetnam TW, Allen CD, Betancourt JL, Chung-MacCoubrey AL (1995) Pinyon-juniper woodlands. United States Department of Agriculture Forest Service General Technical Report RM, pp 95–132

Griffith DM, Osborne CP, Edwards EJ, Bachle S, Beerling DJ, Bond WJ, Gallaher TJ, Helliker BR, Lehmann CE, Leatherman L, Nippert JB (2020) Lineage-based functional types: characterising functional diversity to enhance the representation of ecological behaviour in land surface models. *New Phytol* 228(1):15–23

Grivet D, Climent J, Zabal-Aguirre M, Neale DB, Vendramin GG, González-Martínez SC (2013) Adaptive evolution of Mediterranean pines. Mol Phylogenet Evol 68:555–566

Guadagno CR, Ewers BE, Speckman HN, Aston TL, Huhn BJ, DeVore SB, Ladwig JT, Strawn RN, Weinig C (2017) Dead or alive? Using membrane failure and chlorophyll fluorescence to predict mortality from drought. Plant Physiol 00581.2016

Gutzler DS, DuBois D (2022) Our climate future, climate change in New Mexico over the next 50 years: impacts on water resources. New Mexico Bureau of Geology and Mineral Resources

Haile SA (2019) How does precipitation pattern affect annual tree growth? A multi-year data analysis a case study of semi-arid Yatir Forest, Israel. University of Valladolid, Spain

Hammond WM (2020) A matter of life and death: alternative stable states in trees, from xylem to ecosystems. Front For Glob Change 3:560409

Hammond WM, Adams HD (2019) Dying on time: traits influencing the dynamics of tree mortality risk from drought. Tree Physiol 39:906–909

Hammond WM, Yu K, Wilson LA, Will RE, Anderegg WRL, Adams HD (2019) Dead or dying? Quantifying the point of no return from hydraulic failure in drought-induced tree mortality. New Phytol 223:1834

Hammond WM, Johnson DM, Meinzer FC (2021) A thin line between life and death: radial sap flux failure signals trajectory to tree mortality. Plant Cell Environ 44:1311–1314

Hammond WM, Williams AP, Abatzoglou JT, Adams HD, Klein T, López R, Sáenz-Romero C, Hartmann H, Breshears DD, Allen CD (2022) Global field observations of tree die-off reveal hotter-drought fingerprint for earth's forests. Nat Commun 13:1761

Hartmann H, Adams HD, Hammond WM, Hoch G, Landhäusser SM, Wiley E, Zaehle S (2018) Identifying differences in carbohydrate dynamics of seedlings and mature trees to improve carbon allocation in models for trees and forests. Environ Exp Bot 152:7–18

Hartmann H, Bastos A, Das AJ, Esquivel-Muelbert A, Hammond WM, Martínez-Vilalta J, NG MD, Powers JS, Pugh TAM, Ruthrof KX et al (2022) Climate change risks to global Forest health: emergence of unexpected events of elevated tree mortality worldwide. Annu Rev Plant Biol 73:673–702

Hernández-Serrano A, Verdú M, González-Martínez SC, Pausas JG (2013) Fire structures pine serotiny at different scales. Am J Bot 100:2349–2356

Herrero A, Ruiz-Benito P, Andivia E, Madrigal-González J, Villar-Salvador P, Ratcliffe S, Zavala MA (2021) Mediterranean pine forest distribution: assessing vulnerability and resilience under climate change. In: Ne'eman G, Osem Y (eds) Pines and their mixed forest ecosystems in the Mediterranean Basin, Managing forest ecosystems, vol 38. Springer, Cham, pp 251–277

IPCC, Masson-Delmotte V, Zhai P, Pirani A, Connors SL, Péan C, Berger S, Caud N, Chen Y, Goldfarb L et al (2021) Climate change 2021: the physical science basis. Contribution of working group I to the sixth assessment report of the intergovernmental panel on climate change 2. IPCC

Jacobs BF, Romme WH, Allen CD (2008) Mapping "old" vs. "young" piñon–juniper stands with a predictive topo-climatic model. Ecol Appl 18:1627–1641

Jin Y, Qian H (2022) V.PhyloMaker2: an updated and enlarged R package that can generate very large phylogenies for vascular plants. Plant Divers 44:335–339

Johnson DM, Wortemann R, McCulloh KA, Jordan-Meille L, Ward E, Warren JM, Palmroth S, Domec J-C (2016) A test of the hydraulic vulnerability segmentation hypothesis in angiosperm and conifer tree species. Tree Physiol 36:983–993

Jones MW, Abatzoglou JT, Veraverbeke S, Andela N, Lasslop G, Forkel M, Smith AJ, Burton C, Betts RA, van der Werf GR, Sitch S (2022) Global and regional trends and drivers of fire under climate change. Rev Geophys 60(3):e2020RG000726

Julio Camarero J, Gazol A, Sangüesa-Barreda G, Cantero A, Sánchez-Salguero R, Sánchez-Miranda A, Granda E, Serra-Maluquer X, Ibáñez R (2018) Forest growth responses to drought at short-and long-term scales in Spain: squeezing the stress memory from tree rings. Front Ecol Evol 6:9

Jump AS, Ruiz-Benito P, Greenwood S, Allen CD, Kitzberger T, Fensham R, Martínez-Vilalta J, Lloret F (2017) Structural overshoot of tree growth with climate variability and the global spectrum of drought-induced forest dieback. Glob Chang Biol 23:3742–3757

Kannenberg SA, Driscoll AW, Malesky D, Anderegg WR (2021) Rapid and surprising dieback of Utah juniper in the southwestern USA due to acute drought stress. For Ecol Manag 480:118639

Karavani A, Boer MM, Baudena M, Colinas C, Díaz-Sierra R, Pemán J, de Luis M, Enríquez-de-Salamanca Á, Resco de Dios V (2018) Fire-induced deforestation in drought-prone Mediterranean forests: drivers and unknowns from leaves to communities. Ecol Monogr 88:141–169

Klein T, Di Matteo G, Rotenberg E, Cohen S, Yakir D (2012) Differential ecophysiological response of a major Mediterranean pine species across a climatic gradient. Tree Physiol 33:26–36

Klein T, Cahanovitc R, Sprintsin M, Herr N, Schiller G (2019) A nation-wide analysis of tree mortality under climate change: forest loss and its causes in Israel 1948–2017. For Ecol Manag 432:840–849

Klockow PA, Edgar CB, Moore GW, Vogel JG (2020) Southern pines are resistant to mortality from an exceptional drought in East Texas. Front For Glob Change 3:23

Kröniger K, De Roo F, Brugger P, Huq S, Banerjee T, Zinsser J, Rotenberg E, Yakir D, Rohatyn S, Mauder M (2018) Effect of secondary circulations on the surface–atmosphere exchange of energy at an isolated semi-arid forest. Bound-Layer Meteorol 169:209–232

Leslie AB, Beaulieu J, Holman G, Campbell CS, Mei W, Raubeson LR, Mathews S (2018) An overview of extant conifer evolution from the perspective of the fossil record. Am J Bot 105:1531–1544

Liphschitz N, Biger G (2001) Past distribution of Aleppo pine (Pinus halepensis) in the mountains of Israel (Palestine). The Holocene 11:427–436

Liu Y, Stanturf J, Goodrick S (2010) Trends in global wildfire potential in a changing climate. For Ecol Manag 259(4):685–697

Lloret F, Calvo E, Pons X, Díaz-Delgado R (2002) Wildfires and landscape patterns in the Eastern Iberian Peninsula. Landsc Ecol 17:745–759

Lloret F, Siscart D, Dalmases C (2004) Canopy recovery after drought dieback in holm-oak Mediterranean forests of Catalonia (NE Spain). Glob Chang Biol 10:2092–2099
López R, Cano FJ, Rodríguez-Calcerrada J, Sangüesa-Barreda G, Gazol A, Camarero JJ, Rozenberg P, Gil L (2021) Tree-ring density and carbon isotope composition are early-warning signals of drought-induced mortality in the drought tolerant Canary Island pine. Agric For Meteorol 310:108634
López-Toledo L, Heredia-Hernández M, Castellanos-Acuña D, Blanco-García A, Saénz-Romero C (2017) Reproductive investment of Pinus pseudostrobus along an altitudinal gradient in Western Mexico: implications of climate change. New For 48:867–881
Mantova M, Menezes-Silva PE, Badel E, Cochard H, Torres-Ruiz JM (2021) The interplay of hydraulic failure and cell vitality explains tree capacity to recover from drought. Physiol Plant 172:247–257
Mantova M, Herbette S, Cochard H, Torres-Ruiz JM (2022) Hydraulic failure and tree mortality: from correlation to causation. Trends Plant Sci 27:335–345
Martínez-Vilalta J, Sala A, Piñol J (2004) The hydraulic architecture of Pinaceae—a review. Plant Ecol. (formerly Vegetatio) 171:3–13
Mátyás C (2010) Forecasts needed for retreating forests. Nature 464:1271–1271
Mcdowell NG, Allen CD (2015) Darcy's law predicts widespread forest mortality under climate warming. Nat Clim Chang 5:669–672
McDowell NG, Beerling DJ, Breshears DD, Fisher RA, Raffa KF, Stitt M (2011) The interdependence of mechanisms underlying climate-driven vegetation mortality. Trends Ecol Evol 26(10):523–532. https://doi.org/10.1016/j.tree.2011.06.003
McDowell NG, Williams AP, Xu C, Pockman WT, Dickman LT, Sevanto S, Pangle R, Limousin J, Plaut J, Mackay DS (2016) Multi-scale predictions of massive conifer mortality due to chronic temperature rise. Nat Clim Chang 6:295
McDowell NG, Grossiord C, Adams HD, Pinzón-Navarro S, Mackay DS, Breshears DD, Allen CD, Borrego I, Dickman LT, Collins A (2019) Mechanisms of a coniferous woodland persistence under drought and heat. Environ Res Lett 14:045014
McDowell NG, Allen CD, Anderson-Teixeira K, Aukema BH, Bond-Lamberty B, Chini L, Clark JS, Dietze M, Grossiord C, Hanbury-Brown A et al (2020) Pervasive shifts in forest dynamics in a changing world. Science 368:eaaz9463
McDowell NG, Sapes G, Pivovaroff A, Adams HD, Allen CD, Anderegg WRL, Arend M, Breshears DD, Brodribb T, Choat B et al (2022a) Mechanisms of woody-plant mortality under rising drought, CO2 and vapour pressure deficit. Nat Rev Earth Environ 3:294–308
McDowell NG, Ball M, Bond-Lamberty B, Kirwan ML, Krauss KW, Megonigal JP, Mencuccini M, Ward ND, Weintraub MN, Bailey V (2022b) Processes and mechanisms of coastal woody-plant mortality. Glob Chang Biol 28(20):5881–5900
Meddens AJ, Hicke JA, Ferguson CA (2012) Spatiotemporal patterns of observed bark beetle-caused tree mortality in British Columbia and the western United States. Ecol Appl 22:1876–1891
Mills C, Bartlett MK, Buckley TN (2024) The poorly-explored stomatal response to temperature at constant evaporative demand. Plant Cell Environ 47(9):3428–3446
Morales-Molino C, Colombaroli D, Valbuena-Carabaña M, Tinner W, Salomón RL, Carrión JS, Gil L (2017) Land-use history as a major driver for long-term forest dynamics in the Sierra de Guadarrama National Park (Central Spain) during the last millennia: implications for forest conservation and management. Glob Planet Chang 152:64–75
Muller JD, Rotenberg E, Tatarinov F, Oz I, Yakir D (2021) Evidence for efficient nonevaporative leaf-to-air heat dissipation in a pine forest under drought conditions. New Phytol 232(6):2254–2266
Murray-Tortarolo GN (2021) Seven decades of climate change across Mexico. Atmósfera
Mutke S, Vendramin GG, Fady B, Bagnoli F, González-Martínez SC (2019) Molecular and quantitative genetics of stone pine (*Pinus pinea*). In: Nandwani D (ed) Genetic diversity in horticultural plants, Sustainable development and biodiversity, vol 22. Springer, Cham, pp 61–84

Nishad A, Nandi AK (2021) Recent advances in plant thermomemory. Plant Cell Rep 40:19–27
O'Brien MJ, Leuzinger S, Philipson CD, Tay J, Hector A (2014) Drought survival of tropical tree seedlings enhanced by non-structural carbohydrate levels. Nat Clim Chang 4:710–714
Oertli JJ (1971) Stability of water under tension in the xylem. Z Pflanzenphysiol 65:195–209
Oliver WW, Ryker RA. 1990. Pinus ponderosa Dougl. ex Laws. ponderosa pine. Silvics of North America Burns R. M. Honkala B. H. (Washington, DC: US Department of Agriculture, Forest Service),413
Osem Y, Yavlovich H, Zecharia N, Atzmon N, Moshe Y, Schiller G (2013) Fire-free natural regeneration in water limited Pinus halepensis forests: a silvicultural approach. Eur J For Res 132:679–690
Partelli-Feltrin R, Johnson DM, Sparks AM, Adams HD, Kolden CA, Nelson AS, Smith AM (2020) Drought increases vulnerability of Pinus ponderosa saplings to fire-induced mortality. Fire 3:56
Partelli-Feltrin R, Smith AMS, Adams HD, Thompson RA, Kolden CA, Yedinak KM, Johnson DM (2023) Death from hunger or thirst? Phloem death, rather than xylem hydraulic failure, as a driver of fire-induced conifer mortality. New Phytol 237:1154–1163
Pausas JG, Fernández-Muñoz S (2012) Fire regime changes in the Western Mediterranean Basin: from fuel-limited to drought-driven fire regime. Clim Chang 110:215–226
Pausas JG, Keeley JE (2017) Epicormic Resprouting in fire-prone ecosystems. Trends Plant Sci 22:1008–1015
Perry JP Jr (1991) The pines of Mexico and Central America. Timber Press, Inc
Pozner E, Bar-On P, Livne-Luzon S, Moran U, Tsamir-Rimon M, Dener E, Schwartz E, Rotenberg E, Tatarinov F, Preisler Y, Zecharia N (2022) A hidden mechanism of forest loss under climate change: the role of drought in eliminating forest regeneration at the edge of its distribution. For Ecol Manag 506:119966
Preisler Y, Tatarinov F, Grünzweig JM, Bert D, Ogée J, Wingate L, Rotenberg E, Rohatyn S, Her N, Moshe I et al (2019) Mortality versus survival in drought-affected Aleppo pine forest depends on the extent of rock cover and soil stoniness (A Sala, Ed.). Funct Ecol 33:901–912
Preisler Y, Tatarinov F, Grünzweig JM, Yakir D (2021) Seeking the "point of no return" in the sequence of events leading to mortality of mature trees. Plant Cell Environ 44:1315–1328
Preisler Y, Hölttä T, Grünzweig JM, Oz I, Tatarinov F, Ruehr NK, Rotenberg E, Yakir D (2022) The importance of tree internal water storage under drought conditions (R Oren, Ed.). Tree Physiol 42:771–783
Raz-Yaseef N, Yakir D, Rotenberg E, Schiller G, Cohen S (2010) Ecohydrology of a semi-arid forest: partitioning among water balance components and its implications for predicted precipitation changes. Ecohydrology 3:143–154
Reporte Nacional (2022) Análisis climático y productivo en relación a la mortalidad de Pinus taeda en la Provincia de Corrientes. In: Secretaría de Desarrollo Forestoindustria, Provincia de Corrientes; Consejo Nacional de Investigaciones Científicas y Técnicas; Instituto Nacional de Tecnología Agropecuaria. Secretaría de Agricultura Ganadería y Pesca. Ministerio de Economía de Argentina
Revell LJ (2012) Phytools: an R package for phylogenetic comparative biology (and other things). Methods Ecol Evol 2:217–223
Rice KJ, Matzner SL, Byer W, Brown JR (2004) Patterns of tree dieback in Queensland, Australia: the importance of drought stress and the role of resistance to cavitation. Oecologia 139:190–198
Richardson DM (ed) (2000) Ecology and biogeography of Pinus. Cambridge University Press
Romme WH, Allen CD, Bailey JD, Baker WL, Bestelmeyer BT, Brown PM, Eisenhart KS, Floyd ML, Huffman DW, Jacobs BF (2009) Historical and modern disturbance regimes, stand structures, and landscape dynamics in pinon–juniper vegetation of the western United States. Rangel Ecol Manag 62:203–222
Rotenberg E, Yakir D (2010) Contribution of semi-arid forests to the climate system. Science 327(5964):451–454

Rubiales JM, García-Amorena I, Hernández L, Génova M, Martínez F, Manzaneque FG, Morla C (2010) Late quaternary dynamics of pinewoods in the Iberian Mountains. Rev Palaeobot Palynol 162:476–491

Rubin-Aguirre A, Saenz-Romero C, Lindig-Cisneros R, del-Rio-Mora AA, Tena-Morelos CA, Campos-Bolaños R, del-Val E (2015) Bark beetle pests in an altitudinal gradient of a Mexican managed forest. For Ecol Manag 343:73–79

Rubio-Cuadrado Á, López R, Rodríguez-Calcerrada J, Gil L (2021) Stress and tree mortality in Mediterranean pine forests: anthropogenic influences. In: Pines and Their Mixed Forest Ecosystems in the Mediterranean Basin, pp 141–181

Ruiz-Benito P, Lines ER, Gómez-Aparicio L, Zavala MA, Coomes DA (2013) Patterns and drivers of tree mortality in Iberian forests: climatic effects are modified by competition. PLoS One 8:e56843

Sáenz-Ceja JE, Arenas-Navarro M, Torres-Miranda A (2022) Prioritizing conservation areas and vulnerability analyses of the genus Pinus L. (Pinaceae) in Mexico. J Nat Conserv 67:126171

Sáenz-Romero C (2024) Arriving at a tipping point for worldwide forest decline due to accelerating climatic change. For Chron 100:5–7

Sáenz-Romero C, Mendoza-Maya E, Gómez-Pineda E, Blanco-García A, Endara-Agramont AR, Lindig-Cisneros R, López-Upton J, Trejo-Ramírez O, Wehenkel C, Cibrián-Tovar D et al (2020) Recent evidence of Mexican temperate forest decline and the need for ex situ conservation, assisted migration, and translocation of species ensembles as adaptive management to face projected climatic change impacts in a megadiverse country. Can J For Res 50:843–854

Sáenz-Romero C, Cambrón-Sandoval VH, Hammond W, Méndez-González J, Luna-Soria H, Macías-Sámano JE, Gómez-Romero M, Trejo-Ramírez O, Allen CD, Gómez-Pineda E et al (2023) Abundance of Dendroctonus frontalis and D. Mexicanus (Coleoptera: Scolytinae) along altitudinal transects in Mexico: implications of climatic change for forest conservation (N Aguilar-Rivera, Ed.). *PLOS ONE* 18:e0288067

SAGyP. (2022) Tablero de plantaciones forestales. Secretaría de Agricultura Ganadería y Pesca. Ministerio de Economía de Argentina

Sakes A, van der Wiel M, Henselmans PW, van Leeuwen JL, Dodou D, Breedveld P (2016) Shooting mechanisms in nature: a systematic review. Plos One 11:e0158277

Salazar-Tortosa D, Castro J, Saladin B, Zimmermann NE, Rubio De Casas R (2020) Arid environments select for larger seeds in pines (Pinus spp.). Evol Ecol 34:11–26

Salzer MW, Pearson CL, Baisan CH (2019) Dating the methuselah walk bristlecone pine floating chronologies. Tree-Ring Res 75:61–66

Sánchez-González A (2008) Una visión actual de la diversidad y distribución de los pinos de México. Madera y Bosques 14:107–120

Sánchez-Salguero R, Navarro-Cerrillo RM, Camarero JJ, Fernández-Cancio Á (2012) Selective drought-induced decline of pine species in southeastern Spain. Clim Chang 113:767–785

Sánchez-Salguero R, Camarero JJ, Grau JM, De la Cruz AC, Gil PM, Minaya M, Fernández-Cancio Á (2016) Analysing atmospheric processes and climatic drivers of tree defoliation to determine forest vulnerability to climate warming. Forests 8:13

Savchenko GE, Klyuchareva EA, Abramchik LM, Serdyuchenko EV (2002) Effect of periodic heat shock on the inner membrane system of etioplasts. Russ J Plant Physiol 49:349–359

Schönbeck LC, Schuler P, Lehmann MM, Mas E, Mekarni L, Pivovaroff AL, Turberg P, Grossiord C (2022) Increasing temperature and vapour pressure deficit lead to hydraulic damages in the absence of soil drought. Plant Cell Environ 45(11):3275–3289

Schoonmaker AL, Lieffers VJ, Landhäusser SM (2014) Uniform versus asymmetric shading mediates crown recession in conifers (M Reigosa, Ed.). *PLoS One* 9:e104187

Shaw JD, Menlove J, Witt C, Morgan TA, Amacher MC, Goeking SA, Werstak CE Jr (2018) Arizona's forest resources, 2001–2014. United States Department of Agriculture, Forest Service, Rocky Mountain

Sheffer E (2012) A review of the development of Mediterranean pine–oak ecosystems after land abandonment and afforestation: are they novel ecosystems? Ann For Sci 69:429–443

Smith P, Gregory PJ, Van Vuuren D, Obersteiner M, Havlík P, Rounsevell M, Woods J, Stehfest E, Bellarby J (2010) Competition for land. Philos Trans R Soc Lond B Biol Sci 365:2941–2957

Stephenson NL, Das AJ, Ampersee NJ, Bulaon BM, Yee JL (2019) Which trees die during drought? The key role of insect host-tree selection (D Edwards, Ed.). J Ecol 107:2383–2401

Still CJ, Page G, Rastogi B, Griffith DM, Aubrecht DM, Kim Y, Burns SP, Hanson CV, Kwon H, Hawkins L et al (2022) No evidence of canopy-scale leaf thermoregulation to cool leaves below air temperature across a range of forest ecosystems. Proc Natl Acad Sci 119:e2205682119

Still CJ, Sibley A, DePinte D, Busby PE, Harrington CA, Schulze M, Shaw DR, Woodruff D, Rupp DE, Daly C et al (2023) Causes of widespread foliar damage from the June 2021 Pacific northwest heat dome: more heat than drought (S Sevanto, Ed.). Tree Physiol 43:203–209

Stocker BD, Roth R, Joos F, Spahni R, Steinacher M, Zaehle S, Bouwman L, Prentice IC (2013) Multiple greenhouse-gas feedbacks from the land biosphere under future climate change scenarios. Nat Clim Chang 3:666–672

Swetnam TW, Baisan CH (1996) Historical fire regime patterns in the. In: Fire effects in southwestern forests: Proceedings of the Second La Mesa Fire Symposium, Los Alamos, New Mexico, March 29–31, 1994. USDA Forest Service, Rocky Mountain Forest and Range Experiment Station, 11

Swetnam TW, Betancourt JL (1998) Mesoscale disturbance and ecological response to decadal climatic variability in the American southwest. J Clim 11:3128–3147

Teskey R, Wertin T, Bauweraerts I, Ameye M, Mcguire MA, Steppe K (2015) Responses of tree species to heat waves and extreme heat events. Plant Cell Environ 38:1699–1712

Trugman AT, Anderegg LD, Wolfe BT, Birami B, Ruehr NK, Detto M, Bartlett MK, Anderegg WR (2019) Climate and plant trait strategies determine tree carbon allocation to leaves and mediate future forest productivity. Glob Change Biol 25:3395–3405

US Forest Service (nd) Detection surveys. U.S. Department of agriculture. https://www.fs.usda.gov/science-technology/data-tools-products/fhpmapping-reporting/detection-surveys

Urban J, Ingwers MW, McGuire MA, Teskey RO (2017) Increase in leaf temperature opens stomata and decouples net photosynthesis from stomatal conductance in Pinus taeda and Populus deltoides x nigra. J Exp Bot 68:1757–1767

USGCRP, Reidmiller DR, Avery CW, Easterling DR, Kunkel KE, KLM L, Maycock TK, Stewart BC (2018) Impacts, risks, and adaptation in the United States: the fourth national climate assessment, volume II. U.S. Global Change Research Program

Vadell E, de-Miguel S, Pemán J (2016) Large-scale reforestation and afforestation policy in Spain: A historical review of its underlying ecological, socioeconomic and political dynamics. Land Use Policy 55:37–48

Van Doorn WG (2011) Classes of programmed cell death in plants, compared to those in animals. J Exp Bot 62:4749–4761

Van Doorn WG, Woltering EJ (2005) Many ways to exit? Cell death categories in plants. Trends Plant Sci 10:117–122

Vieira J, Nabais C, Campelo F (2022) Dry and hot years drive growth decline of Pinus halepensis at its southern range limit in the Moroccan High Atlas Mountains. Trees 36:1585–1595

Wagner Y, Feng F, Yakir D, Klein T, Hochberg U (2022) *In situ*, direct observation of seasonal embolism dynamics in Aleppo pine trees growing on the dry edge of their distribution. New Phytol 235:1344–1350

Wahid A, Gelani S, Ashraf M, Foolad M (2007) Heat tolerance in plants: An overview. Environ Exp Bot 61:199–223

Williams AP, Allen CD, Macalady AK, Griffin D, Woodhouse CA, Meko DM, Swetnam TW, Rauscher SA, Seager R, Grissino-Mayer HD et al (2013) Temperature as a potent driver of regional forest drought stress and tree mortality. Nat Clim Chang 3:292–297

Williams AP, Cook BI, Smerdon JE (2022) Rapid intensification of the emerging southwestern North American megadrought in 2020–2021. Nat Clim Chang 12:232–234

Wion AP, Breshears DD, Carroll CJ, Cobb NS, Hart SJ, Law DJ, Meneses N, Redmond MD (2022) Dead again: predictions of repeat tree die-off under hotter droughts confirm mortality thresholds for a dryland conifer species. Environ Res Lett 17:074031

Yanoviak SP, Gora EM, Bitzer PM, Burchfield JC, Muller-Landau HC, Detto M, Paton S, Hubbell SP (2020) Lightning is a major cause of large tree mortality in a lowland neotropical forest. New Phytol 225(5):1936–1944

MIX
Papier aus verantwortungsvollen Quellen
Paper from responsible sources
FSC® C105338

FSC
www.fsc.org

If you have any concerns about our products,
you can contact us on
ProductSafety@springernature.com

In case Publisher is established outside the EU,
the EU authorized representative is:
Springer Nature Customer Service Center GmbH
Europaplatz 3, 69115 Heidelberg, Germany

Printed by Libri Plureos GmbH
in Hamburg, Germany